AF893493

Fritz Vögtle, Gabriele Richardt and Nicole Werner

Dendrimer Chemistry

Further Reading

Sliwa, W., Kozlowski, C.

Calixarenes and Resorcinarenes
Synthesis, Properties and Applications

2009
Hardcover
ISBN: 978-3-527-32263-3

Diederich, F., Stang, P. J., Tykwinski, R. R. (eds.)

Modern Supramolecular Chemistry
Strategies for Macrocycle Synthesis

2008
Hardcover
ISBN: 978-3-527-31826-1

Schalley, C. A. (ed.)

Analytical Methods in Supramolecular Chemistry

2007
Hardcover
ISBN: 978-3-527-31505-5

J. Dodziuk, H. (ed.)

Cyclodextrins and Their Complexes
Chemistry, Analytical Methods, Applications

2006
Hardcover
ISBN: 978-3-527-31280-1

Fritz Vögtle, Gabriele Richardt and Nicole Werner

Dendrimer Chemistry

WILEY-VCH Verlag GmbH & Co. KGaA

Authors

Prof. Dr. Fritz Vögtle
Universität Bonn
Kekulé-Institut
Gerhard-Domagk-Str. 1
53121 Bonn

Dr. Gabriele Richardt
Universität Bonn
Kekulé-Institut
Gerhard-Domagk-Str. 1
53121 Bonn

Dr. Nicole Werner
Universität Bonn
Kekulé-Institut
Gerhard-Domagk-Str. 1
53121 Bonn

Translator
Anthony J. Rackstraw
Weinheim, Germany

Cover
The branched treelike "*Lichtenberg Figure*" has been generated by bombarding a plexiglass block with electrons which in one of the electron generators at the Max-Planck-Institut für Strahlenchemie had been boosted to reach the impact energy necessary to enter the highly isolating material. The electrons are slowed down and immobilized in the plexiglass which then is being negatively charged. When after the electron bombardement the plexiglass block is contacted by an earthed electrode, the negative charges are removed through this contact. The discharge process is accompanied by tiny lightnings which melt the plexiglass and thus generate the finely branched figure. Since the discharge cannot be controlled accurately, a unique figure is formed each time. These figures have been named "*Lichtenberg Figures*", after the physicist and philosopher Georg Christoph Lichtenberg (1742–1799) of the University of Göttingen. In his treatise "Über eine neue Methode, die Natur und die Bewegung der elektrischen Materie zu erforschen", Lichtenberg has described the novel phenomenon using the notion "positive and negative electricity" for the first time. – The photo of the "*Lichtenberg Figure*" in a plexiglass block has been superimposed by a schematic drawing of a poly-bifunctionalized dendritic molecule.

Source: Max-Planck-Institut für Strahlenchemie, Mülheim an der Ruhr. We thank Dr. W.E. Klotzbücher for providing the photograph and related information.

Library of Congress Card No.: applied for

British Library Cataloguing-in-Publication Data
A catalogue record for this book is available from the British Library.

Bibliographic information published by the Deutsche Nationalbibliothek
Die Deutsche Nationalbibliothek lists this publication in the Deutsche Nationalbibliografie; detailed bibliographic data are available in the Internet at http://dnb.d-nb.de.

Originally published in the German language by B.G. Teubner Verlag as "Fritz Vögtle/Gabriele Richardt/Nicole Werner: Dendritische Moleküle".

Typesetting K+V Fotosatz GmbH, Beerfelden
Printing Strauss GmbH, Mörlenbach
Binding Litges & Dopf GmbH, Heppenheim
Cover Design Adam-Design, Bernd Adam, Weinheim

Printed in the Federal Republic of Germany
Printed on acid-free paper

ISBN 978-3-527-32066-0

Contents

Preface *XI*

1 Introduction *1*
1.1 Historical – Cascade molecules and dendrimers *1*
1.2 Dendritic architectures *7*
1.3 Perfection, defects, dispersity *13*
1.4 Definition and classification of dendritic molecules *15*
1.5 Nomenclature of dendritic molecules *16*
1.5.1 Newkome nomenclature *16*
1.5.2 Cascadane nomenclature *17*
Bibliography and Notes for Chapter 1 "Introduction" *22*

2 Synthetic methods for dendritic molecules *25*
2.1 Divergent synthesis *25*
2.2 Convergent synthesis *27*
2.3 Recent synthetic methods *28*
2.3.1 Orthogonal synthesis *28*
2.3.2 Double-stage convergent method *29*
2.3.3 Double-exponential method *29*
2.3.4 Hypermonomer method *30*
2.3.5 Click chemistry *31*
2.4 Solid phase synthesis *31*
2.5 Coordination-chemical synthesis *33*
2.5.1 Metal complex as core unit *33*
2.5.2 Metal complexes as branching unit *34*
2.6 Supramolecular synthesis *35*
2.7 Hyperbranched polymers *38*
2.8 Dendronised linear polymers *39*
2.8.1 Polymer-analogous method *39*
2.8.2 Macromonomer method *41*
2.9 Dendro-Isomers *42*
Bibliography and Notes for Chapter 2 "Synthetic methods for dendritic molecules" *44*

Dendrimer Chemistry. Fritz Vögtle, Gabriele Richardt and Nicole Werner

ISBN: 978-3-527-32066-0

3 Functional dendrimers *49*
3.1 Monofunctional dendrimers *50*
3.1.1 Functional core *50*
3.1.2 Functional periphery *51*
3.1.2.1 Functionalisation of terminal groups *52*
3.1.2.2 Introduction of peripheral groups prior to dendrimer growth *54*
3.1.3 Functional units in the dendrimer scaffold *55*
3.1.3.1 Modification prior to dendrimer growth *55*
3.1.3.2 Internal modification on conclusion of dendrimer growth *58*
3.2 Multifunctional dendrimers *61*
3.2.1 Bifunctionalised molecular periphery *62*
3.2.2 Two different functional units in different parts of the molecule *67*
3.2.3 More than two different functional units *67*
3.2.4 Overview of functional dendrimers and their synthesis *71*
Bibliography and Notes for Chapter 3 "Functional dendrimers" *75*

4 Types of dendrimers and their syntheses *81*
4.1 Achiral dendrimers *81*
4.1.1 POPAM *81*
4.1.2 PAMAM *85*
4.1.3 POMAM *89*
4.1.4 Polylysine dendrimers *90*
4.1.5 Dendritic hydrocarbons *93*
4.1.5.1 Condensed arene components – Iptyceness *93*
4.1.5.2 Dendrimers from arene and multiply bonded building blocks *93*
4.1.5.3 Stilbenoid dendrimers *97*
4.1.5.4 Hyperbranched polybenzenes *102*
4.1.6 Carbon/oxygen-based (and Fréchet) dendrimers *102*
4.1.6.1 Polyether dendrimers *102*
4.1.6.2 Polyester dendrimers *104*
4.1.6.3 Carbohydrate dendrimers (glycodendrimers) *108*
4.1.7 Porphyrin-based dendrimers *109*
4.1.8 Ionic dendrimers *111*
4.1.8.1 Polyanionic dendrimers *111*
4.1.8.2 Polycationic dendrimers *114*
4.1.9 Silicon-based dendrimers *118*
4.1.9.1 Silane dendrimers *118*
4.1.9.2 Carbosilane dendrimers *120*
4.1.9.3 Carbosiloxane dendrimers *123*
4.1.9.4 Siloxane dendrimers *127*
4.1.9.5 Hyperbranched silicon-based polymers *128*
4.1.10 Phosphorus-based dendrimers *130*
4.1.11 Metallodendrimers (and Newkome dendrimers) *133*
Bibliography and Notes for Section 4.1 "Achiral dendrimers" *138*
4.2 Chiral dendrimers *143*

4.2.1 Classification of chiral dendrimers *143*
4.2.2 Studies on the chirality of dendritic molecules *145*
4.2.2.1 Chiroptical studies *145*
4.2.2.2 Possible applications of chiral dendrimers *146*
4.2.3 Dendrimers with chiral core and achiral branching scaffold *146*
4.2.3.1 Chiroptical studies on dendrimers with chiral cores *147*
4.2.3.2 Possible applications of chiral-core dendrimers *153*
4.2.4 Dendrimers with chiral building blocks as spacers or branching units *156*
4.2.4.1 Chiroptical studies on dendrimers with chiral dendrimer scaffold *156*
4.2.4.2 Possible applications of dendrimers with chiral branching scaffold *158*
4.2.5 Chirality in the periphery *158*
4.2.5.1 Chiroptical studies on dendrimers with peripheral chiral units *158*
4.2.5.2 Possible applications of dendrimers with peripheral chiral units *160*
4.2.6 Chiral dendrimers for asymmetric catalysis *161*
4.2.7 Interpretation of the chirality of dendritic molecules *161*
Bibliography and Notes for Section 4.2 "Chiral Dendrimers" *165*

5 Photophysical properties of dendritic molecules *169*
5.1 Luminescence and energy transfer *169*
5.1.1 Luminescence *169*
5.1.2 Energy transfer *172*
5.1.2.1 Dexter mechanism: Energy transfer by radiative emission *172*
5.1.2.2 Förster mechanism: Energy transfer by dipole-dipole interactions *172*
5.1.2.3 Examples from the field of dendritic molecules *173*
5.2 Antenna effect and photoisomerisation of dendrimers *176*
5.2.1 Antenna effect *176*
5.2.2 Photoisomerisation *186*
Bibliography and Notes for Chapter 5 "Photophysical properties of dendritic molecules" *192*

6 (Special) chemical reactions of dendritic molecules *195*
6.1 Covalent chemical reactions *195*
6.1.1 Metathesis *195*
6.1.2 Molecular imprinting *201*
6.1.3 Covalent introduction of functionalities in the interior of dendritic molecules *203*
6.2 Supramolecular (host/guest) interactions *204*
6.2.1 Non-covalent modification of a dendrimer periphery *204*
6.2.2 Self-assembly of dendrimers *207*
6.2.3 Inclusion of guest species in dendritic host molecules *207*

6.2.3.1 Dendrimers with multiple receptor units *207*
6.2.3.2 Guest inclusion by steric compression *212*
6.2.3.3 Guest inclusion by dynamic processes (diffusion) *214*
6.2.4 Self-assembly of dendrimers *217*
6.2.5 Dendritic stopper groups (in rotaxanes) *220*
6.3 Dendritic effects *221*
6.3.1 Dendritic effect on inclusion of guests *223*
6.3.2 Dendritic effects in catalysis *224*
6.3.2.1 Metal-containing dendritic catalysts *224*
6.3.2.2 Metal-free dendritic catalysts *234*
6.3.3 Dendritic effects on electrochemical properties *240*
6.3.3.1 Metal-free dendritic catalysts *240*
6.3.3.2 Redox gradients *242*
6.3.3.3 Redox sensors *243*
6.3.3.4 Redox potential and redox transfer kinetics *244*
6.3.3.5 Charge-separation processes *244*
6.3.4 Summary of the dendritic effect *244*
Bibliography and Notes for Chapter 6 "(Special) chemical reactions of dendritic molecules" *246*

7 Characterisation and analysis *253*
7.1 Chromatography *253*
7.1.1 Liquid chromatography *253*
7.1.1.1 Preparative liquid chromatography *254*
7.1.1.2 High-performance liquid chromatography *254*
7.1.2 Gel permeation chromatography *257*
7.2 Gel electrophoresis *259*
7.3 NMR spectroscopy *260*
7.3.1 (1D)-NMR spectroscopic studies *261*
7.3.2 Multidimensional NMR spectroscopy in dendrimer research *261*
7.3.3 Diffusion NMR spectroscopy *262*
7.3.4 Dynamic NMR spectroscopy *263*
7.4 Mass spectrometry *263*
7.4.1 Gentle ionisation methods: MALDI and ESI *263*
7.4.1.1 Study of dendrimers by MALDI and ESI-MS *264*
7.5 X-ray crystal structure analysis *266*
7.6 Small-angle scattering *266*
7.6.1 Principle of small-angle scattering *267*
7.6.2 Capability of small-angle scattering *267*
7.6.3 Structural analysis of dissolved dendrimers with SANS and SAXS *268*
7.6.3.1 Radial segment density distribution of flexible dendrimers *268*
7.6.3.2 Distribution of end groups *269*
7.6.3.3 Intermolecular interactions of flexible dendrimers in solution *270*
7.7 Scanning probe microscopy *271*

7.7.1 STM and AFM *271*
7.7.1.1 AFM images of dendrimers *273*
7.7.1.2 STM images of dendrimers *274*
7.8 Transmission electron microscopy *274*
7.8.1 TEM *274*
7.8.1.1 TEM images of dendrimers *275*
7.9 Chiroptical methods *276*
7.9.1 Optical rotatory dispersion and circular dichroism *277*
7.9.2 Chiroptical studies on chiral dendritic structures *279*
7.10 Summary *280*
Bibliography and Notes for Chapter 7 "Characterization and analysis" *282*

8 Special properties and potential applications *289*
8.1 Introduction *289*
8.2 Catalysis, membrane technology *290*
8.2.1 Dendrimers as catalyst supports *291*
8.2.2 Catalytic dendrimers for membrane reactors *292*
8.2.3 Dendrimers in enantioselective catalysis *293*
8.2.4 Dendrimers as phase transfer catalysts *294*
8.3 Pigments, adhesives, additives in chemical materials *294*
8.3.1 Dendrimers as additives *294*
8.3.2 Dendritic polymers for printing inks *295*
8.3.3 Dendritic polymers for paints *295*
8.3.4 Dendritic polymers as additives in foam formulation *295*
8.3.5 Network precursors for plastics *296*
8.3.6 Dendrimers as nanocapsules for dyes and for molecular imprinting *296*
8.4 Dendrimers for displays and (opto)electronics *298*
8.4.1 Liquid-crystalline dendrimers *299*
8.5 Biomimetics, sensor technology, diagnostics (fluorescence) *300*
8.5.1 Protein dendrimers *300*
8.5.2 Glycomimetics *302*
8.5.3 Dendrimers in sensor technology *304*
8.5.3.1 Quartz micro balance with dendritic sensor layers *304*
8.5.3.2 Luminescent dendrimers as sensor materials *306*
8.5.3.3 Fluorescing PET sensors *308*
8.6 Dendrimers in medical diagnostics *311*
8.6.1 Magnetic resonance imaging (MRI) processes *311*
8.6.2 DNA dendrimers as biosensors for DNA hybridisation *313*
8.7 Medical applications *313*
8.7.1 Dendrimers as carriers for cytostatic agents *313*
8.7.2 Gene therapy *315*
8.7.3 Photodynamic therapy *316*

8.7.4 Dendrimers in prevention against HIV *317*
8.7.5 Culture of organs and tissue *318*
8.7.5.1 Wound healing *318*
8.7.6 Boron neutron capture therapy *319*
8.8 Dendrimers in nanotechnology *320*
8.8.1 Photoswitchable dendrimers *320*
8.8.2 Dendrimers as impellers *320*
8.8.3 Dendrimers as nanotubes *321*
8.8.4 Dendritic polymers as templates *324*
Bibliography and Notes for Chapter 8 "Special properties and potential applications" *324*

Outlook *331*

Subject Index *333*

Preface

The positive feedback we have received regarding of our book "Dendritische Moleküle" in German gave us good reason to prepare a corresponding international edition in English. Since this fast developing research area of dendritic molecules has gained international recognition and promises to generate more applications in the "high chem" field as well as in technology, we chose the new title "Dendrimer Chemistry". Nevertheless, this three-author book goes far beyond chemistry, touching on important topics in material and life sciences.

We have aimed to provide short introductory explanations for non-experts, e.g. in areas such as spectroscopic methods and new methodologies in adjacent disciplines, in order to make this interdisciplinary book easier to read for students and not only for chemists.

Apart from the modified title and the new cover illustration and legend, many corrections, additions, hints and remarks have been added to the text and the illustrations. Dozens of recent literature references from 2007 and 2008, especially to new books and overviews, have been included in all chapters to keep the book up to date.

It is especially pleasing that the publisher could maintain the style of the illustrations, in particular the red and green colors symbolising host and guest molecules, respectively, throughout the book.

We hope that this most recent overview on dendritic molecules in chemistry (and beyond) will be useful not only as an introduction and source of general information, but also as a basis for future teaching and research in all disciplines of chemistry as well as material and life sciences.

We would like to thank Dr. Heike Nöthe, STM-books, Wiley-VCH-Verlag, for her valuable advice in the course of production of this book and Mr. Anthony Rackstraw for his excellent translation of the German manuscript into English.

Fritz Vögtle
Kekulé-Institut der Rheinischen Friedrich-Wilhelms-Universität Bonn
Bonn, September 2008

Dendrimer Chemistry. Fritz Vögtle, Gabriele Richardt and Nicole Werner

ISBN: 978-3-527-32066-0

1 Introduction

1.1 Historical – Cascade molecules and dendrimers

In 1978 Vögtle *et al.* [1] described a series of synthetic "cascade molecules" [2–4] as the first tangible representatives of compounds exhibiting potentially perpetual branching. Starting from diverse primary monoamines and diamines, "cascade synthesis" was performed to attach spacer units of propylenamine structure whose N atoms served as a $1 \rightarrow 2$ branching point (formal branching of a bond into two new ones) during each subsequent repetitive step [5].

In the simplest case, reaction of a primary *mono*amine *via* a two-fold Michael reaction with acrylonitrile (bis-cyanoethylation) led to the *dinitrile* (Fig. 1.1). Subsequent reduction of the two nitrile functions – by hydrogenation with sodium borohydride in the presence of cobalt(II) ions – afforded the corresponding terminal *diamine*. Repetition (iteration) of this synthetic sequence, consisting in Michael addition followed by reduction, provided the first – structurally variable – access to regularly branched, many-armed molecules.

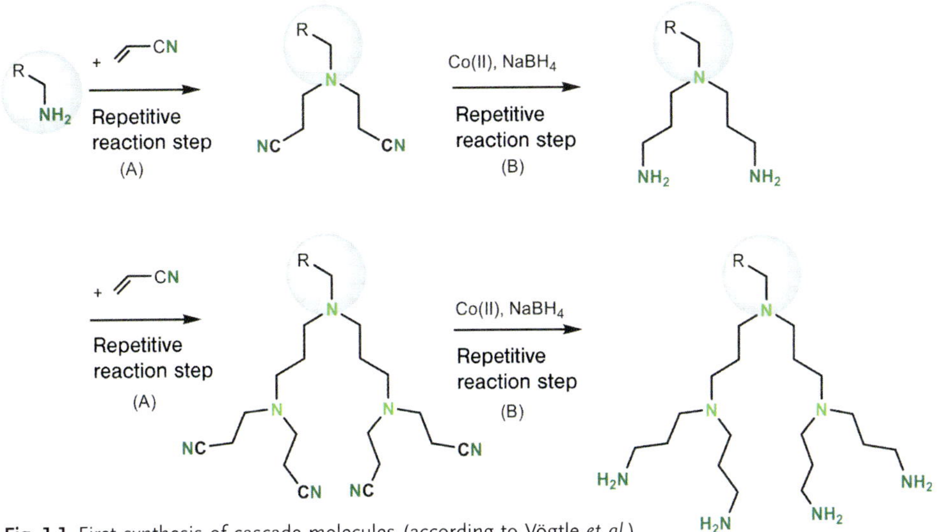

Fig. 1.1 First synthesis of cascade molecules (according to Vögtle *et al.*)

Dendrimer Chemistry. Fritz Vögtle, Gabriele Richardt and Nicole Werner

ISBN: 978-3-527-32066-0

Several years earlier (1974), the same group had already described many-armed, albeit non-branched, molecules as *octopus molecules* [6], whose numerous arms were used for complexation with metal ions (Fig. 1.2). These octopus molecules can be regarded as forerunners of nitrogen-containing propylenamine cascade molecules since they already demonstrated the utility of many adjacent functional arms – all the more readily attainable by branching – for example for host-guest interactions [6].

Initially, further development of highly branched (cascade-type, dendritic) molecules proceeded slowly for many years, probably due to the synthetic and analytical obstacles to be overcome at the time with these species, which occupy an intermediate position at the boundary between low-molecular weight and high-molecular weight compounds.

Denkewalter *et al.* described a pathway to polylysine dendrimers (Fig. 1.3) *via* divergent synthesis in 1981 patents (see Section 2.1) [7].

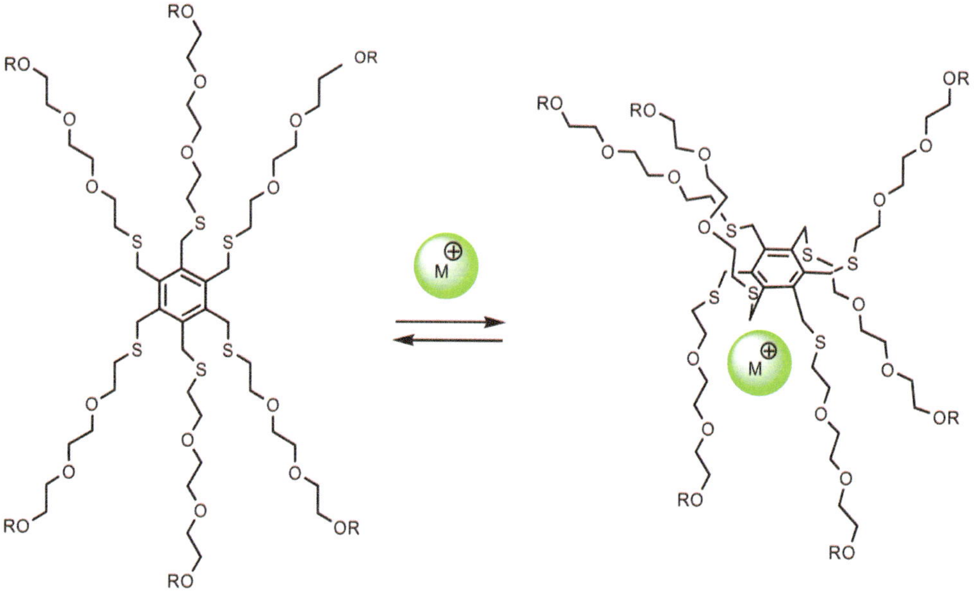

Fig. 1.2 Octopus molecule (left): host-guest interaction with metal ions (schematic; according to Vögtle, Weber)

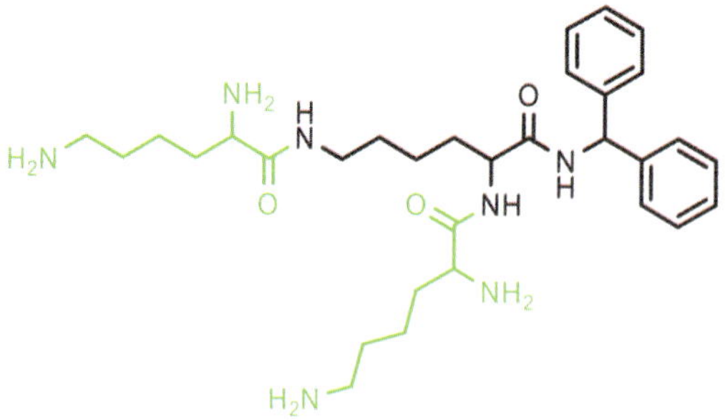

Fig. 1.3 Polylysine dendrimer (according to Denkewalter *et al.*); the two peripheral lysine units are shown in green

In 1982 Maciejewski developed a densest packing concept for polymers having a cascade-like structure [8]. On the basis of statistical considerations, in 1983 de Gennes (1991 Nobel laureate for physics) and Hervet presented limits to the growth of branched molecules (*starburst-limited generation*), paying due attention to the influence of steric hindrance [9]. Further statistical model calculations were combined with the "cascade theory" [10]. According to de Gennes, highly branched molecules are considered as "soft material" [11].

In 1985 Tomalia developed branched poly(*amidoamines*) (PAMAM), which he also designated as "starburst dendrimers" (Fig. 1.4) and generally propagated the name "dendrimer" (from the Greek *dendron* = tree and *meros* = part) [12]. Like the first cascade synthesis, the synthetic route again involved Michael addition (of methyl acrylate to ammonia). The resulting ester was converted into the primary triamine by reaction with an excess of ethylenediamine. Repetition of the reaction sequence (iteration) by analogy with the cascade synthesis led to dendrimers of up to the tenth generation – with decreasing purity and perfec-

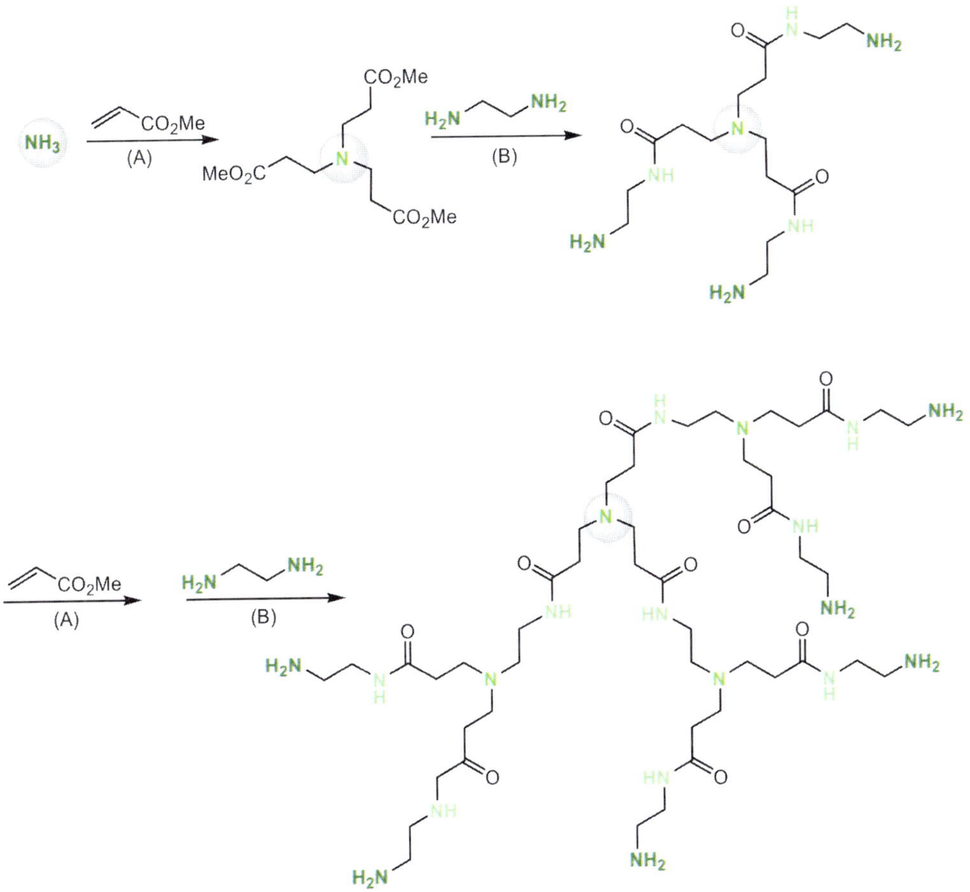

Fig. 1.4 Synthesis of poly(amidoamine) dendrimers (PAMAM; according to Tomalia *et al.*)

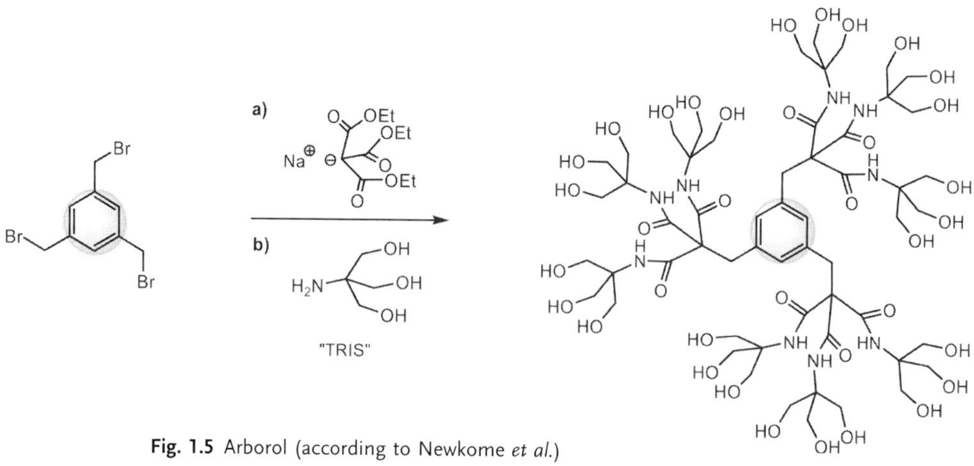

Fig. 1.5 Arborol (according to Newkome *et al.*)

tion (see Section 1.3). Tomalia referred to the individual ester stages as half generations (0.5, 1.5, 2.5) [13].

Also in 1985, Newkome *et al.* presented a divergent synthetic route to water-soluble, highly branched "arborol systems" (Fig. 1.5) with terminal hydroxyl groups, whose name is derived from the Latin *arbor* = tree [14].

Tomalia's exhaustive review paper with coloured illustrations [10] did much to popularise the highly branched compounds and to ensure broad general acceptance of the family name "dendrimers". In the same year, Fréchet and Hawker described the first convergent synthesis of dendrimers. They constructed poly(aryl ether) architectures "from the outside inwards" (Fig. 1.6; see Section 2.2) [15].

Miller and Neenan succeeded in the very same year in producing the first hydrocarbon dendrimers based exclusively on arene units, also using a convergent synthetic strategy [16] (Fig. 1.7).

In this introduction, the history of dendritic molecules is limited to initial developments. It also warrants mention that theoretical considerations of infinite (polymer) networks by Flory can be traced back to 1941 [17].

Further pioneering work and workers in the field of dendrimers will be mentioned in Chapters 2 and 4, which also consider more recent developments relating to synthetic methods and types of dendrimers.

The exponential growth and topicality of research into dendritic molecules almost thirty years after their first synthesis (1978) is apparent not only from the large number of publications (presently totalling more than 10 000, and increasing by more that 1000 per annum, plus about 150 patents), but also from the mere fact that more than 8000 researchers are currently active in this area and more than 150 companies have already applied for patents relating to dendritic compounds (source: IDS-5 Programme).

The Dendrimer Symposia which have been held biannually since 1999 also reflect this development: The first *I*nternational *D*endrimer *S*ymposium (*IDS*-

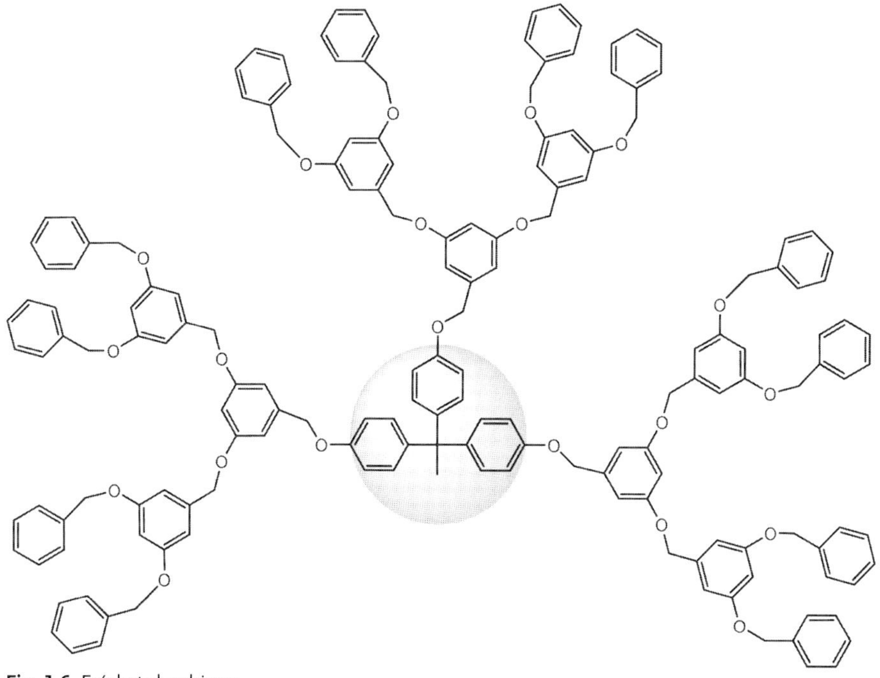

Fig. 1.6 Fréchet dendrimer

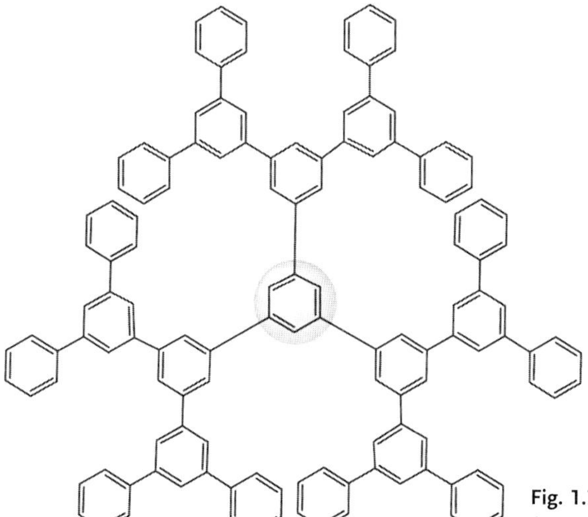

Fig. 1.7 Hydrocarbon dendrimers (according to Miller and Neenan)

Fig. 1.8 Flyers and logos of the first four Dendrimer Symposia held at: *Frankfurt*, 1999 (Organisation: F. Vögtle, K. Müllen, DECHEMA); *Tokio*, 2001 (Organisation: T. Aida, M. Ueda, DECHEMA); *Berlin*, 2003 (Organisation: A. Schlüter, F. Vögtle, E.W. Meijer, A. Hult, DECHEMA); *Michigan*, 2005 (Organisation: D. Tomalia, J.-F. Stoddart, F. Swenson, J. F. Fréchet); (*IDS-5: Toulouse*, 2007 (Organisation: J.-P. Majoral, A.-M. Caminade))

1), initiated by Vögtle and Müllen, was held in 1999 at Frankfurt/Main under the auspices of DECHEMA (Fig. 1.8). 183 participants from 21 countries gathered together to discuss the design, synthesis, structures, analysis, and applications of dendritic molecules [18].

At the second International Dendrimer Symposium emphasis was placed, among other topics, on the multiplication of functional groups on the periphery of dendritic molecules with a view to enhancing physical and chemical effects. Moreover, the inclusion of guest species was also discussed from the standpoint of supramolecular chemistry. The guests considered also included active substances (*e.g.* cytostatics), which can be specifically released from the "embrace" of the dendrimer at certain target sites in the organism (*drug targeting; drug release;* see Chapter 8).

The third International Dendrimer Symposium took place at Berlin Technical University in 2003. Interdisciplinary lectures demonstrated the extent to which dendritic molecules "branch out" into other areas of science, such as physics, biology, medicine, and engineering. The possibilities of functionalisation and resulting applications in industry were at the focus of this symposium. For example, nano-dimensioned dendrimer-based contrast agents were presented as *multilabels* for visualisation of blood vessels (see Chapter 8). Potential applications of dendritic materials as luminescence markers in diagnostics attracted lively interest (see Chapter 8). Consideration of the differences between dendrimers and hyperbranched polymers from the viewpoint of their cost-favourable application was also a topic of discussion [18].

At the fourth International Dendrimer Symposium held in 2005 at Mount Pleasant, Michigan/USA with 81 international speakers, the main emphasis was placed on the potential applications of dendritic molecules. This was apparent not least from the strict categorisation of the lectures according to possible commercial applications as (nano)materials (displays, optical sensors) or in medicine (see Chapter 8). The fifth International Dendrimer Symposium in 2007 at Toulouse was dominated by the topics synthesis, nanomaterials and in particular by the development of dendritic therapeutics [18 b].

1.2 Dendritic architectures

The name dendrimers, which has meanwhile largely displaced the original designation of *cascade molecules*, is derived from the Greek words *dendron* and *meros*, and is meant to underscore the tree-like branched structure of this class of compounds (see Section 1.1).

Multiply branched (dendritic) structures are frequently encountered in nature, science, technology, art, and everyday life. Examples of naturally occurring dendritic structures are seen in the branching of trees and roots (Fig. 1.9), blood vessels, nerve cells, rivers, lightning, corals, and snowflakes [19 a]. Metals deposited on electrodes or on noble metals [19b] often exhibit branching, as do fibres and gels [20] The bottoms of geckos' feet bear millions of multiply split thin hair ends, permitting them to climb up walls and cross absolutely smooth ceilings [21 a, b]. On the basis of this model – and that of flies, spiders, and other animals – a peelable adhesive was developed at the Max Planck Institute for Metals Research in Stuttgart, which could replace refrigerator magnets, for example [21 c]. Evolution itself can be depicted dendritically [21 b, 22–23]. Many-armed deities play a special role in some religions. A room divider that can be individually shaped and joined up to form a network structure has been launched on the market [24]. Zone maps of underground railways show dendritic patterns.

Fractals are mathematically defined self-similar structures (Fig. 1.11) [26]. The *scaffold* of cascade or dendritic molecules is fractal if the atoms are considered to be points and the bonds to be strictly one-dimensional lines. Self-similarity

Fig. 1.9 Dendritic structures in nature: Branching of trees, before and after hosting of "guests" (mistletoe) in the intermediate spaces [25 a]

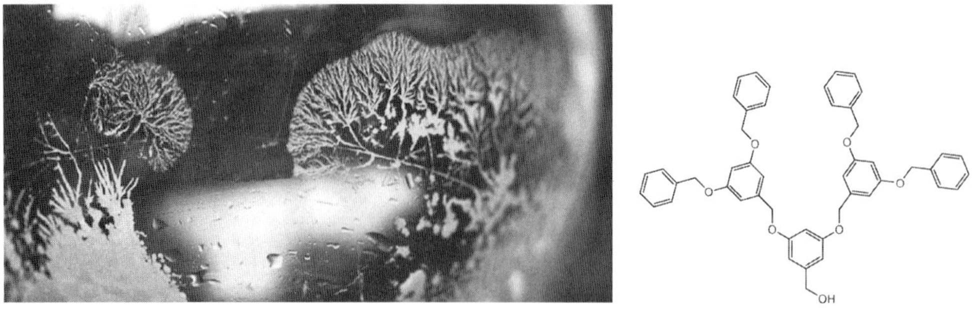

Fig. 1.10 Substance crystallising in dendritic patterns [25 b] (Fréchet dendron of the second generation)

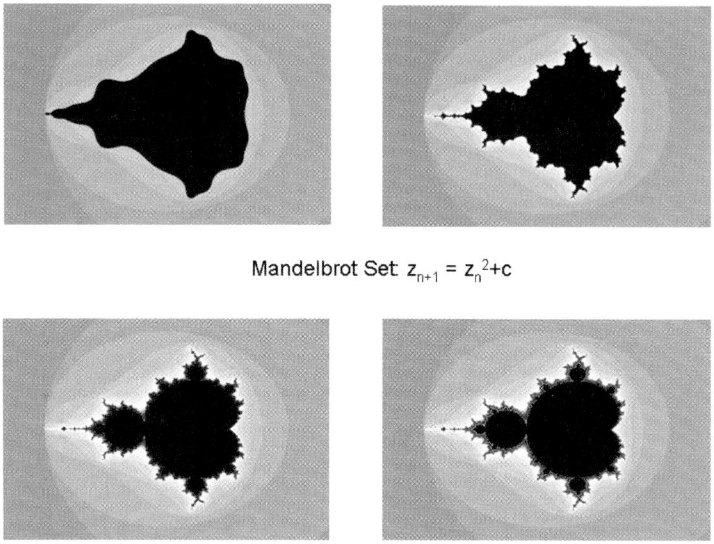

Fig. 1.11 Fractal structures (computer generated) [28]: The four parts of the figure (from left to right) show a series of Mandelbrot sets [26] after 5, 10, 50, and 150 iterations. The intermediate result serves as the starting point for each iteration. (The individual images were produced in colour, but could only be presented in modified form here)

means that structural elements are repeated on different scales. While the self-similarity of the mathematically calculated constructions in Fig. 1.11 may not be immediately apparent, the principle is readily illustrated with the aid of the Sierpinski triangle [27]. Joining the mid-points of the three sides of the equilateral triangle shown on the left of Fig. 1.12 produces another equilateral triangle reduced in size by a factor of 1/4 compared to the original triangle. Removal of this new triangle leaves the middle figure of Fig. 1.12. Application of this procedure to the middle figure affords the right-hand figure. Each repetition of the procedure results in a new generation of the Sierpinski triangle. The starting figure is called the first generation.

Fig. 1.12 Principle of self-similarity, demonstrated by the first three generations of the Sierpinski triangle

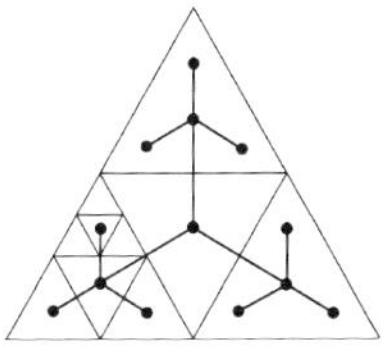

Fig. 1.13 Modified (dendritic) representation of Sierpinski triangle

If the graphical representation of the Sierpinski triangle is modified by drawing the triangles not as geometric bodies but only as lines joining the centroids of the triangles removed, then Fig. 1.13 results. The dendritic branching is now readily recognised and it is apparent that the total length of the branches increases with each new generation.

If a three-dimensional body such as a tetrahedron is substituted for the two-dimensional triangle, the branching sites lie on the surface of an imaginary sphere. This accounts for the shell-like structure of dendrimers.

Dendrimers are molecular (nano)architectures of well-defined size and number of terminal groups. Starting from a multi-functional core unit, the structure branches – often in regular layers (shells) resembling onion skins – in three dimensions from the inside outwards. These *generations* can serve to characterise the molecular size – within a given type of dendrimer. The branched structures linked in the form of segments to the central unit are termed *dendrons*. The end groups, which may in turn be "terminal functional groups", are located on the – formal – surface of the dendrimers, which is often designated as the periphery (Fig. 1.14). Given a conformationally flexible dendrimer with largely aliphatic skeleton, these terminal groups will, however, often be folded back into the interior of the molecule and therefore form neither a homogeneous outer shell nor well-defined free pores (see Section 7.6.3.2).

Depending upon the nature of the terminal groups, the dendrimers will vary in shape, stability, solubility, conformational rigidity/flexibility, and viscosity. The number of end groups of desired functionality increases with increasing number of generations. This can lead to reinforcement of certain phenomena, such as light-harvesting effects (Chapter 5), or also – as described in Chapter 8 – to an *amplification effect*. Within certain limits, appropriate design and synthesis of a dendrimer thus permits certain properties to be modified or possibly tailored to requirements.

Owing to their self-similar (fractal) structure, the number of terminal groups of a dendrimer of any generation can be calculated with the aid of the following equation:

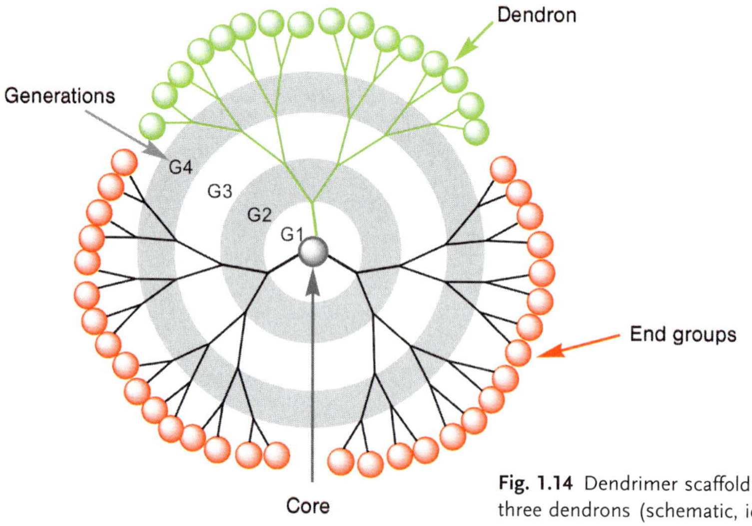

Fig. 1.14 Dendrimer scaffold – with three dendrons (schematic, idealised)

$$n_G = F_k(F_v - 1)^G \tag{1.1}$$

n_G: number of end groups in the G-th generation
F_k: functionality of the core (= number of bonds emanating from the core)
F_v: functionality of branching unit (number of bonds emanating from a branching point)
G: generation of the dendrimer

This equation ultimately expresses no more than the fact that the number of terminal groups increases as a function of the number of (former) functional groups of the core (core multiplicity) and that of the branching units (branching multiplicity) rises exponentially with the number of generations.

The mechanical stability of a dendrimer depends upon the conformational flexibility/rigidity of the branching units and the end groups. Modification of the branching units leads to a change of density in the interior of the dendrimer molecule. This is important for the host/guest chemistry considered in detail in Section 6.2, which makes use of areas of lower density to accommodate guests. The dendrimer skeleton acts here as a kind of – reversible – "dendritic box". Appropriate choice of the branching modules permits selective inclusion of guest molecules – apart from solvent – in a dendrimer, without any need for preformed *vacant* cavities or niches.

The oligo- or multi-functional core unit also plays a role in determining the space occupied by a dendrimer. The core itself can exercise a function, as demonstrated by metallodendrimers (see Section 4.1.11), in which the metal ion core in a supramolecular or coordinatively constructed architecture coordinates with the surrounding branching units – and in this way can influence catalytic and photochemical processes.

The ways in which the individual structural units of a dendrimer can affect its properties are compiled in Table 1.1.

A specific feature of dendritic molecules is that they show a lower viscosity in solution than corresponding compounds with a lower degree of branching. This behaviour is characterised by the Staudinger index η (dimensions mL/g) determined by recording the change of viscosity of a solution with different dendrimer (or polymer) concentrations and then extrapolating to zero concentration with the aid of empirical equations. The molecular mass dependence of the Staudinger index, known as the *intrinsic viscosity* or also as the *limiting viscosity*, is given by the Mark–Houwink relation:

$$[\eta] = KM^{a} \qquad (1.2)$$

η: Staudinger index
K: system-dependent constant
M: molecular mass
a: Exponential index which depends upon the shape of the dissolved dendrimer; it can lie between 0 and 2

Unlike that of linear polymers, the *intrinsic viscosity* of dendrimers does not increase linearly (Fig. 1.15) with molar mass, but reaches a maximum at a certain generation (limiting generation), only to decrease again at high generations

Table 1.1 Influence of the various components of dendrimers

Core	Branching unit	Surface	End group
	Influences		
Shape	Shape	Shape	Shape
Size	Size	Size	Stability
Multiplicity	Density/niches	Flexibility	Solubility
Functions	Guest inclusion	Properties	Viscosity

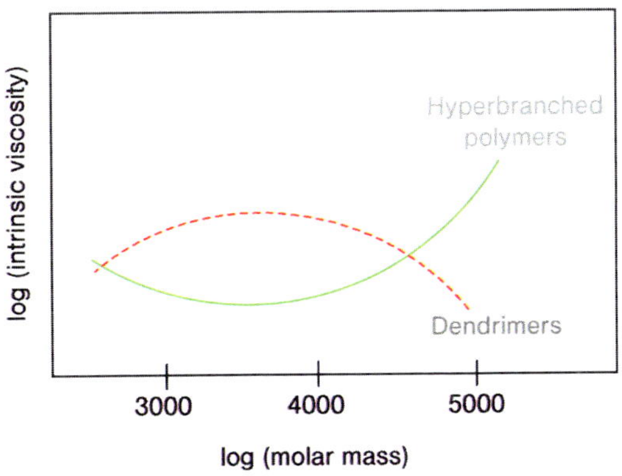

Fig. 1.15 Intrinsic viscosity of dendrimers – compared with that of polymers (schematic) [10, 30]

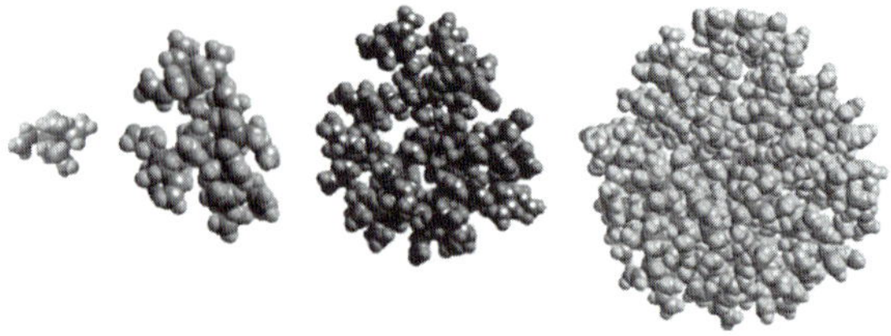

Fig. 1.16 Progressively globular form adopted with increasing number of generations illustrated for carbosilane dendrimers of zeroth to third generation (from left to right; according to Frey *et al.* [31b])

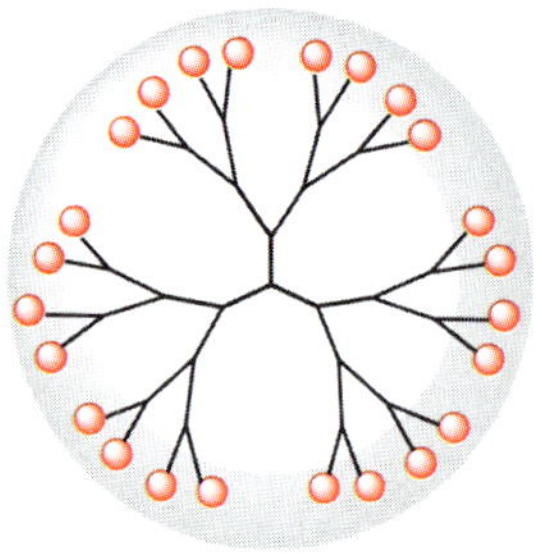

Fig. 1.17 Segment density distribution of a dendrimer molecule according to the "dense-shell model" (schematic)

(*dendrimer effect*). The intrinsic viscosity of hyperbranched polymers (Sections 4.1.5.4 und 2.7) likewise increases with increasing molecular mass [29].

This phenomenon can be explained by the gradual transition from a practically open structure of lower generations of dendrimers to an almost globular form of the higher generations (Fig. 1.16) [31].

This is why dendrimers of higher generations also have a smaller volume than corresponding linear polymers. Moreover, they also show better solubility in organic solvents and generally crystallise only with difficulty.

Depending upon their molecular structure, dendrimers can be classified as tendentially more rigid (*e.g.* "polyphenylene dendrimers"; Section 4.1.5) and tendentially more flexible (*e.g.* POPAM, PAMAM dendrimers; Sections 4.1.1 and 4.1.2).

In their theoretical considerations of molecular structure, de Gennes and Hervet assumed an ideal dendrimer with extended branches with all terminal groups arranged at its periphery in a kind of "outer ring" around the dendrimer core [9]. According to this model, dendrimers should exhibit a lower segment density at the core, which increases to a maximum value on moving to the periphery. This concept is known as the "dense-shell model" (Fig. 1.17).

In their postulated "dense-core model", Lescanec and Muthukumar advocated an opposing segment density profile [32]. Thus a maximum density is postulated at the core of the dendrimer and a decrease in segment density proportional to the distance from the core. The decrease in segment density towards

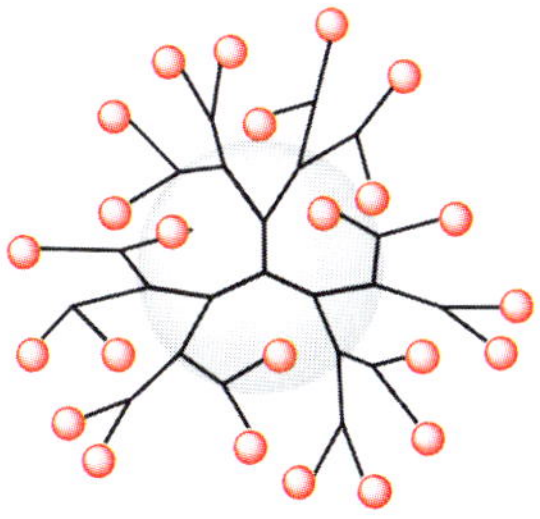

Fig. 1.18 Segment density distribution in the "dense-core model" (schematic)

the periphery is caused by a partial *backfolding* of end groups into the interior of the dendrimer (Fig. 1.18).

Most studies performed partly on molecular models [33] but also on real POPAM and PAMAM dendrimers support the latter model concept [34]. Careful studies on the three-dimensional structure of flexible dendrimers in solution were performed by Ballauff *et al.* by means of *SANS* (*S*mall *A*ngle *N*eutron *S*cattering) [35] (see Section 7.6).

1.3 Perfection, defects, dispersity

The *degree of branching DB* defined by Fréchet *et al.* provides a criterion for the classification of dendritic molecules with regard to their perfection [36].

$$DB = (I_T + I_D)/(I_T + I_D + I_L) \qquad (1.3)$$

I_T: number of terminal monomer units
I_D: number of dendritic monomer units
I_L: number of linear monomer units

In contrast to perfect dendrimers with a degree of branching of 100%, hyperbranched polymers (Sections 2.7 and 4.1.5.4) will have degrees of branching between 50 and 85%, depending upon the monomer – for example, whether AB_2 or AB_8 monomers were used as starting materials [37].

Above a certain dendrimer size, a limiting generation is reached beyond which a dendrimer of perfect structure is no longer possible. If dendrimer construction takes place from the inside outwards (divergent; see Section 2.1), the space needed for the terminal groups increases with the square of the dendrimer radius *r*. However, the number of terminal groups increases exponentially with $(F_v-1)^G$ (see Eq. 1.1). This means that with each new generation there is a – formal – increase in the occupancy of the surface with terminal groups, leading to an increased density of the outer shell(s) of the dendrimer.

$$\text{Area per end group} = (\text{dendrimer surface area})/(\text{number of end groups}) \approx 4\pi r^2/(F_k \times F_v^G) \quad (1.4)$$

F_k: functionality of the core
F_v: functionality of the branching unit
G: generation of the dendrimer

Even if the above-mentioned backfolding of peripheral groups is taken into consideration, beyond a certain limiting generation (Fig. 1.19) further reaction – *e.g.* a quantitative chemical reaction of terminal groups – will be hindered by steric effects, resulting in growth defects. This phenomenon is also known as the *starburst limit effect* [9]. According to Tomalia, ideally branched 5th generation polyethyleneimine dendrimers (PEI) [38] are ruled out by "*starburst dense packing*" [10].

The *polydispersity* of dendritic molecules, expressed in the form of their *polydispersity index* (PDI), is directly related to their structural perfection. The PDI is a measure of molecular weight distribution.

$$\text{PDI} = M_w/M_n \quad (1.5)$$

M_w corresponds to the average molar mass (*e.g.* by sedimentation equilibrium measurements)
M_n represents the number average of the molar mass (*e.g.* by determination of terminal groups) [14 b, 39].

If the polydispersity index (PDI) has a value of unity, the substance is designated as *monodisperse. Monodispersity* is considered to be a property of the *cascadanes* (defect-free dendritic molecules; cf. Section 1.4) and almost perfect dendrimers. Since these compounds are synthesised *via* an iterative approach, monodispersity has so far generally been limited to lower generations. Should it prove possible to repeatedly remove all reactants and by-products of the individual synthetic steps during the construction of a dendrimer, then structurally perfect dendrimers will result.

In contrast, *polydispersity* is a characteristic property of hyperbranched dendritic polymers, and results from the formation of by-products as a result of cycli-

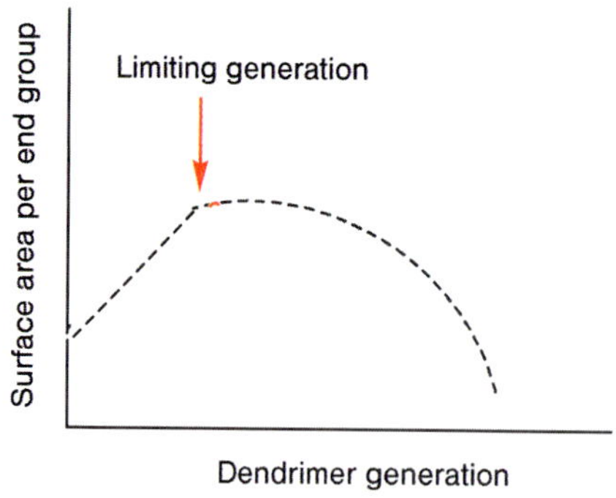

Fig. 1.19 Attainment of the limiting generation in dendrimer growth as a result of the starburst limit effect (schematic; idealised)

sation and steric hindrance during polymerisation. A *monodisperse* substance always consists of molecules of homogeneous size, whereas in *polydisperse* compounds the individual molecules have different (heterogeneous) masses. Branching defects make a minor contribution to polydispersity, which arises mainly from coiling, bridging (ring formation), and irregular growth.

1.4 Definition and classification of dendritic molecules

The first correctly dendritically branched molecules were termed *cascade molecules* and could be prepared divergently by a *cascade synthesis* (Section 1.1).

Dendritic molecules (*cascade molecules*) are repetitively branched compounds. This collective term embraces the various *dendrimers*. The latter generally exhibit "almost perfect" structures and display properties characteristic of monodisperse compounds (see also Section 1.3). With regard to their molecular masses, dendrimers range from low-molecular to high-molecular chemistry.

Cascadanes consist exclusively of molecules of the same kind and the same weight with correct, regularly branched, or perfect, defect-free structure [40].

In contrast, *hyperbranched* compounds, which do not have perfect structures owing to their method of synthesis, show polydisperse properties because they contain molecules of different masses.

If parts of dendrimers and cascadanes acts as substituents or functional groups of molecules, then they are called *dendrons*, or – if defect-free – *cascadons.*

If cascadanes form host/guest complexes, for example by inclusion of smaller guest molecules, then they are designated as *cascadaplexes*; corresponding dendrimer-based complexes are known as *dendriplexes* (Fig. 1.20) [40].

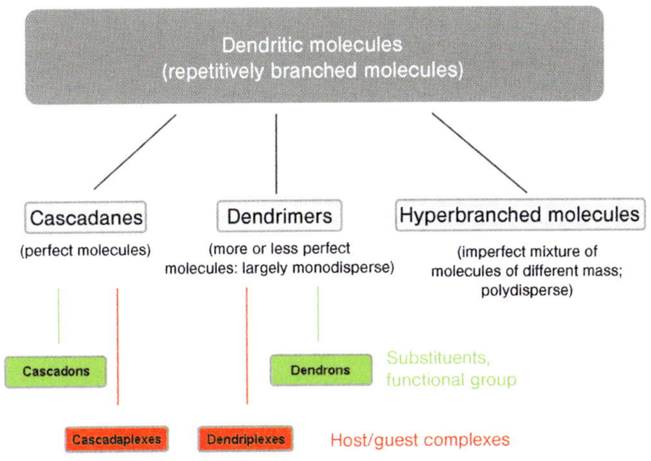

Fig. 1.20 Classification of dendritic molecules [40, 45]

1.5 Nomenclature of dendritic molecules

Like other known (macro)molecules (*e.g.* molecular knots [41], catenanes, rotaxanes [42]), dendritic molecules can be named in accordance with the *IUPAC Rules of Nomenclature*. However, these rules are not always sufficiently comprehensive to adequately, unequivocally, and clearly name such complex structures as dendritic molecules [43].

1.5.1 Newkome nomenclature

In 1993 Newkome [44] constructed a dendrimer nomenclature on the basis of the family names of the cascade molecules. This modular naming procedure for dendritic molecules and their fragments (dendrons, dendryl-/cascadyl substituents) begins with a statement of the number of peripheral terminal groups, so that the multiplicity is already clear from the beginning of the name. After the class designation "cascade", the individual branches are enumerated ("divergently") starting from the core (carbon and hetero atoms; number of branches as superscript), with the individual generations separated by colons. The terminal groups are then characterised. The names constructed thus take the form:

$$Z \text{ cascade: core building block } [N_{\text{core}}]\text{: (branching unit)}^{G}\text{: end groups} \qquad (1.6)$$

Z: number of terminal groups
N_{core}: core multiplicity
G: number of generations with branching building blocks

By way of illustration, Fig. 1.21 shows a POPAM dendrimer of the 2nd generation (cf. Fig. 1.14), which, according to Newkome, bears the name 16-Cascade:1,4-diaminobutane[4-*N*,*N*,*N'*,*N'*]:(1-azabutylidene)^2:aminopropane.

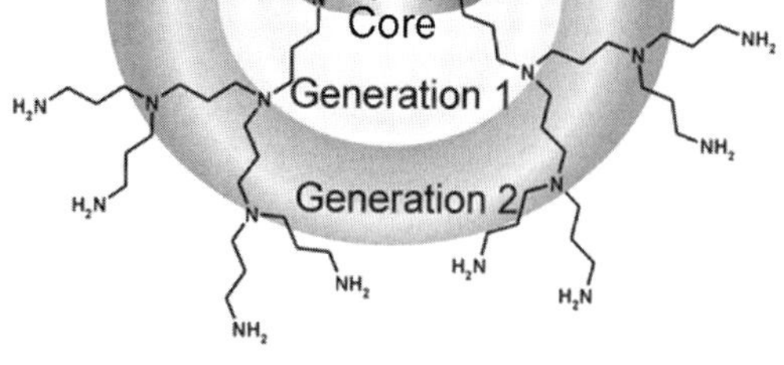

Fig. 1.21 POPAM dendrimer according to the Newkome nomenclature: 16-Cascade:1,4-diaminobutane[4-*N*,*N*,*N'*,*N'*]:(1-azabutylidene)^2:aminopropane [45]

1.5.2
Cascadane nomenclature

In the case of complex dendrimers with differing branches or differing dendryl substituents on a non-dendritic scaffold it is necessary to include further details in the name. For this reason, a more detailed *"cascadane nomenclature"* [45] has been developed, according to which the above POPAM would be called:

$$\text{1,4-Diaminobutane}[N,N,N',N']:\{\text{4-azabutyl(4,4)}\}^{G_1,G_2}{}_{4n,8n}$$
$$:\text{3-aminopropyl}_{16}\text{-cascadane.}$$

Here the generations (G_1 and G_2) with the corresponding number of branchings (4 and 8) are clearly indicated as superscripts and subscripts. The number of terminal groups (16) is also given as a subscript. The class designation cascadane appears at the end of the name. The numbers (locants 4,4) of the two branching atoms are given in parentheses.

As a consequence of the complex molecular structures, such nomenclatures are not without complications and require numerous rules. However – unlike the *IUPAC* [46] or *nodal nomenclature* [47] – owing to their modular structure they quickly reveal important individual characteristics (number of generations, number of terminal groups), which is of benefit in the laboratory and also in computer searches.

The individual rules are reduced here to their bare essentials; more detailed information will be found in the original literature [45].

Rule 1 Dendritic structures consist of self-similar units (*fractals*).

Rule 2 A dendritic structure consists of a core unit and a dendritic unit.

Rule 3 If a molecule contains one dendritic structure part or only different ones, the dendritic structures (=dendrons) are treated as substituents and are assigned the suffix *-cascadyl*. If at least two dendrons in a molecule are of the same kind, then the name is given the suffix *-cascadane* (Fig. 1.22).

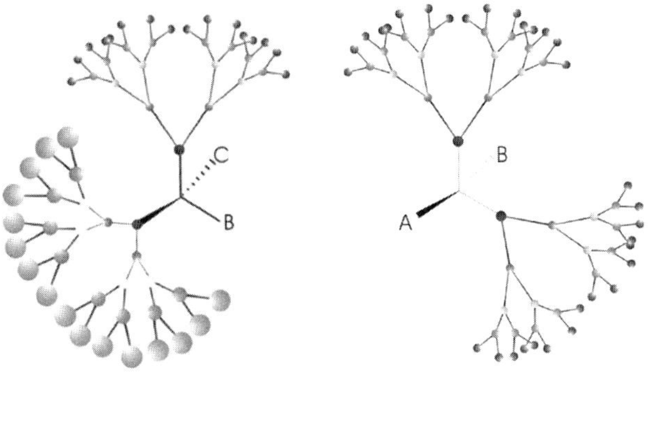

Fig. 1.22 *Cascadyl* (left), *cascadane* (right) [45]

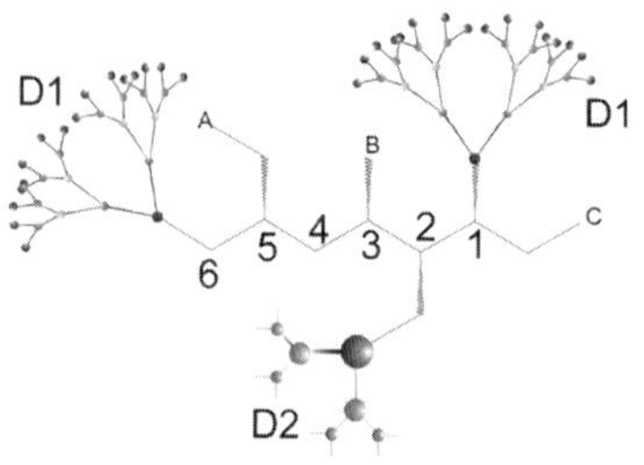

Fig. 1.23 5-(A-methyl)-3-B-1-(C-methyl)-2-(D2-cascadyl)-hexane(1,6) [45]

Rule 4 The enumeration of several different dendrons begins with that dendron which has the longest chain in the first generation. If the chain lengths are the same, then the order is decided by the next generation.

Rule 5 The name of the core unit in dendrimers is derived from the shortest unbranched chain connecting dendrons of the same kind. The positions at which the dendrons are bound to the core unit are placed in parentheses after the name of the core unit (Fig. 1.23).

Rule 6 The atoms in the branches of the scaffold are numbered from the inside outwards, excluding inner branching. The length of the chain represents the shortest connection between the branching points.

Rule 7 Every bifurcation represents the end of a generation. The terminal groups [suffix (end)] do not form a generation.

Rule 8 The name of the core unit is followed by the names of the scaffold units. These are placed in curly brackets. The term in curly brackets precedes a superscript indicating the ordinal number of the generation, and a subscript giving the total number of framework units in the pertinent generation followed by an "*n*". The individual generations are separated from one another and from the end groups by colons. The number of end groups is indicated by a subscript.

The name of this example (Fig. 1.24) is thus (in general):

"Core(1,1):{branch A (A,A)}$^{G_1}{}_{2n}$:{branch B (B,B)}$^{G_2}{}_{4n}$:end$_8$-cascadane"

Rule 9 If units are repeated in several generations, they are listed in the superscript after the *curly brackets* and separated by commas. The number of framework units in the corresponding generations is written in analogous order.

Rule 10 If repetitions of units occur in generations which do not immediately follow each other they are indicated as described in rule 9. The term listing the generation completing the dendritic structure immediately precedes the end groups.

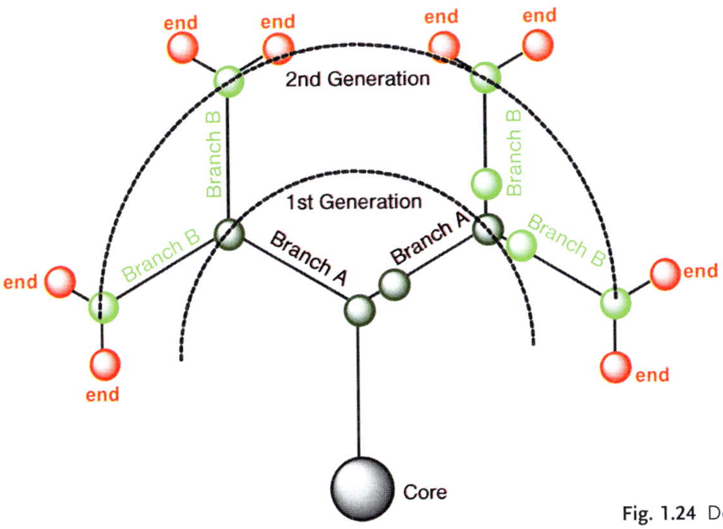

Fig. 1.24 Dendritic unit

Rule 11 If different scaffold units are distributed symmetrically in a generation, the term in curly brackets gives their names placed in angle brackets. They are ordered according to increasing chain length. After the angle brackets a subscript shows the number of repetitions followed by an "*n*". The subscript after the curly brackets gives the total number of scaffold units in the generation. If the framework units are distributed non-symmetrically in the scaffold, the scaffold has to be divided into smaller dendrons. If it is necessary to know the exact location of the scaffold unit, superscripts are added to the branching positions. Equal branching units are separated by obliques. These branching positions are placed in front of the names of the framework units of the next generation to which they are joined. If the same branching unit occurs several times, all branching positions to the preceding generation are listed in front of the name. The number of repetitions followed by an "*n*" is added as a subscript after the angle brackets.

Rule 12 Arene units are counted as a chain member by analogy with *nodal nomenclature* [47]. Ring atoms are numbered according to the *IUPAC rules*.

The names thus have the following appearance:

$$\text{Core(a,b)}:\{\langle\text{branch A(c,d)}\rangle\langle\text{branch B(e,f)}\rangle\}^{G_1}{}_{2n}$$
$$:\{\text{c,d,e}\langle\text{branch C(g,h)}\rangle_{3n}\ \text{f}\langle\text{branch D(x,x)}\rangle\}^{G_2}{}_{4n}:\text{end}_8\text{-cascadane}$$

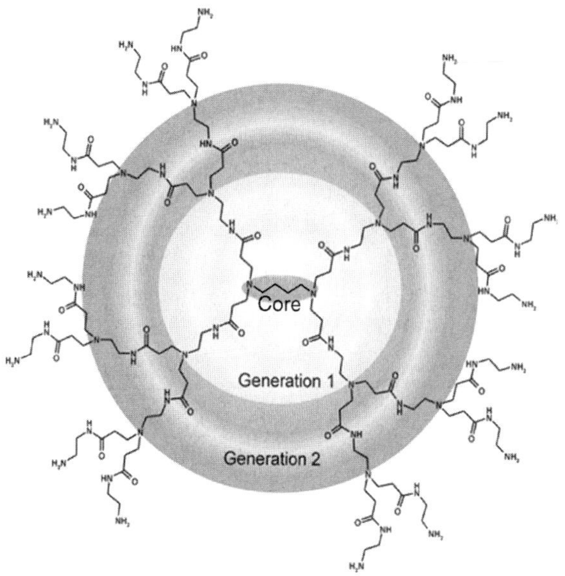

Fig. 1.25 Second generation of a PAMAM dendrimer [45]

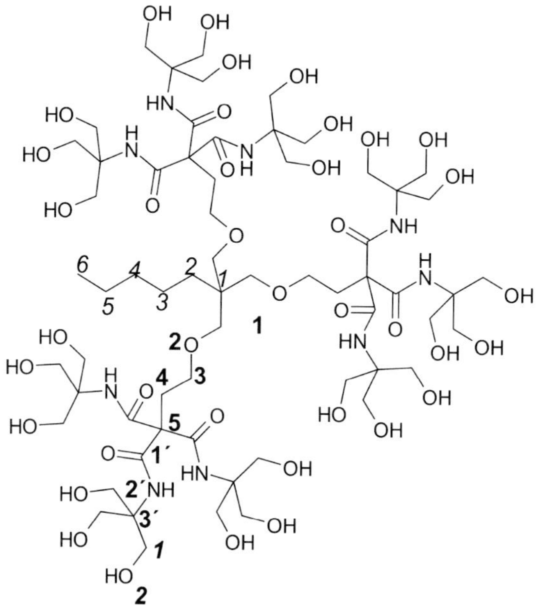

Fig. 1.26 Hexane(1,1,1):{2-oxapentyl(3,3,3)}$^{G_1}{}_{3n}$:
{1-oxo-2-azapropyl(3,3,3)}$^{G_2}{}_{9n}$:hydroxymethyl$_{27}$-cascadane [45]

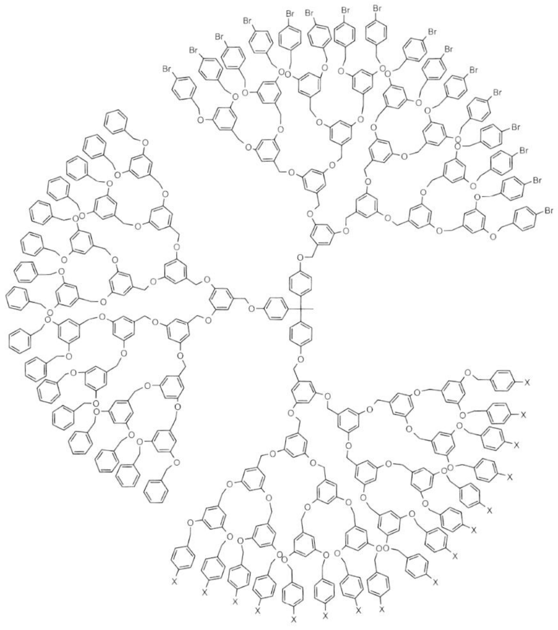

Fig. 1.27 For X=H: 1-(Phenyl (4′):{1-oxa-3-(phenyl-3′,5′-diyl) propyl(3′,5′)}$^{G_1,G_2,G_3,G_4}{}_{1n,2n,4n,8n}$: (1-oxa-4′-brombenzyl)$_{16}$-cascadyl) ethane-1,1-dibenzenol(4″,4‴):{1-oxa-3-(phenyl-3′,5′-diyl) propyl(3′,5′)}$^{G_1,G_2,G_3,G_4}{}_{2n,4n,8n,16n}$: (1-oxabenzyl)$_{32}$-cascadane [45]

Further examples of cascadane nomenclature are given below for the sake of illustration. The reader is referred to the original literature for further explanations and rules [45].

According to cascadane nomenclature, the second-generation PAMAM dendrimer shown in Fig. 1.25 has the name:

1,4-Diaminobutane[N,N,N',N']: {4,7-diaza-3-oxoheptyl(7,7)}$^{G_1,G_2}{}_{4n,8n}$
: 3-aminoethyl$_{16}$-cascadane

Bibliography and Notes for Chapter 1 "Introduction"

Review articles are indicated by the words "Review(s)" or "Book/Books" in bold-faced type.

1 E. Buhleier, W. Wehner, F. Vögtle, *Synthesis* **1978**, 155–158.

2 The term "cascade" was chosen because it evokes analogies to the repetitive branching of the arms, for example the water cascade of a fountain, in which the number of "waterfalls" increases on proceeding from the top small basin to the progressively larger ones below to an extent determinable by the designer.

3 With some justification, this type of molecule could have been called dendritic at the time (see the title of this book), in allusion to corresponding nerve cells whose branching is, however, somewhat more random in nature. Several years later Tomalia coined the term "dendrimers" from the branching of trees [12]. We later used the new designation synonymously with "cascade molecules", because it had been popularised by the review published in *Angewandte Chemie* [10]. Although there is a tendency to reserve the name dendrimers preferentially for high-molecular substances [39a], we use the two terms interchangeably: In our opinion dendritic molecules are by no means limited to polymer chemistry, as the last syllable of the name dendrimer might suggest, but play an important role in lower-molecular chemistry which will undoubtedly increase in the future when synthesis and analysis have advanced further. Nevertheless, Newkome 1993 [44] and we 2006 [45] use the more suitable root "cascade" for construction of a nomenclature for all dendritic compounds.

4 Remarkably, this publication attracted little attention and generated little response at the time and for several years afterwards, nor were any priority claims made. The macrocyclic iterative variant *"non-skid-chain like molecules"* also mentioned in reference [1] shares this fate to the present day.

5 ***Review*** of repetitive/iterative syntheses: N. Feuerbacher, F. Vögtle, *Top. Curr. Chem.* **1998**, *197*, 2–18; Y. Zhao, R. R. Tykwinski, *J. Am. Chem. Soc.* **1999**, *121*, 458–459; P. A. Jacobi, H. Liu, *J. Am. Chem. Soc.* **1999**, *121*, 1958–1959; A. Boydston, Y. Yin, B. L. Pagenkopf, *J. Am. Chem. Soc.* **2004**, *126*, 10350–10354.

6 F. Vögtle, E. Weber, *Angew. Chem.* **1974**, *86*, 896–898; *Angew. Chem. Int. Ed.* **1974**, *13*, 814–816; F. Vögtle, H. Sieger, W. M. Müller, *J. Chem. Research (S)* **1978**, 398–399; ***Review***: F. M. Menger, *Top. Curr. Chem.* 1986, *136*, 1–16.

7 R. G. Denkewalter, J. F. Kolc, W. J. Lukasavage, in *US Pat. 4.360.646*, **1979**; R. G. Denkewalter, J. F. Kolc, W. J. Lukasavage, in *US Pat. 4.289.872*, **1981**; R. G. Denkewalter, J. F. Kolc, W. J. Lukasavage, in *US Pat. 4.410.688*, **1983**.

8 M. Maciejewski, *Macromol. Sci. Chem.* **1982**, *A17*, 689–703.

9 P.-G. de Gennes, H. Hervet, *J. Phys. Lett. Fr.* **1983**, *44*, L351–L361.

10 ***Review***: D. A. Tomalia, A. Naylor, W. A. Goddard III, *Angew. Chem.* **1990**, *102*, 119–157; *Angew. Chem. Int. Ed.* **1990**, *29*, 138–175.

11 P.-G. de Gennes, *Angew. Chem.* **1992**, *104*, 856–857, *Angew. Chem. Int. Ed.* **1992**, *31*, 842–845.

12 D. A. Tomalia H. Baker, J. Dewald, M. Hall, G. Kallos, S. Martin, J. Roeck, J. Ryder, P. Smith, *Macromolecules* **1986**, *19*, 2466–2468.

13 D. A. Tomalia, H. Baker, J. R. Dewald, M. Hall, G. Kallos, S. Martin, J. Roeck, J. Ryder, P. Smith, *Polym. J.* **1985**, *17*, 117–132.

14 a) G. R. Newkome, Z.-Q. Yao, G. R. Baker, V. K. Gupta, *J. Org. Chem.* **1985**, *50*, 2003–2004; b) ***Book***: G. R. Newkome, C. N. Moorefield, F. Vögtle, *Dendrimers and Dendrons*, Wiley-VCH, Weinheim 2001, 1st Edit.

15 C. Hawker, J. M. J. Fréchet, *J. Chem. Soc., Chem. Commun.* **1990**, 1010–1013.

16 T. M. Miller, T. X. Neenan, *Chem. Mater.* **1990**, *2*, 346–349.

17 P. J. Flory, *J. Am. Chem. Soc.* **1941**, *63*, 3091–3100; **1952**, *74*, 2718–2723.

18 a) ***Short review***: M. Freemantle, *Science/Technology* **1999**, *77*, 27–35;

b) J. P. Majoral, A.-M. Caminade (eds.), *Book of Abstracts*, 5th International Dendrimer Symposium, Tolouse **2007**.

19 ***Books***: a) For further examples of branching in nature, see also G. Pölking, *Schöpfungsdesign*, Tecklenburg, Steinfurt **2006**; b) B. H. Kaye, *A Random Walk through Dimensions*, VCH-Verlagsgesellschaft, Weinheim **1994**.

20 ***Reviews***: M. Žinić, F. Vögtle, F. Fages, *Top. Curr. Chem.* **2005**, *256*, 39–76; F. Fages, F. Vögtle, M. Žinić, *Top. Curr. Chem.* **2005**, *256*, 77–131, and further literature cited therein.

21 a) K. Autumn, V. A. Liang, S. T. Hsieh, W. P. Chan, T. W. Kenny, R. Fearing, R. J. Full, *Nature* **2000**, *405*, 681–685; b) E. Pennisi, *Science* **2000**, *288*, 1717–1718; c) S. Gorb, *J. Royal Soc. Interface* (DOI: 10.1098/rsif.2006.0164); c) ***Short overview***: *GIT Labor-Fachzeitschrift* **2008**, *7*, 629; d) K. Takahashi, J. O. L. Berengueres, K. J. Obata, S. Saito, *Int. J. of Adhesion and Adhesives* **2006**, *26*, 639–643; cf. *Nachr. Chem.* **2008**, *56*, 406.

22 R. Haag, F. Vögtle, *Angew. Chem.* **2004**, *116*, 274–275; *Angew. Chem. Int. Ed.* **2004**, *43*, 272–273.

23 K. Hien, *Laborjournal 1–2*/**2004**, 32–34; K. Autumn, R. Full, *GEO Magazin* 10/**2000**; J. Kahn, M. Thiessen, K. Eward, *National Geographic Deutschland*, June **2006**.

24 Design: Roman & Erwan Bouroullec.

25 a) Photo taken by F. Vögtle; see also *Generalanzeiger*, Bonn, 12. 03. 2007; b) U. Hahn, G. Pawlitzki, F. Vögtle, June **2001**; DFG Calendar, published by Deutsche Forschungsgemeinschaft, Bonn **2003**; c) ***Overview***: D. Singh, G. Fernandes, R. Sinha, A. Jerath, M. Lindsay, *Bäume der Welt*, Dorling Kindersley, Starnberg **2006**.

26 ***Books***: B. B. Mandelbrot, *The Fractal Geometry of Nature*, W. H. Freeman and Company, New York, **1982**; G. Binnig, *Aus dem Nichts*. Piper, München, 4th Edit. **1992**; H.-O. Peitgen, P. H. Richter, *The Beauty of Fractals*. Springer, Berlin **1986**; H.-O. Peitgen, H. Jürgens, D. Saupe, *Chaos and Fractals*. Springer, Berlin **1992**; B. Kaye, *Chaos and Complexity*. VCH, Weinheim **1993**.

27 W. Sierpinski, *C. R. Acad. Paris* **1915**, *160*, 302–305; J.-L. Giavitto, S. Spicher, in N. Krasnogor, S. Gustafson, D. A. Pelta, J. L. Verdegay (eds.), *Systems Self-Assembly*: Multidisciplinary Snapshots. *Studies in Multidisciplinarity*, Elsevier, Amsterdam **2008**, *5*, 199–243.

28 We thank Dr. Jörg Friedhofen, Kekulé Institute of the University of Bonn, for these graphics.

29 ***Review***: C. J. Hawker, J. M. J. Fréchet, *Step-Growth-Polymers for High Performance Materials* (Eds. J. L. Hedrick, J. W. Labadie), Oxford Press, Oxford **1996**, Chapter 7; M. Seiler, *Chem. Eng. Technol.* **2002**, *3*, 237–253; comparison of thermosensitivity properties of POPAM dendrimers and linear analogues: Y. Haba, C. Kojima, A. Harada, K. Kono, *Angew. Chem.* **2007**, *119*, 238–241; *Angew. Chem. Int. Ed.* **2007**, *46*, 234–237.

30 E. M. M. de Brabander-van den Berg, E. W. Meijer, *Angew. Chem.* **1993**, *105*, 1370–1372; *Angew. Chem. Int. Ed.* **1993**, *32*, 1308–1311; T. H. Mourey, S. R. Turner, M. Rubinstein, J. M. J. Fréchet, C. J. Hawker, K. L. Wooley, *Macromolecules* **1992**, *25*, 2401–2406.

31 a) ***Reviews***: A. W. Bosman, H. M. Janssen, E. W. Meijer, *Chem. Rev.* **1999**, *99*, 1665–1688; R. Hourani, A. Kakkar, M. A. Whitehead, *J. Mater. Chem.* **2005**, *15*, 2106–2113; b) H. Frey, K. Lorenz, C. Lach, *Chem. unserer Zeit* **1996**, 75–85.

32 L. Lescanec, M. Muthukumar, *Macromolecules* **1990**, *23*, 2280–2288.

33 M. L. Mansfield, L. I. Klushin, *Macromolecules* **1993**, *26*, 4262–4268; M. L. Mansfield, *Polymer* **1994**, *35*, 1827–1830; D. Boris, M. Rubinstein, *Macromolecules* **1996**, *29*, 7251–7260; N. W. Suek, M. H. Lamm, *Macromolecules* **2006**, *39*, 4247–4255.

34 A. M. Naylor, W. A. Goddard III, G. E. Kiefer, D. A. Tomalia, *J. Am. Chem. Soc.* **1989**, *111*, 2339–2341; R. Scherrenberg, B. Coussens, P. van Vlief, G. Edouard, J. Brackmann, E. de Brabander, K. Mortensen, *Macromolecules* **1998**, *31*, 5892–5897.

35 M. Ballauff, C. N. Likos, *Angew. Chem.* **2004**, *116*, 3060–3082; *Angew. Chem. Int. Ed.* **2004**, *43*, 2998–3020; ***Review***: M. Ballauff, (Volume editor F. Vögtle), *Top.*

Curr. Chem. **2001**, *212*, 177–194; C. N. Likos, M. Schmidt, H. Löwen, M. Ballauff, D. Pötschke, P. Lindner, *Macromolecules* **2001**, *34*, 2914–2920; S. Rosenfeldt, N. Dingenouts, M. Ballauff, P. Lindner, N. Wemer, F. Vögtle, *Macromolecules* **2002**, *35*, 8098–8105, reference to work by other authors cited therein; S. Rosenfeldt, E. Karpuk, M. Lehmann, H. Meier, P. Lindner, L. Harnau, M. Ballauff, *ChemPhysChem.* **2006**, *7*, 2097–2104.

36 C. J. Hawker, R. Lee, J. M. J. Fréchet, *J. Am. Chem. Soc.* **1991**, *113*, 4583–4588.

37 E. Malmström, M. Johansson, A. Hult, *Macromolecules* **1995**, *28*, 1698–1703; Y. Ishida, A. C. F. Sun, M. Jikai, M. Kakimoto, *Macromolecules* **2000**, *33*, 2832–2838; P. Bharathi, J. S. Moore, *Macromolecules* **2000**, *33*, 3212–3218.

38 Cf. PPI in Chapter 4, Fig. 4.1; replacement of propano by ethano groups in the latter formula yields the corresponding third-generation PEI.

39 ***Books***: a) J. M. J. Fréchet, D. A. Tomalia, *Dendrimers and Other Dendritic Polymers*, Wiley, Chichester **2001**; b) G. R. Newkome, C. N. Moorefield, F. Vögtle, *Dendrimers and Dendrons: Concepts, Syntheses, Applications*, Wiley-VCH, New York, Weinheim **2001**.

40 In our opinion an adequate distinction between ideal, defect-free dendritic molecules, for which we propose the term "cascad*anes*", and the more or less defect-free (monodisperse) "dendrimers", and polydisperse, imperfect "hyperbranched compounds" has so far been lacking in the area between polymer chemistry and small-molecule chemistry. Other reasons for such a distinction are because highly pure compounds have always been especially important in small-molecule chemistry, and because of the need to name an individual molecule.

41 O. Safarowsky, B. Windisch, A. Mohry, F. Vögtle, *J. Prakt. Chem.* **2000**, *342*, 337–342; ***Review***: F. Vögtle, O. Lukin, *Angew. Chem.* **2005**, *117*, 2–23; *Angew. Chem. Int. Ed.* **2005**, *44*, 2–23.

42 New IUPC recommendations for rotaxanes: A. Harada, W. V. Metanomski, G. P. Moss, E. S. Wilks, A. Yerin, "*Nomenclature of Rotaxanes*", International Union of Pure and Applied Chemistry, Project 2002-007-1-800.

43 It would be beyond the scope of this textbook to dive further into the details of dendrimer nomenclature. Instead, basic rules are given to create an awareness of the problems and complexity. Nevertheless, with the aid of the literature cited (see ref. [45]) it is possible to develop names for certain formulae – and *vice versa*.

44 G. R. Newkome, G. R. Baker, J. K. Young, J. G. Traynham, *J. Polym. Sci. A., Polym. Chem.* **1993**, *31*, 641–651; G. R. Newkome, G. R. Baker, *Polym. Preprints* **1994**, *35*, 6–9.

45 J. H. Friedhofen, F. Vögtle, *New J. Chem.* **2006**, *30*, 32–42.

46 ***Review***: *International Union for Pure and Applied Chemistry, Nomenclature of Organic Chemistry Sections A, B & C*, Butterworth, London **1971**.

47 ***Review***: N. Lozac'h, A. L. Goodson, W. H. Powell, *Angew. Chem.* **1979**, *91*, 951–1032; *Angew. Chem. Int. Ed.* **1979**, *18*, 887–899, N. Lozac'h, A. L. Goodson, *Angew. Chem.* **1984**, *96*, 1–15; *Angew. Chem. Int. Ed.* **1984**, *23*, 33–46.

2
Synthetic methods for dendritic molecules

Dendrimer architectures can be constructed in a wide variety of ways. Principal goals in the past were to assure general availability of routine methods for readily accessible dendrimer structures, as has been accomplished with POPAM and PAMAM dendrimers as well as Fréchet dendrons (see Chapter 4). Although other types of dendrimers have not received so much attention from other groups of workers, they nevertheless have the potential to do so. On the other hand, it is both important and desirable to construct new branching generations on such readily accessible and in part commercially available dendrons, to endow them with new functionalities, with pores, or with specific properties (solubility, aggregation behaviour, philicity, luminescence, rigidity, backfolding, chirality, guest inclusion, gel formation, *etc.*). A cherished pipe dream continues to be a readily available "modular synthesis toolkit" for dendrimers with "drawers" filled with readily accessible starting materials and intermediates which can be selected and processed in well-described standard steps and combined to form new molecular structures with new properties. Despite the many contributions on the synthesis of dendrimers published so far, much remains to be desired in this respect. There is still a need for yet higher generations of all types of dendrimers and dendrons and for high-yield syntheses of certain dendrimer families and for their targeted mono-functionalisation, (regioselective) oligo-functionalisation, and multi-functionalisation – also with differing substituents – in the individual dendrimer molecule.

In this chapter general synthetic strategies are presented in schematic form; details are provided later with the aid of specific examples in Chapter 4. The majority of the syntheses considered will be discussed in terms of a general "CFP concept": The coupling sites are designated by a *C* (in a red sphere), the functional end groups by *F* (in a green sphere). If the relevant groups are made to react in their protected form, the corresponding colour coding is retained and the protected functionality subsequently marked with a *P*.

2.1
Divergent synthesis

Synthesis of a dendrimer according to the divergent method (Fig. 2.1) proceeds stepwise starting from a multi-functionalised core building block, to whose reac-

Dendrimer Chemistry. Fritz Vögtle, Gabriele Richardt and Nicole Werner

ISBN: 978-3-527-32066-0

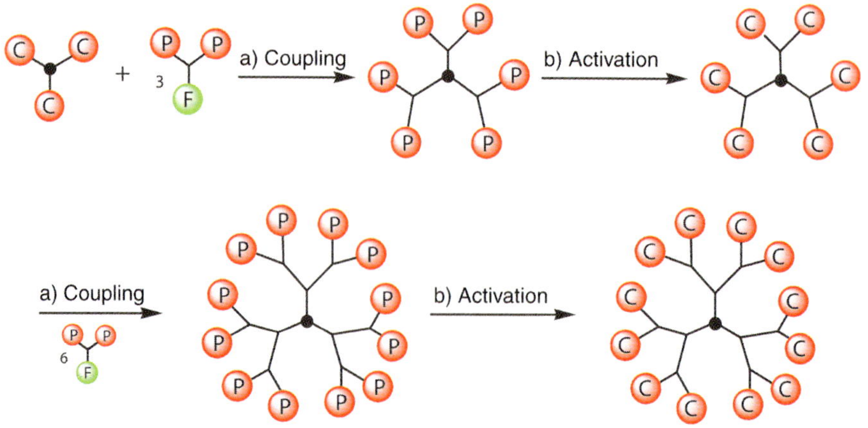

Fig. 2.1 Divergent synthesis. *C*=coupling points; *F*=active, unprotected functional group; *P*=protected, inactive (protective group) functionality. The core of the dendrimer is shown as a black central point. Steps a) coupling and b) activation stand for a repetitive pair of steps for construction of the next generation

tive coupling sites *C* new branching units in the form of dendritic branches are attached *via* a reactive terminal functionality *F*. During reaction, other functional groups of the branching unit are protected as indicated by the letter *P*. After the first reaction step, the protected functional groups *P* are deprotected (activated) and then serve as new reactive coupling sites *C* for further branching units. A new dendrimer generation arises with each branching unit. The *repetitive* (iterative) synthetic sequence, consisting of both the construction step, in which coupling of a branching unit to two further units ($1 \rightarrow 2$ branching) takes place, as well as the activation step, progressively yields higher generations and permits the dendrimer to grow from the inside outwards.

An advantage of the divergent method – which was the first to be developed [1] – is the attainable high-molecular (nano)scaffold architecture as well as the possibility of automation of the repetitive steps. The divergent method is therefore the method of choice for – commercially available – *POPAM* and *PAMAM dendrimers* (see Section 4.1).

One disadvantage of this synthetic methodology is seen in the exponentially increasing number of functional terminal groups ($K \times M^n$; see Section 1.2), since they cannot always be made to react quantitatively and thus give rise to structural defects. Such defects cannot always be avoided, even on addition of large excesses of reactants. Moreover, purification and separation of structurally perfect from defective dendrimers are problematic because the compounds have very similar properties.

The first known syntheses of dendritic molecules were divergent and were used by the working groups of Vögtle [1], Denkewalter [2], Tomalia [3], and Newkome [4].

2.2
Convergent synthesis

The convergent synthesis strategy proceeds in the opposite direction, from the periphery to the core, that is from the outside inwards. (Functionalised) dendrimer components ("dendrons") are bonded to the reactive terminal groups linked to a focal point of a multifunctional core unit. The CFP symbolism is again used to illustrate the principle of the synthesis (Fig. 2.2).

Coupling of an active (unprotected) functional terminal group *F*, bearing two further protected linkage sites *P* (in a red sphere), with a branching unit, consisting of two active coupling sites *C* as well as an inactive (protected) functional group *P* (in a green sphere), leads to a first-generation dendrimer/dendron (step a). For further build-up of the next dendron generation, the inactive group of this G1 dendron can be activated (step b), and a further branching unit formed with two active sites *C* and an inactive functional group *P* – not shown in Fig. 2.2. The steps can be repeated until all segment-shaped dendrons of the desired generation react – like pieces of a pie – with an oligo-functional core module (*e.g.* "C_3" in Fig. 2.2) to form the desired higher-generation dendrimer.

Because of the small number of reactive terminal groups involved, this type of synthesis has the advantage that it does not produce the structural defects (*e.g.* missing branches in higher generations) often observed in the case of the divergent synthetic route. Moreover, this synthetic method can be performed with equimolar quantities – without any need for large excesses – thus facilitating preparative work-up. The by-products formed – owing to incomplete reaction of bulky dendrons with the branching unit – differ drastically in molecular mass and can be removed more easily after each step than in the case of divergent synthesis.

The dimensions of dendrimer growth are subject to limitations set by steric hindrance during reaction of the dendrons at the periphery. This is the reason

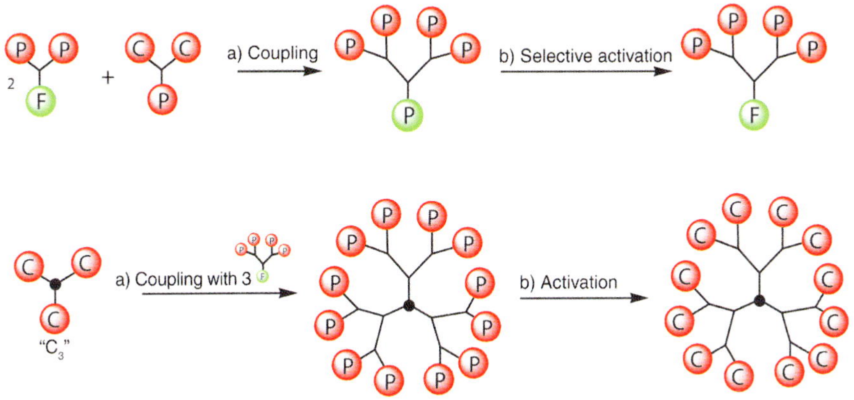

Fig. 2.2 Convergent synthesis (*C*=coupling group; *F*=functional group; *P*=protective group)

why this synthetic strategy is used mainly for the preparation of lower-generation dendrimers [5]. Hence divergent and convergent syntheses are in a way complementary.

The convergent synthetic strategy is well suited for the production of macromolecular architectures such as *segmented-block dendrimers,* which bear either the same or different generations of dendrons, but with different molecular scaffolds linked to a core unit. This type of dendrimer is of interest owing to its multifunctionality.

In the case of *surface-block dendrimers,* the dendrimer periphery exhibits different functionalities in specific molecular segments. They are formed by coupling of dendrons differing in the nature of their terminal functionalities to a common core unit (cf. Section 3.2.1) [6, 10].

2.3
Recent synthetic methods

2.3.1
Orthogonal synthesis

In an orthogonal synthesis [7] two different branching units with complementary coupling functions are used alternatingly and no activation step is employed [8]. The selected reactants as well as the resulting coupling product must be inert towards the subsequent reaction conditions. The term orthogonal means that the functionalities are initially inert towards the coupling conditions, but can be activated *in situ* for the desired subsequent reaction or coupling (Fig. 2.3). If this condition is met, the dendrimer can be constructed divergently or convergently in just a few steps [9].

However, the method of orthogonal coupling has still not been very widely adopted because the building blocks used have to meet very stringent structural requirements [10]. Spindler and Fréchet were the first to prepare a third-generation polyether carbamate dendron – starting from 3,5-diisocyanatobenzyl chloride and 3,5-dihydroxybenzyl alcohol – in a one-pot synthesis [8]. Zimmerman *et al.* reported the first application of orthogonal coupling to the synthesis of higher-generation dendrimers [9].

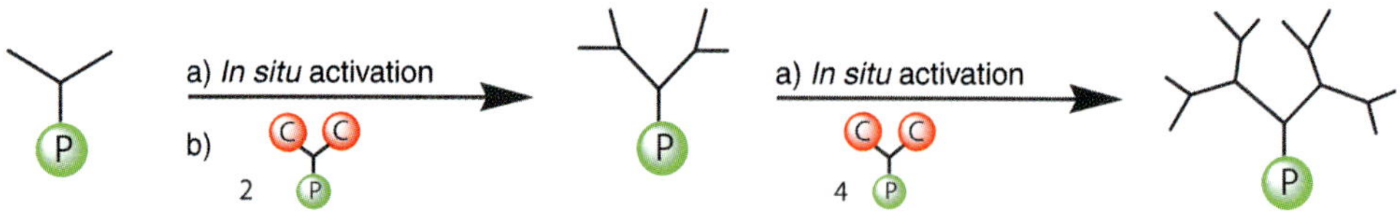

Fig. 2.3 Orthogonal synthesis (*C*=coupling group; *P*=protected group). On *in-situ* activation, *P* is temporarily transformed into *F* (= functional group), which undergoes spontaneous further reaction

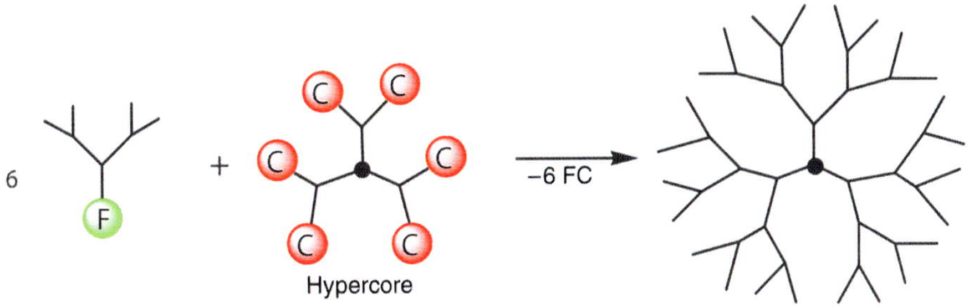

Fig. 2.4 Double-stage convergent method (*C*=coupling group, *F*=functional group)

2.3.2
Double-stage convergent method

This new variant is a combination of the convergent and the divergent method. Its significant step consists in the coupling of a small dendron with an active functional group *F* at the focal point, produced by convergent synthesis, to a low-generation multifunctional dendrimer with peripheral coupling sites *C*, the *hypercore*, produced by divergent synthesis (Fig. 2.4) [11,12].

Compared to conventional convergent synthesis, the two-stage method promises fast access to higher-generation monodisperse dendrimers, combined with a jump in terminal functionalities, since the surface functionalities of the hypercore are subject to less steric hindrance than in the case of simple non-dendritic core building blocks of conventional convergent synthesis. Moreover, this approach permits the formation of dendrimers with different inner and outer branching units (*layer-block dendrimers*) [10, 13].

2.3.3
Double-exponential method

The double-exponential method can basically be regarded as a convergent growth strategy for a dendron [14]. It involves synthesis in two directions, towards the periphery and towards the focal point: Starting from a completely protected branching unit, consisting of two protected coupling sites *C* and a protected functionality *F*, selective deprotection is performed such that the functional group *F* is activated in one of the branching units and the two coupling units are activated in the other. If two branching units with active groups *F* (marked green in Fig. 2.5) now react with such a group having two coupling sites *C* (marked red), a second-generation dendron is formed. Iteration of the synthetic sequence leads to the corresponding fourth-generation dendron.

Access to a fourth-generation Fréchet dendron can thus be shortened by one stage, since no first-generation dendron can be formed by this method – *i.e.* by doubling of the numbers of generations.

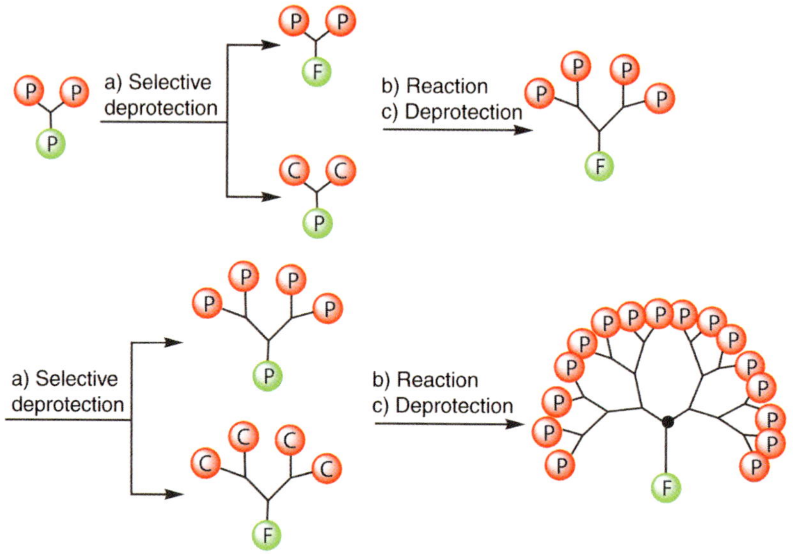

Fig. 2.5 Double-exponential method (*C*=coupling group, *F*=functional group; *P*=protected group)

2.3.4
Hypermonomer method

Compared to conventional FC_2 or FC_3 *monomers*, the number of terminal groups increases faster ($1 \rightarrow 4$ branching) with monomeric FC_4 synthetic building blocks (*i.e. hypermonomers*). However, the number of synthetic steps required for construction of a dendrimer remains the same as in the case of preparation by conventional methods. On reduction to the simple CFP scheme, the course of synthesis can be described as follows (Fig. 2.6): Four branching units with an active functional group *F* are made to react with active coupling groups *C* of the hypermonomer FC_4, where the functional group *F* in the hypermonomer itself is deactivated and therefore (in Fig. 2.6) labelled *P* for "protected" [15].

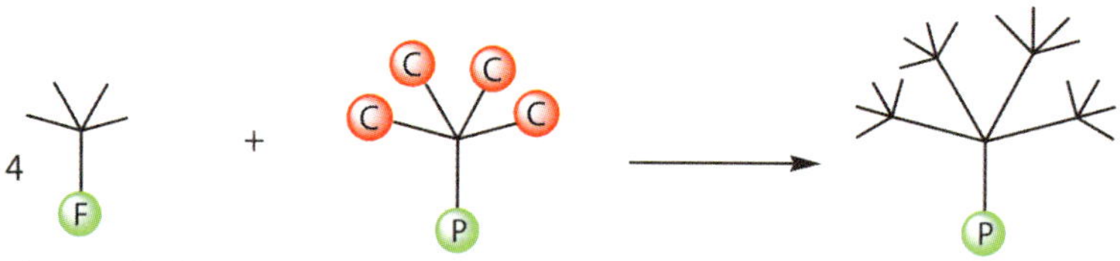

Fig. 2.6 The hypermonomer method (*C*=coupling unit, *F*=functional group, *S*=protective group)

2.3.5
Click chemistry

In 2001 Sharpless proposed this method – in which smaller units are joined together by way of heteroatom bonds (C–X–C; X=heteroatom) – as a fast track route to compounds [16]. The criteria for a "click reaction" were recently summarised: broad applicability with high yields; readily accessible starting compounds; readily separable by-products or none at all; straightforward reaction conditions; easy product isolation; stereospecific.

Typical reactions applied are 1,3-dipolar cycloadditions, nucleophilic substitutions for ring opening of strained electrophilic heterocycles, as well as additions to carbon-carbon multiple bonds (*e.g.* epoxidation). Just one representative example will be mentioned here since click chemistry in itself does not represent a new type of synthesis, but merely facilitates the course of synthesis and work-up of the products through favourable choice of reaction parameters mentioned above: Wooley and Hawker *et al.* [17] prepared second and third-generation dendrimers with a "divergent click strategy". For this purpose, a first-generation azido-dendrimer was transformed in a Cu(I)-catalysed reaction with an alkynylated monomer into a triazole dendrimer with terminal hydroxyl groups, which in turn were transformed in a second step into azido functions, in order to again undergo *repetitive* reaction with fresh alkynylated monomers [18 a, b].

Unprotected glycol-dendrimers, peptido- and redox-dendrimers and dendronized polymer organogels can also be prepared by click chemistry [18 c].

2.4
Solid phase synthesis

Solid-phase methodology was established in 1963 in pioneering work conducted by Merrifield in the area of peptide synthesis [19]. Interest in this synthetic strategy continues unabated to this day, particularly in connection with the production of new active components for drugs, since the repetitive amide bond formation performed in automated synthesisers lends itself ideally to the construction of extensive substance libraries by combinatorial chemistry [20].

The synthetic scheme shown in Fig. 2.7 proceeds in the first step as a covalent coupling of substrate (*B*) *via* a linker to the functional group *A* (*e.g.* $-NH_2$) attached in a prior step to a commercially available solid phase. This solid phase usually consists of an insoluble polymeric material, with polystyrene most frequently serving as the polymeric support – generally in the form of beads. Of critical importance for optimum reaction are good swelling properties of the support material: the greater the swelling, the greater the surface area available for chemical reactions. Reaction of the new substrate C with the previously coupled substrate (*solid phase–A–B*) is followed either by (generally hydrolytic) cleavage of the product *B–C* formed on the solid phase from the support, or by further reaction with substrates *D, E, F,* to give a linear sequence *B–C–D–E–F* –

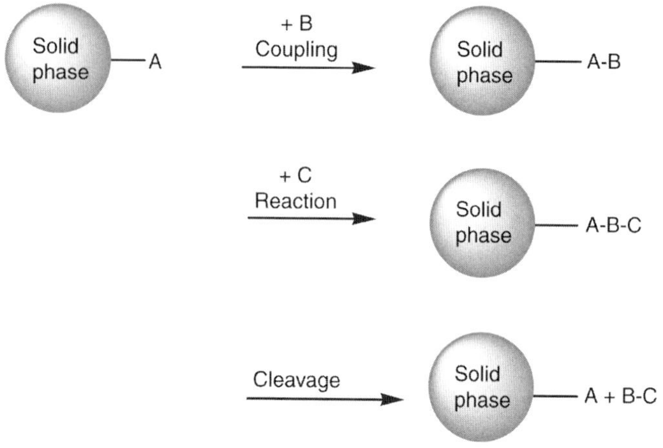

Fig. 2.7 Solid-phase synthesis (schematic); for the sake of simplicity only one of the many functional groups (A) attached to a solid phase bead is shown

or isomers thereof with the letters in a different order – should the substrate solutions be added in a different order.

Recent solid-phase syntheses utilise light-sensitive linkers which can subsequently be cleaved photochemically. And oxidative and reductive cleavage concepts are also used. The support material is subsequently filtered off and reused after washing.

In contrast to reactions in solution, solid-phase synthesis has the advantage that excess amounts of reactants can be used and the yields thus increased. Work-up and purification processes – which often prove difficult in homogeneous solution – are rationalised as straightforward washing or filtration. Recycling of the support material after cleavage of the product from the solid phase also has cost benefits.

Notwithstanding all its advantages, the principle of solid-phase synthesis cannot be applied to all kinds of chemical reactions. Although reactants are used in excess, reaction is not always quantitative. The resulting impurities cannot be separated on the solid phase, giving rise to separation problems particularly in multistep systems. Moreover, only limited use can often be made of conventional analytical methods (NMR, MS). Recent methods of ^{13}C-NMR spectroscopy on solid phases [21] or in gel phases [22] are ideally suited for solid-phase synthesis, but are not universally available owing to the expensive instrumentation.

In the present context, solid-phase synthesis has been used primarily for the preparation of peptide and glycopeptide dendrimers [23]. For example, a second-generation dendrimer could be prepared by successive addition of branched polyproline building blocks to a solid phase [24]. The divergent synthesis of polyamide dendrimers on polystyrenes was accomplished [25] by Fréchet *et al.* 1991 (cf. Fig. 2.8) [26]. PAMAM dendrons could be grown up to the fourth-

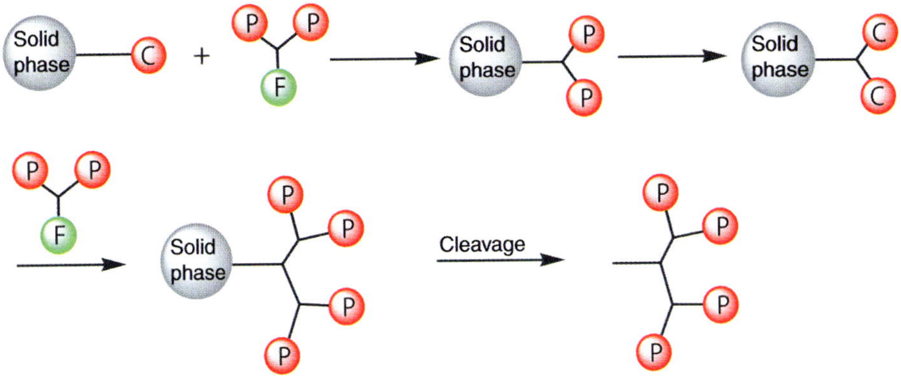

Fig. 2.8 Divergent synthesis of dendritic molecules on a solid phase (*C*=coupling group, *F*=functional group; *P*=protective group)

generation [27]. Solid-phase synthesis was also employed for polylysine dendrimers, whose basic structures are used as "multiple antigen peptides" (MAP) [28]. Resin-bonded polylysine dendrimers have meanwhile become commercially available (*Novabiochem, Colcom, Sigma-Aldrich*).

2.5 Coordination-chemical synthesis

Dendritic building blocks having the nature of complex ligands can coordinate with a central metal unit [29]. Such "self-assembly" opens up direct access to metallodendrimers (see also Section 4.1.7). It should, however, not go unmentioned in this context that the strong Ru^{2+} bond to the bipyridine nitrogen – although non-covalent – is not regarded as "supramolecular" by some chemists because of its lack of reversibility. That is why we consider it from a coordination-chemical viewpoint [30].

2.5.1 Metal complex as core unit

There are basically two routes available for the preparation of dendrimers with a metal complex as core unit. One of these approaches starts from a preconstructed metal complex, whose ligand framework is covalently substituted with dendritic groups. Aida [31], Diederich [32], and Kaifer [33] were among those to use this strategy.

In a second variant, growth of metallodendrimers can proceed *via* complexation of a metal cation with dendritic ligands. In this way, Balzani, Vögtle, De Cola *et al.* [34] obtained photoactive ruthenium complexes by spontaneous self-assembly of the components starting from various dendritically substituted bipyridines. Fig. 2.9 shows a representative example (see Sections 5.1.2.3 and 5.2.1 for other photoactive complexes).

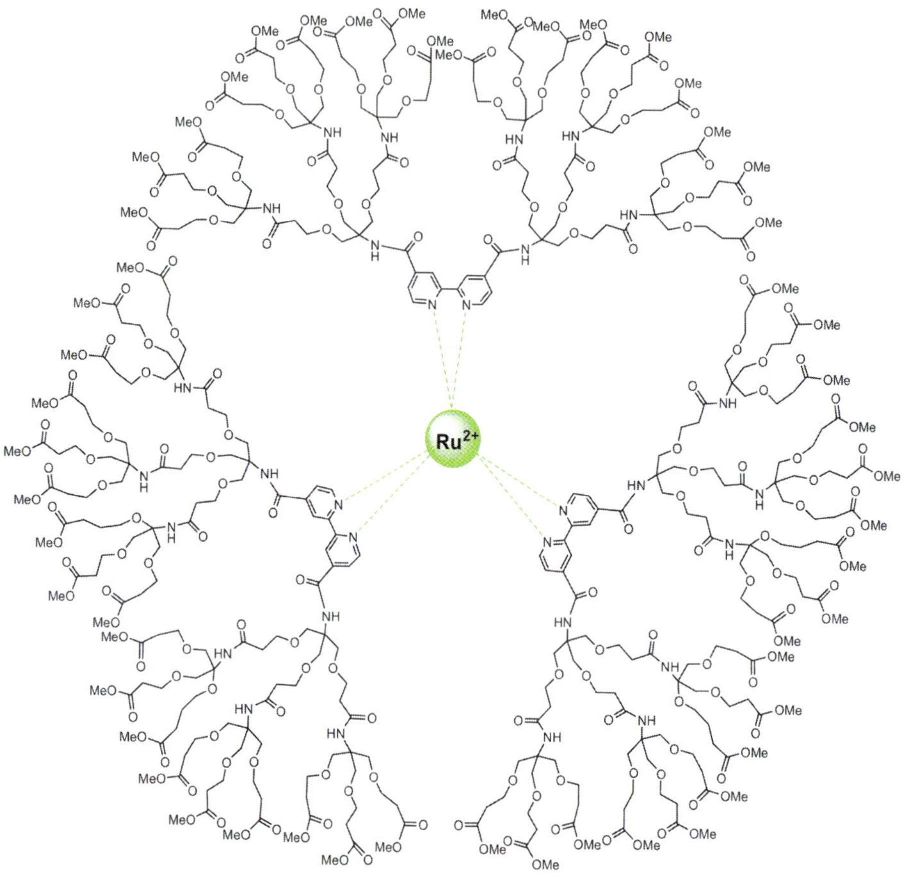

Fig. 2.9 Ru^{2+}-tris-bipy complex with dendritic periphery

An in-depth consideration of the product variety offered by this strategy will be found in Section 4.1.11 (metallodendrimers) and in the pertinent literature [31, 32, 35].

2.5.2
Metal complexes as branching unit

Balzani *et al.* prepared dendrimers with metal complexes serving both as core [36] and as branching unit. The metallodendrimer in Fig. 2.10 is constructed solely from polypyridine ligands and transition metal ions. Such dendritic transition metal complexes can be synthesised both convergently and divergently and different transition metal ions (ruthenium/osmium) can be incorporated. This provides a means of influencing the luminescence properties of the dendrimer. Thus the energy transfer process proceeds from the inside outwards in

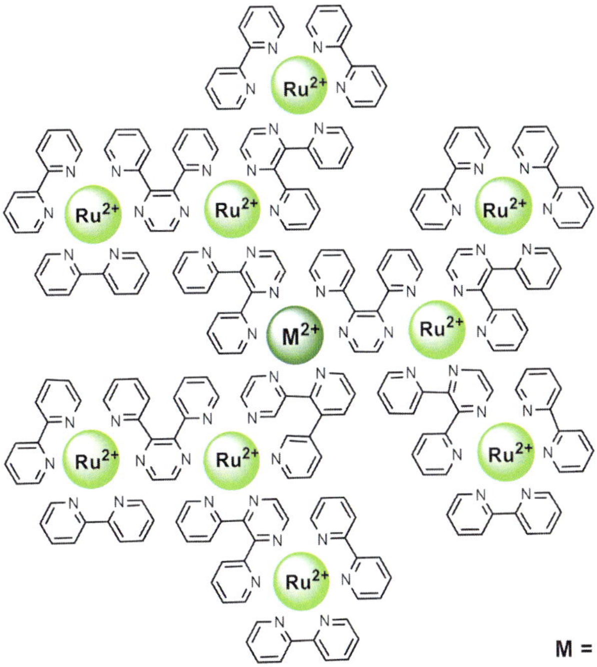

Fig. 2.10 Metallodendrimers (according to Balzani *et al.*)

a dendrimer with a ruthenium ion as central metal and peripheral osmium. If the central atom is osmium, then energy transfer proceeds in the opposite direction, from the outside inwards [37].

Newkome *et al.* prepared metallodendrimers with a ligand/metal/ligand architecture allowing separate construction of the dendrons. Two polyamide dendrons were preconstructed and linked to a ruthenium complex [38] (see Section 4.1.11).

Further complex-chemical dendrimer syntheses (*"metal-directed self-assembly utilising metal ions as convex templates"*) are to be found in the literature [39].

2.6 Supramolecular synthesis

In contrast to the methods already presented, the supramolecular synthesis [40b] of dendrimers [40b] does not involve covalent bond formation, but instead exploits *non-covalent interactions.*

Fréchet *et al.* were able to coordinate polyether dendrons having carboxylate functional groups at the focal point with lanthanide ions up to the fourth generation (Fig. 2.11) [41]. Preparation by straightforward ligand exchange starting from lanthanide triacetates with dendron carboxylates was made possible by the

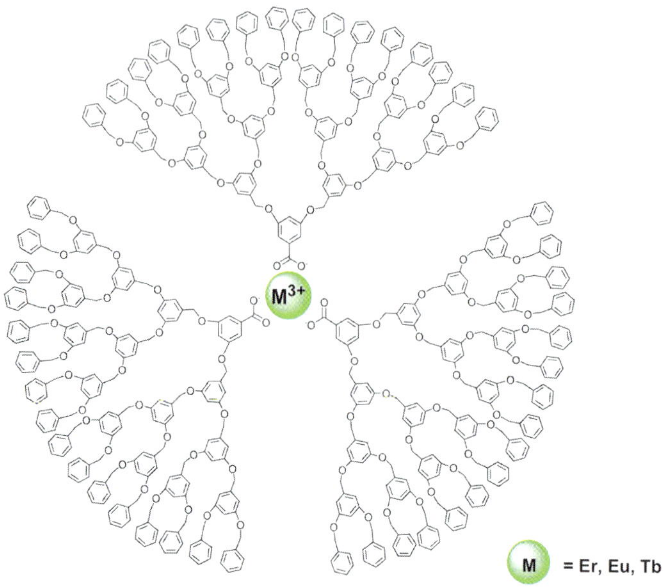

Fig. 2.11 Lanthanide ion as core unit of a dendritic metal complex

predominantly ionic – and reversible – interactions between the lanthanide ion bearing a triple positive charge and carboxylate groups.

A dendritic "two-component gelator" was synthesised by Smith *et al.* on the basis of self-assembling acid-base/hydrogen bond interactions. Dendritic lysine building blocks serve as dendrons, and an aliphatic diamine as core (Fig. 2.12). Depending upon the choice of building blocks, the supramolecular complex forms fibrous gel phases by hierarchic self-organisation.

The dendritic peptides used with D- and L-lysine building blocks each contain three stereocentres. The chirality of D- and L-lysine exerts a controlling effect on structure during self-assembly in the gel fibres and hence on the morphology and macroscopic properties of the product. LLL- or DDD enantiomeric units lead to fibres, whereas the corresponding racemic gels are more prone to form planar structures. The chirality accordingly affects the pattern of hydrogen bonding during the formation of molecular aggregates [42].

Lüning *et al.* reported on attempts at supramolecular linkage of branching units and core units to form dendrimers [43]. To this end, the building blocks are provided with recognition units (Fig. 2.13), which can undergo spontaneous selective self-assembly with other non-identical components *via* hydrogen bonds. In order to assure the desired strength of bonding and stability, the recognition units must have several positions suitable for hydrogen bonding in the molecule. Moreover, certain sequences of acceptor (carbonyl oxygen) and donor properties (HN groups) generate a "supramolecular regioselectivity", in that only those molecular building blocks which have adequate complementary bonding sites dock onto one another.

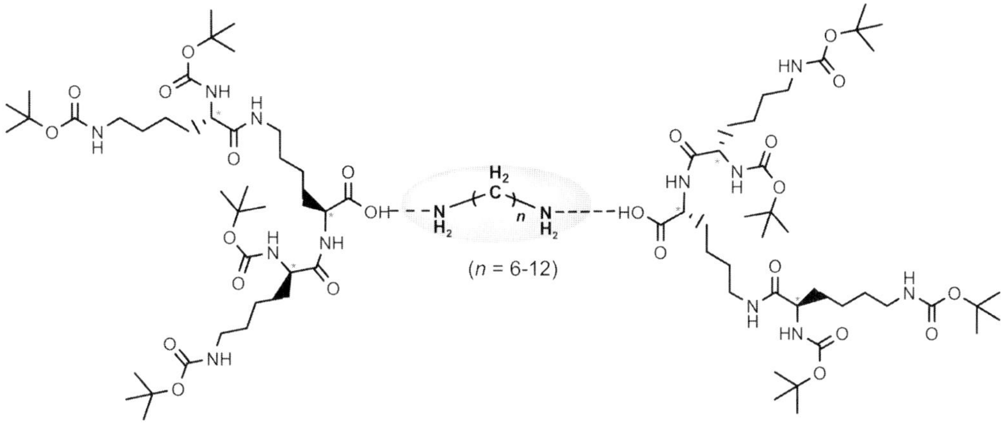

Fig. 2.12 Dendritic two-component gelator produced by supramolecular synthesis (according to Smith *et al.*); stereocentres are marked red

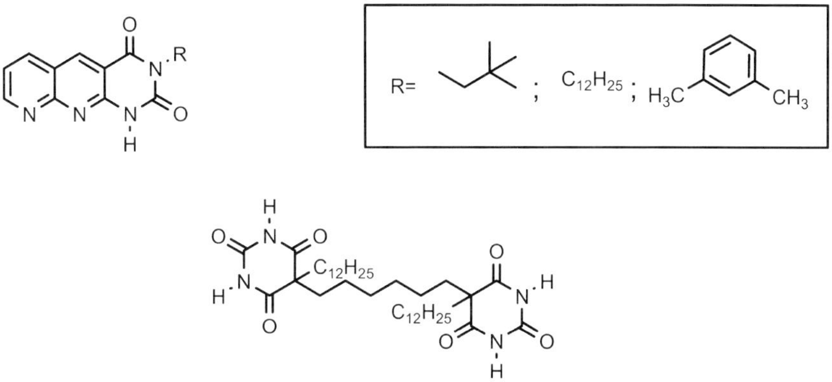

Fig. 2.13 Examples of recognition and core units with designed H-donor- and acceptor units (according to Lüning *et al.*)

This concept could be used for self-organised generation-wise growth of dendrimers by controlled shell-by-shell construction around the core utilising various recognition units. However, the poor solubility of the multiple amide and urea structural elements has so far made isolation of pure oligomeric products difficult.

Complete self-construction of dendrimers according to a modular principle was accomplished by Hirsch *et al.* The synthetic strategy adopted will be considered only briefly here because it is described in greater detail in Chapter 6: Starting from a core unit with three recognition domains and two recognition units (identical with those of the core unit), a branching unit is bound to a complementary recognition domain *via* hydrogen bonding and built up to the desired generation, whereupon appropriate end groups are attached (Fig. 2.14) [44].

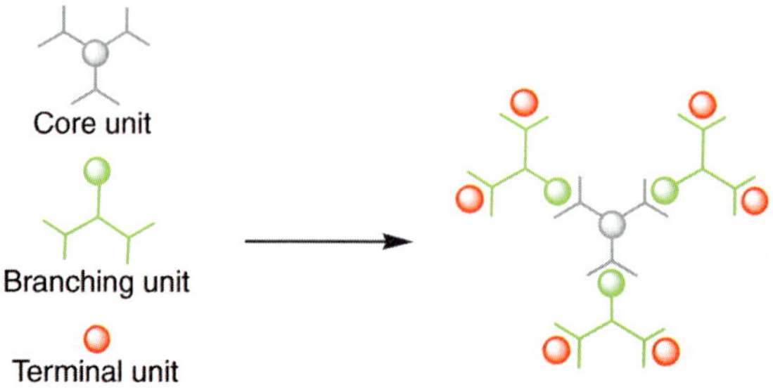

Fig. 2.14 Supramolecular self-construction of dendrimers (schematic; according to Hirsch *et al.*)

Further supramolecular dendrimer syntheses and syntheses of dendrimers with metal complex or polyglycerol structure will be found in the literature [45, 39b].

2.7
Hyperbranched polymers

Preparation of dendrimers requires a high degree of purity of the starting materials and high yields of the individual synthetic steps, all of which generally increases the effort involved. Polydisperse, hyperbranched compounds, which admittedly show defects yet often display properties similar to their ideally perfect dendritic relations, can readily be synthesised.

Synthesis of hyperbranched polymers [46] proceeds in a single-stage process *via* polyaddition, polycondensation, radical polymerisation, and so forth, of an FC_n monomer (Fig. 2.15). Reaction of the functional *F* groups with the functional *C* (coupling) groups of a second monomer molecule gives rise to randomly branched molecules. Since the *C* groups are present in excess ($n \geq 2$), crosslinking reactions are avoided from the outset. Reaction can be brought to a standstill by addition of stopper components. Since the synthesis of hyperbranched polymers does not involve coupling to a core molecule, but only FC_n monomers react with one another, both branched molecules and linear sequences may be formed [47]. If reactive groups are present during the synthesis of hyperbranched polymers, then a protective group technique is required be-

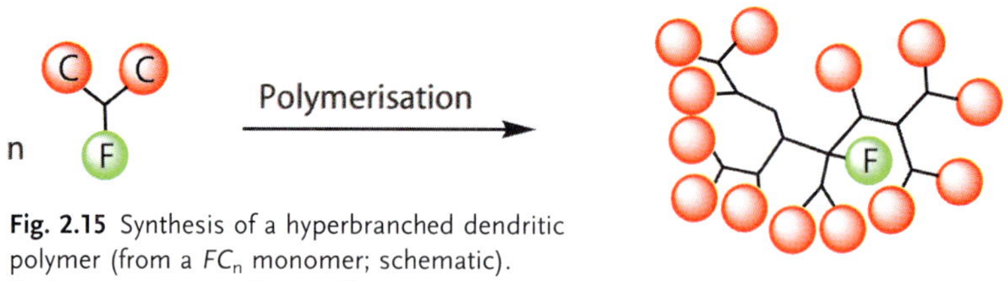

Fig. 2.15 Synthesis of a hyperbranched dendritic polymer (from a FC_n monomer; schematic). *F*=functional group, *C*=coupling site

cause the desired molecular architecture would otherwise not be formed (appropriate protective groups enhance the selectivity of selected groups for bond formation).

Owing to their molecular structures and their properties, hyperbranched polymers represent a transition between linear polymers and highly branched dendrimers.

2.8
Dendronised linear polymers

Dendronised linear polymers are polymers which bear dendrons at more or less regular intervals along their polymeric backbone (cf. Fig. 2.16). They can be assigned to the comb polymers [48], since the arrangement of dendrons resembles that of the teeth of a comb. Apart from polymer-analogous "graft-to-" and "graft-from-" approaches, the commonest synthetic route to dendronised linear polymers is the macromonomer method [49]:

2.8.1
Polymer-analogous method

a) "Graft-to" method

The "*graft-to*" method of attaching dendrons numbers among the polymer-analogous synthetic strategies which start from a functionalised polymer backbone to which convergently preconstructed dendritic units of the desired generation can be fixed in a dense sequence (Fig. 2.16).

This variant can be used for the preparation of dendronised poly(*p*-phenylene) polymers [50].

b) "Graft-from" method

The "graft-from" strategy is a variant of the "graft-to" method since the starting material is again a functionalised polymer, on which the dendritic unit is assembled generation by generation. Specifically, a first-generation dendron is linked to a functionalised polymer and then undergoes divergent growth (Fig. 2.17). Amidoamine dendrimers could be prepared in this way starting from a polyethylenimine (*PEI*) backbone [51].

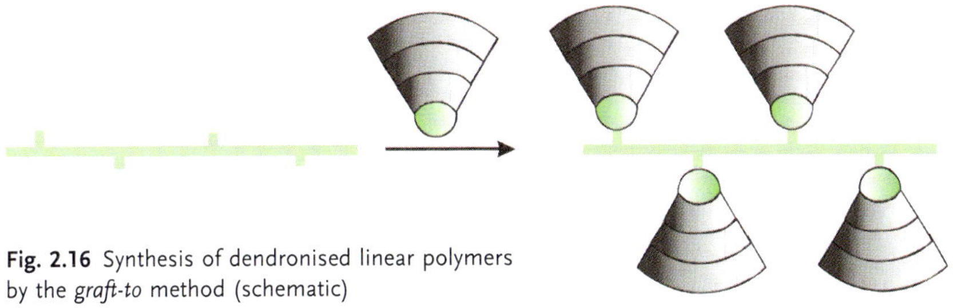

Fig. 2.16 Synthesis of dendronised linear polymers by the *graft-to* method (schematic)

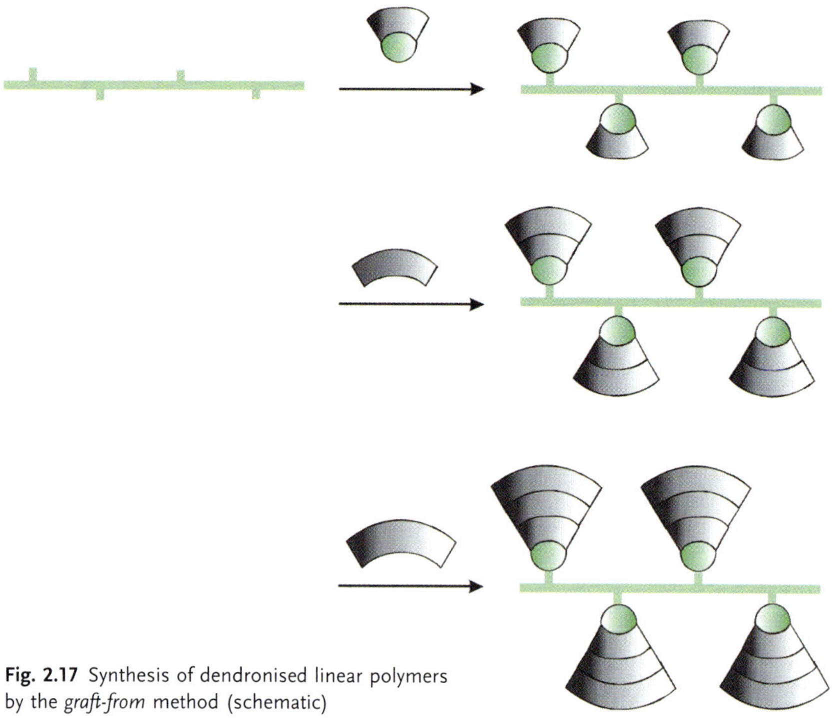

Fig. 2.17 Synthesis of dendronised linear polymers by the *graft-from* method (schematic)

Polymer-analogous variants reach their limits in the face of insufficient rigidity of the functionalised linear polymers which can lead to undesired coiled structures. These have to re-assume a linear structure, with an attendant loss of entropy, in order to assure reaction of all the dendrons with the functional groups attached to the backbone. The dendrons are mostly added in excess in order to facilitate complete reaction, which in turn necessitates tedious purification of the products. The additional steric hindrance occurring on attachment of larger dendrons if higher generation dendrons are already located in the close

Table 2.1 Possible coupling sites on a polymer backbone

Polymer	**Reactive group as linkage sites**
Cellulose	–OH
Poly(vinyl alcohol)	–OH
Poly(vinyl chloride)	–Cl
Poly(acrylic acid)	–COOH
Polyamide	–COOH
Partly hydrolysed poly(vinyl acetate)	–OH
Styrene-divinylbenzene copolymer	vinyl group

vicinity of the polymer backbone can slow down the reaction even to the point of non-reaction. A low reaction rate can lead to side reactions which do not take place when the reaction is fast.

Table 2.1 lists a number of reactive coupling sites – for attachment of dendrons – on various polymer structures.

2.8.2
Macromonomer method

An alternative approach uses macromonomers, that is monomeric building blocks already bearing dendrons, which subsequently undergo polymerisation (Fig. 2.18). Percec *et al.* described the synthesis of dendronised methacrylate monomers and their radical polymerisation [52]. They introduced the term tapered side chains for polymers bearing one dendron per repeating unit. It is essential for the success of the concluding polymerisation step that the dendronised monomers bear polymerisable functionalities such as vinyl, acryl, or oxiran end groups. Apart from radical polymerisation [53], ring-opening metathetic polymerisation [54], Suzuki polycondensation [55], as well as Heck coupling [56] were also employed, to mention just a few methods.

The macromonomer route assures uniform distribution of dendrons along the polymer backbone. Acrylates and styrenes with pendant dendrons proved to be well suited for polymerisation [57]. However, steric hindrance between monomers bearing higher-generation dendrons and the spatial requirements of the end of the chain cause problems: The monomer reacts with the chain end only in the case of slight, if any, steric hindrance. Thus the method only provides access to dendronised linear polymers with relatively low molar masses

To summarise, a common feature of all methods for the preparation of dendronised linear polymers is the change of shape of a formerly flexible filamentary molecule by attachment of dendrons to the backbone of the polymer. As the extent of dendronisation increases, the coil structure extends and acquires greater rigidity, until the polymer assumes a linear extended form with a stiffened backbone. Stiffening of the polymer molecules to this extent has hitherto only been accom-

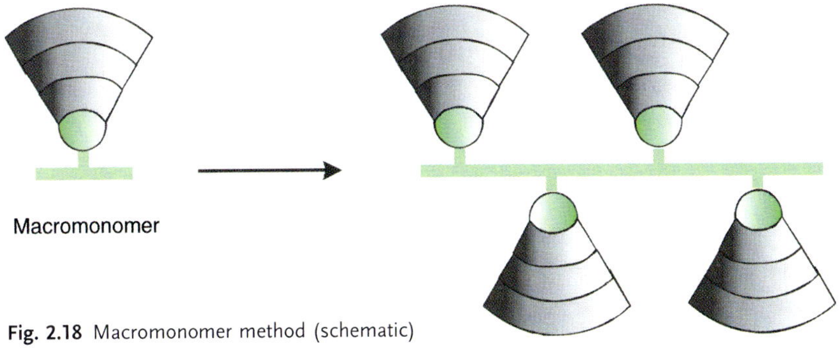

Fig. 2.18 Macromonomer method (schematic)

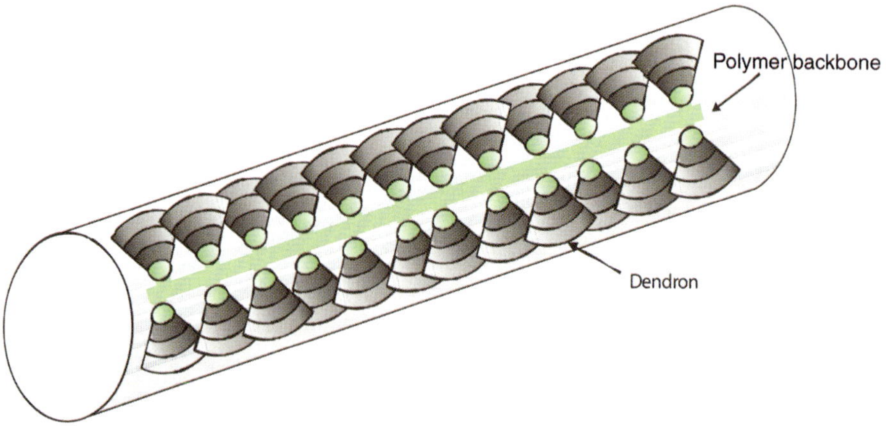

Fig. 2.19 Cylindrical shape of a *den*dronised linear *pol*ymer molecule (*denpol*; schematic)

plished by such attachment of dendrons. Convincing proof was provided by Schlüter, Rabe *et al.* on the basis of STM/AFM images of individual polymer molecules [58]. These molecules could also be moved and manipulated on the surface with the tip of the STM/AFM probe.

Extreme loading of the *polymer* with *dendrons* results in a cylindrical shape of the *"denpol" polymer* (Fig. 2.19) with polydisperse properties. It is thus possible to influence or even control the size of the coupled dendron and the density of its coverage on the polymer backbone through choice of the type of polymer (*e.g.* polyacrylate or polystyrene; see Table 2.1).

The dimensions of the dendritic cylinder depend on the one hand upon the degree of polymerisation, which determines the length, and on the other hand upon the generation of the dendron, which determines the diameter (about twice the size of the attached dendron). Conventional polymers have diameters in the Ångström range, whereas the dendronised linear polymers described here have nanometre diameters.

Such well-defined architectures could be useful for nanoscale applications in, for example, catalysis or as carrier materials for chemical transport.

The parallel arrangement of dendrons on the polymer backbone is also attractive for surface orientation in liquid crystal displays.

2.9 Dendro-Isomers

Selective formation and elimination of sulphonamides permit synthesis of constitutional isomers ("dendro isomers") of dendritic architectures of the kind shown in Fig. 2.20.

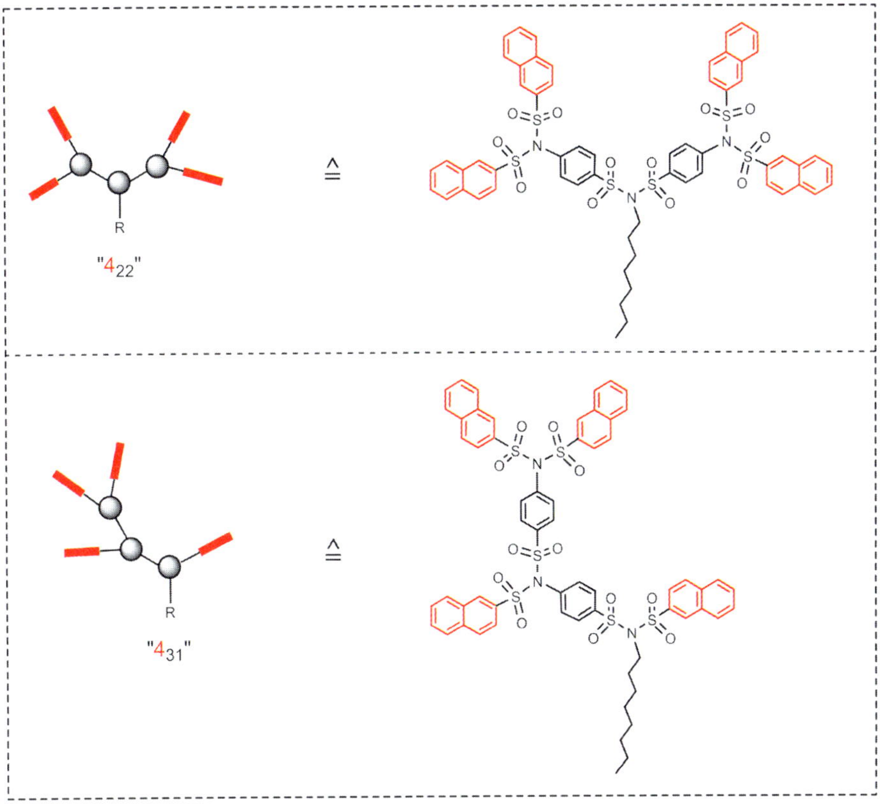

Fig. 2.20 Dendro-isomers – with four end groups (according to Lukin *et al.*). Left: schematic; right: specific example

They have the same molecular mass, but differ in their manner of linkage, their branching, and in the order (sequence) of their atoms (constitutional isomers). The number of possible isomers increases rapidly in higher generations – by analogy with the classical case of the alkanes.

Fig. 2.21 shows examples of isomeric dendrimers with a large number of peripheral groups. The melting points of such isomers are seen to fluctuate widely owing to hydrogen bonding and donor/acceptor effects of the benzene and naphthalene rings involved and depending upon the extent of branching. Moreover, the dendritic sulphonimides crystallise so well – even in the case of higher generations – that information could be obtained about their fine structure from a series of X-ray crystal structure analyses [59].

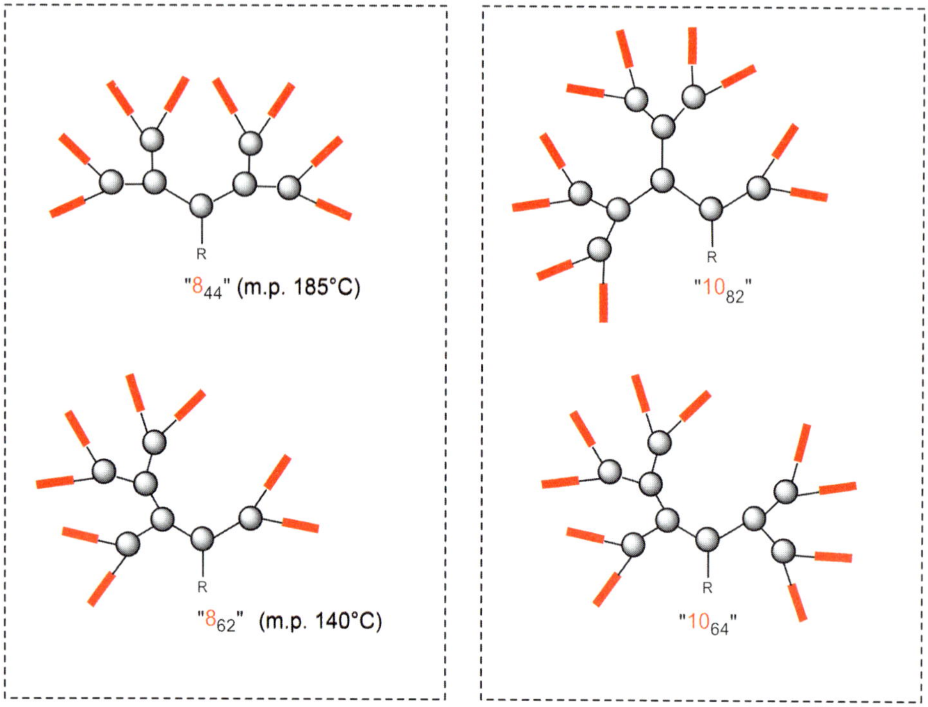

Fig. 2.21 Dendro-isomers – with eight (left) and ten end groups (right; according to Lukin *et al.*); melting points are given as examples for the isomers with eight end groups

Bibliography and Notes for Chapter 2 "Synthetic methods for dendritic molecules"

Review articles are indicated by the words "Review(s)" or "Book/Books" in bold-faced type.

1 E. Buhleier, W. Wehner, F. Vögtle, *Synthesis* **1978**, 155–158.

2 R. G. Denkewalter, J. F. Kolc, W. J. Lukasavage, *US Pat. 4 360 646,* **1979**; R. G. Denkewalter, J. F. Kolc, W. J. Lukasavage, *US Pat. 4 289 872,* **1981**; R. G. Denkewalter, J. F. Kolc, W. J. Lukasavage, *US Pat. 4 410 688,* **1983**.

3 D. A. Tomalia, H. Baker, J. R. Dewald, M. Hall, G. Kallos, S. Martin, J. Roeck, J. Ryder, P. Smith, *Polym. J.* **1985**, *17,* 117–132.

4 G. R. Newkome, Z.-Q. Yao, V. K. Gupta, G. R. Baker, *J. Org. Chem.* **1985**, *50,* 2003–2004.

5 C. J. Hawker, J. M. J. Fréchet, *J. Chem. Soc., Chem. Commun.* **1990**, 1010–1013; C. J. Hawker, J. M. J. Fréchet, *Macromolecules* **1990**, *23,* 4726–4729; *J. Am. Chem. Soc.* **1990**, *112,* 7638–7647; C. J. Hawker, K. L. Wooley, J. M. J. Fréchet, *J. Chem. Soc., Perkin Trans. 1* **1993**, 1287–1297; J. M. J. Fréchet, C. J. Hawker, K. L. Wooley, *J. Macromol. Sci., Pure Appl. Chem.* **1994**, *A31,* 1627–1645; J. S. Moore, Z. Xu, *Macromolecules* **1991**, *24,* 5893–5894; Z. Xu, J. S. Moore, *Angew. Chem.* **1993**, *105,* 261–264; *Angew. Chem. Int. Ed.* **1993**, *32,* 246–248.

6 V. Maraval, R. Laurent, B. Donnadieu, M. Mauzac, A.-M. Caminade, J.-P. Majoral, *J. Am. Chem. Soc.* **2000**, *122,* 2499–2511.

7 ***Books** and **Reviews***: G. R. Newkome, C. N. Moorefield, F. Vögtle, *Dendrimers and Dendrons: Concepts, Syntheses, Applications*, Wiley-VCH, Weinheim **2001**; J. M. J. Fréchet, D. A. Tomalia, *Dendrimers and Other Dendritic Polymers*, Wiley, New York **2002**; I. In, S. Y. Kim, *Macromolecules* **2005**, *38*, 9399–9401.
8 R. Spindler, J. M. J. Fréchet, *J. Chem. Soc., Perkin Trans. 1* **1993**, 913–918.
9 F. Zeng, S. C. Zimmerman, *J. Am. Chem. Soc.* **1996**, *118*, 5326–5327.
10 ***Review***: S. M. Grayson, J. M. J. Fréchet, *Chem. Rev.* **2001**, *101*, 3819–3867.
11 K. L. Wooley, C. J. Hawker, J. M. J. Fréchet, *J. Am. Chem. Soc.* **1991**, *113*, 4252–4261.
12 T. M. Miller, T. X. Neenan, R. Zayas, H. E. Bair, *J. Am. Chem. Soc.* **1992**, *114*, 1018–1025; Z. F. Xu, M. Kahr, K. L. Walker, J. S. Moore, *J. Am. Chem. Soc.* **1994**, *116*, 4537–4550; H. Ihre, A. Hult, J. M. J. Fréchet, I. Gitsov, *Macromolecules* **1998**, *31*, 4061–4068.
13 V. Maraval, R. Laurent, B. Donnadieu, M. Mauzac, A.-M. Caminade, J.-P. Majoral, *J. Am. Chem. Soc.* **2000**, *122*, 2499–2511.
14 R. Klopsch, P. Franke, A. D. Schlüter, *Chem. Eur. J.* **1996**, *2*, 1330–1334; T. Kawaguchi, K. L. Walker, C. L. Wilkins, J. S. Moore, *J. Am. Chem. Soc.* **1995**, *117*, 2159–2165.
15 K. L. Wooley, C. Hawker, J. M. J. Fréchet, *Angew. Chem.* **1994**, *106*, 123–126; *Angew. Chem. Int. Ed.* **1994**, *33*, 82–85; B. Forier, W. Dehaen, *Tetrahedron* **1999**, *55*, 9829–9846; G. L'abbé, B. Forier, W. Dehaen, *Chem. Commun.* **1996**, 2143–2144.
16 H. C. Kolb, M. G. Finn, K. B. Sharpless, *Angew. Chem.* **2001**, *113*, 2056–2075; *Angew. Chem. Int. Ed.* **2001**, *40*, 2004–2021.
17 P. Wu, A. K. Feldman, A. K. Nugent, C. J. Hawker, A. Scheel, B. Voit, J. Pyun, J. M. J. Fréchet, K. B. Sharpless, V. V. Fokin, M. J. Joralemon, R. K. O'Reilly, J. B. Matson, A. K. Nugent, C. J. Hawker, K. L. Wooley, *Macromolecules* **2005**, *38*, 5436–5443.
18 a) M. Malkoch, K. Schleicher, E. Drockenmuller, C. J. Hawker, T. P. Russell, P. Wu, V. V. Fokin, *Macromolecules* **2005**, *38*, 3663–3678; b) P. Wu, A. K. Feldman, A. K. Nugent, C. J. Hawker, A. Scheel, B. Voit, J. Pyun, J. M. J. Fréchet, K. B. Sharpless, V. V. Fokin, *Angew. Chem.* **2004**, *116*, 4018–4022; *Angew. Chem. Int. Ed.* **2004**, *43*, 3928–3932; c) E. Fernandez-Megia, J. Correa, I. Rodriguez-Meizoso, R. Riguera, *Macromolecules* **2006**, *39*, 2113–2120; K.-N. Lau, H.-F. Chow, M.-C. Chan, K.-W. Wong, *Angew. Chem.* **2008**, *120*, 5606–5624; *Angew. Chem. Int. Ed.* **2008**, *47*, 6912–6916; ***Overview***: P. Wu, V. Fokin, *Aldrichimica Acta*, **2007**, *40*, 7–17.
19 R. B. Merrifield, *J. Am. Chem. Soc.* **1963**, *85*, 2149–2154. ***Reviews***: R. B. Merrifield, *Angew. Chem.* **1985**, *97*, 801–812; *Angew. Chem. Int. Ed.* **1985**, *24*, 799–810; R. B. Merrifield, *Science* **1986**, *232*, 341–347; ***Book***: F. Zaragoza Dörwald, *Organic Synthesis on Solid Phase*, 2nd Edit., Wiley-VCH, Weinheim **2002**.
20 ***Books***: N. K. Terret, *Kombinatorische Chemie*, Springer, Berlin **2000**; J. Eichler, *Kombinatorische Chemie*, Teubner-Verlag, Stuttgart **2003**; W. Bannwarth, B. Hinzen, *Combinatorial Chemistry*, Wiley-VCH, Weinheim **2006**.
21 ***Review***: *Top. Curr. Chem.* (*New Techniques in Solid-State NMR;* Volume editor: J. Klinowski), **2005**, *246*.
22 P. Haberz, J. Farjon, C. Griesinger, *Angew. Chem.* **2005**, *117*, 431–433; *Angew. Chem. Int. Ed.* **2005**, *44*, 427–429; J. C. Freudenberger, S. Knorr, K. Kobzar, D. Heckmann, D. Paululat, H. Kessler, B. Luy, *Angew. Chem.* **2005**, *117*, 427–430; *Angew. Chem. Int. Ed.* **2005**, *44*, 423–426.
23 P. Vepřek, J. Ježek, *J. Peptide Sci.* **1999**, *5*, 5–23; J. Ježek, J. Velek, P. Vepřek, V. Velková, T. Trnka, J. Pecka, M. Ledvina, J. Vondrášek, M. Pisačka, *J. Peptide Sci.* **1999**, *5*, 46–55; S. Monaghan, D. Griffith-Johnson, I. Matthews, M. Bradley, *ARKIVOC* **2001**, 46–53; A. Esposito, E. Delort, D. Lagnoux, F. Djojo, J.-L. Reymond, *Angew. Chem.* **2003**, *115*, 1419–1421; *Angew. Chem. Int. Ed.* **2003**, *42*, 1381–1383; A. Clouet, T. Darbre, J.-L. Reymond, *Angew. Chem.* **2004**, *116*, 4712–4715; *Angew. Chem. Int. Ed.* **2004**, *43*, 4612–4615.
24 G. Sanclimens, L. Crespo, E. Giralt, M. Royo, F. Albericio, *Peptide Sci.* **2004**, *76*,

283–297; N. J. Wells, A. Basso, M. Bradley, *Peptide Sci.* **1999**, *47*, 381–396.
25 H.-F. T. K.-K. Mong, M. F. Nongrum, C.-W. Wan, *Tetrahedron* **1998**, *54*, 8543–8660.
26 K. E. Uhrich, S. Boegemann, J. M. J. Fréchet, S. R. Turner, *Polym. Bull.* **1991**, *25*, 551–558.
27 V. Swali, N. J. Wels, G. J. Langley, M. Bradley, *J. Org. Chem.* **1997**, *62*, 4902–4903.
28 D. N. Posnett, H. McGrath, J. P. Tam, *J. Biol. Chem.* **1988**, *263*, 1719–1725; J. P. Tam, *Proc. Natl. Acad. Sci. USA* **1988**, *85*, 5409–5413; ***Review***: K. Sadler, J. P. Tam, *Rev. Mol. Biotechnol.* **2002**, *90*, 195–229.
29 E. Constable, *J. Chem. Soc., Chem. Commun.* **1997**, 1073–1080; C. B. Gorman, *Adv. Mater.* **1998**, *10*, 295–309.
30 a) This does not rule out the possibility that – less stable, reversibly dissociable – complexes of bipyridine, phenanthroline, or other complex ligands with other cations (alkali metal, zinc, silver) can be categorised under supramolecular aspects/supramolecular synthesis. However, such a categorisation of metal-complexing dendrimers on the basis of a measurable quantity (complex stability constant) has the drawback that – in borderline cases – it will depend upon the measuring conditions (solvent, temperature, pH, *etc.*), showing a certain analogy to the situation with keto/enol equilibria. b) We thank Prof. Dr. J.-M. Lehn and Prof. Dr. J.-P. Sauvage, Strasbourg, for fruitful discussions of these problems; c) B. Champin, V. Sartor, J.-P. Sauvage, *New. J. Chem.* **2006**, *30*, 22–25; d) regarding the definition of "supramolecular" see also J.-M. Lehn, *Supramolecular Chemistry: Concepts and Perspectives*, Wiley/VCH, Weinheim **1995**, p. 90; J. S. Hannam, S. M. Lacy, D. A. Leigh, C. G. Saiz, A. M. Z. Slawin, S. G. Stitchel, *Angew. Chem.* **2004**, *116*, 3270–3277; *Angew. Chem. Int. Ed.* **2004**, *43*, 3260–3264, reference [13].
31 Y. Tomoyose, D. L. Jiang, R. H. Jin, T. Aida, T. Yamashita, K. Horie, E. Yashima, Y. Okamoto, *Macromolecules* **1996**, *29*, 5236–5238.
32 P. J. Dandliker, F. Diederich, M. Gross, C. B. Knobler, A. Louati, E. M. Sanford, *Angew. Chem.* **1994**, *106*, 1821–1824; *Angew. Chem. Int. Ed.* **1994**, *33*, 1739–1741; P. J. Dandliker, F. Diederich, J. P. Gisselbrecht, A. Louati, M. Gross, *Angew. Chem.* **1995**, *107*, 2906–2909; *Angew. Chem. Int. Ed.* **1995**, *34*, 2725–2728.
33 C. M. Cardona, A. E. Kaifer, *J. Am. Chem. Soc.* **1998**, *120*, 4023–4024.
34 M. Plevoets, F. Vögtle, L. De Cola, V. Balzani, *New. J. Chem.* **1999**, 63–69; J. Issberner, F. Vögtle, L. De Cola, V. Balzani, *Chem. Eur. J.* **1997**, *3*, 706–712.
35 J.-F. Nierengarten, D. Felder, J.-F. Nicoud, *Tetrahedron Letters* **1999**, *40*, 273–276; P. Weyermann, J.-P. Gisselbrecht, C. Boudon, F. Diederich, M. Gross, *Angew. Chem.* **1999**, *111*, 1716–1721; *Angew. Chem. Int. Ed.* **1999**, *38*, 3215–3219; D.-L. Jiang, T. Aida, *Chem. Commun.* **1996**, 1523–1524; J. P. Collman, L. Fu. A. Zingg, F. Diederich, *Chem. Commun.* **1997**, 193–194; ***Reviews***: M. Momenteau, C. A. Reed, *Chem. Rev.* **1994**, *94*, 659–698; V. Balzani, F. Vögtle, *C. R. Chimie* **2003**, 867–872.
36 ***Review***: V. Balzani, P. Ceroni, A. Juris, M. Venturi, S. Campagna, *Coord. Chem. Rev.* **2001**, *219–221*, 545–572.
37 G. Denti, S. Serroni, S. Campagna, V. Ricevuto, V. Balzani, *Inorg. Chim. Acta* **1991**, *182*, 127–129; S. Campagna, S. Serroni, A. Juris, V. Balzani, *Inorg. Chem.* **1992**, *31*, 2982–2984; S. Campagna, S. Serroni, V. Balzani, G. Denti, A. Juris, M. Venturi, *Acc. Chem. Res.* **1998**, *31*, 29–34; ***Review***: S. Campagna, S. Serroni, V. Balzani, A. Juris, M. Venturi, *Chem. Rev.* **1996**, *96*, 756–833.
38 G. R. Newkome, R. Güther, C. N. Moorefield, F. Cardullo, L. Echegoyen, E. Perez-Cordero, H. Luftmann, *Angew. Chem.* **1995**, *107*, 2159–2162; *Angew. Chem. Int. Ed.* **1995**, *34*, 2023–2026; G. R. Newkome, X. Lin, *Macromolecules* **1991**, *24*, 1443–1444.
39 a) G. R. Newkome, C. N. Moorefield, G. R. Baker, A. L. Johnson, R. K. Behera, *Angew. Chem.* **1991**, *103*, 1205–1207; *Angew. Chem. Int. Ed.* **1991**, *30*, 1176–1178; b) G. R. Newkome, J. M. Keith, G. R. Baker, G. H. Escamilla, C. N. Moorefield,

Angew. Chem. **1994**, *106*, 701–703; *Angew. Chem. Int. Ed.* **1994**, *33*, 666–668.

40 ***Books***: *Comprehensive Supramolecular Chemistry* (11 volumes, Eds. J. L. Atwood, J. E. D. Davies, D. D. MacNicol, F. Vögtle), Pergamon Press, Oxford **1996**; F. Zeng, S. C. Zimmerman, *Dendrimers in Supramolecular Chemistry: From Molecular Recognition to Self-Assembly, Chem. Rev.* **1997**, *97*, 1681–1712; J.-M. Lehn, *Supramolecular Chemistry*, VCH, Weinheim **1995**; F. Vögtle *Supramolekulare Chemie*, Teubner-Verlag, Stuttgart, **1992**; *Supramolecular Chemistry*, Wiley, Chichester, **1993**; M. Venturi, S. Serroni, A. Juris, S. Campagna, V. Balzani, *Top. Curr. Chem.* (Volume editor F. Vögtle), **1998**, *197*, 193–228; K. Ariga, T. Kunitake, *Supramolecular Chemistry – Fundamentals and Applications*, Springer-Verlag, Berlin, Heidelberg **2006**.

41 M. Kawa, J. M. J. Fréchet, *Chem. Mater.* **1998**, *10*, 286–296.

42 A. R. Hirst, D. K. Smith, M. C. Feiters, H. P. M. Geurts, *Chem. Eur. J.* **2004**, *10*, 5901–5910; A. R. Hirst, D. K. Smith, J. P. Harrington, *Chem. Eur. J.* **2005**, *11*, 6552–6559; K. S. Partridge, D. K. Smith, G. M. Dykes, P. T. McGrail, *Chem. Commun.* **2001**, 319–320; A. R. Hirst, D. K. Smith, M. C. Feiters, H. P. M. Geurts, *Langmuir* **2004**, *20*, 7070–7077; A. R. Hirst, D. K. Smith, *Org. Biomol. Chem.* **2004**, *2*, 2965–2971; A. R. Hirst, D. K. Smith, *Langmuir* **2004**, *20*, 10851–10857; A. R. Hirst, D. K. Smith, M. C. Feiters, H. P. M. Geurts, A. C. Wright, *J. Am. Chem. Soc.* **2003**, *125*, 9010–9011.

43 PhD Theses by Stefan Brammer (**2001**) and Ahmet Dogan (**2005**), University of Kiel. We thank Prof. Dr. U. Lüning for supplying these theses and further information.

44 A. Franz, W. Bauer, A. Hirsch, *Angew. Chem.* **2005**, *117*, 1588–1592; *Angew. Chem. Int. Ed.* **2005**, *44*, 1564–1597.

45 a) T. Nagasaki, M. Ukon, S. Arimori, S. Shinkai, *J. Chem. Soc., Chem. Commun.* **1992**, 608–610; T. Nagasaki, O. Kimura, M. Ukon, S. Arimori, I. Hamachi, S. Shinkai, *J. Chem. Soc., Perkin Trans. 1* **1994**, 75–81; b) S. C. Zimmermann, J. R. Quinn, E. Burakowska, R. Haag, *Angew. Chem.* **2007**, *119*, 8312–8315; *Angew. Chem. Int. Ed.* **2007**, *46*, 8164–8176.

46 W. H. Hunter, G. H. Woollett, *J. Am. Chem. Soc.* **1921**, *43*, 135–142; ***Overviews***: H. Frey (ed.), Special Issue "Branched Polymers", *Macromolecular Chemistry and Physics* **2007**, *208*, issue 15; N. Tirelli (ed.), Special Issue "Branched Macromolecular Structures and their Bio-applications", *Macromolecular Bioscience* **2007**, *7*, issue 8.

47 ***Book***: P. J. Flory, *Principles of Polymer Chemistry*, Cornell University Press, Ithaca **1952**. Y. H. Kim, O. W. Webster, *J. Am. Chem. Soc.* **1990**, *112*, 4592–4593; Y. H. Kim, *J. Polym. Sci. Part A: Polym. Chem.* **1998**, *36*, 1685–1698; B. Voit, *J. Polym. Sci. Part A: Polym. Chem.* **2000**, *38*, 2505–2525; M. Seiler, *Chem. Eng. Technol.* **2002**, *25*, 237–253.

48 ***Books***: H.-G. Elias, *Makromoleküle*, 5th Edit., Hüthig & Wepf, Basel **1990**; H.-G. Elias, *Macromolecules*, Volume 1, Wiley-VCH, Weinheim **2005**; H.-G. Elias, *Macromolecules*, Volume 2, Wiley-VCH, Weinheim **2006**.

49 J. L. Mynar, T.-L. Choi, M. Yoshida, V. Kim, C. J. Hawker, J. M. J. Fréchet, *Chem. Commun.* **2005**, 5169–5171; A. D. Schlüter, J. P. Rabe, *Angew. Chem.* **2000**, *122*, 860–880; *Angew. Chem. Int. Ed.* **2000**, *39*, 864–883; A. Zang, L. Shu, Z. Bo, A. D. Schlüter, *Macromol. Chem. Phys.* **2003**, *204*, 328–339; C. C. Lee, J. M. J. Fréchet, *Macromolecules* **2006**, *39*, 476–481.

50 B. Karakaya, W. Claussen, K. Gessler, W. Saenger, A. D. Schlüter, *J. Am. Chem. Soc.* **1997**, *119*, 3296–3301; W. Stocker, B. L. Schürmann, J. P. Rabe, S. Förster, P. Lindner, I. Neubert, A. D. Schlüter, *Adv. Mater.* **1998**, *10*, 793–797; A. Wenzel, I. Neubert, A. D. Schlüter, *Macromol. Chem. Phys.* **1998**, *199*, 745–749; L. Shu, A. Schäfer, A. D. Schlüter, *Macromolecules* **2000**, *33*, 4321–4328; A. D. Schlüter, J. P. Rabe, *Angew. Chem.* **2000**, *112*, 860–880; *Angew. Chem. Int. Ed.* **2000**, *39*, 864–883.

51 D. A. Tomalia, P. M. Kirchhoff, *US Pat. 4 694 064*, **1987**; R. Yin, Y. Zuh, D. Tomalia, H. Ibuki, *J. Am. Chem. Soc.* **1998**, *120*, 2678–2679.

52 V. Percec, J. Heck, D. Tomazos, F. Falkenberg, H. Blackwell, G. Ungar, *J. Chem. Soc. Perkin. Trans. 1* **1993**, 2799–2811.

53 G. Draheim, H. Ritter, *Macromol. Chem. Phys.* **1995**, *196*, 2211–2222; V. Percec, C. H. Ahn, G. Ungar, D. J. P. Yeardley, M. Möller, S. S. Sheiko, *Nature* **1998**, *391*, 161–164; Y. M. Chen, C. F. Chen, W. H. Liu, F. Xi, *Macromol. Rapid Commun.* **1996**, *17*, 401–407; I. Neubert, R. Klopsch, W. Claussen, A. D. Schlüter, *Acta Polym.* **1996**, *47*, 455–459; I. Neubert, A. D. Schlüter, *Macromolecules* **1998**, *31*, 9372–9378; L. Shu, A. D. Schlüter, *Macromol. Chem. Phys.* **2000**, *201*, 239–245; A. Zistler, S. Koch, A. D. Schlüter, *J. Chem. Soc. Perkin Trans. 1* **1999**, 501–508; I. Neubert, E. Amoulong-Kirstein, A. D. Schlüter, *Macromol. Rapid Commun.* **1996**, *17*, 517–527.

54 ***Books***: K. J. Ivin, *Olefin Metathesis*, Academic Press, London **1983**; V. Percec, D. Schlüter, J. C. Ronda, G. Johansson, G. Ungar, J. P. Zhou, *Macromolecules* **1996**, *29*, 1464–1472; V. Percec, D. Schlüter, *Macromolecules* **1997**, *30*, 5783–5790; R. H. Grubbs, *Handbook of Metathesis*, Wiley-VCH, Weinheim **2003**, *Vol. 1–3*; ***Reviews***: (Nobel lectures): Y. Chauvin, *Angew. Chem.* **2006**, *118*, 3824–3831; *Angew. Chem. Int. Ed.* **2006**, *45*, 3740–3747; R. R. Schrock, *Angew. Chem.* **2006**, *118*, 3832–3844; *Angew. Chem. Int. Ed.* **2006**, *45*, 3748–3759; R. H. Grubbs, *Angew. Chem.* **2006**, *118*, 3845–3850; Angew. *Chem. Int. Ed.* **2006**, *45*, 3760–3765.

55 A. D. Schlüter, G. Wegner, *Acta Polymer.* **1993**, *44*, 59–69; W. Claussen, N. Schulte, A. D. Schlüter, *Macromol. Rapid Commun.* **1995**, *16*, 89–94; B. Karakaya, W. Claussen, K. Gessler, W. Saenger, A. D. Schlüter, *J. Am. Chem. Soc.* **1997**, *119*, 3296–3301; Z. Bo, J. P. Rabe, A. D. Schlüter, *Angew. Chem.* **1999**, *111*, 2540–2542; *Angew. Chem. Int. Ed.* **1999**, *38*, 2370–2372; Z. Bo, C. Zhang, N. Severin, J. P. Rabe, A. D. Schlüter, *Macromolecules* **2000**, *33*, 2688–2694.

56 Z. Bao, K. R. Amundson, A. J. Lovinger, *Macromolecules* **1998**, *31*, 8647–8649.

57 L. Shu, A. D. Schlüter, *Macromol. Chem. Phys.* **2000**, *201*, 139–142; see also W. Li, A. Zhang, A. D. Schlüter, *Macromolecules* **2008**, *41*, 43–49.

58 I. Shu, A. D. Schlüter, C. Ecker, N. Severin, J. P. Rabe, *Angew. Chem.* **2001**, *113*, 4802–4805; *Angew. Chem. Int. Ed.* **2001**, *40*, 4666–4669.

59 O. Lukin, V. Gramlich, R. Kandre, I. Zhun, T. Felder, C. Schalley, G. Dolgonos, *J. Am. Chem. Soc.* **2006**, *128*, 8964–8974; G. Bergamini, P. Ceroni, V. Balzani, M. Villavieja, R. Mandre, I. Zhun, O. Lukin, *Chem. Phys. Chem.* **2006**, *7*, 1980–1984; cf.: V. Vicinelli, G. Bergamini, P. Ceroni, V. Balzani, F. Vögtle, O. Lukin, *J. Phys. Chem. B* **2007**, 6620–6627; for Janus-type poly(ethylene oxide) dendrimers see X. Feng, D. Taton, E. Ibarboure, E. L. Chaikof, Y. Gnanou, *J. Am. Chem. Soc.* **2008**, *130*, 11662–11676.

3
Functional dendrimers

The fascination of dendrimers is due essentially to the unique architecture of these molecules and the attendant possibilities of designing well-defined "functional (macro)molecules" [1], whose functional groups can participate in chemical/physical processes [2]. In discussions about "functional dendrimers" a distinction is made between dendrimers bearing more complex, function-bearing units (such as catalytically active, photoactive units, liquid-crystalline groups) and those whose functionalities serve to control simple chemical/physical molecular properties (such as solubility, viscosity). Since both fulfil a function, the word "functional" is used as a generic term in the following sections.

The many possibilities of directing the physical/chemical properties of a dendrimer by introduction of selected functional groups into the molecule lead to an great variety of potential applications of dendrimers. Depending upon the position and the nature of the functional units within the dendrimer structure, it is possible to distinguish between different types of functional dendrimers:

1. Monofunctional dendrimers
   - with functional core unit
   - with functional molecular periphery
   - with the same kind of functional units within the dendrimer framework, *i.e.* in the inner shells (generations)

2. Multifunctional dendrimers
   - with two or more different functional units in the molecular periphery
   - with functional core unit and functional molecular periphery
   - with functional core unit and functional dendrimer scaffold

The following Sections give a detailed account of the strategies most commonly used for synthesis of the various dendrimer types and the synthetic utility for targeted design of functional dendrimers is evaluated.

Dendrimer Chemistry. Fritz Vögtle, Gabriele Richardt and Nicole Werner

ISBN: 978-3-527-32066-0

3.1 Monofunctional dendrimers

3.1.1 Functional core

As a rule, dendrimers with a functional core are synthesised by the convergent growth method (Fig. 3.1, Route a); see also Section 2.2) by covalent linkage of pre-synthesised dendrons with a function-bearing core unit. In some cases, core-functionalised dendrimers could also be obtained by a supramolecular synthetic strategy (Fig. 3.1; Route b); see also Section 2.6) [3]. Fresh orientation in this field came from the method of convergent self-assembly of polypyridine ligands by means of transition metal ions frequently used by Balzani *et al.* [4]. The advantage of the convergent (Section 2.2) and the supramolecular synthetic strategy (Section 2.6) over the divergent method (Section 2.1) is that the functional core unit is not introduced until the last synthetic step. This enables the dendrimer chemist to introduce sensitive functional core units. Only if this core unit is stable under the conditions of the subsequent activation and coupling

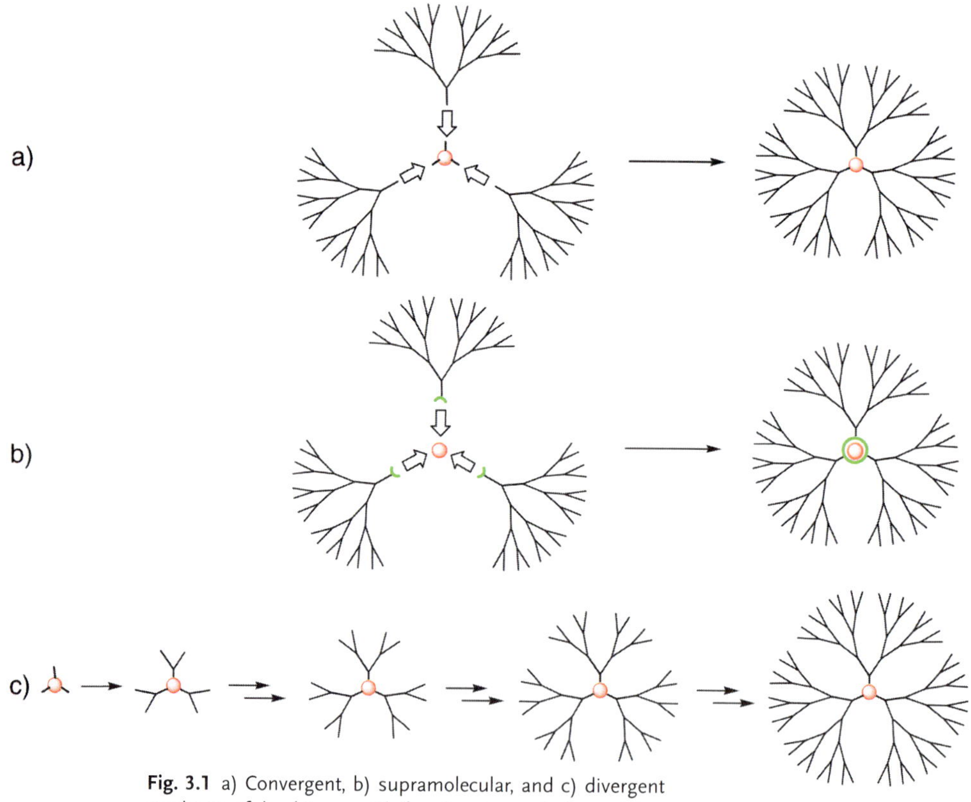

Fig. 3.1 a) Convergent, b) supramolecular, and c) divergent synthesis of dendrimers with functional core (schematic)

steps will it also be possible to synthesise dendrimers of this type through divergent growth of a dendritic shell around the core unit (Fig. 3.1; Route c), see also Section 2.1) [5]. The convergent synthetic strategy for dendritic shell formation around functional core components is considered the most versatile method, not least because it yields dendritic materials of greater perfection and homogeneity [6].

A wide range of dendrimers with functional core is described in the literature. Thus chromophores [7], electrochemically active, redox active [8], and catalytically active [9] or also self-associating and chiral units as well as polymerisable monomers and polymers have been successfully introduced into the centre of dendrimers. However, the core unit not only has a determining effect on the function, but also has a decisive influence on the multiplicity, size, and shape of the dendrimer.

Studies have shown the dendrimer shell to exert a significant influence on the properties of a functional core unit. Thus the reduced accessibility of the core unit leads in some cases to establishment of a separate microenvironment in the immediate vicinity of the function-bearing central unit, which is dependent upon the chemical and electronic properties of the dendrimer shell. Depending upon the polarity and the electron density of the resulting microenvironment, characteristic properties of active sites, such as their redox behaviour, can deviate from those of their freely accessible analogues [10]. In addition, steric shielding [6b] by dendritic substituents can be utilised in the case of redox-active and photoactive core units to prevent undesired intermolecular interactions between active sites. The grouping of large dendritic branches around a catalytically active unit can influence access of substrates to the catalytic site and generate substrate, regio-, or enantioselectivity. Moreover, the increase in molecular size associated with the attachment of dendrons opens up new possibilities of catalyst recovery in homogeneous catalysis [11].

The long-term goal is to be able to tune the properties of the functional core unit to the technical requirements of specific applications by attachment of appropriate dendrons. Numerous publications devoted to the elucidation of these interrelationships are bringing the day when this goal will be realised closer and closer, in spite of the many factors which still have to be considered.

3.1.2
Functional periphery

Studies of very different kinds have shown that the chemical and physical properties of dendritic (macro)molecules are extensively dictated by the nature of their functional peripheral groups [12]. Thus, for example, the stability, solubility, viscosity, aggregation, and chemical reactivity of a dendrimer can be controlled *via* the peripheral groups and its spatial and surface structure as well as its conformational flexibility influenced in the same way. Here, the effect of the end groups on the properties of the molecule increases with increasing number of generations since the number of end groups of an ideally growing dendrimer increases exponentially with the addition of further generations as a function of the multiplicity of the core and the branching units. This is why functionalisation of the periphery of the molecule represents the most promising and the simplest method of designing properties and opening up new areas of application. In certain cases, the reproduction of identical functional units can lead to a dramatic enhancement of the effect exerted by these units *(dendritic effect;* see Section 6.3). In addition to reinforcing desired functions, good accessibility of the functional units or functionalities on the periphery of the molecule is of decisive advantage for many of the possible applications of dendrimers (*e.g.* in catalysis).

Various strategies for the synthesis of peripherally functionalised dendrimers are presented and evaluated in the following sections. The expression "peripheral group" is used here as a neutral term. It can refer both to a peripheral function-bearing unit and to a peripheral functionality. Terminal groups linked covalently to the periphery can also be – partly – backfolded (see Sections 1.2 and 7.6.3).

3.1.2.1 **Functionalisation of terminal groups**

The simplest synthetic approach to the preparation of monofunctional dendritic systems is to functionalise the periphery of existing molecular scaffolds, generally constructed by divergent or convergent growth (Fig. 3.2, Route a) or b), respectively).

Commercially available POPAM and PAMAM dendrimers (see Chapter 4) bearing terminal primary amino groups are currently the most commonly used divergently grown dendrimer structures, followed by poly(benzyl ether) dendrons or poly(benzyl ether) dendrimers (Fréchet type) as representatives of convergently synthesised dendritic molecules.

Such subsequent "end group functionalisation" requires just a single synthetic step and yields dendrimers bearing a single kind of peripheral group in the corresponding generation-dependent multiplicity, and which are therefore designated according to Newkome *et al.* as "peripherally homogeneous dendrimers" [13]. However, synthesis of monodisperse dendrimers of this type requires selective and quantitative reaction of the substrates at the periphery (see Section 2.1). The number of suitable coupling reactions and the types of end group reagents are therefore limited.

Functionalisation of the end groups of divergently synthesised dendrimers: Several examples of the functionalisation of peripheral NH_2–, CO_2H–, CHO–, $P(X)Cl_2$– (X = S, O) and Si–H groups [14] of divergently grown dendrimer scaffolds are to be found in the literature. A common feature of all these end groups is that they react with a large number of readily accessible or even commercially available reagents. In addition, they are all regenerated in the course of the normal iterative growth of common types of dendrimers (*e.g.* POPAM, PAMAM, phosphorus-containing dendrimers, carbosilane dendrimers) at each generation stage.

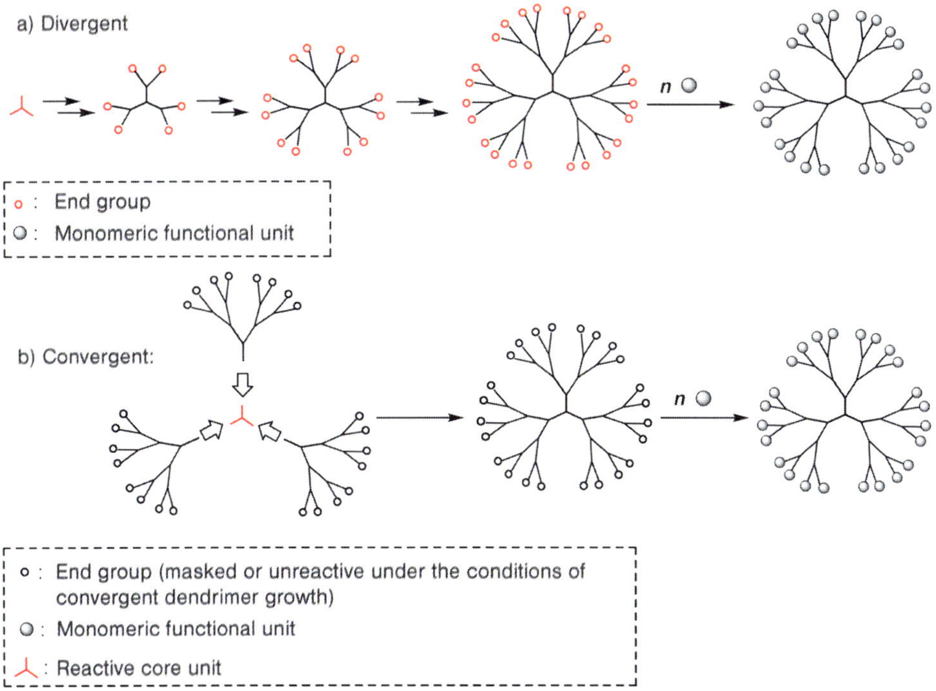

Fig. 3.2 Molecular periphery on conclusion of a) divergent and b) convergent dendrimer growth (schematic). The step shown on the right generates the final periphery in each case

Functionalisation reactions of peripheral NH_2 groups are already well established in dendrimer chemistry, especially in the case of POPAM and PAMAM dendrimers. These include coupling with activated carboxylic acid derivatives such as acid chlorides [15], fluorides [16], anhydrides [17], and *N*-hydroxysuccinimide esters [18] to form the corresponding oligo-/polyamides. Reactions of oligo-/polyamine dendrimers with sulphonyl chlorides to give the corresponding sulphonamides [19] and reaction with isothiocyanates or isocyanates to give thiourea [20] or urea derivatives [21] have often proved of value. Other attractive reactions of polyamine dendrimers with epoxides to yield -amino alcohols [22], by double phosphinomethylation *via* hydroxymethyldiphenylphosphine formed *in situ* to give the corresponding bis(methyldiphenylphosphin)amines [23], or with *ortho*-hydroxybenzaldehyde derivatives to give imines [24] have so far remained limited to individual cases.

Follow-on functionalisation with molecular moieties of biological importance, such as anti-tumour agents (*e.g.* 5-fluorouracil) currently appears particularly attractive. In this context, acetylation is particularly suitable as functionalisation reaction for oligo/polyamine dendrimers. It renders the dendrimers more water soluble, which is essential for biomedical applications [25]. The toxicity of dendritic systems also plays a decisive role here. Studies conducted so far on dendrimers (including POPAM and PAMAM dendrimers) with various peripheral groups have shown that the cell toxicity of dendrimers can be significantly modified by end group functionalisation [26].

The principle of "*click chemistry*" presented by Hawker *et al.* (see Section 2.3.5) offers an efficient and versatile method of functionalisation. For example, it utilises [3+2]-cycloaddition of azide-functionalised reagents with ethyne end groups of a dendrimer precursor to prepare dendrimers with triazole-functionalised end groups. The mild reaction conditions, almost quantitative reaction, and not least the tolerance towards numerous functional groups permit the use of widely differing molecular frameworks (*e.g.* poly(benzyl ether), POPAM dendrimer structures or hyperbranched polyesters) and different functionalised azides [27].

Functionalisation of the end groups of convergently grown dendrimers/dendrons: The desired functional groups can also be introduced into the periphery after convergent growth of a dendron or dendrimer (Fig. 3.2, Route b)), provided that the end groups present at the beginning of the synthesis remain masked throughout molecule growth (*i.e.* stay unreactive), but can subsequently be modified.

Thus the alkyl ester end groups of convergently grown poly(benzyl ether) dendrimers can be subjected to post-synthetic modification by hydrolysis [28], reduction [29], transesterification/amidation [28] in a variety of ways. Subsequent modification of poly(benzyl ether) dendrons bearing *p*-bromobenzyl end groups by palladium-catalysed coupling reactions permitted preparation of dendrons with phenyl, pyridinyl, or thiophenyl end groups [30].

The concept of "postmodification" of the molecule's periphery is thus not only appropriate for the introduction of "simple" functionalities determining the physical and chemical properties of the molecule (*e.g.* solubility), but also allows the introduction of "complex" function-bearing units (*e.g.* catalytically active units, anti-tumour agents), which could be too sensitive for convergent dendrimer synthesis.

3.1.2.2 **Introduction of peripheral groups prior to dendrimer growth**

A convergent growth strategy starting from the peripheral groups of the later dendrimer (see Section 2.2) offers another approach to peripherally homogeneously functionalised dendrimers. The advantage over the "postmodification" variant is that significantly fewer coupling steps are required to introduce the desired peripheral functional moieties or functionalities. As already mentioned, only those peripheral groups which exhibit adequate solubility and which neither cause side reactions nor undergo degradation in the course of subsequent activation and coupling steps can be introduced in this way.

The relatively mild synthetic conditions during iterative growth of poly(benzyl ether) dendrons are tolerated by many peripheral groups (*e.g.* cyano, bromide functionality, alkyl ester, alkyl ether, perfluoroalkyl ether, oligo(ethylene glycol) ether). The peripheral groups of such Fréchet-type dendrons can therefore be varied over a wide range and adapted to meet specific requirements. In particular the solubility can be tuned within very wide limits – from solubility in water to solubility in petroleum ether. With their individual functionality at the focal point, such dendrons can serve as modification reagents to solubilise poorly soluble substances sufficiently for spectroscopic purposes (*e.g.* NMR) or to stabilise them towards reagents. From an applications standpoint, particular interest attaches to Fréchet-type dendrons or dendrimers with functional molecular periphery, which were grown starting from organoruthenium units [31], redox-active ferrocenyl groups [32], catalytically aktive TADDOL units [33], liquid-crystalline groups or chromophores [34]. On a smaller scale, other dendron structures with peripheral functional units, such as sugars [35], ferrocenes, or fullerenes [36] could also be prepared in this way.

3.1.3 Functional units in the dendrimer scaffold

In the synthesis of functional dendrimers, interest has hitherto been focussed on variation of the functional core unit or peripheral groups and the resulting effects on the properties of the dendrimer. For a long time, the only function ascribed to the dendritic branches and their repeating units was that of a scaffold linking periphery and core. It was overlooked that, in the interior of the dendrimer scaffold, an individual characteristic (nano)environment can arise which is largely dependent upon the chemical characteristics and the polarity of the repeating units used to construct the dendrimer. Moreover, they can facilitate cascade processes and serve as a platform for cooperative effects between dendrimer branches [37].

To make maximum use of the potential offered by the unique architecture of dendrimers, synthetic strategies are required which permit internal manipulation of dendrimers through incorporation of special functional units (chemically functional or physically active groups) in the inner shells. Depending upon the synthetic goal, functionalisation of the dendrimer interior can again take place either prior to dendrimer construction or on the ready-synthesised dendrimer scaffold [38]. Since functionalisation of the interior framework after completion of dendrimer growth ("postmodification") requires the introduction of subsequently modifiable functionalities during dendrimer growth the two strategies cover some common ground, as do the two following sections.

3.1.3.1 Modification prior to dendrimer growth

Incorporation of a large number of internal functional units is desirable for certain applications (*e.g.* catalysis). This can be accomplished by use of appropriately pre-functionalised monomeric branching units during growth of the dendrimer scaffold. On adopting this strategy, it should be assured that the functional groups of these monomeric building blocks are insensitive to the reaction conditions prevailing during dendrimer growth. The use of two differently functionalised monomereric branching units of the type AB_2FG (FG = functional group) permits layerwise growth of dendrimer structures in a divergent or convergent fashion. Given appropriate combinability of the functionalised monomeric building blocks, the number and arrangement of the internal functionalities can be controlled in accordance with the synthetic goal (see Fig. 3.3). Use of just one kind of functionalised AB_2FG-type branching unit, on the other hand, will lead to dendrimers with identical functionalities in every generation.

The *parallel monomer-combination method* developed by Fréchet *et al.* offers rapid access to poly(benzyl ether) dendrimers having a layered arrangement of internal functional units while maintaining precise control over the number and geometry of these groups in the internal dendrimer. This synthetic strategy was developed with the aim of combining the synthetic utility and the speed of convergent *"two-stage" synthesis* with the throughput and versatility of classical *parallel synthesis*. The internal functionalities are again introduced with the aid

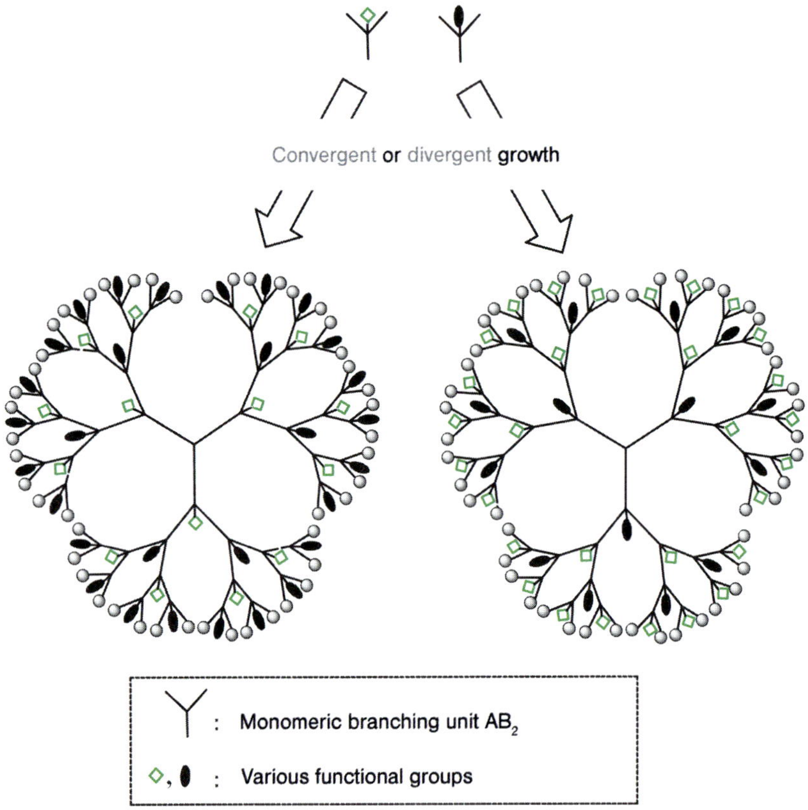

Fig. 3.3 "Premodification": Growth of dendrimers starting from pre-functionalised monomeric branching units permits preparation of dendrimers with different internal functionalities in subsequent layers

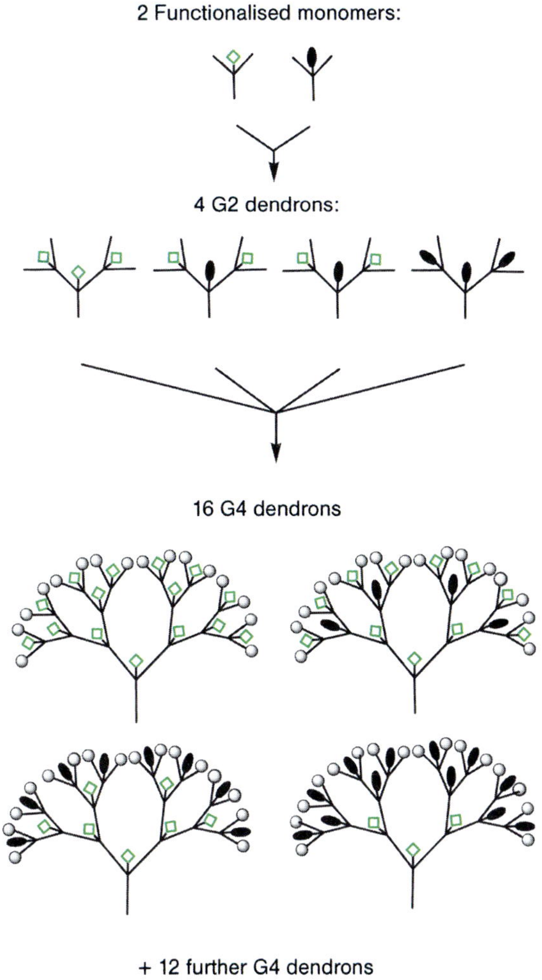

Fig. 3.4 The parallel monomer-combination method according to Fréchet *et al.* leads to 2^G dendrons (G = number of generations; simplified illustration)

of two differently functionalised monomeric building blocks AB_2FG (FG = desired functionality, which may also be H), each of which must be combinable in parallel. If all possible combinations are exhausted, then 2^G (G = generation) different dendrons will be obtained (see Fig. 3.4). Fréchet *et al.* demonstrated the power of this method by synthesis of a series of fourth-generation poly(benzyl ether) dendrons with one to fifteen postmodified internal allyloxy groups.

A dendrimer containing azobenzene groups in every repeating unit also numbers among the dendrimers to have had a large number of internal functional groups introduced in a targeted manner during dendrimer growth. On UV irradiation, typical azobenzene isomerisation was observed in combination with significant shrinkage of the molecule (see Section 5.2) [39]. Other functional units apart from azobenzene to have been used as functional components of monomeric building blocks in dendrimer growth include fullerenes, quaternary ammonium salts, pyridines, triarylamines, carbazoles, liquid-crystalline groups [40], and, in connection with chiral recognition, also chiral groups [41].

In some cases, however, attention is focussed not on a large number of internal functionalities, but on generation-specific introduction of a single functional group. This is the case, for example, when the influence of the microenvironment formed by the dendrimer structure on the internal functional group is to be investigated. Moreover, this group can serve as linkage site for functional units such as chromophores, electrophores, or catalytically active units and for internal modification of the dendrimer at predetermined positions.

Schlüter *et al.* successfully accomplished generation-specific incorporation of an individual aryl bromide functionality into poly(benzyl ether) dendrimers. Whatever its position in the dendrimer framework, this bromide function could also be utilised for further chemical modification because it can react with arylboronic acid *via* Suzuki cross-coupling [42].

3.1.3.2 **Internal modification on conclusion of dendrimer growth**

As an alternative to the functionalisation strategy just outlined, in the "postmodification" approach monomeric repeating units (branching building blocks) with an orthogonally protected (see Section 2.3.1) or selectively activatable functionality are used for convergent or divergent dendrimer construction. All internal functionalities of the dendrimer precursor can then be modified in a single step (Fig. 3.5).

Subsequent modification of the branching units in the interior of a dendrimer framework has so far received comparatively little attention. From a synthetic standpoint, however, this strategy offers significantly more design scope than its counterpart. Thus dendrimers also become accessible whose branching units are hard to synthesise or are either not sufficiently reactive or too sensitive to be used in iterative dendrimer synthesis. However, restricted accessibility of internal groups compared to peripheral functionalities can prove problematic.

By oxidation of internal dibenzyl sulphide moieties with hydrogen peroxide, Chow *et al.* converted first- to third-generation oligo(dibenzyl sulphide) dendrimers into the corresponding oligo(dibenzyl sulphone) dendrimers, which are inaccessible by divergent or convergent synthetic strategies. Subsequent transformation of the sulphone units into stilbene groups could be accomplished up to the second dendrimer generation and was based, respectively, on three and nine Ramberg–Bäklund rearrangements in the interior of the dendrimer frameworks. Since the rearrangements lead to a new kind of dendrimer with a totally different molecular framework, Chow *et al.* designated this synthetic strategy as "dendrimer metamorphosis" (Fig. 3.6) [43].

The literature contains mainly references to subsequent internal functionalisation of Fréchet dendrimers [44]. The suitability of this type of dendrimer comes

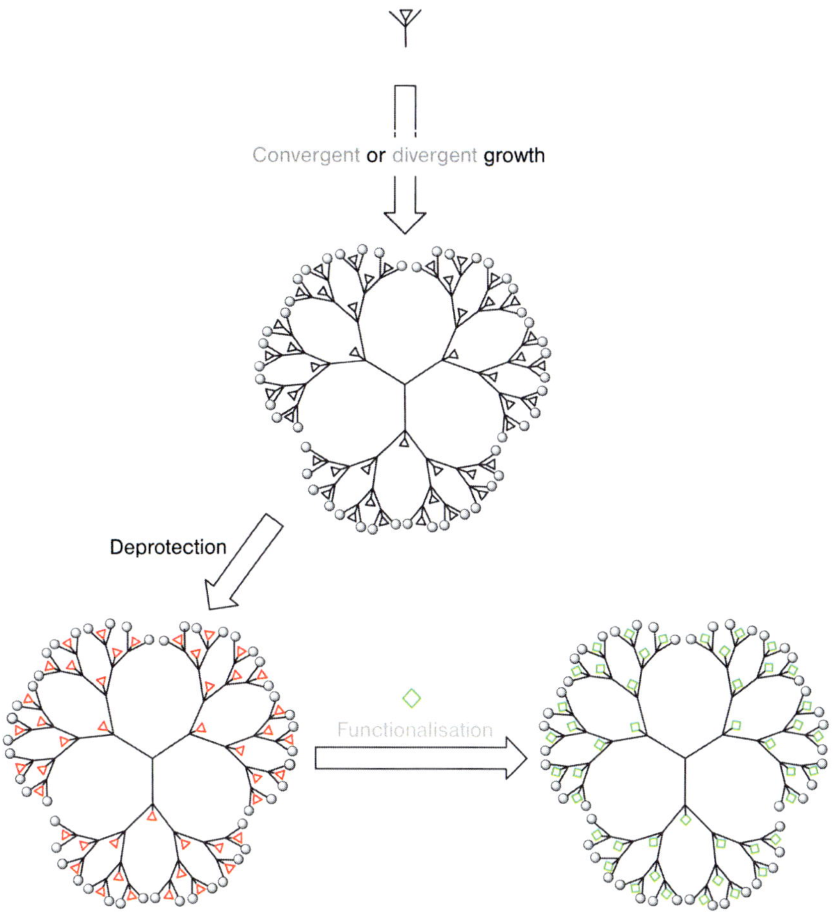

Fig. 3.5 "Postmodification": Dendrimer growth with protected monomeric branching units permits subsequent introduction of functionalities into the scaffold of the dendrimer precursor

about because the convergent growth of poly(benzyl ether) dendrimers [45] proceeds under relatively mild conditions tolerated by many functional groups, purification is comparatively easy in convergent synthesis, and a wide selection of commercially available or at least readily accessible arene building blocks is at hand.

Lochmann *et al.* were the first to attempt postmodification of the internal molecular framework of Fréchet-type dendrons [46]. Multifunctionalisation of the interior of the dendron *via* metallation (superbase followed by reaction with various electrophiles) proved to be of a somewhat random nature owing to a lack of regioselectivity. It is admittedly suitable for engineering the solubility properties of the dendrons, but not for producing dendritic structures having functional groups in defined positions of the internal scaffold of the molecule.

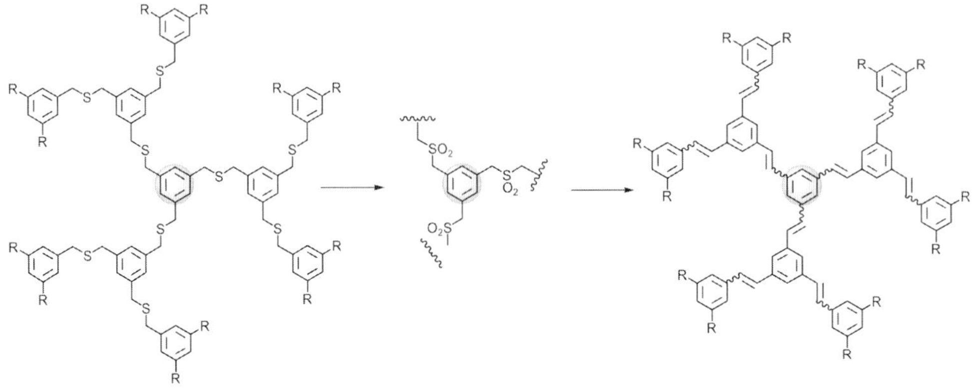

Fig. 3.6 Conversion of internal functional groups and "metamorphosis" of a second-generation oligo(dibenzyl sulphone) dendrimer to a corresponding oligo(phenylene-vinylene) dendrimer (The central benzene core is marked grey for the sake of clarity)

There are only a few examples of a follow-on regioselective and covalent functionalisation of the interior scaffold of a dendrimer [47].

The first controlled internal functionalisation of dendrimers was accomplished by Newkome *et al.* They synthesised dendrimer-type water-soluble polycarboranes for possible use in boron neutron capture therapy by utilising the generation-specific reaction of decaboranes with the internal alkyne functionalities of a polyalkyl dendrimer and subsequently introduced polar peripheral groups [48]. Müllen *et al.* successfully performed controlled "postmodification" of second-generation polyphenylene dendrimers with a defined number of keto groups in the dendrimer scaffold. After transformation of the internal keto groups of the precursor dendrimer into hydroxyl groups with the aid of organolithium reagents, these could be regioselectively and quantitatively functionalised with functional units of various natures and sizes – including reaction partners as large as pyrene [49]. The introduction of functional groups into the interior dendrimer scaffold also permits intramolecular bond formation through reaction between functional groups within the same molecule on conclusion of dendrimer construction. Thus incorporation of aromatic spacers with two allyl groups each into a poly(benzyl ether) dendrimer framework permits subsequent controlled covalent coupling of the individual dendron subunits *via* ring closing methathesis [50].

Majoral *et al.* undertook a systematic investigation of subsequent generation-specific introduction of various functional groups (allyl, propargyl, isothiocyanates, primary amines, aldehydes [51], crown ethers, fluorescence labels [52]) as well as charged groups into phosphorus-containing dendrimers and even studied the divergent growth of dendrons in the interior of large, flexible phosphorus-containing dendrimers.

As shown in the above sections, internal functionalisation greatly increases the degree of complexity of such molecular structures and hence also the chance of developing new dendritic materials for future applications, for instance in nanotechnology or biotechnology. The long-term goal is to develop versatile new synthetic strategies permitting control over targeted positioning of various functional units, both in the scaffold and in the periphery of dendrimers.

3.2 Multifunctional dendrimers

The properties of a dendrimer are determined not only by the specific properties of its functional units, but also by their number and structural variety, and by cooperative effects between different functional units. A number of special applications and the development of substances suitable for mimicking biological systems [53] require such multifunctional dendritic systems with more than one kind of functionality.

The following sections are devoted primarily to the introduction of two different types of functional units. A distinction is made between dendrimers with bifunctionalised molecular periphery (Fig. 3.7; Types A, B, C, D) and those in which one function is located in the core and the other in the branching units or in the periphery (Types E, F). Multifunctional dendrimers of type G with different functional units in the core, scaffold, and periphery have so far played only a minor role and will therefore only be treated briefly here, particularly since compounds of this type will be considered in greater depth in Chapter 6.

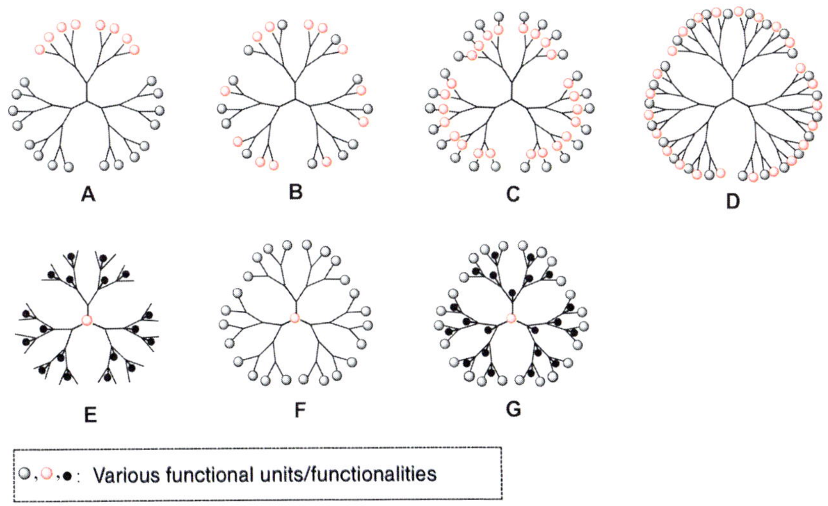

Fig. 3.7 Various bifunctionalisation (A–F) and multifunctionalisation concepts (G)

3.2.1
Bifunctionalised molecular periphery

The convergent synthetic strategy appears to be the fundamentally superior method for introducing a controlled number of functional units, also of different type, into the periphery of a dendrimer. Thus, within the framework of a convergent synthesis, two kinds of dendrons differing in their terminal groups can be joined together to form a dendrimer molecule or coupled to a common core unit (Fig. 3.8). Dendrimers with two different peripheral groups in separate molecular segments, known as *surface-block dendrimers*, can be obtained in this way [54]. The numerical ratio of the two peripheral groups to each other depends upon the respective generation of the two dendron types and on the relative amounts of each type of dendron used (see also Section 2.2).

An amphiphilic poly(benzyl ether) dendrimer described by Fréchet *et al.*, with terminal COOH groups in one half of the molecule and lipophilic terminal alkyl chains in the other, corresponds to this type of peripherally heterogeneous dendrimer [55]. Other surface-block dendrimers formed by coupling of two dendrons with strong π-donor or strong π-acceptor groups showed inter- and intramolecular charge-transfer interactions and an amphoteric redox behaviour [56].

Phosphorus-containing surface-block dendrimers can be prepared by a method developed by Majoral *et al.* [57] (Fig. 3.9).

Convergent synthesis using dendrons having two different terminal groups offers a further route to dendrimers with bifunctionalised molecular periphery [58]. Unsymmetrical dendrons of this kind can be prepared, for example, by stepwise incorporation of two different functionalities starting from an AB_2 monomer. Synthesis proceeds without random monfunctionalisation, and is thus far less problematic, if it is possible to distinguish between the two "B" functionalities during convergent dendron growth either because one of the functionalities is protected (ABB_P monomer, subscript P stands for protected) or because the two of them react differently under specific reaction conditions (ABB′ monomer; Fig. 3.10).

Newkome *et al.* were the first to present a combinatorial approach to the synthesis of peripherally heterogeneously functionalised POPAM dendrimers of type B (see Fig. 3.7), which is based on functionalisation of end groups with a

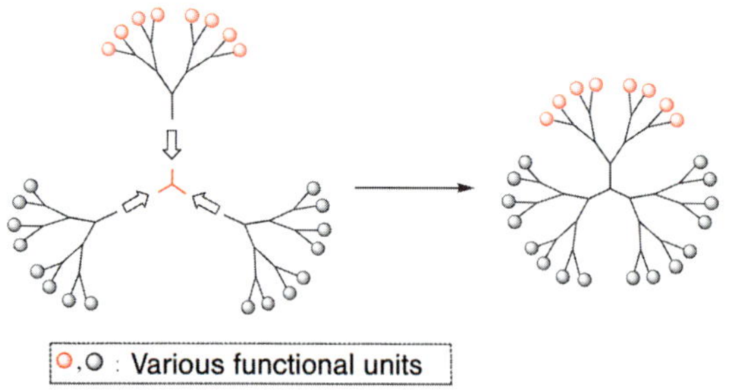

Fig. 3.8 Convergent synthesis of a bifunctionalised dendrimer with two different peripheral groups in separate molecular segments (surface-block dendrimers)

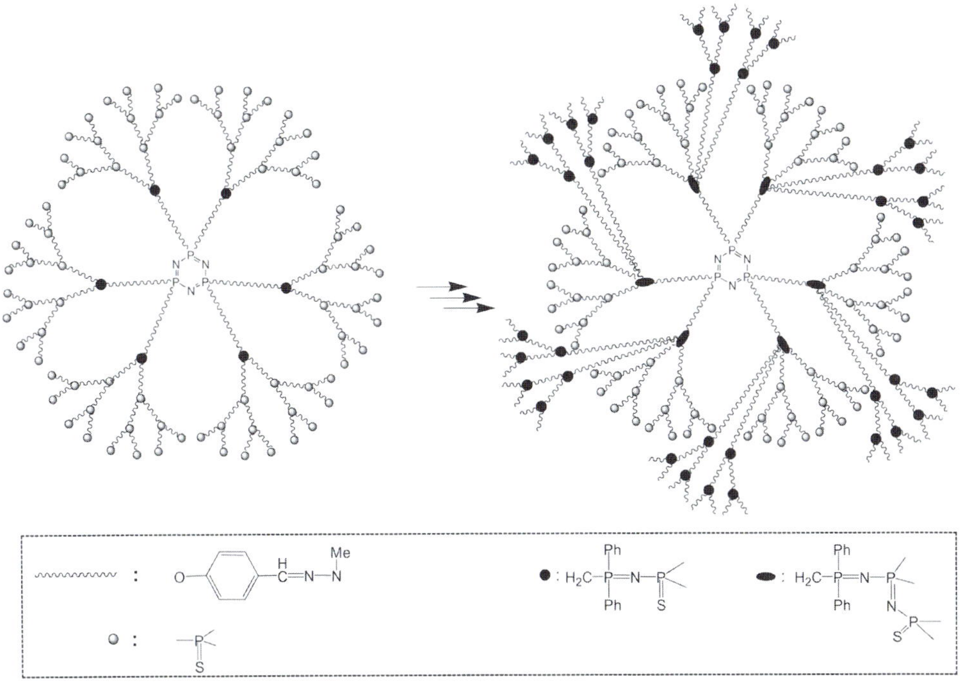

Fig. 3.9 On bifunctionalisation according to Majoral *et al.* "daughter dendrons" are produced divergently starting from specific functions in the interior of the dendrimer scaffold

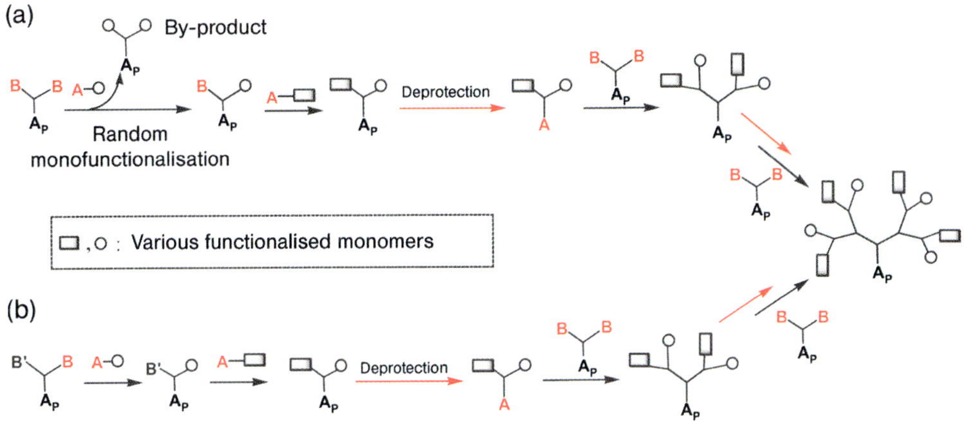

Fig. 3.10 Synthesis of unsymmetrical dendrons with the aid of AB_2 monomers or ABB′ monomers (the functionalities B and B′ show different reactivities under the necessary synthesis conditions)

mixture of different but similarly functionalised substrates. The drawback of this method is that the resulting dendrimers have an unpredictable distribution of the various functional units in their periphery [13]. Meijer *et al.* functionalised POPAM dendrimers with a 1:1 mixture of pentafluorophenyl ester-activated azobenzene units and correspondingly activated alkyl chains. They obtained amphiphilic switchable dendrimers whose peripheries contain the two functional units in equal proportions but in a random distribution (Fig. 3.11) [59].

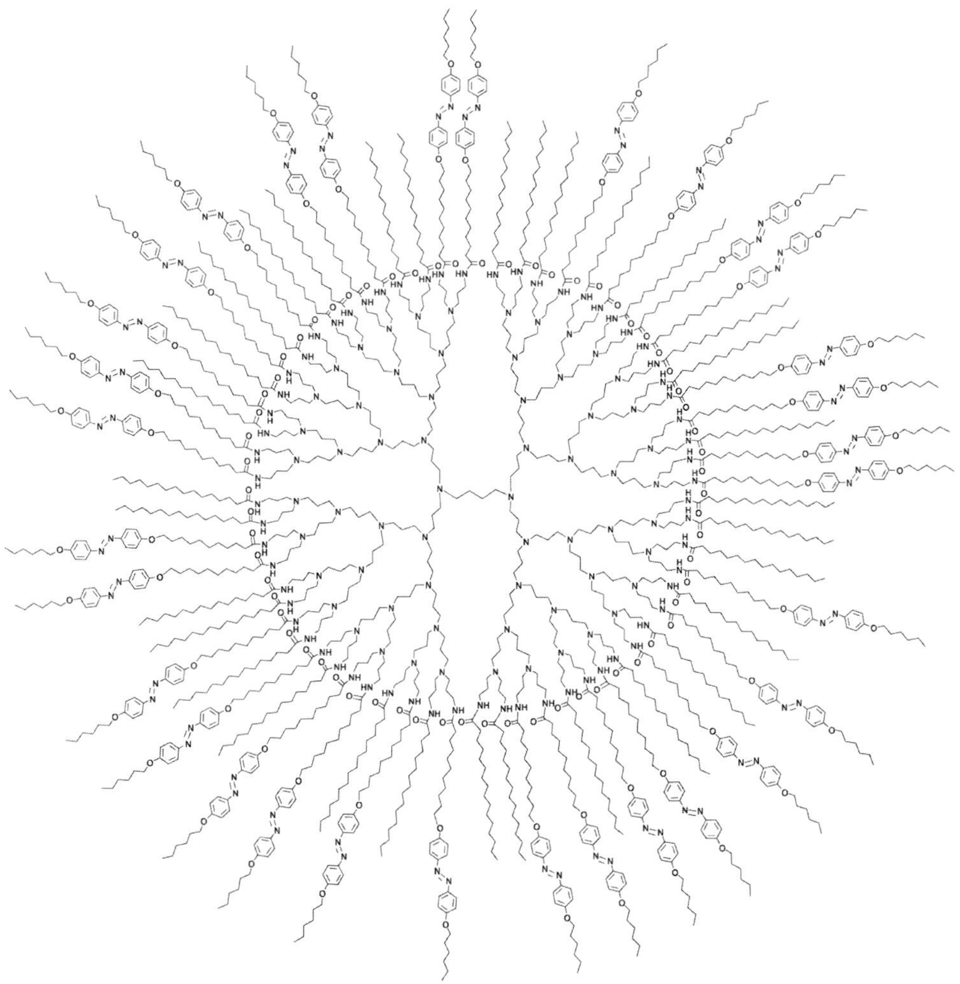

Fig. 3.11 Amphiphilic switchable POPAM dendrimer (according to Majoral *et al.*), whose periphery contains the two functional units in equal proportions but randomly distributed

A bifunctionalisation method generally leading to peripherally heterogeneous dendrimers of defined composition consists in targeted coupling of two different functional units to the individual end groups of a dendrimer scaffold [60]. Initial concepts for such local bifunctionalisation of a dendrimer periphery were presented by Shinkai *et al.* and by Kim *et al.* Preparation of a dendritic saccharide sensor with locally bifunctionalised molecular periphery was accomplished by Shinkai *et al.* through reductive amination of a second-generation PAMAM dendrimer with eight anthracene moieties as signalling groups and subsequent reaction with boronic acid reagent (Fig. 3.12) (see also Section 8.5.3.4) [61].

Kim *et al.* first reacted dendrimers bearing hydroxy terminal groups with amine-protected α-amino acids; after basic cleavage of the protective group they were able to introduce a second functionality by means of carboxylic acids [62].

However, these bifunctionalisation methods are comparatively laborious and applicable only in special cases, since the monofunctionalisation step is limited to substrates possessing an additional coupling site in protected form for the second functional unit. A more versatile method of local bifunctionalisation, which has no need of a deprotection step and also utilises commercially available dendrimer scaffolds, consists in the functionalisation of POPAM dendrimers bearing amine terminal groups with sulphonyl chlorides and subsequent substitution of the sulphonamide proton with other sulphonyl chlorides [63] or with alkyl- or (dendritic) benzyl bromides [64] (see Fig. 3.13).

The preparation of peripherally bifunctionalised dendrimers by the method of *non-covalent local bifunctionalisation* has hitherto remained restricted to individual cases. This method of bifunctionalisation can be applied to peripherally

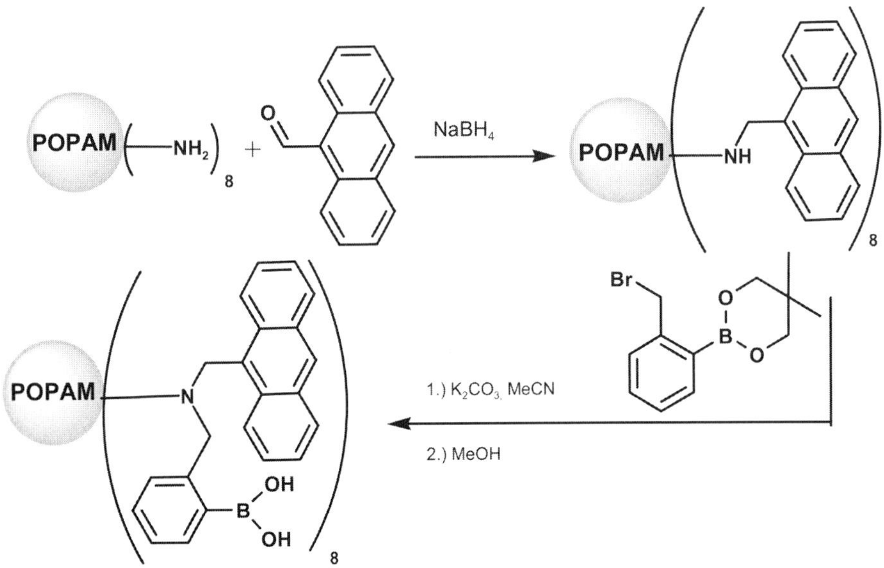

Fig. 3.12 Synthesis of a locally bifunctionalised dendrimer (according to Shinkai *et al.*)

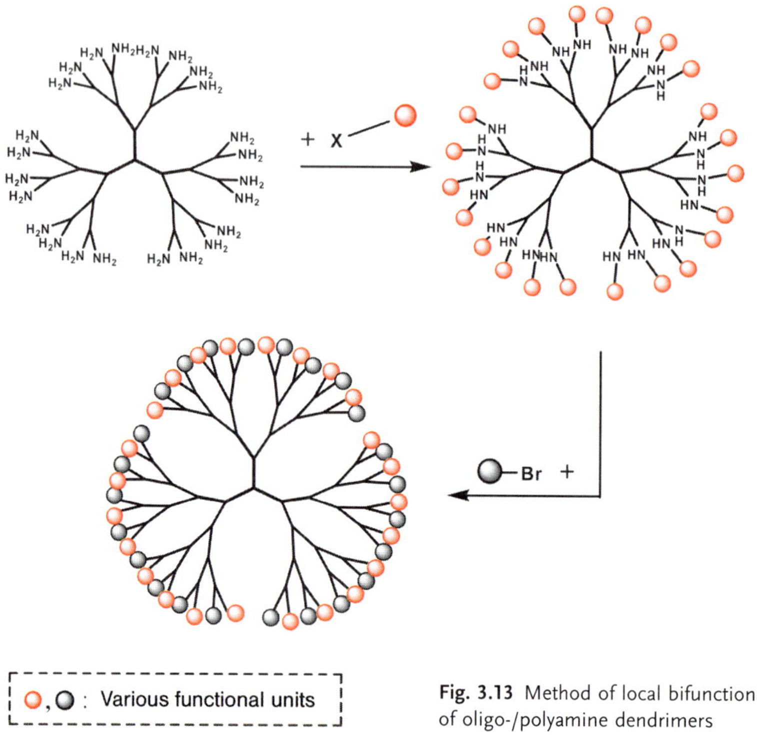

Fig. 3.13 Method of local bifunctionalisation of oligo-/polyamine dendrimers

monofunctionalised dendrimers, which can act as host dendrimers since they possess end groups capable of undergoing supramolecular interactions. In such cases it is possible to incorporate a second function-bearing unit, serving as a functional guest, non-covalently into the periphery of the molecule. This is illustrated by the example of POPAM dendrimers with peripheral adamantylurea groups, which could be functionalised with catalytically active, urea-substituted phosphine ligands by hydrogen bonding [65].

The method of local bifunctionalisation of end groups offers the advantage that the two different functional units are not only readily accessible, but are also spatially proximate. Such bifunctionalised dendrimers are ideal model compounds for the study of cooperative and allosteric interactions between various peripheral groups. This kind of spatial geometry permits intramolecular interactions between different neighbouring functionalities which could be of interest with a view to possible applications. Thus the chromophore group coupled to the ionophore ligand in "chromoionophores" reacts sensitively to its complexation [66]. The mechanism of action of enzymes is similarly influenced by the spatial proximity of several functional groups [67].

3.2.2
Two different functional units in different parts of the molecule

Examples of dendrimers possessing both a functional core and a periphery bearing functional units are relatively well represented in the literature. In many cases, the dendrimer shell serves to shield and solubilise the functional core unit (*e.g.* photoactive or catalytically active core unit) and is furnished with solubility-promoting peripheral groups for this very purpose. Depending upon the solvent, the introduction of alkyl chains is recommended for enhancing the lipophilicity, perfluoroalkyl chains for raising the hydrophobicity, or hydrophilic groups such as sulphonates [68], carboxylates [69, 28], phosphonates [70], quaternary ammonium groups [71], sugars [72], or polyethylene glycols [5 b, 73] for improving solubility.

Energy gradients can arise if two or more different kinds of functional units capable of interacting in suitable manner are incorporated in different parts (core or focal point, scaffold, periphery) of a dendrimer or dendron [74]. Such entities will frequently be dendritic structures, in which a structurally generated gradient enables energy- or electron-transfer reactions from the molecular periphery to the photoactive or redox-active central unit. Fréchet *et al.* convergently assembled a bifunctionalised poly(benzyl ether) dendron in which light is harvested by the terminal chromophore (coumarin 2) and focussed on the fluorescing focal point (coumarin 343) (light harvesting antenna; Fig. 3.14; see also Section 5.2) [75]. Surrounding a central tris(bipyridine)ruthenium(II) complex with naphthalene-functionalised Fréchet type dendrons also leads to dendrimers exhibiting effective energy transfer from terminal naphthyl units *via* the dendritic scaffold to the central ruthenium complex [76].

3.2.3
More than two different functional units

Whereas the bifunctionalisation strategies presented so far are relatively straightforward, the synthesis of multifunctional dendrimers with more than two kinds of functional units requires considerable synthetic effort. Preparation of dendrimers with a functional core and additional functional units in the dendrimer scaffold and in the periphery requires *de novo* synthesis of the entire dendrimer scaffold, with the synthesis conditions having to be tolerable for all groups (Fig. 3.15).

Müllen *et al.* presented a multi-chromophoric dendrimer which absorbs over the entire visible spectrum. Three different dyes are positioned in a rigid polyphenylene dendrimer scaffold in such a manner that an energy gradient is generated between the periphery and the centre of the core and efficient energy transfer to the central chromophore (A) takes place on excitation of the peripheral chromophore (C1, C2) (Fig. 3.16) [77].

Starting from ABB′ monomers, a multifunctionalised third-generation poly (benzyl ether) dendron could be prepared in which all eight terminal groups are

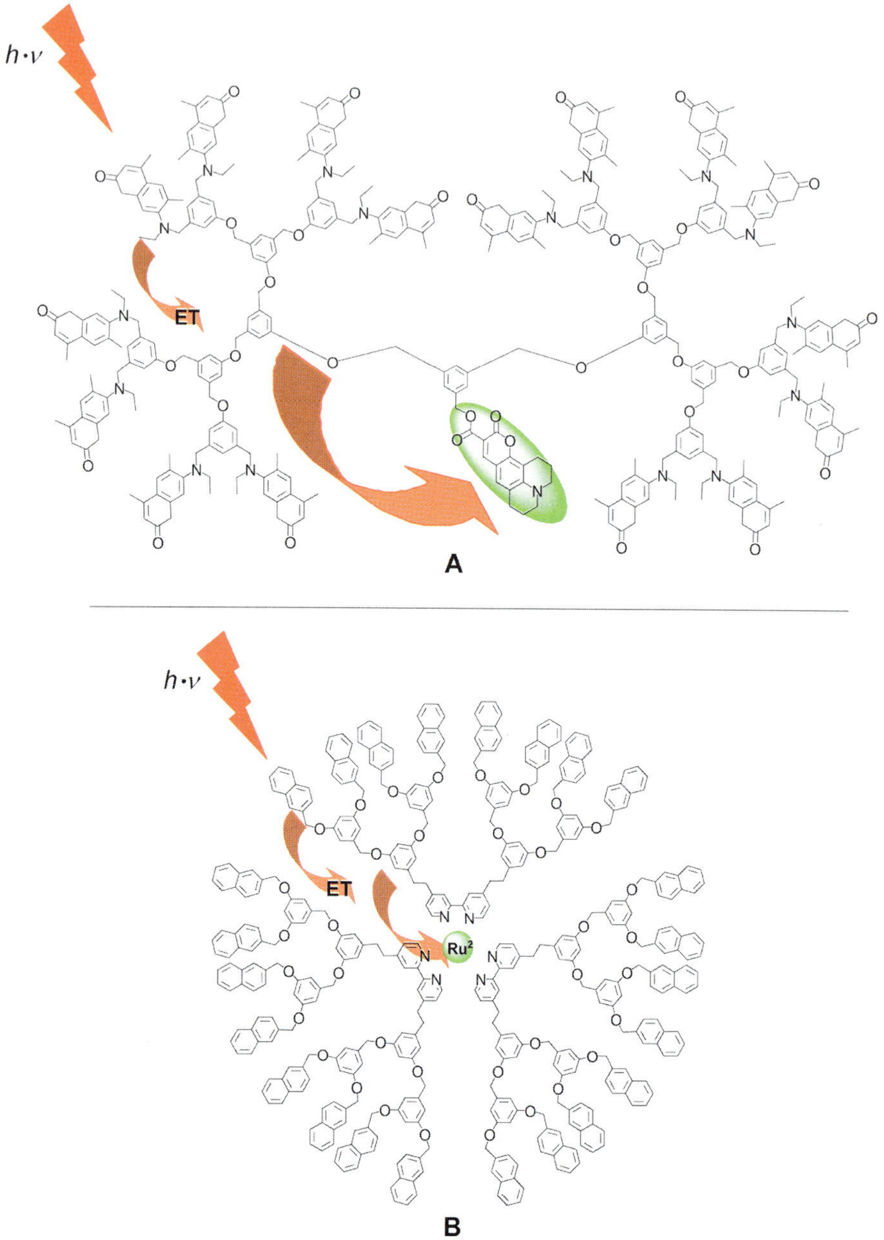

Fig. 3.14 Two different dendritic light-harvesting antennae **A** and **B**

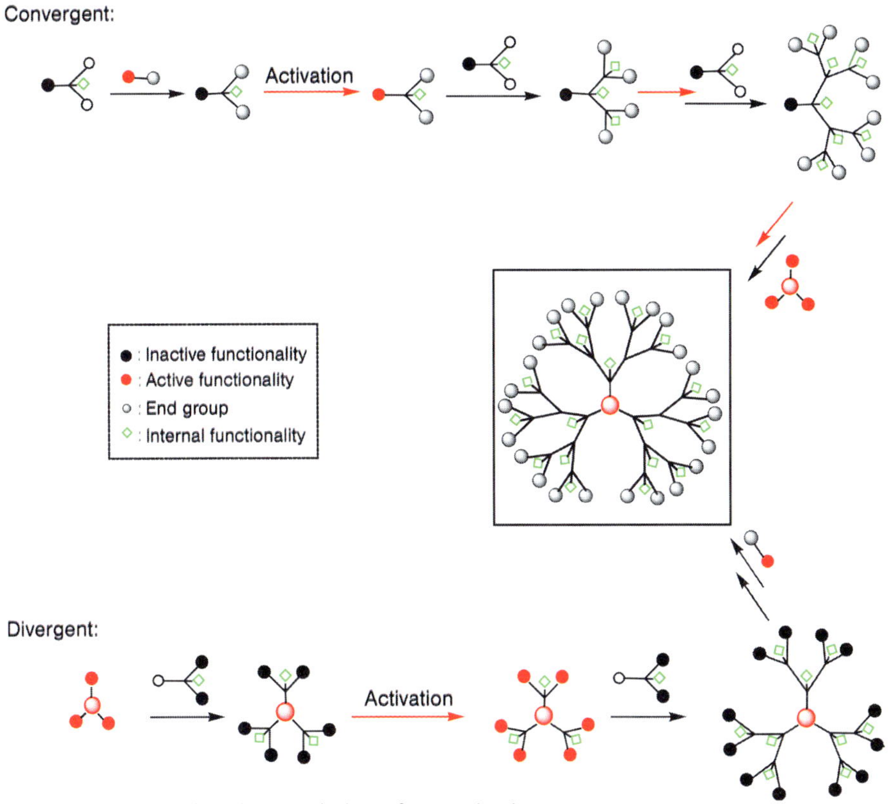

Fig. 3.15 Synthesis of dendrimers which are functionalised in the core, in the branching scaffold, and in the periphery

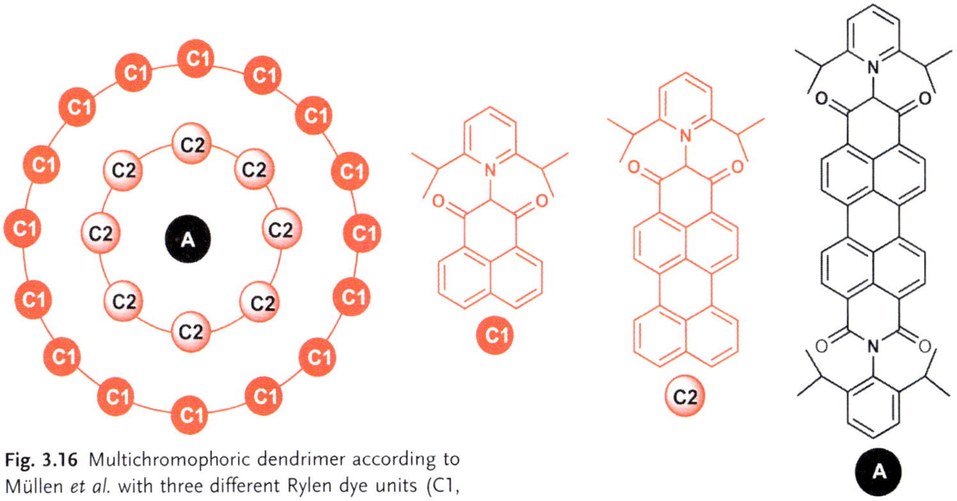

Fig. 3.16 Multichromophoric dendrimer according to Müllen *et al.* with three different Rylen dye units (C1, C2, A) in the periphery, the scaffold, and the core (schematic)

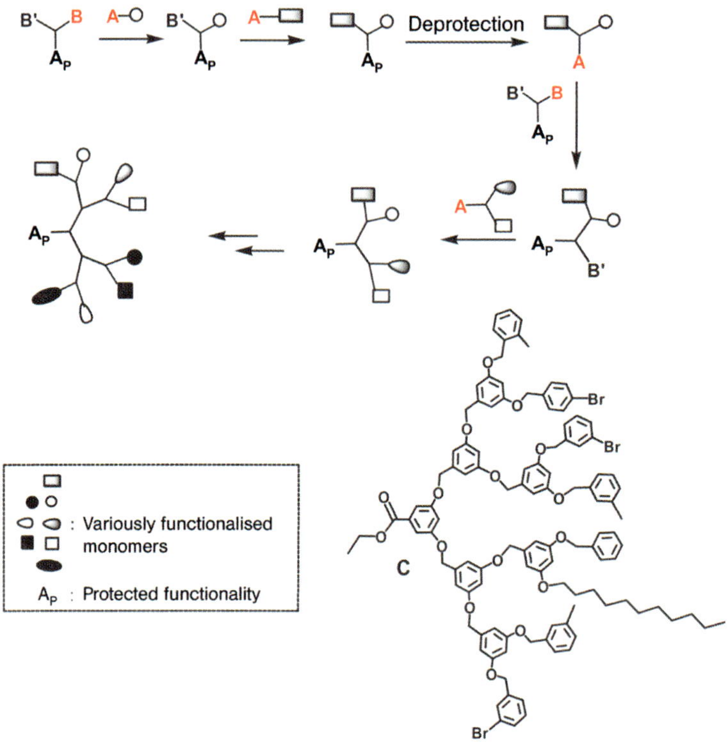

Fig. 3.17 Synthetic strategy for non-symmetrical dendrons. The procedure exploits the fact that functionality A reacts better with functionality B than with B′ (A_P = protected functionality)

different (Fig. 3.17) [78]. Coupling of these dendrons to a core unit would yield dendrimers with a multifunctionalised periphery of predictable composition. The advantages of a pronounced functional variety in the periphery of the molecule justifies the large number of synthetic and purification steps compared to classical convergent synthesis.

The above sections demonstrate that the dendrimer chemist meanwhile has recourse to an extensive range of possibilities for introducing specific functional units in one or more selected regions of the molecule (centre, periphery, interior branching scaffold) in order to customise dendrimers or dendrons for projected applications. It should, however, be borne in mind that the synthetic strategies for introduction of two or more different functional units (bifunctionalisation, multifunctionalisation) demand considerable synthetic effort as well as time and expense and their use is limited to certain dendrimers and dendrons, because special conditions have to be fulfilled (*e.g.* special end groups such as primary amino functions).

3.2.4
Overview of functional dendrimers and their synthesis

To conclude this Chapter, the characteristics of, and relationships between, the different types of functional dendrimers and the hitherto applied, and hence recommendable, synthetic strategies are compiled in tabular form:

Type of dendrimer Monofunctional dendrimer	Structure	Synthetic strategy
a) with a functional core unit	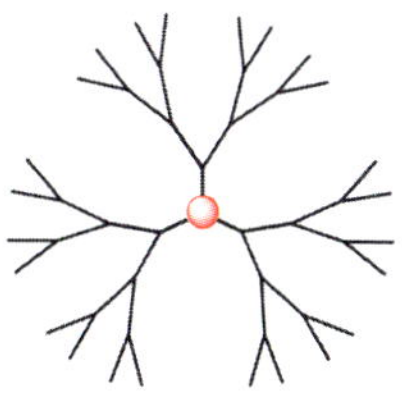	Convergent, supramolecular or more rarely divergent dendrimer assembly using a functional core unit
b) with a functional molecular periphery	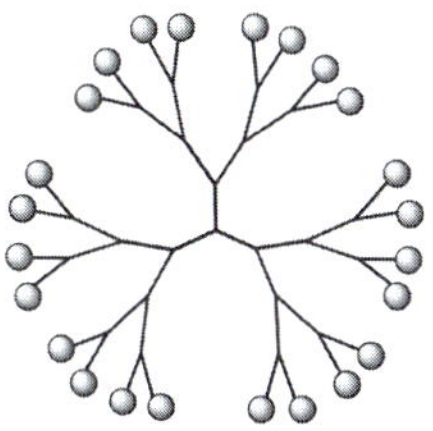	a) Premodification: Convergent dendrimer assembly starting from the future peripheral groups b) Postmodification: End group functionalisation on conclusion of divergent or convergent dendrimer assembly
c) with internal functional units	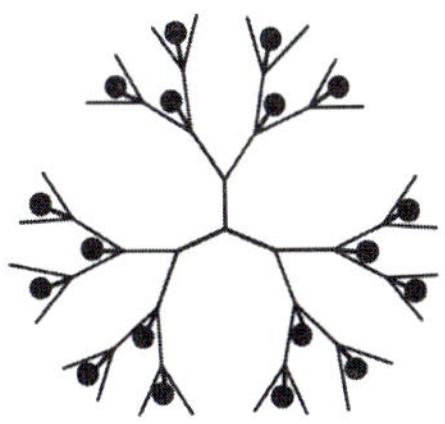	a) Premodification: Divergent, convergent, or supramolecular dendrimer assembly using monomeric building blocks (*e.g.* AB_2FG), already containing the desired function. b) Postmodification: Convergent or divergent assembly of a dendrimer precursor using monomeric building blocks with orthogonally protected or selectively activatable functionalities. Introduction of the desired functional unit by subsequent functionalisation of the internal functionalities of the dendrimer precursor.

Type of dendrimer Multifunctional dendrimer	Structure	Synthetic strategy
	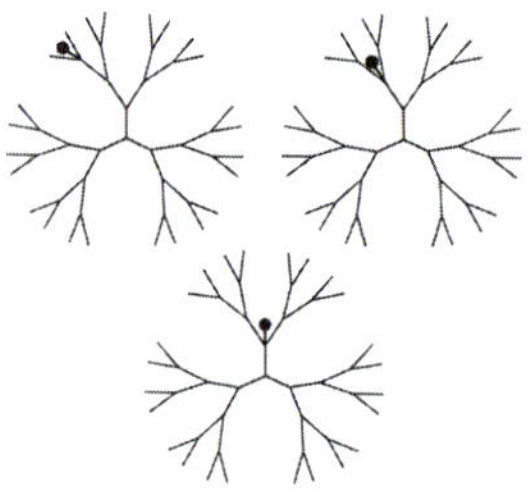 Dendrimers with a single, generation-specifically incorporated functional unit.	Convergent assembly using a monofunctionalised AB_2FG-type monomeric branching building block. Random monofunctionalisation of dendritic branches on monomeric building blocks is a necessary feature of the synthesis.
a) with globally bifunctionalised periphery	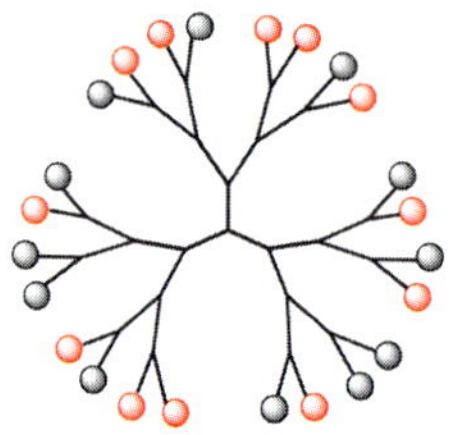 Random distribution of different functional units in the periphery of the molecule.	Postmodification: Reaction of the end groups of an existing dendrimer scaffold with a mixture of two differently functionalised substrates.
	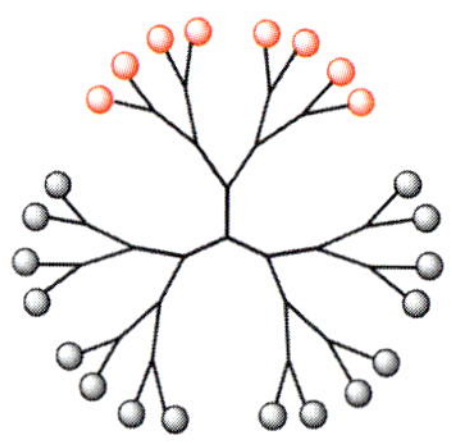 Surface-block dendrimer with two different peripheral groups in separate segments of the molecule.	a) Coupling of two kinds of dendrons, differing in their peripheral groups. b) Coupling of such dendrons to a common core unit.

Type of dendrimer Multifunctional dendrimer	Structure	Synthetic strategy
b) with locally bifunctionalised molecular periphery	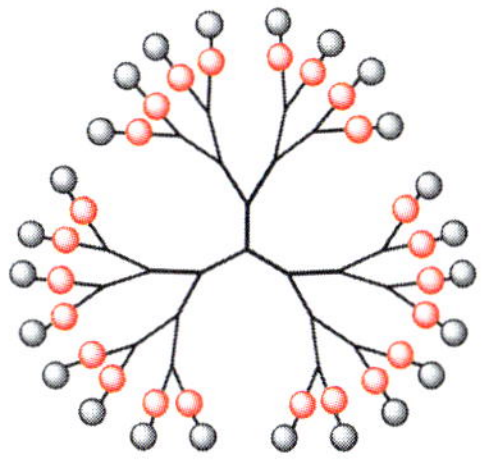	Monofunctionalisation of the end groups with a substrate containing an additional coupling site in protected form. Subsequent introduction of another functional unit *via* the activated coupling site.
	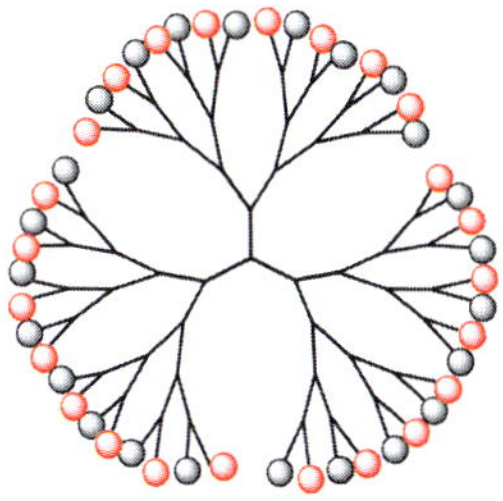 Bifunctionalised dendrimer with perfectly alternating molecular periphery.	Monofunctionalisation of polyamine dendrimers with functionalised sulphonyl chlorides. Subsequent substitution of the sulphonamide proton with other sulphonyl chlorides or with alkyl- or (dendritic) benzyl bromides.
c) with functional core and internal functional units	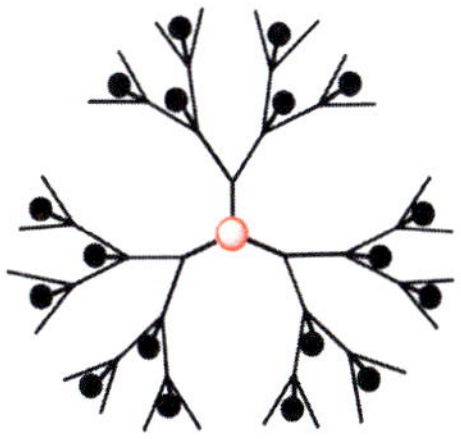	a) Convergent or divergent synthesis using the desired functional core units and correspondingly prefunctionalised monomeric building blocks.

Type of dendrimer Multifunctional dendrimer	Structure	Synthetic strategy
d) with functional core, internal functional units, and functionalised molecular periphery	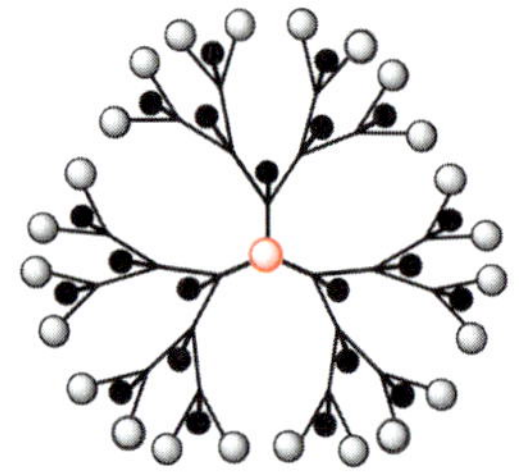	*De novo* synthesis of dendrimer scaffold necessary: a) Convergent synthesis: Dendron assembly using appropriately prefunctionalised monomeric building blocks (*e.g.* AB_2FG) starting from the peripheral functional units of the dendrimer to be formed. Subsequent coupling of the dendrons to the desired functional core unit. b) Divergent synthesis (rare): Starting from the functional core unit, the dendrimer scaffold is assembled using monomeric building blocks already containing the desired internal functional unit. There follows functionalisation of the end groups for introduction of the desired peripheral functional units. c) Special case of metallodendrimers: Divergent or convergent assembly from transition-metal complexes (frequently with polypyridine ligands) as core and branching units.

Bibliography and Notes for Chapter 3 "Functional dendrimers"

Review articles are indicated by the words "Review(s)" or "Book(s)" in bold-faced type.

1 Dendrimers are not, as could be mistakenly assumed, limited to macromolecules, polymers, but are also of importance in the chemistry of low-molecular weight compounds. For example, low-generation dendrimers do not generally number among the macromolecules, and functionalised dendrons and dendritic substituents will be sure to play an important role in the chemistry of small molecules, both in organic and in bio-organic and inorganic chemistry, for control of steric effects, philicity effects, and solubilities. Thus, in spite of what their name may imply, dendrimers are of relevance throughout the entire field of chemistry.

2 ***Reviews***: a) H.-F. Chow, T. K. K. Mong, M. F. Nongrum, C.-W. Wan, *Tetrahedron* **1998**, *54*, 8543–8660; b) F. Diederich, *Chem. Eur. J.* **1998**, *4*, 1353–1361; c) A. Archut, F. Vögtle, *Chem. Soc. Review* **1998**, 27, 227–240; d) F. Vögtle, S. Gestermann, R. Hesse, H. Schwierz, B. Windisch, *Prog. Polym. Sci.* **2000**, *25*, 987–1041.

3 a) M. Kawa, J. M. J. Fréchet, *Chem. Mater.* **1998**, *10*, 286–296; b) H. F. Chow, I. Y. K. Chan, D. T. W. Chan, R. W. M. Kwok, *Chem. Eur. J.* **1996**, *2*, 1085–1091; c) M. Plevoets, F. Vögtle, L. De Cola, V. Balzani, *New J. Chem.* **1999**, *3*, 63–69; d) F. Vögtle, M. Plevoets, M. Nieger, G. C. Azzellini, A. Credi, L. De Cola, V. Marchis, M. Venturi, V. Balzani, *J. Am. Chem. Soc.* **1999**, *121*, 6290–6298.

4 a) ***Review***: V. Balzani, A. Juris, M. Venturi, S. Campagna, S. Serroni, *Chem. Rev.* **1996**, *96*, 759–833; b) G. Denti, S. Campagna, S. Serroni, M. Ciano, V. Balzani, *J. Am. Chem. Soc.* **1992**, *114*, 2944–2950; c) ***Review***: V. Balzani, S. Campagna, G. Denti, A. Juris, S. Serroni, M. Venturi, *Acc. Chem. Res.* **1998**, *31*, 26–34.

5 a) P. J. Dandliker, F. Diederich, M. Gross, C. B. Knobler, A. Louati, E. M. Sanford, *Angew. Chem.* **1994**, *106*, 821–1824; *Angew. Chem. Int. Ed.* **1994**, *33*, 1739–1742; b) P. Dandliker, F. Diederich, A. Zingg, J.-P. Gisselbrecht, M. Gross, A. Louati, E. Sanford, *Helv. Chim. Acta* **1997**, *80*, 1773–1801; c) M. Kimura, Y. Sugihara, T. Muto, K. Hanabusa, H. Shirai, N. Koboayashi, *Chem. Eur. J.* **1999**, *5*, 3495–3450.

6 a) J. M. J. Fréchet, *Science* **1994**, *263*, 1710–1715; b) ***Review***: S. Hecht, J. M. J. Fréchet, *Angew. Chem.* **2001**, *113*, 76–94; *Angew. Chem. Int. Ed.* **2001**, *40*, 74–91.

7 *E.g.* dendritically substituted photoactive metal complexes: ***Review***: V. Balzani, P. Ceroni, A. Juris, M. Venturi, S. Campagna, F. Puntoriero, S. Serroni, *Coord. Chem. Rev.* **2001**, *219–221*, 545–572.

8 C. B. Gorman, J. C. Smith, M. W. Hager, B. L. Parkhurst, H. Sierzputowska-Gracz, C. A. Haney, *J. Am. Chem. Soc.* **1999**, *121*, 9958–9966.

9 A. V. Davis, M. Driffild, D. K. Smith, *Org. Lett.* **2001**, *3*, 3075–3078.

10 D. L. Stone, D. K. Smith, P. T. McGrail, *J. Am. Chem. Soc.* **2002**, *124*, 856–864.

11 a) T. Marquardt, U. Lüning, *Chem. Commun.* **1997**, 1681–1682; b) D. A. Tomalia, P. R. Dvornic, *Nature* **1994**, *372*, 617–618; c) N. Brinkman, D. Giebel, G. Lohmer, M. T. Reetz, U. Cragl, *J. Catal.* **1999**, *183*, 163–168; d) N. J. Hovestad, E. B. Eggeling, H. J. Heidbüchel, J. T. B. H. Jastrzebski, U. Kragl, W. Keim, D. Vogt, G. van Koten, *Angew. Chem.* **1999**, *111*, 1763–1765; *Angew. Chem. Int. Ed.* **1999**, *38*, 1655–1658; e) D. de Groot, E. B. Eggeling, J. C. de Wilde, H. Kooijman, R. J. van Haaren, A. W. van der Made, A. L. Spek, D. Vogt, J. N. H. Reek, P. C. J. Kramer, P. W. N. M. van Leeuwen, *Chem. Commun.* **1999**, 1623–1624; f) M. Albrecht, N. J. Hovestad, J. Boersma, G. van Koten, *Chem. Eur. J.* **2001**, *7*, 1289–1294.

12 a) S. C. Zimmerman, I. Zharov, M. S. Wendland, N. A. Rakow, K. S. Suslik, *J. Am. Chem. Soc.* **2003**, *125*, 13504–13518; b) S. Kimata, D. L. Jiang, T. Aida, *J. Polym. Sci., Polym. Chem.* **2003**, *41*, 3524–3530, c) A. Dahan, M. Portnoy, *Macromolecules* **2003**, *36*, 1034–1038; d) E. M. Harth, S. Hecht, B. Helms,

E. Malmstrom, J. M. J. Fréchet, C. J. Hawker, *J. Am. Chem. Soc.* **2002**, *124*, 3926–3938; e) D. J. Pochan, L. Pakstis, E. Huang, C. J. Hawker, R. Vestberg, J. Pople, *Macromolecules* **2002**, *35*, 9239–9242; f) M. E. Mackay, Y. Hong, M. Jeong, S. Hong, T. P. Russel, C. J. Hawker, R. Vestberg, J. Douglas, *Langmuir* **2002**, *18*, 1877–1882.

13 G. R. Newkome, C. D. Weis, C. N. Moorefield, G. R. Baker, B. J. Childs, J. Epperson, *Angew. Chem.* **1998**, *110*, 318–321; *Angew. Chem. Int. Ed.* **1998**, *37*, 307–310.

14 P. Marchand, L. Griffe, A.-M. Caminade, J.-P. Majoral, M. Destarac, F. Leising, *Org. Lett.* **2004**, *6*, 1309–1312.

15 S. Stevelmans, J. C. M. van Hest, J. F. G. A. Jansen, D. A. F. J. van Boxtel, E. M. M. de Brabander-van den Berg, E. W. Meijer, *J. Am. Chem. Soc.* **1996**, *118*, 7398–7399; b) M. W. P. L. Baars, P. E. Froehling, E. W. Meijer, *J. Chem. Soc., Chem. Commun.* **1997**, 1959–1960; c) C. Valério, J.-L. Fillaut, J. Ruiz, J. Guittard, J.-C. Blais, D. Astruc, *J. Am. Chem. Soc.* **1997**, *119*, 2588–2589; d) J. H. Cameron, A. Facher, G. Lattermann, S. Diele, *Adv. Mater.* **1997**, *9*, 398–403.

16 A. I. Cooper, J. D. Londono, G. Wignall, J. B. McClain, E. T. Samulski, J. S. Lin, A. Dobrynin, M. Rubinstein, A. L. C. Burke, J. M. J. Fréchet, J. M. DeSimone, *Nature* **1997**, *389*, 368–371.

17 R. Roy, D. Zanini, S. J. Meunier, A. Romanowska, *J. Chem. Soc., Chem. Commun.* **1993**, 1869–1872.

18 a) J. F. G. A. Jansen, E. M. M. de Brabander-van den Berg, E. W. Meijer, *Science* **1994**, *266*, 1226–1229; b) P. R. Ashton, S. E. Boyd, C. L. Brown, S. Nepogodiev, E. W. Meijer, H. W. I. Peerlings, J. F. Stoddart, *Chem. Eur. J.* **1997**, *3*, 974–984; c) A. P. H. J. Schenning, C. Elissen-Román, J.-W. Weener, M. W. P. L. Baars, S. J. van der Gaast, E. W. Meijer, *J. Am. Chem. Soc.* **1998**, *120*, 8199–8208.

19 a) H. W. Peerlings, S. A. Nepogodiev, J. F. Stoddart, E. W. Meijer, *Eur. J. Org. Chem.* **1998**, 1879–1886; b) J. F. G. A. Jansen, H. W. I. Peerlings, E. M. M. Brabander-van den Berg, E. W. Meijer, *Angew. Chem.* **1995**, *107*, 1312–1324; *Angew. Chem. Int. Ed.* **1995**, *34*, 1206–1209; c) A. Archut, F. Vögtle, L. De Cola, G. C. Azzellini, V. Balzani, P. S. Ramanujam, R. H. Berg, *Chem. Eur. J.* **1998**, *4*, 699–706.

20 a) D. Pagé, S. Aravind, R. Roy, *J. Chem. Soc., Chem. Commun.* **1996**, 1913–1914; b) T. K. Lindhorst, C. Kieburg, *Angew. Chem.* **1996**, *108*, 2083–2086; *Angew. Chem. Int. Ed.* **1996**, *35*, 1953–1956; c) C. Kieburg, T. K. Lindhorst, *Tedrahedron Lett.* **1997**, *38*, 3885–3888; d) D. Zanini, R. Roy, *J. Org. Chem.* **1998**, *63*, 3486–3489; e) U. Boas, A. J. Karlsson, B. F. W. de Waal, E. W. Meijer, *J. Org. Chem.* **2001**, *66*, 2136–2145.

21 a) G. R. Newkome, C. D. Weis, C. N. Moorefield, G. R. Baker, B. J. Childs, J. Epperson, *Angew. Chem.* **1998**, *110*, 318–321; *Angew. Chem. Int. Ed.* **1998**, *37*, 307–310; b) A. P. H. J. Schenning, C. Elissen-Román, J. W. Weener, M. W. P. L. Baars, S. J. van der Gaast, E. W. Meijer, *J. Am. Chem. Soc.* **1998**, *120*, 8199–8208; c) M. W. P. L. Baars, A. J. Karlsson, V. Sorokin, B. F. W. de Waal, E. W. Meijer, *Angew. Chem.* **2000**, *112*, 4432–4436; *Angew. Chem. Int. Ed.* **2000**, *39*, 4262–4266; d) S. Rosenfeldt, N. Dingenouts, M. Ballauff, N. Werner, F. Vögtle, P. Lindner, *Macromolecules* **2002**, *35*, 8098–8105.

22 M. S. T. H. Sanders-Hovens, J. F. G. A. Jansen, J. A. J. M. Vekemans, E. W. Meijer, *Polym. Mater. Sci. Eng.* **1995**, *73*, 338–339.

23 M. T. Reetz, G. Lohmer, R. Schwickardi, *Angew. Chem.* **1997**, *109*, 1559–1562; *Angew. Chem Int. Ed.* **1997**, *36*, 1526–1529.

24 J. Issberner, M. Böhme, S. Grimme, M. Nieger, W. Paulus, F. Vögtle, *Tetrahedron: Asymmetry* **1996**, *7*, 2223–2232.

25 I. J. Majoros, B. Keszler, S. Woehler, T. Bull, J. R. Baker, Jr., *Macromolecules* **2003**, *36*, 5526–5529.

26 a) N. Malik, R. Wiwattanapatapee, R. Klopsch, K. Lorenz, H. Frey, J. W. Weener, E. W. Meijer, W. Paulus, R. Duncan, *J. Controlled Release* **2000**, *65*, 133–148; b) S. Fuchs, T. Kapp, H. Otto, T. Schöneberg, P. Franke, R. Gust, A. D.

Schlüter, *Chem. Eur. J.* **2004**, *10*, 1167–1192.
27 M. Malkoch, K. Schleicher, E. Drockenmuller, C. J. Hawker, T. P. Russel, P. Wu, V. V. Fokin, *Macromolecules* **2005**, *38*, 3663–3678; RAFT-Copolymerisation: A. P. Vogt, S. R. Gondi, B. S. Sumerlin, *Aust. J. Chem.* **2007**, *60*, 396–399.
28 J. W. Leon, M. Kawa, J. M. J. Fréchet, *J. Am. Chem. Soc.* **1996**, *118*, 8847–8859.
29 M. R. Leduc, W. Hayes, J. M. J. Fréchet, *J. Polym. Sci., Part A* **1998**, *36*, 1–10.
30 L. Groenendaal, J. M. J. Fréchet, *J. Org. Chem.* **1998**, *63*, 5675–5679.
31 Y.-H. Liao, J. R. Moss, *J. Chem. Soc., Chem. Commun.* **1993**, 1774–1777.
32 C. F. Shu, H. M. Shen, *J. Mater. Chem.* **1997**, *7*, 47–52.
33 D. Seebach, R. E. Marti, T. Hintermann, *Helv. Chim. Acta* **1996**, *79*, 1710–1740.
34 M. Serin, D. W. Brousmiche, J. M. J. Fréchet, *Chem. Commun.* **2002**, 2605–2607.
35 a) N. Jayaraman, S. A. Nepogodiev, J. F. Stoddart, *Chem. Eur. J.* **1997**, *3*, 1193–1199; b) P. R. Ashton, E. F. Hounsell, N. Jayaraman, T. M. Nilsen, N. Spencer, J. F. Stoddart, M. Young, *J. Org. Chem.* **1998**, *63*, 3429–3437.
36 J.-F. Nierengarten, D. Felder, J.-F. Nicoud, *Tetrahedron Lett.* **1999**,*40*, 269–272.
37 a) L. A. Baker, R. M. Crooks, *Macromolecules* **2000**, *33*, 9034–9039; b) C. Francavilla, M. D. Drake, F. V. Bright, M. R. Detty, *J. Am. Chem. Soc.* **2001**, *123*, 57–67; c) D. Lagnoux, E. Delort, C. Dout-Casassus, A. Espito, J. L. Reymond, *Chem. Eur. J.* **2004**, *10*, 1215–1226; d) A. W. Kleij, R. A. Gossage, J. T. B. H. Jastrzebski, J. Boersma, G. van Koten, *Angew. Chem.* **2000**, *112*, 179–181; *Angew. Chem. Int. Ed.* **2000**, *39*, 176–178.
38 Synthetic concepts for functionalisation of the interiors of dendrimers: S. Hecht, *J. Polym. Sci. Part A* **2003**, *41*, 1047–1058.
39 T. Nagasaki, S. Tamagaki, K. Ogini, *Chem. Lett.* **1997**, 717–718.
40 V. Percec, P. Chu, G. Ungar, J. Zhou, *J. Am. Chem. Soc.* **1995**, 117, 11441–11454; b) J. F. Li, K. A. Krandall, P. Chu, V. Percec, R. G. Petschek, C. Rosenblatt, *Macromolecules* **1996**, *29*, 7813–7819; c) ***Review***: V. Percec, *Pure Appl. Chem.* **1995**, *67*, 2031–2038.
41 *E.g.* a) H. F. Chow, C. C. Mak, *J. Chem. Soc., Perkin Trans 1* **1994**, 2223–2228; b) L. J. Twyman, A. E. Beezer, J. C. Mitchell, *Tetrahedron Lett.* **1994**, *35*, 4423–4424; c) D. Seebach, G. F. Herrmann, U. D. Lengweiler, W. Amrein, *Helv. Chim. Acta* **1997**, *80*, 989–1026; d) P. Murer, D. Seebach, *Helv. Chim. Acta* **1998**, *81*, 603–631.
42 Z. Bo, A. Schäfer, P. Franke, A. D. Schlüter, *Org. Lett.* **2000**, *2*, 1645–1648.
43 H.-F. Chow, M.-K. Ng, C. W. Leung, G.-X. Wang, *J. Am. Chem. Soc.* **2004**, *126*, 12907–12915.
44 C. O. Liang, J. M. J. Fréchet, *Macromolecules* **2005**, *38*, 6276–6284.
45 a) C. J. Hawker, J. M. J. Fréchet, *Macromolecules* **1990**, *23*, 4726–4729; b) C. J. Hawker, J. M. J. Fréchet, *J. Am. Chem. Soc.* **1990**, *112*, 7638–7647.
46 L. Lochmann, K. L. Wooley, P. T. Ivanova, J. M. J. Fréchet, *J. Am. Chem. Soc.* **1993**, *115*, 7043–7044.
47 a) S. Hecht, *J. Polym. Sci., Part A* **2003**, *41*, 1047–1058; b) L. G. Schultz, Y. Zhao, S. C. Zimmerman, *Angew. Chem.* **2001**, *113*, 2016–2020; *Angew. Chem. Int. Ed.* **2001**, *40*, 1929–1932. c) M. F. Ottavania, F. Monalti, N. J. Turro, D. A. Tomalia, *J. Phys. Chem. B* **1997**, *101*, 158–166; d) G. J. M. Koper, M. H. P. van Genderen, C. Elissen-Román, M. Baars, E. W. Meijer, M. Borcovec, *J. Am. Chem. Soc.* **1997**, *119*, 6512–6521.
48 G. R. Newkome, J. M. Keith, G. R. Baker, G. H. Escamilla, C. N. Moorefield, *Angew. Chem.* **1994**, *106*, 701–703; *Angew. Chem. Int. Ed.* **1994**, *33*, 666–668.
49 S. Bernhardt, M. Baumgarten, M. Wagner, K. Müllen, *J. Am. Chem. Soc.* **2005**, *127*, 12392–12399.
50 L. G. Schultz, Y. Zhao, S. C. Zimmerman, *Angew. Chem.* **2001**, *113*, 2016–2020; *Angew. Chem. Int. Ed.* **2001**, *40*, 1962–1966.
51 C. Larre, D. Bressolles, C. Turrin, B. Donnadieu, A. M. Caminade, J. Majoral, *J. Am. Chem. Soc.* **1998**, *120*, 13070–13082; Polyethylene imines are commercially available under trademarks Lupasol® and Polymin® (BASF Company,

Polymer Research, Ludwigshafen, Germany).
52 L. Brauge, A. M. Caminade, J. P. Majoral, S. Slomkowski, M. Wolszczak, *Macromolecules* **2001**, *34*, 5599–5606.
53 a) R. Sadamoto, N. Tomioka, T. Aida, *J. Am. Chem. Soc.* **1996**, *118*, 3978–3979; b) G. R. Newkome, A. Nayak, R. K. Behara, C. N. Moorefield, G. R. Baker, *J. Org. Chem.* **1992**, *57*, 358–362; c) G. R. Newkome, X. Lin, J. K. Young, *Synlett* **1992**, 53–54.
54 C. J. Hawker, K. L. Wooley, J. M. J. Fréchet, *Macromol. Symp.* **1994**, *77*, 11–20.
55 a) E. M. Sandford, J. M. J. Fréchet, K. L. Wooley, C. J. Hawker, *J. Polym. Prepr.* **1993**, *34*, 654–655; b) C. J. Hawker, K. L. Wooley, J. M. J. Fréchet, *J. Chem. Soc., Perkin Trans. 1* **1993**, 1287–1297; J. M. J. Fréchet, C. J. Hawker, K. L. Wooley, *Pure Appl. Chem.* **1994**, *A31*, 1627–1645.
56 M. R. Bryce, P. de Miguel, W. Devonport, *J. Chem. Soc., Chem. Commun.* **1998**, 2565–2566.
57 V. Maraval, R. Laurent, B. Donnadieu, M. Mauzac, A.-M. Caminade, J.-P. Majoral, *J. Am. Chem. Soc.* **2000**, *122*, 2499–2511.
58 P. Furuta, J. M. J. Fréchet, *J. Am. Chem. Soc.* **2003**, *125*, 13173–13181.
59 a) J.-W. Weener, E. W. Meijer, *Adv. Mater.* **2000**, *12*, 741–746; b) M. W. Baars, M. C. V. Boxtel, D. J. Broer, S. H. Söntjens, E. W. Meijer, *Adv. Mater.* **2000**, *12*, 715–719; c) K. Tsuda, G. C. Dol, T. Gensch, J. Hofkens. L. Latterini, J. W. Weener, E. W. Meijer, F. C. De Schryver, *J. Am. Chem. Soc.* **2000**, *122*, 3445–3452; ***Review***: d) T. Gensch, K. Tsuda, G. C. Dol, L. Latterini, J. W. Weener, A. P. Schenning, J. Hofkens, E. W. Meijer, F. C. De Schryver, *Pure Appl. Chem.* **2001**, *73*, 435–441.
60 a) T. D. James, H. Shinmori, M. Takeuchi, S. Shinkai, *J. Chem. Soc., Chem. Commun.* **1996**, 705–706; b) A. Archut, S. Gestermann, R. Hesse, C. Kauffmann, F. Vögtle, *Synlett* **1998**, 546–548.
61 T. D. James, H. Shinmori, M. Takeuchi, S. Shinkai, *Chem. Commun.* **1996**, 705–706.
62 R. M. Kim, M. Manna, S. H. Hutchins, P. R. Griffen, N. A. Yates, A. M. Bernick, T. Chapman, *Proc. Natl. Acad. Sci. USA* **1996**, *93*, 10012–10017.
63 a) F. Vögtle, H. Fakhrnabavi, O. Lukin, *Org. Lett.* **2004**, *6*, 1075–1078; b) F. Vögtle, H. Fakhrnabavi, O. Lukin, S. Müller, J. Friedhofen, C. A. Schalley, *Eur. J. Org. Chem.* **2004**, 4717–4724.
64 a) A. Archut, S. Gestermann, R. Hesse, C. Kauffmann, F. Vögtle, *Synlett* **1998**, 546–548; b) U. Hahn, M. Gorka, F. Vögtle, V. Vicinelli, P. Ceroni, M. Maestri, V. Balzani, *Angew. Chem.* **2002**, *114*, 3747–3750; *Angew. Chem. Int. Ed.* **2002**, *41*, 3595–3598.
65 D. de Groot, B. F. M. de Waal, J. N. H. Reek, A. P. H. J. Schenning, P. C. J. Kamer, E. W. Meijer, P. W. N. M. van Leeuwen, *J. Am. Chem. Soc.* **2001**, *123*, 8453–8458.
66 ***Reviews***: a) A. P. de Silva, H. Q. N. Gunaratne, T. Gunnlaugsson, A. J. M. Huxley, C. P. McCoy, J. T. Rademacher, T. E. Rice, *Chem. Rev.* **1997**, *97*, 1515–1566; b) L. Fabrizzi, A. Poggi, *Chem. Soc. Rev.* **1995**, 197–202.
67 ***Book***: L. Stryer, *Biochemie*, 3rd Edit., Spektrum Akad. Verlag, Heidelberg, Berlin, New York **1991**; L. Stryer, *Biochemistry*, 4th Edit., W. H. Freeman and Company, New York **1995**.
68 L. Howe, J. Z. Zhang, *J. Phys. Chem. A* **1997**, *101*, 3207–3213.
69 a) R. Sadamoto, N. Tomioko, T. Aida, *J. Am. Chem. Soc.* **1996**, 118, 3978–3979; b) C. J. Hawker, K. L. Wooley, J. M. J. Fréchet, *J. Chem. Soc., Perkin Trans. 1* **1993**, 1287–1297; c) S. Mattei, P. Seiler, F. Diederich, *Helv. Chim. Acta* **1995**, *78*, 1904–1912; c) B. Kenda, F. Diederich, *Angew. Chem.* **1998**, *110*, 3357–3358; *Angew. Chem. Int. Ed.* **1998**, *37*, 1531–1534; d) N. Tomioko, D. Takasu, T. Takahashi, T. Aida, *Angew. Chem.* **1998**, *110*, 1611–1614; *Angew. Chem. Int. Ed.* **1998**, *37*, 3154–3157; e) I. B. Rietveld, E. Kim, S. A. Vinogradov, *Tetrahedron* **2003**, *59*, 3821–3831.
70 a) R. W. Boyle, J. E. van Lier, *Synlett* **1993**, 351–352; b) R. W. Boyle, J. E. van Lier, *Synthesis* **1995**, 1079–1080; c) W. M. Sharman, S. V. Krudevich, J. E. van Lier, *Tetrahedron Lett.* **1996**, *37*, 5831.

71 a) P. R. Ashton, K. Shibata, A. N. Shipway, J. F. Stoddart, *Angew. Chem.* **1997**, *109*, 2902–2904; *Angew. Chem. Int. Ed.* **1997**, *36*, 2781–2783; b) S. W. Krska, D. J. Seyferth, *J. Am. Chem. Soc.* **1998**, *120*, 3604–3612.

72 a) D. Zanini, R. Roy, *J. Am. Chem. Soc.* **1997**, *119*, 2088–2095; b) D. Pagé, R. Roy, *Bioconjugate Chem.* **1997**, *8*, 714–723; c) P. R. Ashton, E. F. Hounsell, N. Jayaraman, T. M. Nielsen, N. Spencer, J. F. Stoddart, M. Young, *J. Org. Chem.* **1998**, *63*, 3429–3437; d) H. W. I. Peerlings, S. A. Nepogodiev, J. F. Stoddart, E. W. Meijer, *Eur. J. Org. Chem.* **1998**, 1879–1886.

73 a) P. J. Dandliker, F. Diederich, J.-P. Gisselbrecht, A. Louati, M. Gross, *Angew. Chem.* **1996**, *107*, 2906–2909; *Angew. Chem. Int. Ed.* **1996**, *34*, 2725–2728; b) M. Liu, K. Kono, J. M. J. Fréchet, *J. Polym.* Sci. *Part A* **1999**, *37*, 3492–3503.

74 M. S. Choi, T. Aida, T. Yamazaki, I. Yamazaki, *Angew. Chem.* **2001**, *113*, 3294–3298; *Angew. Chem. Int. Ed.* **2001**, *40*, 3627–3629.

75 a) S. L. Gilat, A. Adronov, J. M. J. Fréchet, *Angew. Chem.* **1999**, *111*, 1519–1524; *Angew. Chem. Int. Ed.* **1999**, *38*, 1422–1427; b) A. Adronov, S. L. Gilat, J. M. J. Fréchet, K. Otha, F. V. R. Neuwahl, G. R. Fleming, *J. Am. Chem. Soc.* **2000**, *122*, 1175–1185; c) ***Review*** of light harvesting dendrimers: A. Adronov, J. M. J. Fréchet, *Chem. Commun.* **2000**, 1701–1710.

76 M. Plevoets, F. Vögtle, L. De Cola, V. Balzani, *New J. Chem.* **1999**, 63–69.

77 T. Weil, E. Reuter, C. Beer, K. Müllen, *Chem. Eur. J.* **2004**, *10*, 1398–1414.

78 K. Sivanandan, D. Vutukuri, S. Thayumanavan, *Org. Lett.* **2002**, *4*, 3751–3753.

4
Types of dendrimers and their syntheses

Chapters 2 and 3 have already introduced the reader to the general principles of the architecture, synthesis, and functionalisation of dendritic molecules – including hyperbranched and dendronised (linear) polymers (denpols). This chapter will now consider specific molecular scaffolds and syntheses of important types of dendrimers and their individual properties. More specialised and applications-relevant properties of particular dendrimers are compiled in Chapter 8.

4.1
Achiral dendrimers

4.1.1
POPAM

The designation *poly*(*propyleneimine*), like the corresponding acronym *PPI*, is systematically correct. However, the prefix *oligo* would actually be more appropriate for the low-molecular-weight representatives. And since the term imine suggests the presence of an imino group (>C=N–), this family names is not ideal. In order to emphasise the exclusively amine nature, the name *polypropylenamine* has also been used for this group of compounds, hence the acronym *POPAM*. This type of dendrimer (Fig. 4.1) belongs to the most widely used representatives because – like the poly(amidoamine) dendrimers (see the following Section 4.1.2) – it is commercially available [1]. POPAM dendrimers contain only AB_2 branching units ($1 \rightarrow 2$-type branching) owing to the trivalence of the amine nitrogen atoms.

Although samples of such dendrimers are assigned a high degree of purity by electrospray ionisation mass spectrometry (ESI) [2], owing to divergent synthesis structural defects can occur here through incomplete reaction during Michael addition – as already described in Section 1.3. The high degree of branching inhibits generation growth from the fifth generation onwards as a result of steric hindrance ("starburst limit effect", see Section 1.3). Some "growth data" are compiled in Table 4.1; extrapolated to the tenth generation, such POPAM dendrimers would already possess 2048 terminal amino groups.

Dendrimer Chemistry. Fritz Vögtle, Gabriele Richardt and Nicole Werner

ISBN: 978-3-527-32066-0

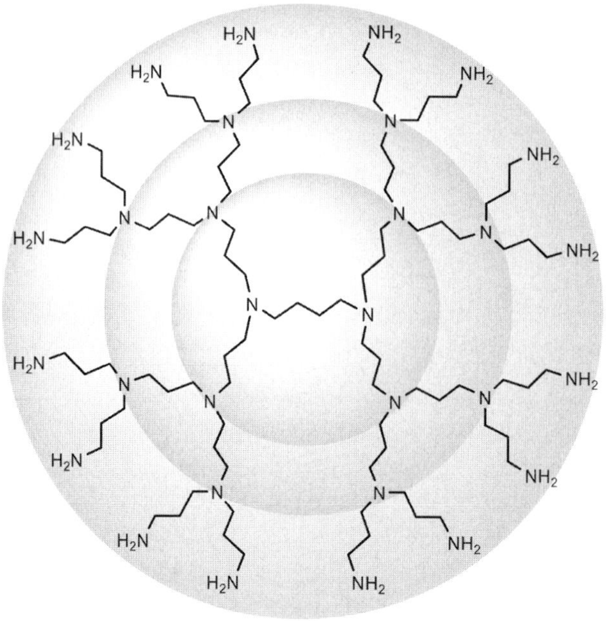

Fig. 4.1 Third-generation (G3) POPAM dendrimer – with tetrafunctional core unit and AB_2 branching (the generations are shown in shades of grey)

Table 4.1 Increase of POPAM parameters with the number of generations [3]

Generation	End groups [number]	Diameter [Å]	Molar mass [g]
1	4	4.4	317
2	8	6.9	773
3	16	9.3	1687
4	32	11.6	3514
5	64	13.9	7168

Divergent synthesis of POPAM dendrimers is based on the concept developed by Vögtle *et al.* in 1978 (see Section 1.1) [4]. Acrylonitrile is added to a primary mono- or oligo-diamine in a Michael reaction (Fig. 4.2). Suitable reducing agents are Co(II)-borohydride complexes or diisobutylaluminium hydride [5]. Subsequent iterative reaction cycles permit repeated addition of acrylonitrile followed by reduction until the limiting generation is reached.

In 1993 Meijer *et al.* and Mülhaupt *et al.* almost simultaneously accomplished the preparative synthesis of higher generation monodisperse POPAM dendrimers on the basis of this principle [6]. Thanks to their terminal amino groups, POPAM dendrimers can readily be *per*-functionalised. Thus Balzani, De Cola,

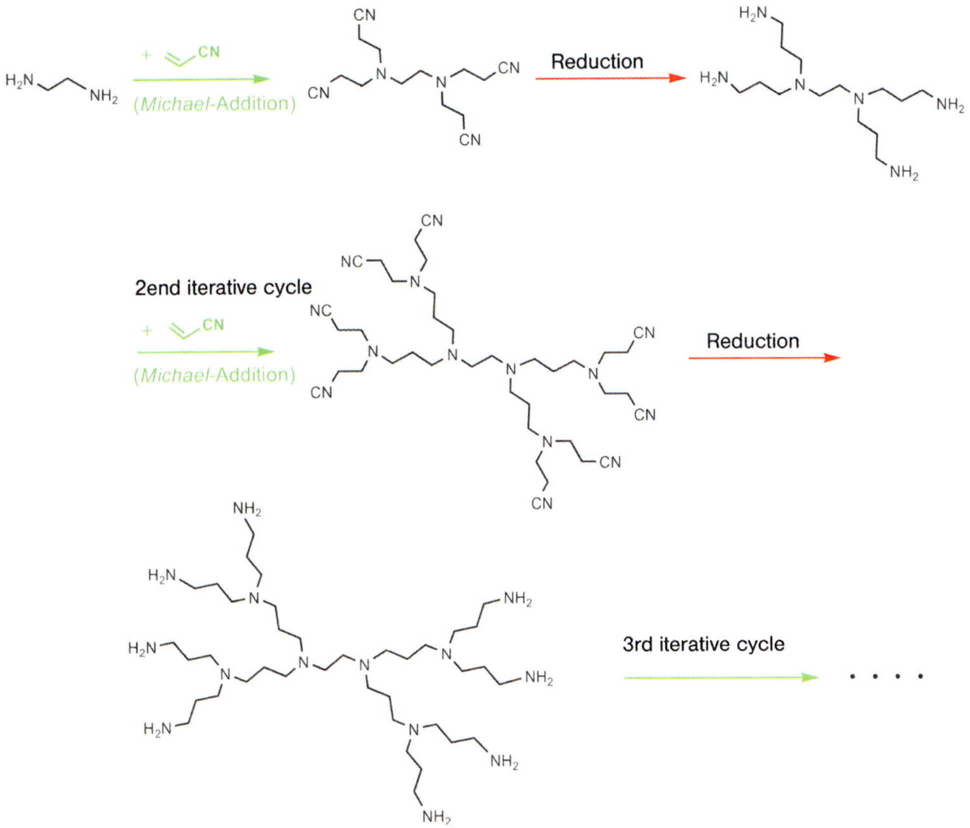

Fig. 4.2 Synthesis of POPAM dendrimers (the key step is Michael addition according to Vögtle *et al.*)

and Vögtle *et al.* succeeded in obtaining pure highly fluorescent poly-dansyl-"decorated" POPAM dendrimers up to the fifth generation with up to 64 peripheral dansyl functions (Fig. 4.3) [7a]. Their fluorescence is found to increase strongly with the number of generations, but not in strict dependence upon the number of fluorophores (dansyl groups). Apparently the overall fluorescence is partly quenched through collisions resulting from the close proximity of the fluorophores in a small space [7b] (see Chapter 5).

For example, Kaifer *et al.* "decorated" the periphery of a POPAM dendrimer with ferrocene and cobaltocenium units. Such a fourth-generation dendrimer (Fig. 4.4) proved to be a suitable guest system – with numerous docking sites – for *β*-cyclodextrin as host compound (see Section 8.3.6) [8].

It also proved possible to attach aza-crown ethers to the periphery of a POPAM dendrimer *via* amidoferrocenyl units (Fig. 4.5) [9a] and also to decorate the outer shell of a second-generation POPAM dendrimer alternately with eight *E*-stilbene and eight 4-*tert*-butylbenzenesulphonic acid units [9b]. Various other

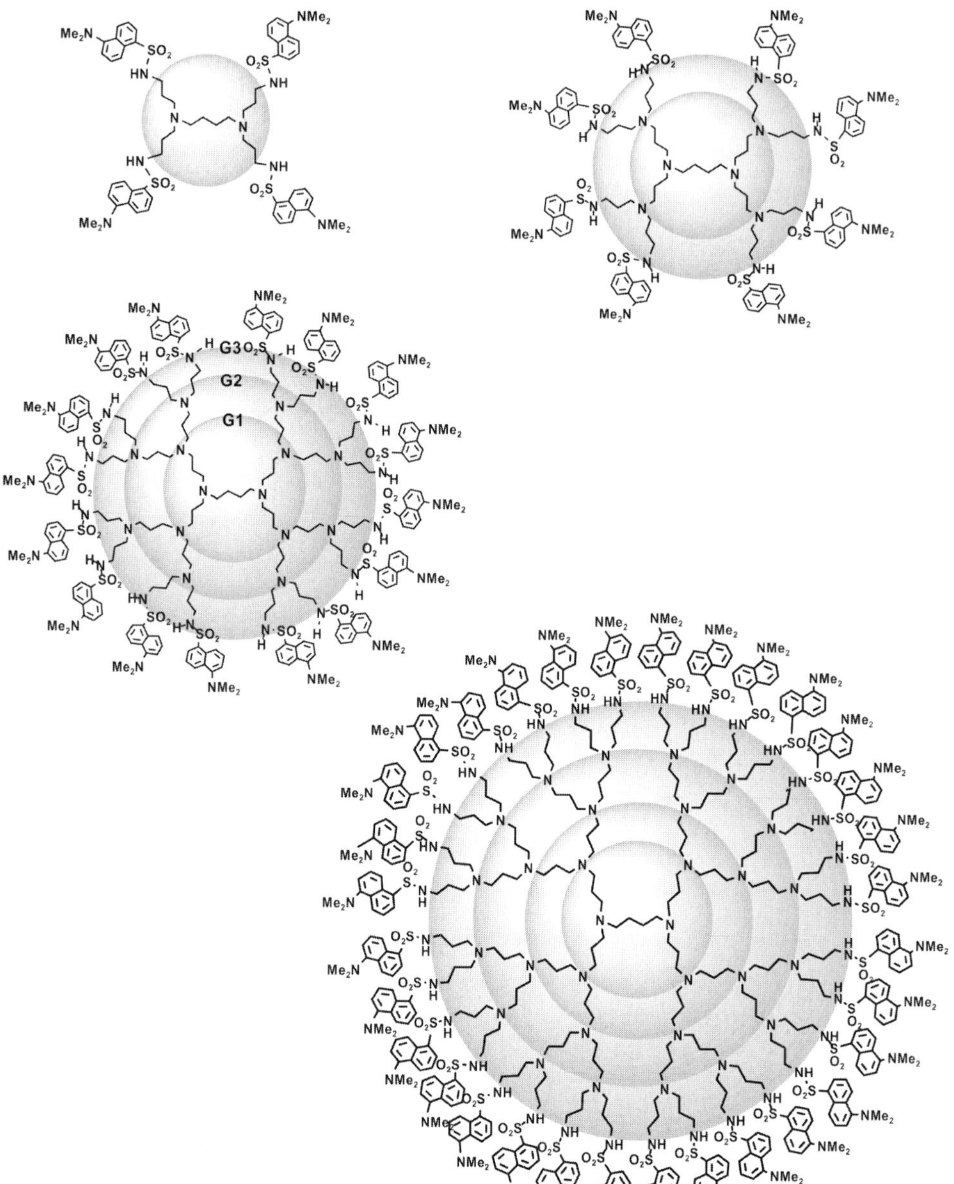

Fig. 4.3 First- to fourth-generation *per-mono-substituted* dansyl-POPAM dendrimers (the fifth generation omitted here was also isolated)

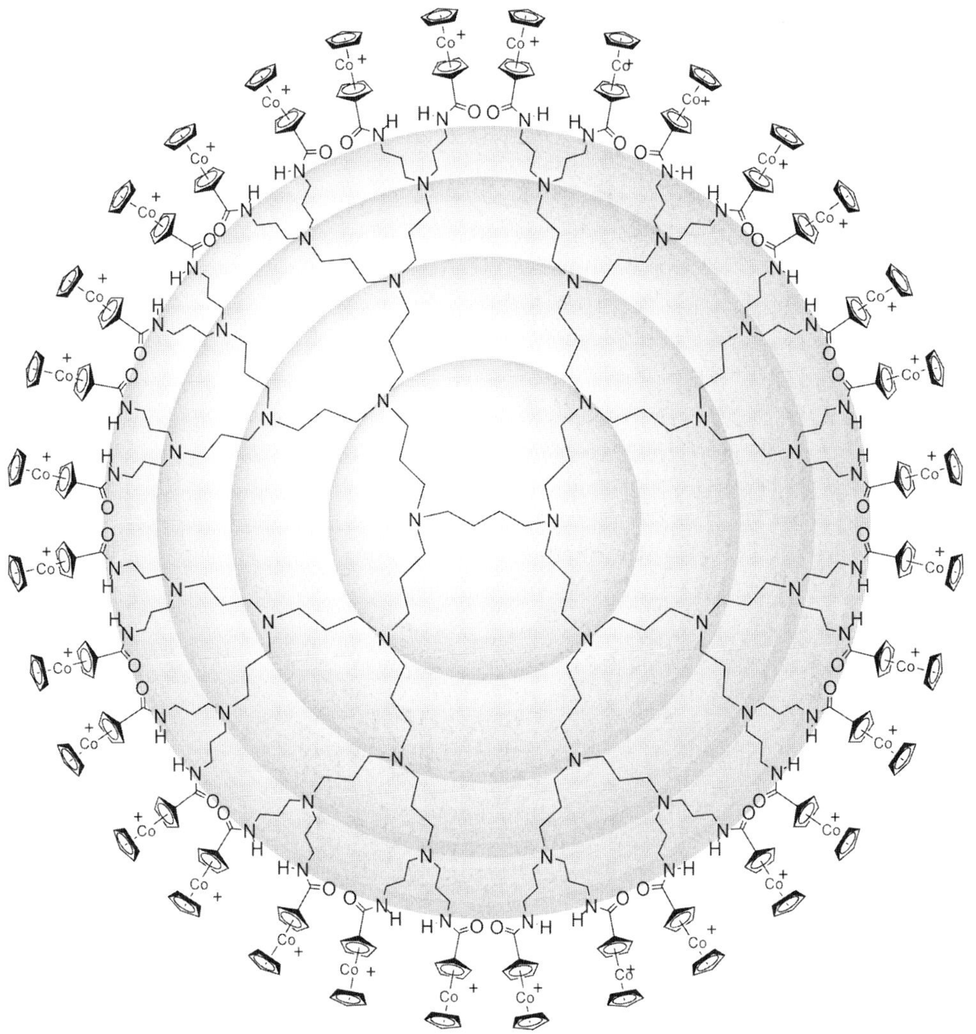

Fig. 4.4 Fourth-generation POPAM dendrimer with terminal cobaltocenium units

examples of functionalisation of POPAM dendrimers will be found in Chapters 5 to 8 of this book.

4.1.2
PAMAM

The acronym *PAMAM* for *po*ly*am*ido*am*ine dendrimers (Fig. 4.6) was intended to emphasise the amide bonds present in the molecule to distinguish them from *POPAM* dendrimers (from *po*ly(*p*ropylen*am*ine)). They are commercially

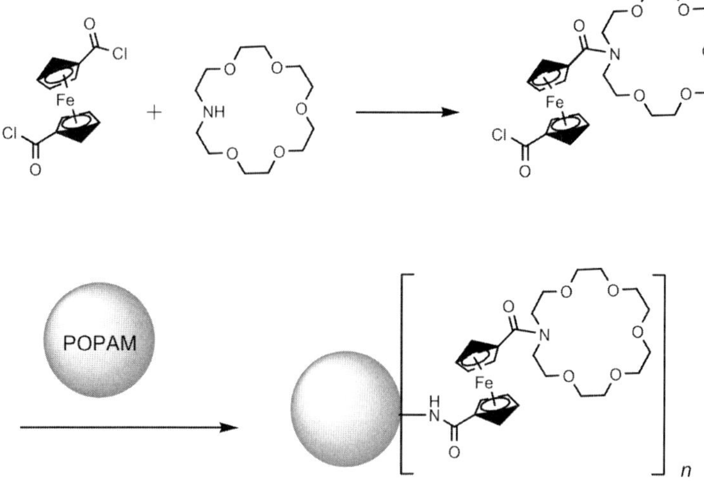

n = 4, 8,16

Fig. 4.5 POPAM dendrimer with terminal aza-crown ether and ferrocene units

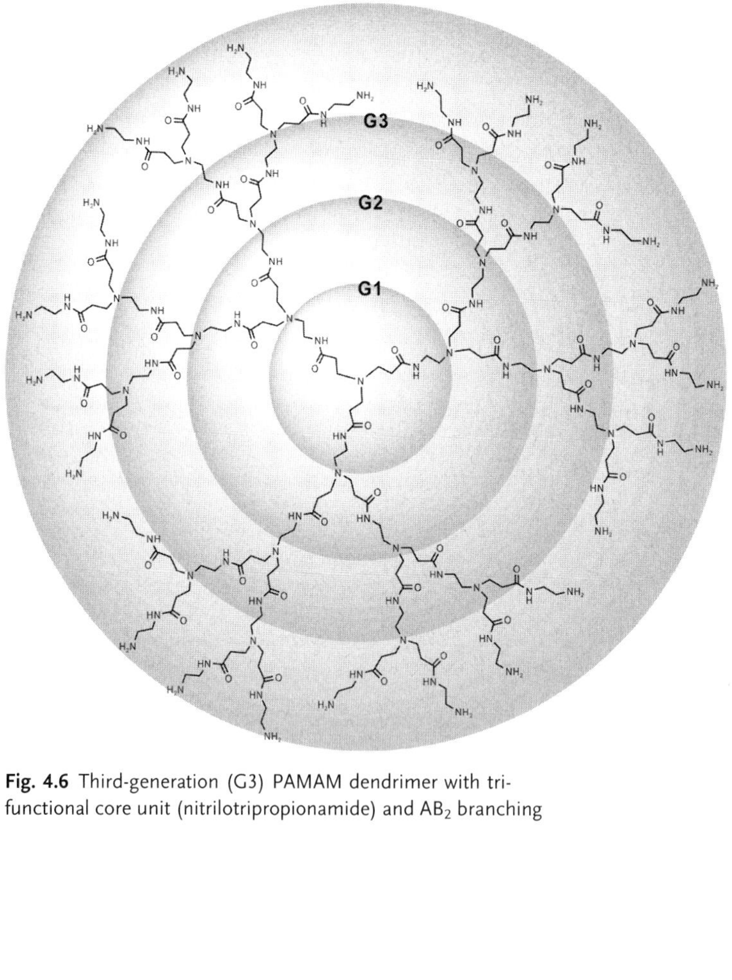

Fig. 4.6 Third-generation (G3) PAMAM dendrimer with tri-functional core unit (nitrilotripropionamide) and AB_2 branching

Table 4.2 Increase of PAMAM parameters with the number of generations (dendrimer type, see Fig. 4.6) [3]. The second column lists the number of terminal groups of 1→3-branched aliphatic Newkome type dendrimers (see Figs. 1–5 and 4.60 for typical formulae)

Generation	Number of end groups		Diameter [Å]	Molar mass [g]
	PAMAM	Newkome type		
0	4		15	512
1	8	12	22	1430
2	16	36	29	3256
3	32	108	36	6909
4	64		45	14215
5	128		54	28826
6	256		67	58048
7	512		81	116493
8	1024		97	233383
9	2048		114	467162
10	4096		135	934720

available as "starburst dendrimers" [10]. The usual PAMAM dendrimers – like the POPAM dendrimers – are made up solely of AB_2 branching units (cf. also Chapter 1, Fig. 1.4).

As shown in Tables 4.1 and 4.2, PAMAM dendrimers (with nitrilotripropionamide core unit; cf. Fig. 4.6) are assembled from more voluminous molecules than POPAM dendrimers (with 1,4-butanediamine core unit; cf. Fig. 4.4). The zeroth generation of such a PAMAM dendrimer (with only 4 end groups) already occupies a greater volume than the fifth generation of a POPAM dendrimer (with 64 peripheral groups). Accordingly, the first generation of the above PAMAM dendrimer has more than four times the molar mass of a first-generation POPAM dendrimer. However, this direct comparison of generations is flawed by the lack of agreement in the literature regarding the definition of generation number when different core units are present. It would therefore appear advisable in many cases to limit comparisons of dendrimers with different core and branching units to representatives with the same number of terminal groups, even if the generation numbers differ.

The space occupied by a dendrimer and its ability to form intramolecular "cavities" depend – as already discussed in general terms in Section 1.2 – upon its conformational flexibility and this, in turn, upon the nature of the branching units. According to Tomalia *et al.*, the "internal surface" of the "cavities" in a sixth-generation PAMAM dendrimer is larger than the external surface [11]. Apart from other organic molecules [12], PAMAM molecules could accommodate dyes [13]. This formal consideration in no way means that cavities or pores (after the manner of vacuoles) are actually preformed in the dendrimer molecule in solution. Nevertheless, solvent or large guest molecules are included in

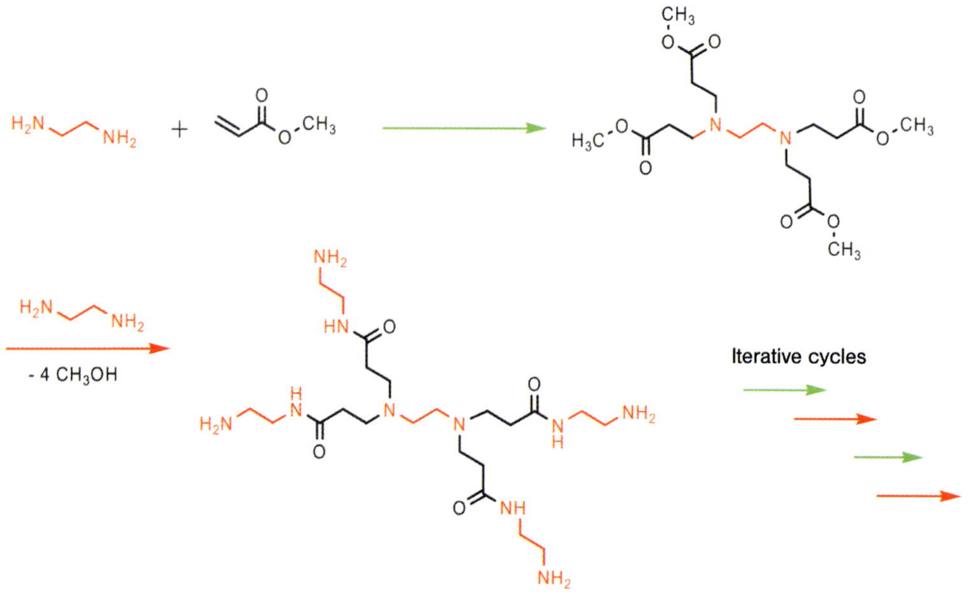

Fig. 4.7 Synthesis of PAMAM dendrimers – with tetrafunctional ethylenediamine core (according to Tomalia *et al.*)

the dendrimer molecule, particularly when attractive forces come into play between host and guest, possibly owing to hydrogen bonding interactions or an acid/base relationship.

In 1985, Tomalia modified the synthetic methodology of Vögtle *et al.* for the likewise divergent preparation (Fig. 4.7) of PAMAM dendrimers by replacing the acrylonitrile reagent used in Michael addition by methyl acrylate (see also Section 1.1). The individual ester stages were also designated as generation 0.5 (cf. Fig. 4.7 above right), generation 1.5, and so forth [14].

Thanks to the amide bonds formed in the course of their synthesis, PAMAM dendrimeres can be readily solvated in polar solvents such as dichloromethane, ethers, as well as short-chain alcohols; they are highly stable to hydrolysis.

The convergent route also leads to symmetrical and unsymmetrical PAMAM dendrimers [15]. However, they are often less easy to prepare in such high purity as POPAM dendrimers, since backfolding and defects in the dendrimer scaffold occur despite the extended branches resulting from the presence of ethylenediamine (as flexible spacer) – clearly owing to hindered rotation around the amide bonds. These can be propagated from generation to generation until a certain "limiting generation" is reached (Chapter 1). The PAMAM limiting generation is only reached at almost twice the generation number as in the case of POPAM dendrimers. From the tenth generation onward, however, the POPAM surface has become so "dense" that further reactions run into problems.

Kono *et al.* compared the thermosensitive properties of *N*-isopropylamide-substituted (NIPAM-)PAMAM dendrimers (G5) with those of linear polymers (poly(*N*-isopropylacrylamide, PNIPAAm). The former show a much smaller endothermic peak around the critical mixing temperature. Apparently the spherical shape of the dendrimer molecules hinders efficient hydration and dehydration of the NIPAM groups in aqueous phosphate solutions in this temperature range [16] (see Section 6.3).

4.1.3
POMAM

Dendritic hybrid architectures of the two dendrimer types – POPAM and PAMAM – designated by Majoros *et al.* as "*POMAM dendrimers*" [17] are structural rarities. One such dendrimer of this type was assembled with PAMAM branching units starting from a POPAM core unit [18]. Vögtle *et al.* developed POPAM/PAMAM hybrid dendrimers up to the third generation (Fig. 4.8) [19].

Starting from a zeroth-generation POPAM dendrimer (Fig. 4.8), the protective group is removed by palladium-catalysed hydrogenation to activate the amino groups. Subsequent reaction with a succinimide ester as bearer of the AB_2 coupling unit leads to a first-generation "hybrid dendrimer". Repetition of this synthetic sequence provides an entry to higher-generation POMAM dendrimers, as depicted in Fig. 4.9.

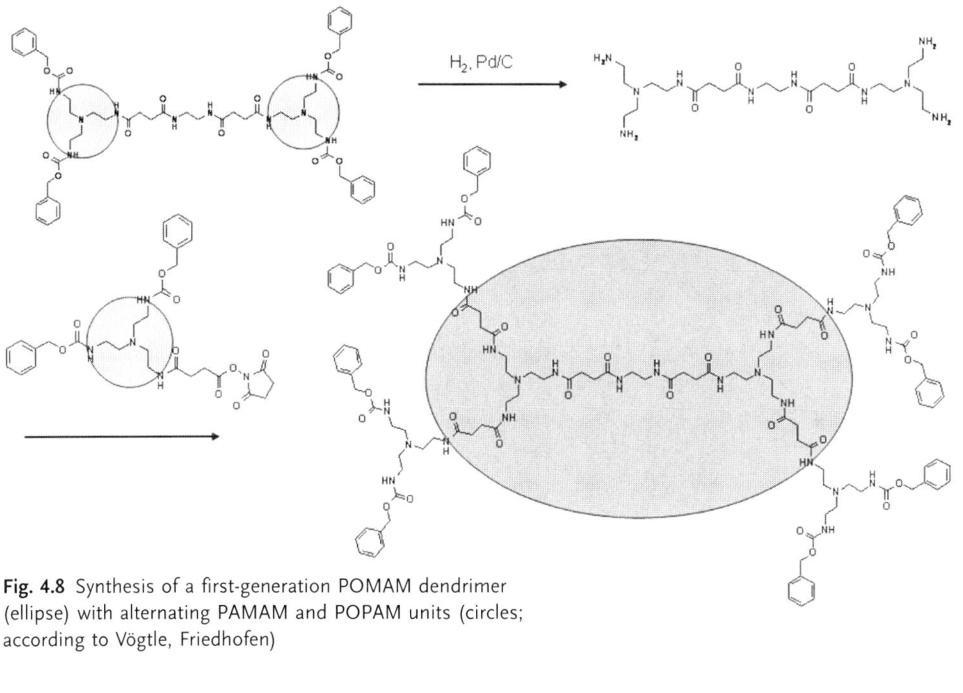

Fig. 4.8 Synthesis of a first-generation POMAM dendrimer (ellipse) with alternating PAMAM and POPAM units (circles; according to Vögtle, Friedhofen)

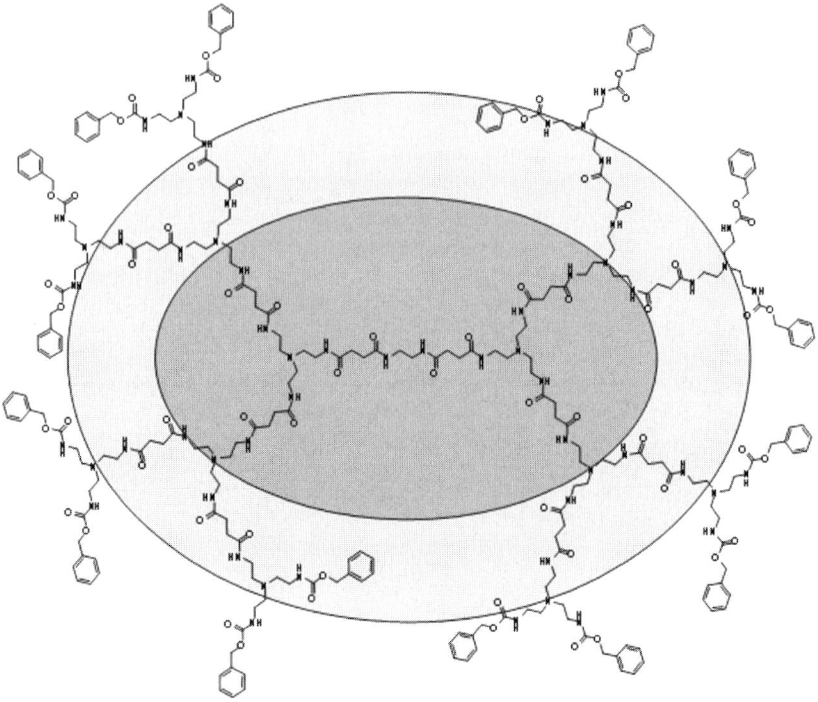

Fig. 4.9 Second-generation POMAM dendrimer analogous to that in Fig. 4.8 (according to Vögtle, Friedhofen)

4.1.4
Polylysine dendrimers

Like PAMAM dendrimers, polylysine dendrimers are largely characterised by amide bonds. They attracted interest as potential therapeutic agents for use in boron neutron capture therapy and in magnetic resonance imaging (MRI) (see Chapter 8), since dendritic polylysines apparently have a lower toxicity than their linear counterparts. The polylysine shown in Fig. 4.10 with a total of 80 terminal boron atoms in the carborane units and a dansyl group was designed specifically for boron neutron capture therapy [20].

Synthesis of these dendrimers is performed by condensation of the amino acid lysine, whose amino functions have previously been protected with *tert*-butyloxycarbonyl groups (Boc), onto an (activated) L-lysine *p*-nitrophenyl ester. The resulting coupling product (Fig. 4.11) is then deprotected with trifluoroacetic acid and thus activated for renewed reaction. Iteration of the assembly and activation step ultimately led to a polylysine dendrimer with 1024 terminal butyloxycarbonyl groups [21].

Balzani and Vögtle *et al.* studied the complexation of neodymium(III) and other rare earth metal ions by a luminescent second-generation polylysine den-

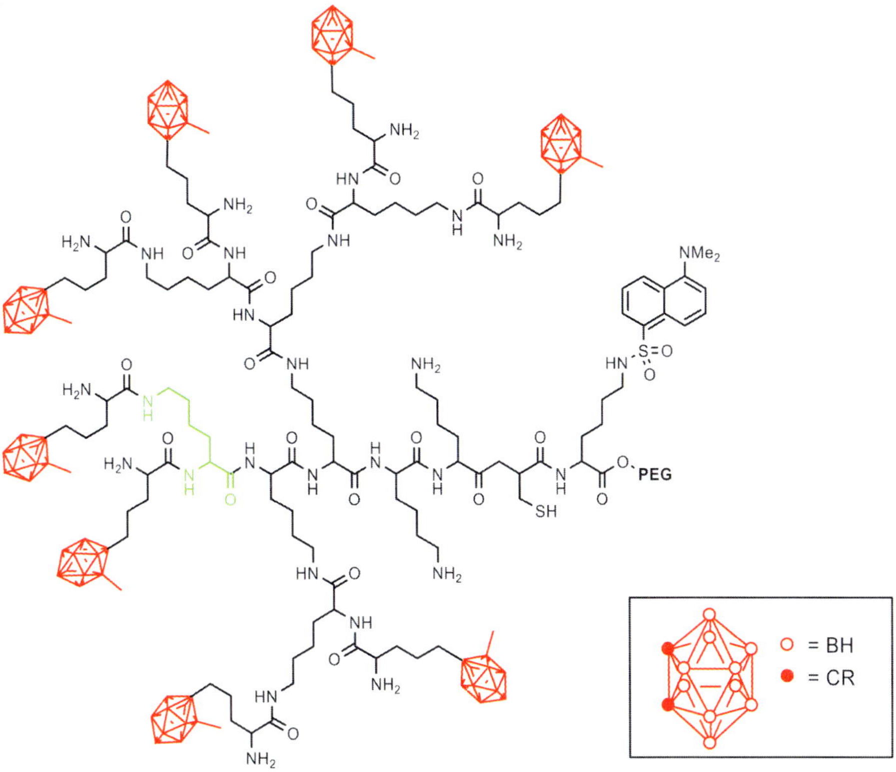

Fig. 4.10 Polylysine with a total of 80 boron atoms in the periphery (according to Qualmann *et al.*). For the sake of clarity a – peptide bonded – lysine building block in one of the dendrimer branches is shown in green (PEG = polyethylene glycol)

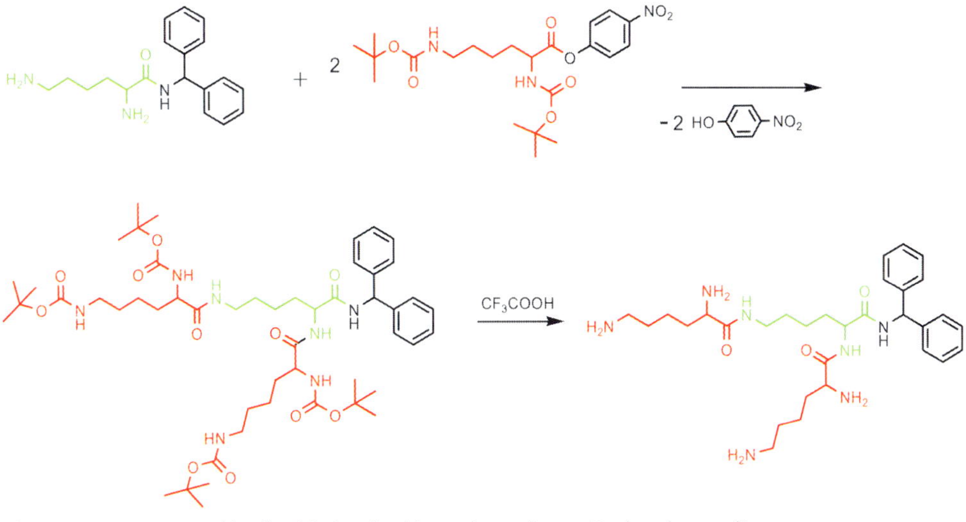

Fig. 4.11 Stepwise assembly of polylysine dendrimers (according to Denkewalter *et al.*)

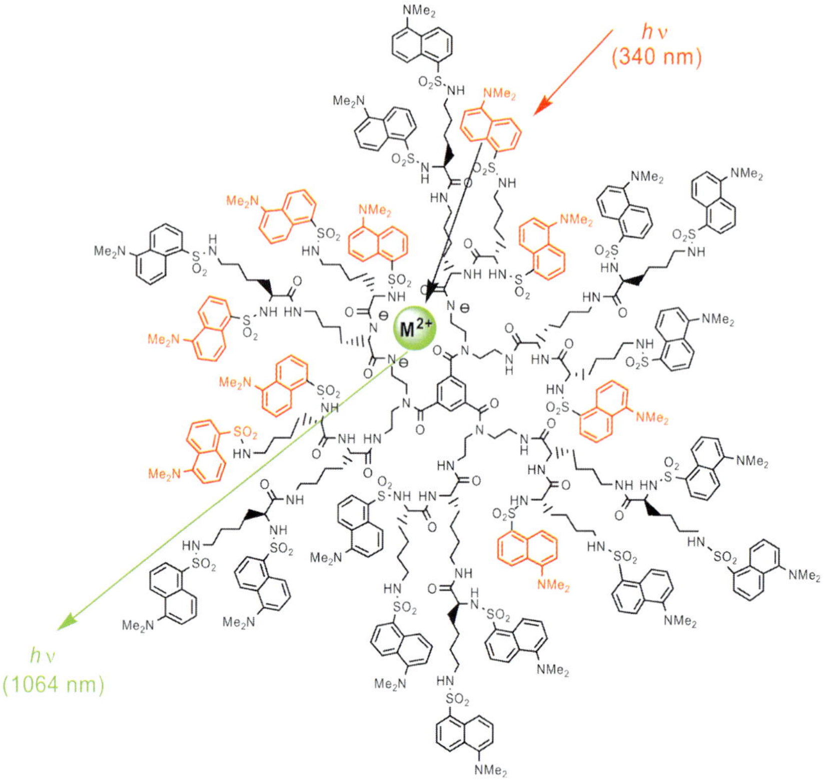

Fig. 4.12 Dansyl-decorated oligo-lysine dendrimer (2nd generation) as dendritic antenna with sensor properties towards metal ions (according to Balzani, Vögtle *et al.*)

drimer with 21 amide and 24 dansyl units (Fig. 4.12). The likewise luminescent lanthanide ions are coordinated in the interior of the dendrimer scaffold, leading to remarkable photophysical properties of the material (see Section 5.2) [22]. Metal ions as well as protons have a pronounced influence on the fluorescence of dansyl groups in such dendrimers (see Sections 6.2 and 3.3). Quantitative measurements showed that one cobalt ion quenches the fluorescence of nine dansyl units. This means that the dendrimer exhibits sensor properties towards Co^{II+} and other metal ions. Such "intradendrimer quenching effects" are also observed with luminescent cations such as Nd^{III+}, Eu^{III+}, Gd^{III+}, Tb^{III+}, Er^{III+}, Yb^{III+}, *i.e.* they quench the fluorescence of some of the dansyl groups, specifically of those which can come into the vicinity of the coordinating metal ions within the dendrimer. Apparently, nine dansyl units can constitute this neighbourhood, shielding the space around the metal ion and thus precluding any negative effect on the fluorescence of the other peripheral dansyl groups. In the case of Nd^{III+} excitation at $\lambda = 340$ nm leads to long-wave near infrared (NIR) emission at $\lambda = 1064$ nm [22]. The dendrimer therefore acts as a dendritic antenna for optical excitation of neodymium ions. We thus have a nanoscale system in which the interaction between a fluorescent host and a fluorescent guest can be studied. Regarding immune properties of Dendri-graft-polylysines see lit. [22 d].

4.1.5 Dendritic hydrocarbons

Hardly any purely aliphatic hydrocarbon dendrimers have hitherto been prepared (see also Section 6.2.3.3), probably owing to fundamental synthetic problems in the formation of unsymmetrical C–C bonds. The dendritic hydrocarbons described so far are nearly all made up of condensed or coupled arene or/ and multiply bonded components, which is easier to achieve synthetically. As a rule, such molecules are generally much more conformationally rigid than their aliphatic analogues made up of $[CH_2]_n$ units.

4.1.5.1 Condensed arene components – Iptycenes

The synthetic route to "supertriptycene" ($C_{104}H_{62}$) based on repeated Diels-Alder reaction is depicted in Fig. 4.13. Remarkably, this molecule possesses a symmetrical arrangement of three relatively rigid "niches" for possible inclusion of guests and enormous thermostability (in a nitrogen atmosphere it does not decompose until 580 °C) [23].

4.1.5.2 Dendrimers from arene and multiply bonded building blocks

Convergent syntheses developed by Moore *et al.* provided access to rigid higher-generation phenylacetylene dendrimers (Fig. 4.14). Dendrons with peripheral *p*-*tert*-butylphenyl units used in initial work showed only limited solubility and were therefore unsuitable for assembly of higher generations. However, minimal modification of the periphery with 3,5-di-*tert*-butylphenyl groups improves the solubility so dramatically that it even proved possible to prepare a dendrimer with 94 "monomeric" units [24].

Such dendrons were derivatised at the focal point to create an energy gradient within a kind of "molecular antenna" (see Chapter 5 for more information) [25]. Percec *et al.* reported a building block library of eleven phenylacetylenes [27] substituted (dendronised, dendrylated) [26] with – self-assembling – dendrons.

Meier *et al.* described hydrocarbon dendrimers with *trans*-stilbene chromophores in the core and in the periphery. Owing to the flexible nature of the arms intra- and intermolecular C–C bonds could be formed by irradiation [28] (see Section 4.1.5.3 for further information).

The monodisperse polybenzene dendrimers reported by Miller and Neenan *et al.* are arranged symmetrically around a 1,3,5-substituted benzene core unit; up to 46 benzene rings were coupled convergently (Fig. 4.15) [29].

Müllen‘s rigid polyphenylene dendrimers (actually polyarene dendrimers) [30] can be assembled both convergently and divergently. Based on Diels-Alder cy-

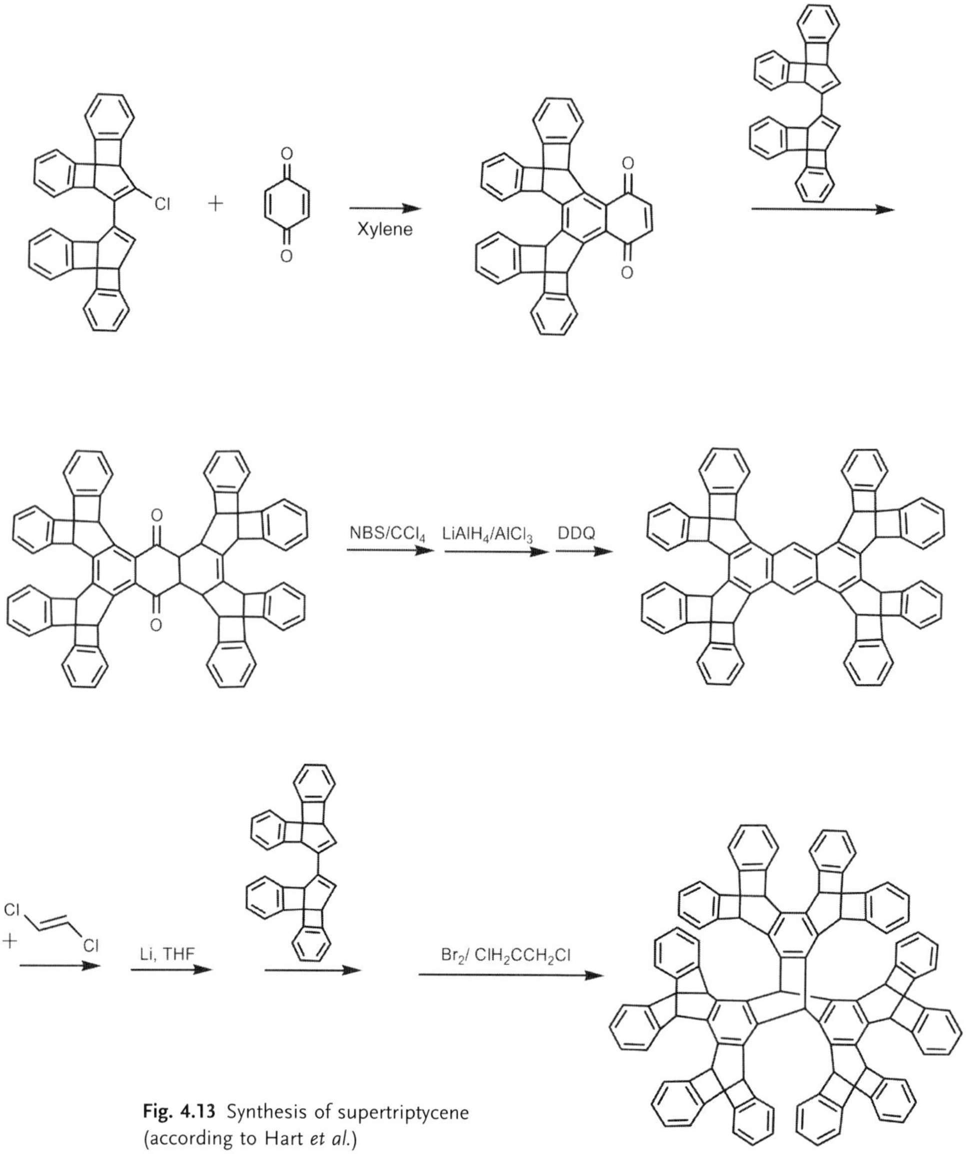

Fig. 4.13 Synthesis of supertriptycene (according to Hart *et al.*)

cloaddition, a core unit with terminal ethyne groups is added to a cyclopentadienone, which may also bear protected triple bonds. The scope of variation is apparent from Fig. 4.16. On removal of the protective groups (with ammonium fluoride), acetylene groups again become available for assembly of higher dendrimer generations.

Although the solubility of (unsubstituted) linear oligophenyls, starting from *p*-terphenyl, becomes poorer and poorer also in lipophilic solvents, polybenzene

3

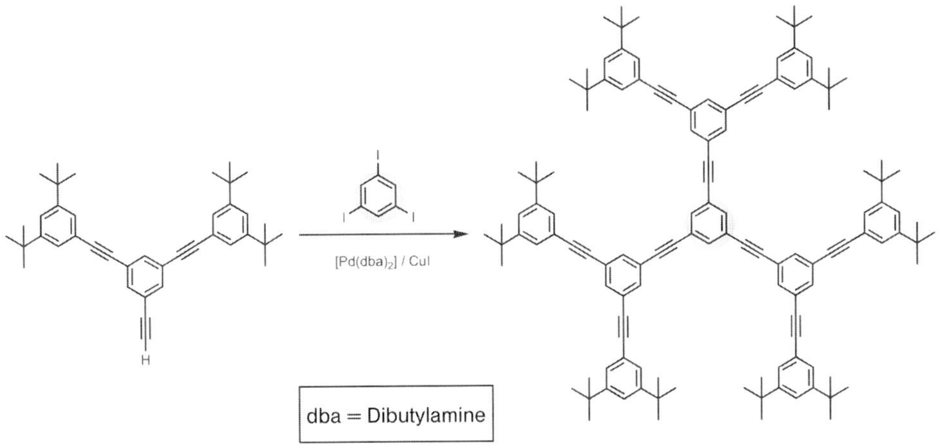

Fig. 4.14 Phenylacetylene dendrimer (according to Moore *et al.*) with three monomeric units

dendrimers including those of higher generations are generally easily handled in this respect, particularly if they bear peripheral substituents (*tert*-butyl-, oligomethylene-). This is because in reality the bulk of the benzene rings causes the molecules to twist out of the plane – in contrast to our simplified flattened representation (*e.g.* Figs. 4.15 and 4.17) – thus precluding aggregation of the kind recently encountered with disc-like species.

In the case of the divergent route (Fig. 4.17), an unprotected tetraethynylated tetraphenylmethane core unit (as dienophile) and a doubly protected ethynylated cylopentadienone (as diene) undergo cycloaddition to give a protected (first-generation) polybenzene dendrimer. Subsequent removal of the triisopropylsilyl protective groups (with ammonium fluoride) generates new dienophilic functions on the periphery. Iteration of the two steps – Diels-Alder reaction [31] and deprotection – provides access to "polyphenylene dendrimers" of up to the fourth generation. However, in the last step of assembling the final generation, a non-functionalised tetracyclone is used as diene – instead of a functionalised one [32]. Incorporation of 2′,5′-dimethyl-*p*-terphenyl spacers in each branching unit permitted the synthesis of fifth-generation oligobenzene dendrimers. Since the spacers expand the macromolecule, and the number of the limit generation can thus be "pushed upwards", no spatial compression occurs through excessive crowding of the end groups [33].

Convergent synthesis (Fig. 4.18) also uses cyclopentadienone for construction of the dendrons. However, in this case two molecules of the diene are added *via* Diels-Alder reaction to 4,4′-diethynylbenzil. The resulting second-generation dendron is reacted with dibenzylacetone in a twofold Knoevenagel reaction. The newly functionalised cyclopentadienone dendron obtained again undergoes (fourfold) Diels-Alder addition to the tetraethynylated tetraphenylmethane core unit [34a].

The synthetic building block of cyclopentadienones, prepared *via* appropriately substituted benzils, places certain limitations on the convergent method

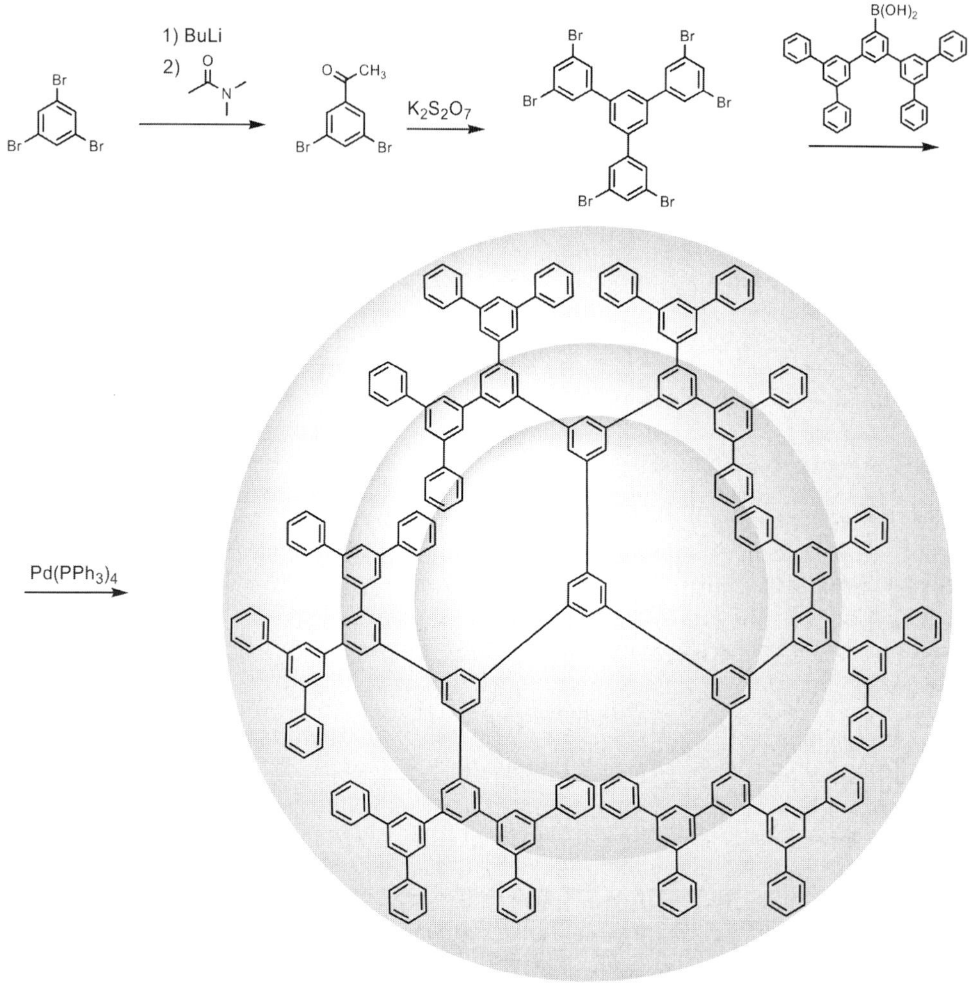

Fig. 4.15 Polybenzene dendrimers (according to Miller, Neenan, *et al.*). For the sake of clarity, some of the C–C bonds are drawn with greatly exaggerated lengths to avoid overlapping in this (energetically unfavourable) planar conformation

for this type of dendrimer. In the case of larger substituents, the benzil derivative exists largely in the *s-trans* conformation for steric reasons. This makes it difficult to perform the twofold Knoevenagel reaction, which requires the *s-cis* conformation (Fig. 4.18). For this reason, convergent synthesis is only used for low generations [34b].

In order to illustrate the divergent and convergent synthetic concepts already presented in Chapter 2 in the case of a specific example, we have used the same core units for both routes in the above examples. In both cases, however, other

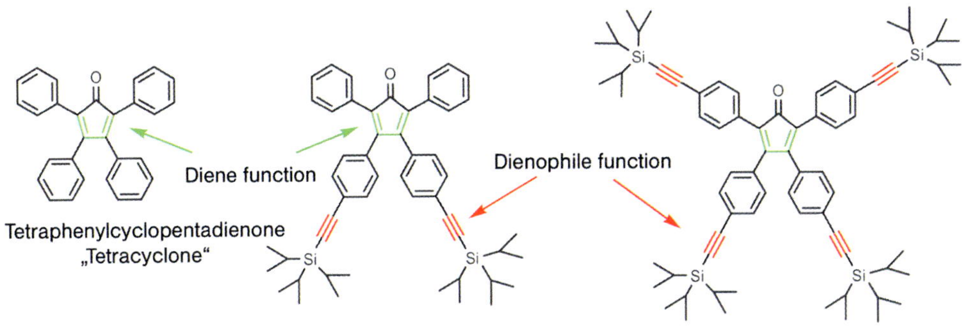

Fig. 4.16 Dienes and dienophiles – and corresponding "zwitter" species with both functions – for assembly of polyphenylene dendrimers (according to Müllen *et al.*)

core units can also be used (Fig. 4.19), leading to different rigid dendrimer scaffolds which tend to assume either an extended or a spherical form, as evidenced by their crystal structures [32, 35]. Because of the large number of benzene rings also present in the interior of the scaffold, neither the dendrimers constructed by Moore nor those synthesised by Miller and Neenan are capable of supramolecular inclusion of guests since the former do not form cavities and the latter will hardly permit entry of guest molecules through the dense surface.

Polybenzene dendrimers can be peripherally functionalised, for example with peptide units, leading to a dendrimer with up to 16 terminal lysine units [36].

4.1.5.3 **Stilbenoid dendrimers**

Meier and Lehmann succeeded in incorporating stilbenoid chromophores into dendrimers [37]. The *E*-configured double bonds were formed by Wittig–Horner reaction. The readily accessible benzenetris(methanephosphonic acid diethyl ester) **1** was used as initiator core (Fig. 4.20). A trisubstituted aldehyde (**2**) and a bisphosphonic acid derivative (**3**) with protected aldehyde function served for construction of the arms; higher aldehydes such as **4** were obtained *via* the latter. Tris(dodecyloxy)phenyl groups attached to the periphery take care of solubility – and lead to formation of liquid-crystalline phases. The *E*-selectivity of the Wittig–Horner reaction suffices to provide configurationally pure *all-E*-isomeric dendrimers **5** – within the detection limits of NMR. The ^{1}H- and ^{13}C-NMR spectra confirm the threefold symmetry of the dendrimers, whose chemical shifts are accordingly similar within the series of compounds. The MALDI–TOF technique is well suited for determination of the molecular masses.

Stilbenoid dendrimers are able to undergo aggregation. Depending upon the generation number, some of the pure substances form liquid-crystalline phases (D_{hd}: discotic hexagonal disordered phase; D_{rd}: discotic rectangular disordered phase; D_{ob}: discotic distorted phase). Differential scanning calorimetry (DSC) revealed phase transitions between 99°C and 0°C.

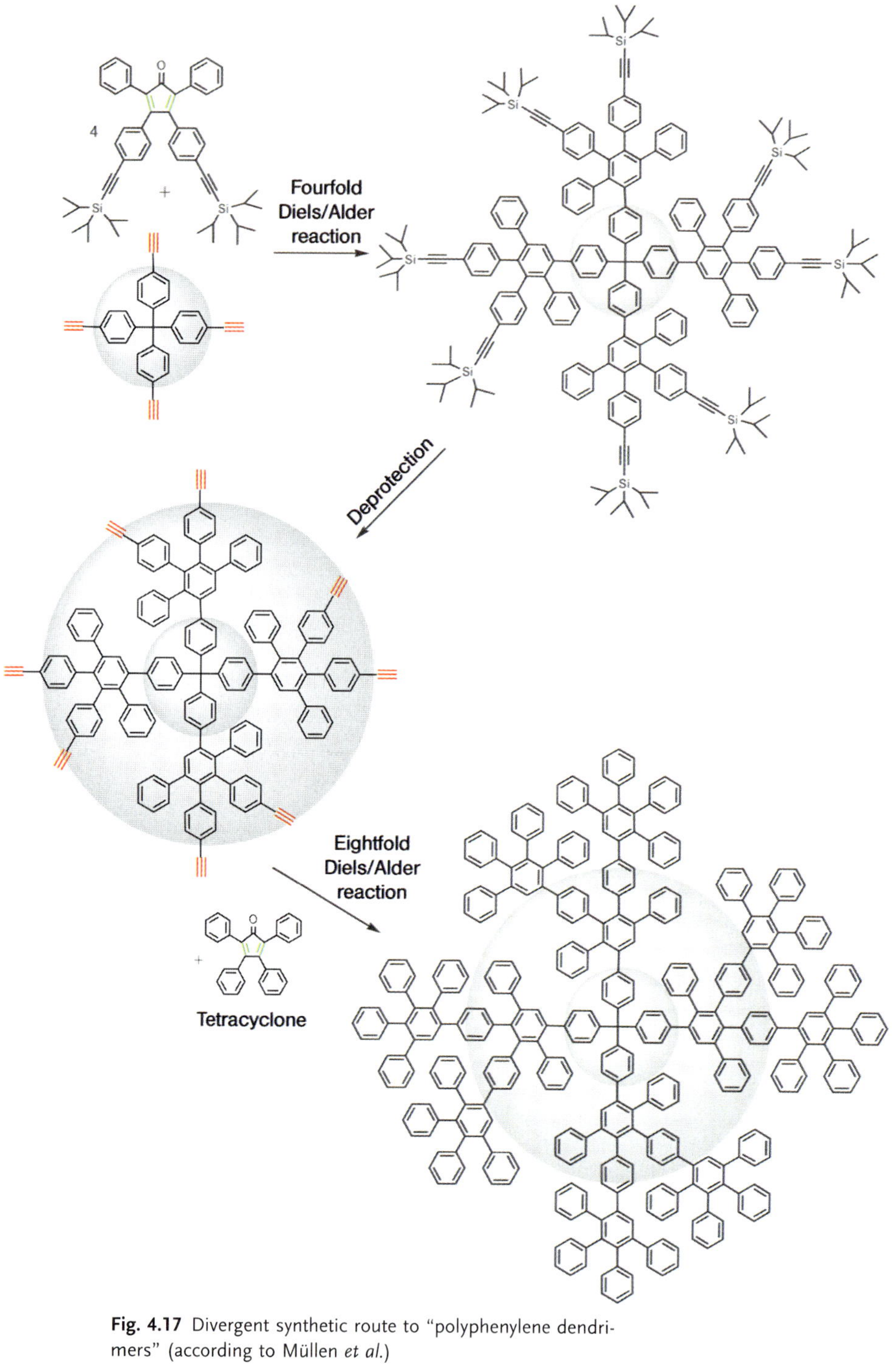

Fig. 4.17 Divergent synthetic route to "polyphenylene dendrimers" (according to Müllen *et al.*)

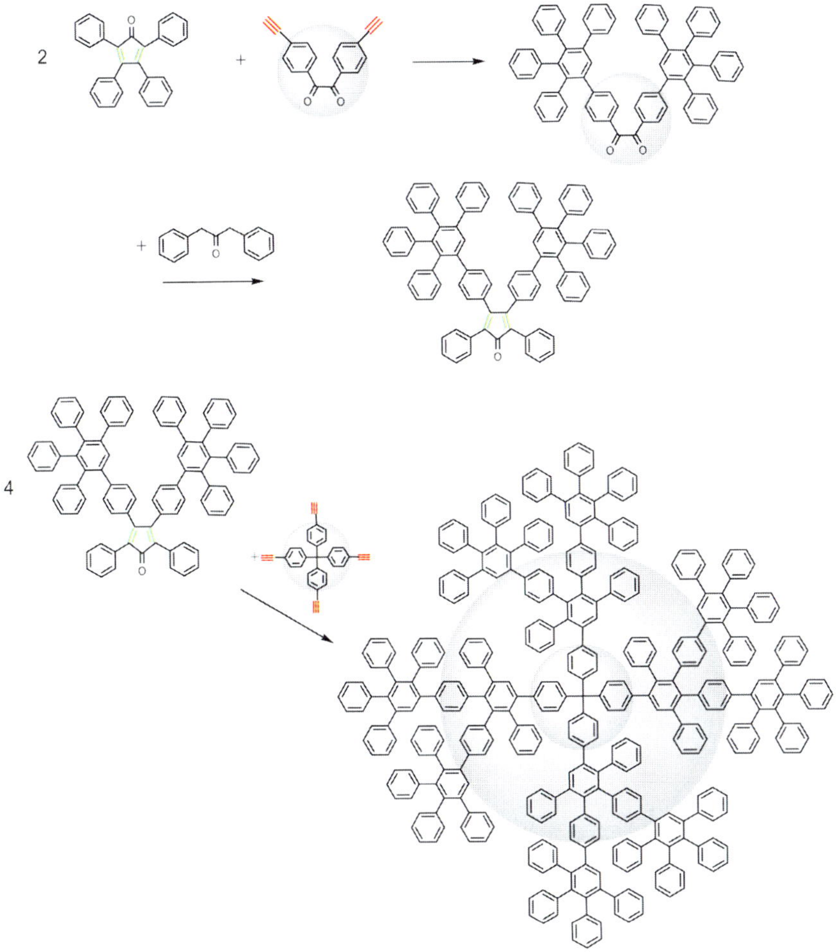

Fig. 4.18 Convergent synthetic route to polybenzene dendrimers (according to Müllen *et al.*). Because of the mutual steric hindrance of the numerous benzene rings, the molecule shown will tend to be spherical rather than planar

In **5** (Fig. 4.20) steric hindrance in the peripheral region appears to be too high for formation of a liquid-crystalline phase. Mesophases were characterised by polarisation microscopy and X-ray diffraction. Presumably the LC properties cease as a result of segment mobility with increasing number of stilbene building blocks; in principle, the number of conformers should double with each double bond although the maximum number of 2^n (*e.g.* 2^{21} for the third generation) is unattainable for symmetry reasons.

The tendency to undergo aggregation in cyclohexane increases with the generation number. It causes strong signal broadening attributable to hindrance of segment mobility. With **5** the effect is so pronounced that a sharply resolved

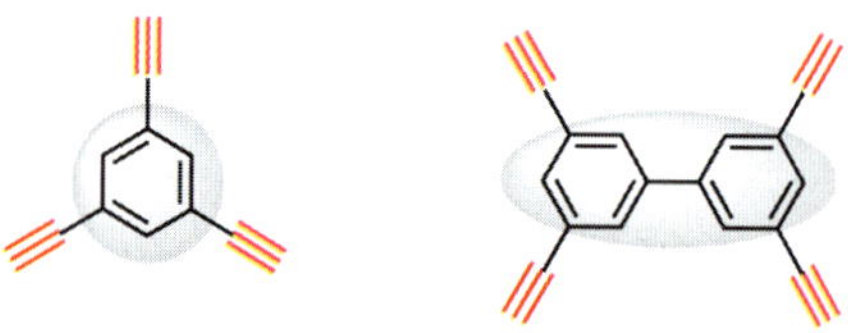

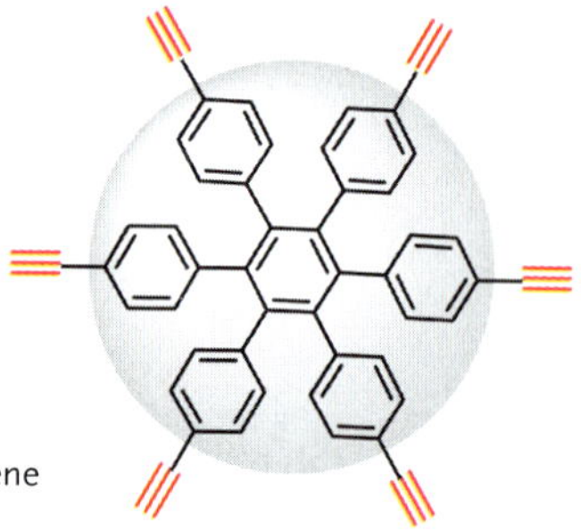

Fig. 4.19 Further core units for the synthesis of polybenzene dendrimers

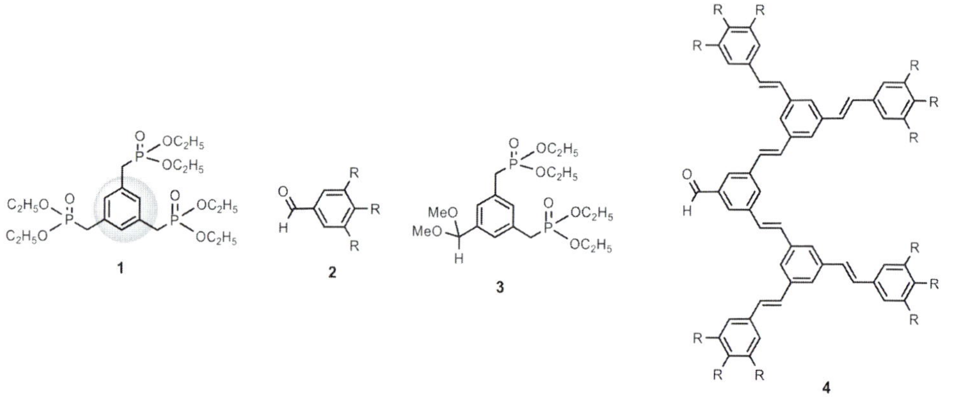

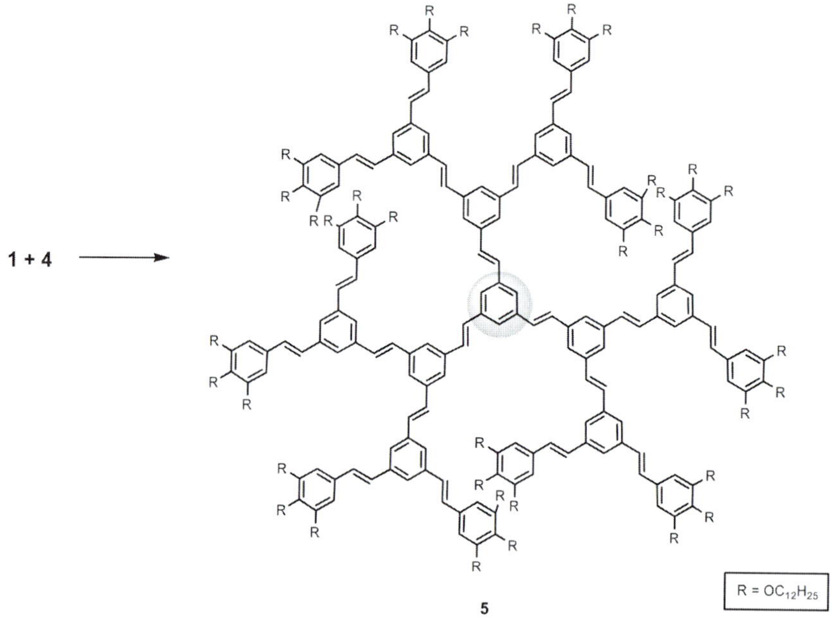

Fig. 4.20 Stilbenoid dendrimers (according to Meier and Lehmann)

1H-NMR spectrum is no longer obtained in C_6D_{12}, even on warming. Aggregation in non-polar media can also be followed as a function of temperature and concentration in the fluorescence spectra.

Structurally, dendrimers of type **5** are best compared with rigid dendrimers made up of tolan building blocks. However, neither the form of aggregation observed in stilbenoid dendrimers, nor the phase behaviour or the photochemistry of these compounds have so far been found to occur in similar form in tolan systems.

The stilbene carbon unit has also been peripherally bound to POPAM cores. Although it does not strictly belong to the hydrocarbon dendrimers, the formula of a G2 dendrimer of this type (Fig. 4.21) is depicted here as an example. It was obtained by alkylation of the corresponding eightfold mono-sulphonamide with 4-(bromomethyl)stilbene. Its fluorescence, *E*/*Z* isomerisation, photoisomerisation (see Section 5.2.2), and excimer formation were compared with those of non-dendritic stilbenes. The quantum yields of photoisomerisation (0.30) and fluorescence of the *E* isomer (0.014) of the dendrimer proved to be substantially lower [38].

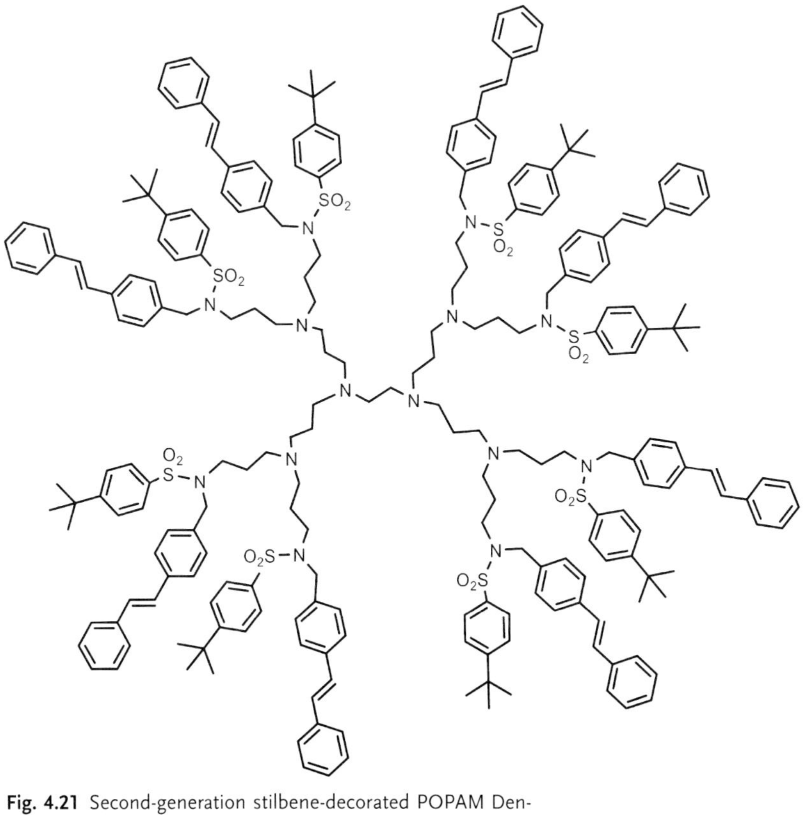

Fig. 4.21 Second-generation stilbene-decorated POPAM Dendrimers (according to Balzani, Vögtle, *et al.*)

The interesting photophysics and photochemistry [37, 38] of stilbenoid compounds make them suitable for applications in materials science. In addition to well-established applications, such as in optical brighteners, others arise in light-emitting diodes (LED), in non-linear optics (NLO), and in optical imaging, storage, and switching techniques.

4.1.5.4 **Hyperbranched polybenzenes**

Kim and Webster laid the foundation for the first hyperbranched polyarenes ("polyphenylenes") in the form of Suzuki-analogous aryl–aryl coupling. Micellar structures developed on this basis [39] are described in Section 4.1.8 (ionic dendrimers).

Initial work based on the Diels-Alder reaction provided access to linear polybenzenes *via* copolymerisation (Fig. 4.22) [40]. Diethynylbenzene acts as dienophile, and bis(cyclopentadienone) as diene.

Since [4+2] cycloaddition is not very susceptible to steric hindrance, this synthetic concept has meanwhile been increasingly used for the assembly of hyperbranched compounds [41]. A diene combining two dienophile functions in one molecule was used as AB_2 monomer for polymerisation (Fig. 4.23). The diene function can then react with a triple bond of a second AB_2 monomer to form pentaphenylbenzene units [30].

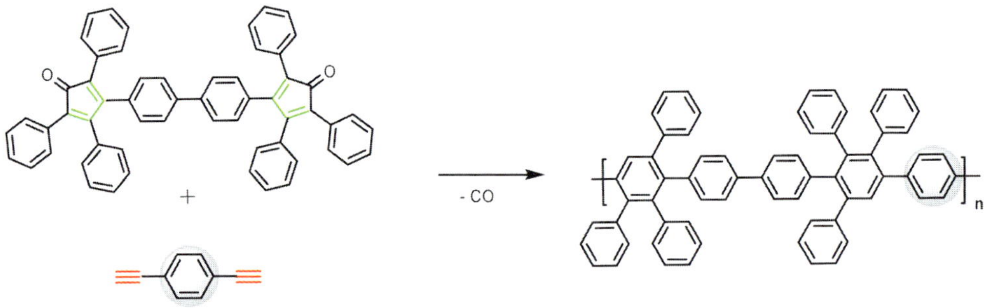

Fig. 4.22 Linear polyphenylenes (according to Stille *et al.*)

4.1.6
Carbon/oxygen-based (and Fréchet) dendrimers

4.1.6.1 **Polyether dendrimers**

Fréchet-type polyether dendrons number among the compounds frequently used in dendrimer synthesis. Fréchet *et al.* first used convergent synthesis in that they assembled polyaryl ether scaffolds from the outside inwards (Fig. 4.24). The attraction of this type of dendrimer is due not least to the reliable and highly perfect architectures obtained so far – in spite of the sensitive-looking benzyl ether units. Unlike in divergent syntheses, the reactive centre for assembly lies not in the end groups, whose increasing density prevents further reaction from a certain limiting generation onwards – or causes defects in the

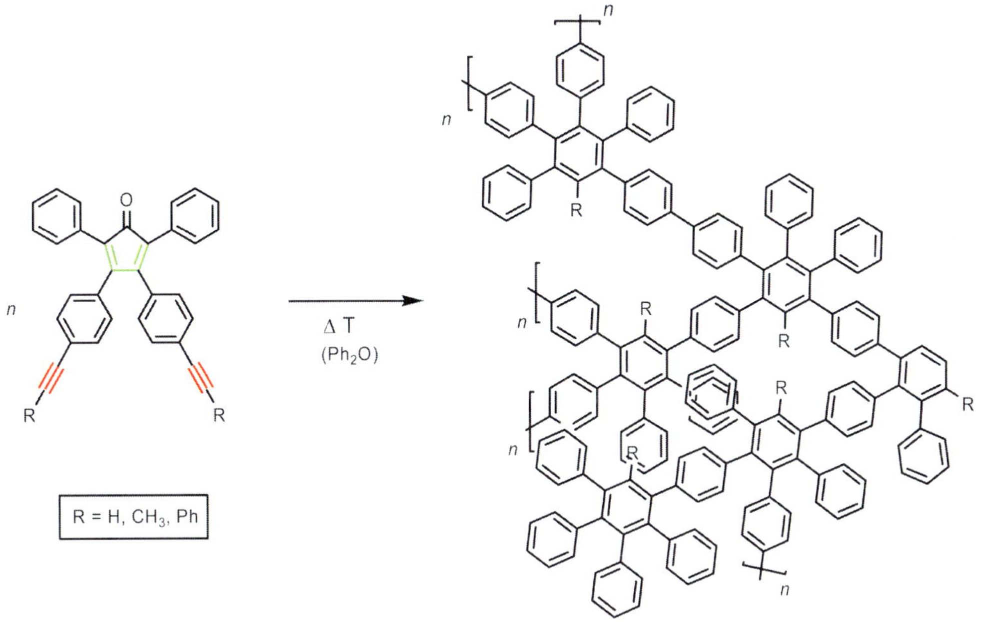

Fig. 4.23 Combination of diene and dienophile in an AB_2 "monomer" (left); reaction with itself to form a polyphenylene dendrimer (right)

dendrimer through backfolding or non-quantitative reactions; instead, reaction takes place exclusively at the respective focal point. This type of dendrimer is used as multifunctional platform of diverse dendritic molecules and thus recurs – like POPAM and PAMAM dendrimers – in almost all chapters of this book.

Repetitive two-stage synthesis of Fréchet dendrons is based in the first stage on the reaction of a 3,5-dihydroxybenzyl alcohol with benzyl bromide. Transformation of the lengthened benzyl alcohols into the corresponding lengthened benzyl bromide permits iteration of the reaction with the monomer [42].

An alternative synthesis of polyether dendrimers proceeds *via* the "hypercores" (Fig. 4.25) already mentioned in Chapter 2 [43]. Such a species is formed by reaction of an activated dendron (or precursor such as **1**) with the core unit 1,1,1-tris(4′-hydroxyphenyl)ethane (**2**). The resulting protected hypercore **3** is activated through removal of its benzyl protective groups by catalytic hydrogenation. Together with the original activated "dendron" **1**, the thus activated hypercore can then grow to the second-generation hypercore **5**. This method has hitherto been extended to dendritic architectures of up to the third generation.

Molecular recognition processes between amphiphilic biaryl dendrimers **2** (with carboxylic acid moieties) as host compounds and the protein chymotrypsin as guest were compared with those involving corresponding benzyl ether dendrons **1** (Fig. 4.26) [44]. In the biaryl dendrimers both the carboxylic acid groups of the periphery and those located in the inner "shells" are available for

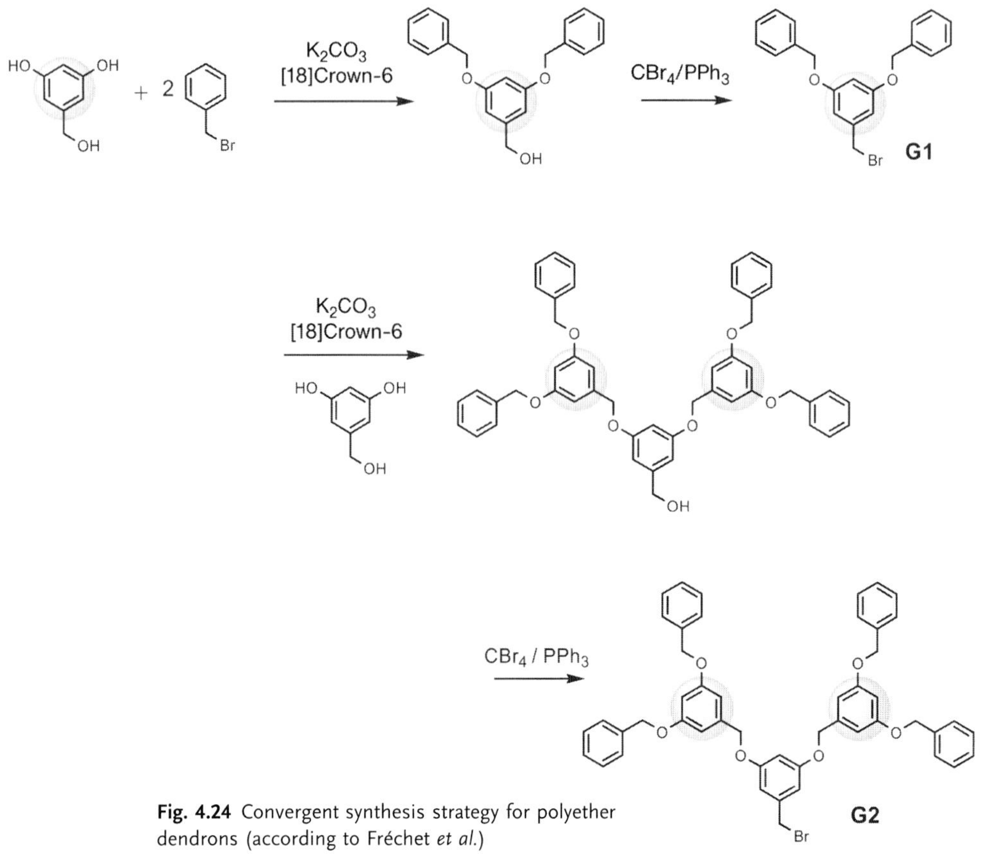

Fig. 4.24 Convergent synthesis strategy for polyether dendrons (according to Fréchet *et al.*)

molecular recognition. In contrast to the amphiphilic benzyl ethers with the same number (eight) of peripheral carboxyl groups, the biaryl dendrimers showed a pronounced affinity for guest molecules, regardless of generation. This comparison also showed that third-generation biaryls can bind six molecules of the proteins, while the corresponding third-generation benzene dendron can only bind three.

4.1.6.2 **Polyester dendrimers** [45]

Owing to the scope offered by their architecture for modification and functionalisation, dendrimers have been considered for therapeutic applications and as delivery systems for active ingredients, although precisely the plethora of possible variation and optimisation measures places demands on time and effort.

Fréchet *et al.* investigated oligoester dendrimers with regard to this potential [46]. These dendrimers were made up of two covalently coupled dendrons, whose "division of labour" consists on the one hand in formation of a "plat-

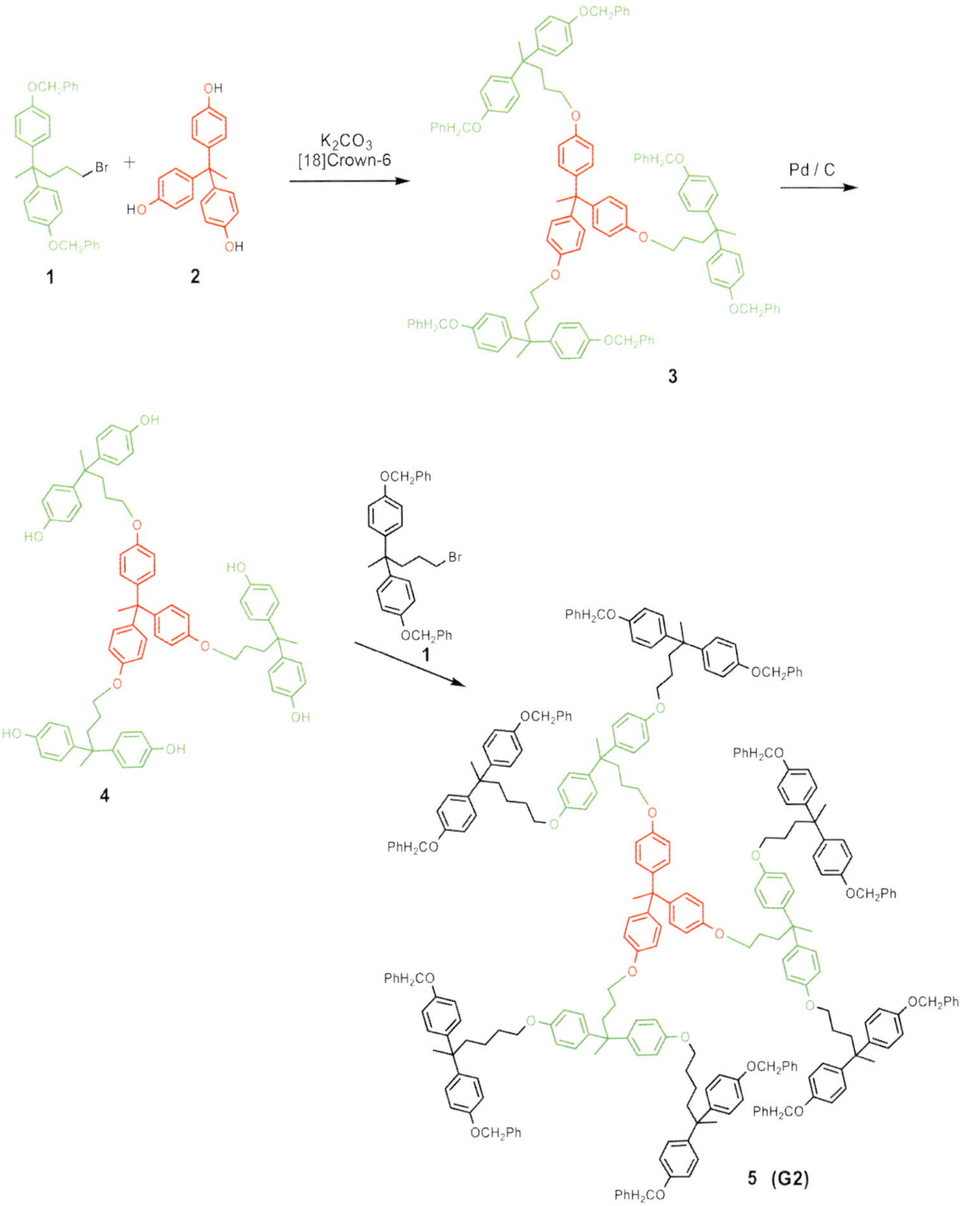

Fig. 4.25 Synthesis of polyether dendrimers starting from a "hypercore" (according to Wooley, Fréchet *et al.*)

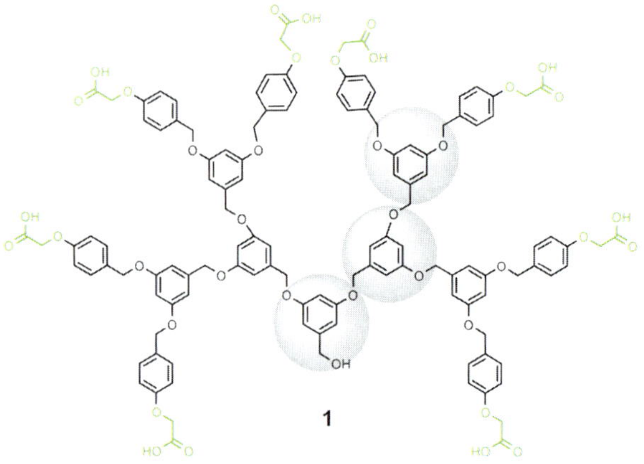

1

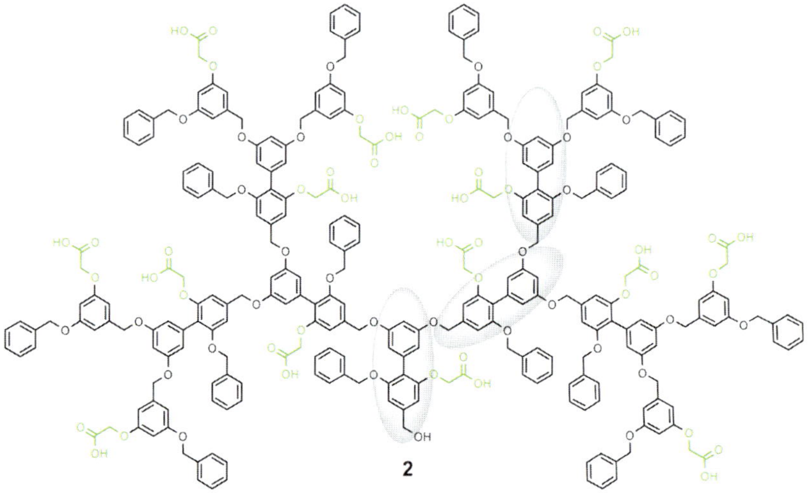

2

Fig. 4.26 Amphiphilic biphenyl dendrimers **2** compared with the corresponding benzene-based Fréchet amphiphiles (**1**; according to Thayumanavan *et al.*). Some of the benzene and biphenyl units are highlighted with a grey background for the sake of clarity

form" for therapeutically active structures while the other dendron serves for attachment of oligoethylene glycol ether chains responsible for solubility. Thus polyester dendrimers could be prepared *via* both convergent and divergent routes as potential active ingredient delivery systems with molecular masses of 20 to 160 kDa.

To this end, the singly protected triol **1** was convergently coupled to the isopropylidene-protected carboxylic acid **2** shown in Fig. 4.27. The resulting protected allyl carbonate **3** was deprotected with $Pd(PPh_3)_4$ in the presence of mor-

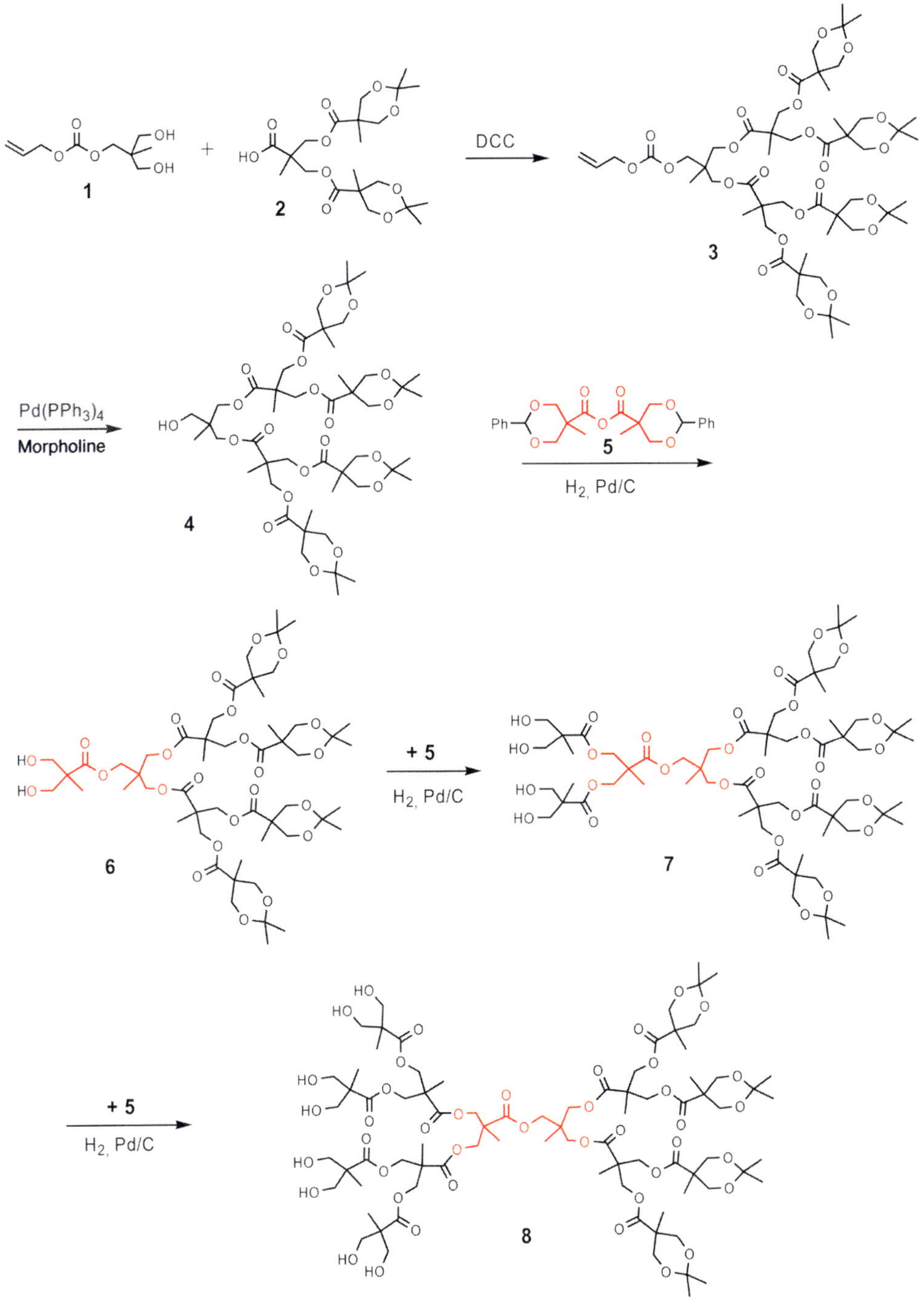

Fig. 4.27 Synthesis of polyester dendrimers (according to Fréchet *et al.*)

pholine to give the mono-alcohol **4**. The free hydroxyl group produced at the focal point was allowed to react with dibenzylidene-2,2-bis(oxymethyl)propionic anhydride **5** to give the dendron **6** with two hydroxyl groups at the focal point. The sequence of coupling and deprotection ultimately afforded the new polyester dendrimer **8** (Fig. 4.27).

Well-defined aliphatic polyesters and dendrons based on 2,2-bis(methylol)propionic acid (bis-MPA) are commercially available from the Swedish company *Polymer Factory*. Their stability on storage, solubility, and low toxicity make them attractive for the medical sector.

4.1.6.3 **Carbohydrate dendrimers (glycodendrimers)**

The molecular scaffolds of carbohydrate dendrimers can be varied in many different ways: carbohydrates can act as core unit, serve for branching, or function as terminal groups [47]. Moreover, they are also suitable for supramolecular bonding and for the transport of active ingredients.

As shown below in Section 4.1.9 (silicon-based dendrimers), there are meanwhile many examples of carbohydrate-coated dendrimers (glycodendrimers) [48], not least because these terminal groups confer better solubility upon the molecules because they are, so-to-speak, “sugar coated”.

The first glycodendrimer was pepared by Roy *et al.* [49]. Access to dendrimers bearing carbohydrate terminal groups is provided by the triglycosylated trihydroxyamine **3** [50] obtained in a several-stage synthesis, which can serve as dendron. The amino dendrons could be attached to the tricarboxylic acid core by DCC coupling (dicyclohexylcarbodiimide) (Fig. 4.28) [51].

Glycopeptide dendrimers with carbohydrate units both in the core and in the branching units were prepared by Lindhorst *et al. via* a generalisable synthetic route [52]. The iterative synthetic sequence is based on the orthogonally protected carbohydrate building block **1** (AB_3 type monomer). After appropriate activation, this species can be combined both divergently as well as convergently with carbohydrate units to form a dendrimer. The actual coupling of core unit and dendrons proceeds *via* peptide bond formation between **2** and **3** (see red and green functional groups in Fig. 4.29). This permits execution as a solid-phase synthesis. The coupling reagents used were HATU (= *O*-(7-*a*zabenztriazol-1-yl)-*N*,*N*,*N*′,*N*′-*t*etramethyl*u*ronium hexafluorophosphate) and DIPEA (= *di*isopropyl*e*thyl*a*mine).

The synthesis of a carbohydrate dendrimer consisting solely of mannopyranosides was accomplished *via* a convergent route [53].

Less has so far been published about carbohydrate core units [54] than about dendrimers with carbohydrate terminal groups. The first PAMAM dendrimer with a carbohydrate core unit was prepared from the α-D-glycoside **1** as core unit precursor. Up to second-generation PAMAM dendrimers with a D-glucose core could be prepared by iteration; Fig. 4.30 shows the first generation [55].

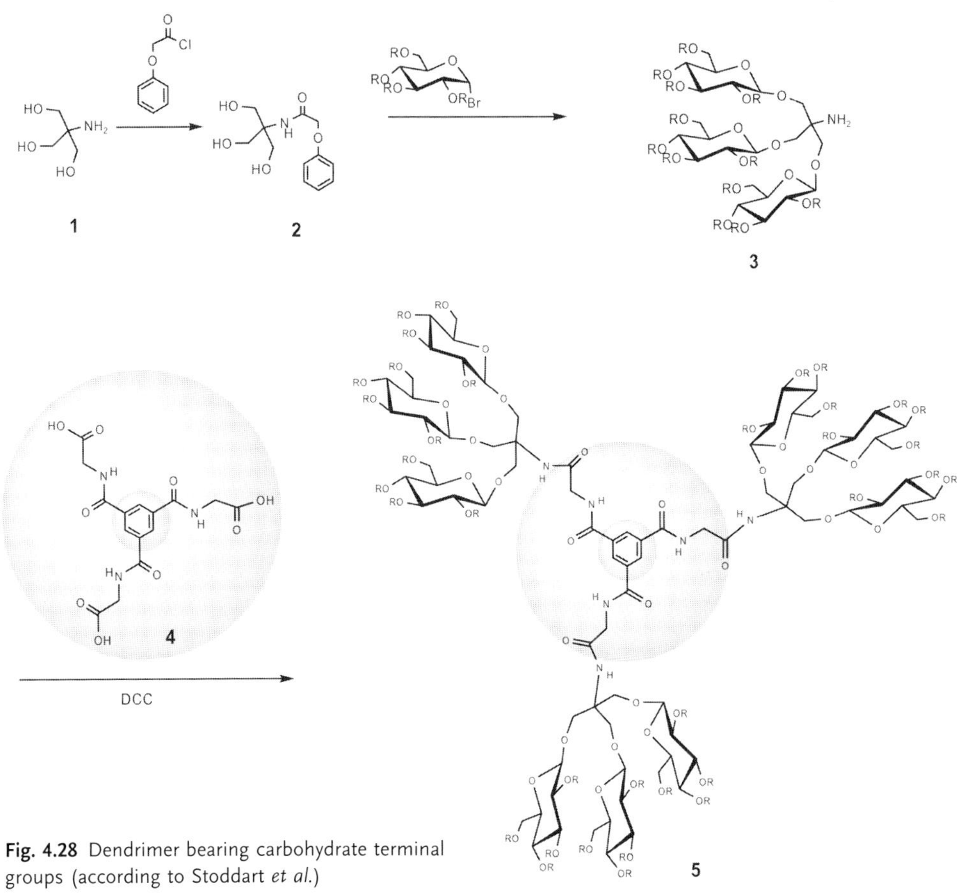

Fig. 4.28 Dendrimer bearing carbohydrate terminal groups (according to Stoddart *et al.*)

4.1.7
Porphyrin-based dendrimers

Sakata *et al.* succeeded in synthesising and characterising a defect-free highly symmetrical, *para*-phenylene-spaced porphyrin-heneicosameric nickel complex ($C_{1244}H_{1350}N_{84}Ni_{20}O_{88}$; molar mass 20061 Da; Fig. 4.31) and corresponding lower generations; they also obtained an STM image of a single molecule. The molecule contains 21 porphyrin units. The synthesis utilises the condensation of formyl-substituted phenyl end groups with pyrrole units, with these reaction steps giving almost quantitative yields. The overall yield of all 17 steps was 17%.

An STM image of individual molecules was obtained on a copper-(111)-surface and shows nanometer-sized flat "mandala"-type squares measuring 6.5 nm (sides) and 8.3 nm (diagonals). The nickel–nickel distance between two adjacent molecules is 13–18 nm. The molar absorption coefficient lies at around

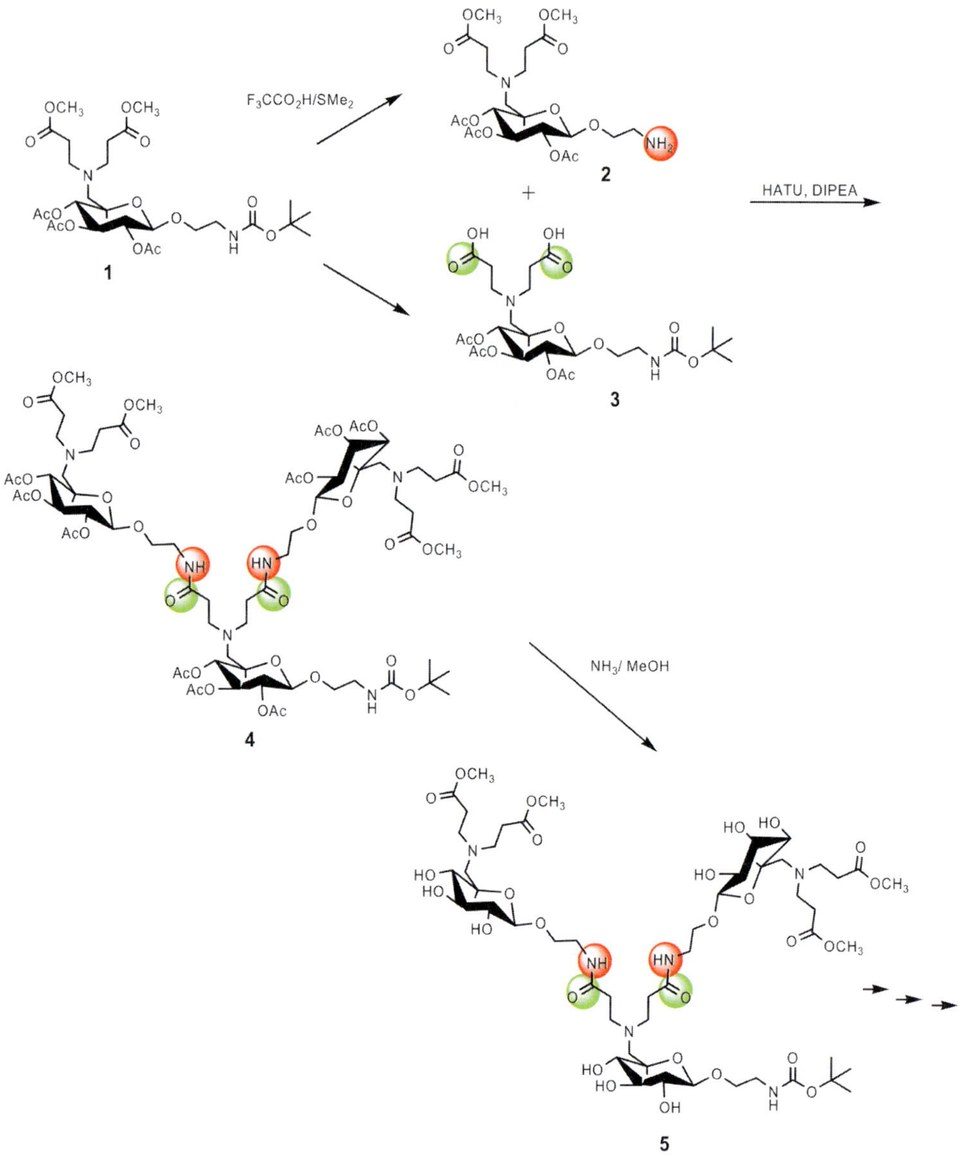

Fig. 4.29 Synthesis of a first-generation glycopeptide dendrimer (5) (according to Lindhorst *et al.*)

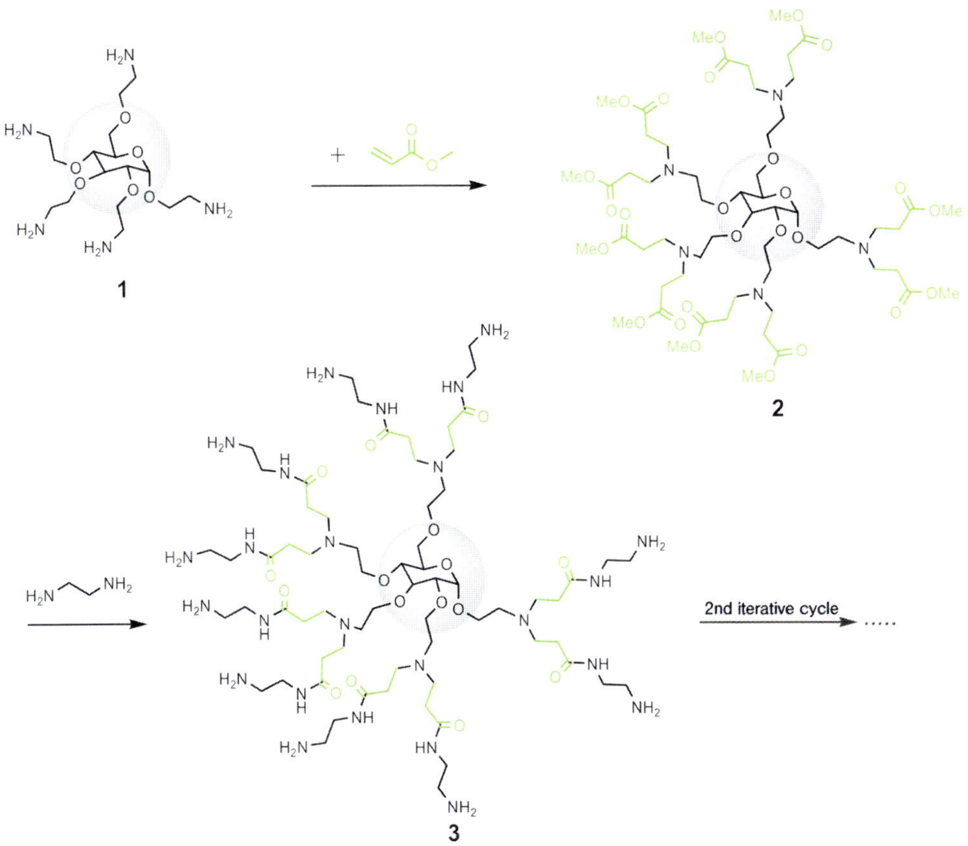

Fig. 4.30 First-generation PAMAM dendrimer with D-glucose core (according to Lindhorst *et al.*)

3.56 million, corresponding to the 13-fold value of the monomeric porphyrin complex itself [56].

Other porphyrin-based dendrimers will be found in Section 5.2.

4.1.8
Ionic dendrimers

Ionic dendrimers [57] can bear positive or negative electric charges in the interior of the molecule (core, branching units) or on terminal (functional) groups.

4.1.8.1 Polyanionic dendrimers

Some polyanionic dendrimers act as "micellar molecules" [58]. Newkome *et al.* first synthesised water-soluble dendrimers, whose hydrophobic molecular niches can accommodate guests. The periphery of such dendrimers was furnished with

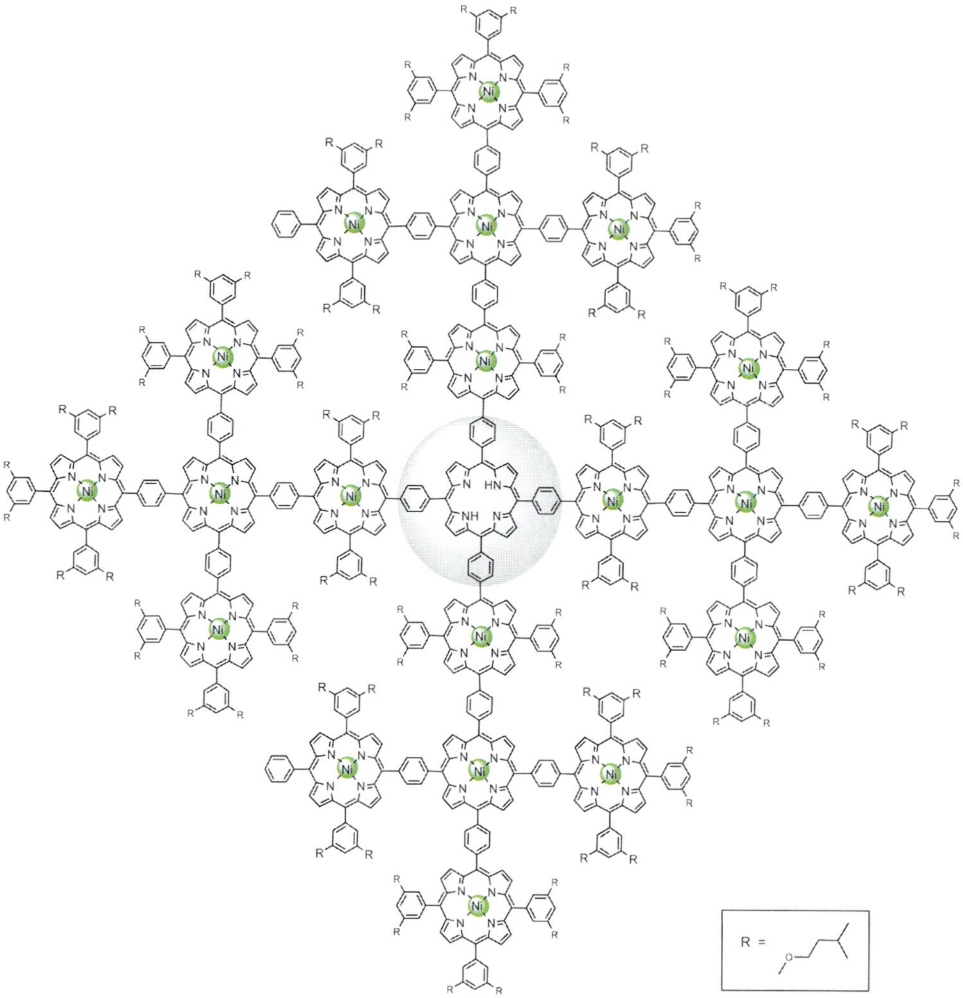

Fig. 4.31 "Mandala dendrimer" (heneicosamer; according to Sakata *et al.*)

carboxylate functions (see Chapter 6, Fig. 6.20) [59]. Corresponding dendritic polyethers are also rendered water-soluble by terminal carboxylate groups [60].

Hirsch and Guldi *et al.* used polyanionic dendrimers with a fullerene core in order to examine *p*hotoinduced *e*lectron *t*ransfer (*PET* see Chapter 5) and their complexing behaviour towards Zn-cytochrome C as guest. Partially deprotonated, negatively charged monolayers of the "amphifullerene" shown in Fig. 4.32 bind cytochrome C, a polycationic redox protein, at the air/water interface. This binding results from electrostatic interactions between the oppositely charged partners. Unlike the corresponding mono-malonic ester adducts of fullerene (Fig. 4.33) [61], the hexakis-malonic ester adduct amphifullerene undergoes pronounced changes of its properties as an electron acceptor.

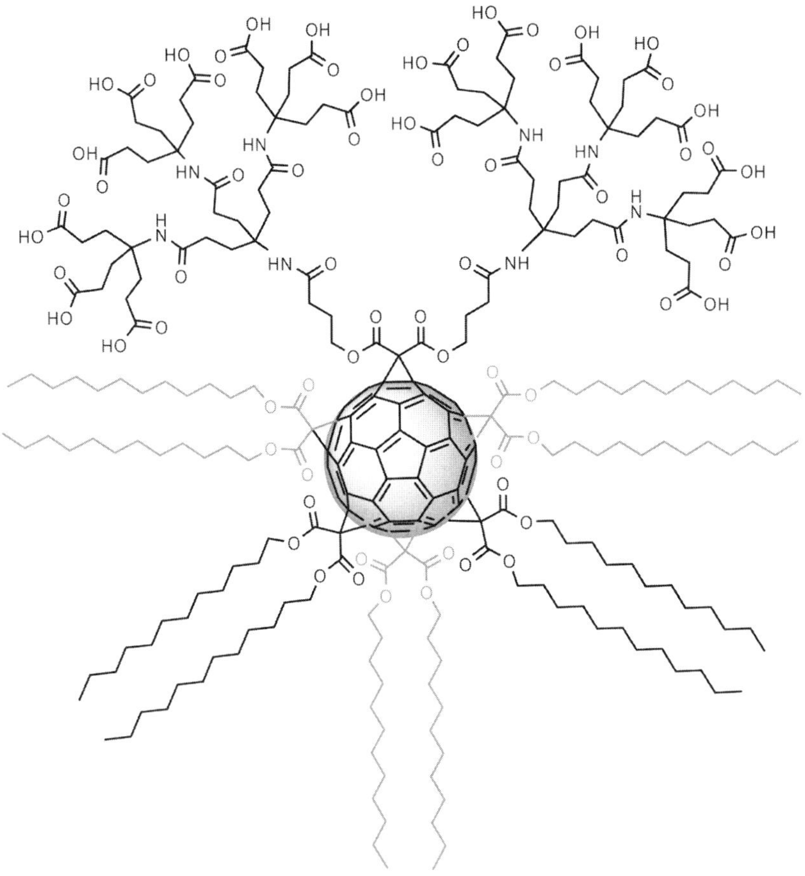

Fig. 4.32 Amphifullerene (according to Hirsch *et al.*)

Preparation of mono-adducts of fullerene – for studies on electrostatic interactions – was undertaken by cyclopropanation of fullerene with appropriately functionalised malonic esters **1** (Bingel reaction) to form **2**. Coupling with the *tert*-butyl protected oligoamide-amino-dendron **3** and subsequent hydrolysis lead to the water-soluble fullerene dendron **5**, which can carry up to nine negative charges after deprotonation. After association with the zinc complex of cytochrome C, photoinduced electron transfer (PET) from the redox protein to the fullerene can be accomplished, which was studied by fluorescence spectroscopy.

Hyperbranched polymers of polybenzene type can be rendered water-soluble like the above-mentioned monodisperse species by introduction of carboxyl groups [62]. Starting from 3,5-dibromophenylboronic acid **1** as AB_2 monomeric building block (Fig. 4.34), the polycondensate **2** was assembled in an aryl–aryl coupling analogous to Suzuki coupling in the presence of Pd(0), and subsequently reacted with carbon dioxide to form the hyperbranched compounds **3** with terminal carboxylate groups:

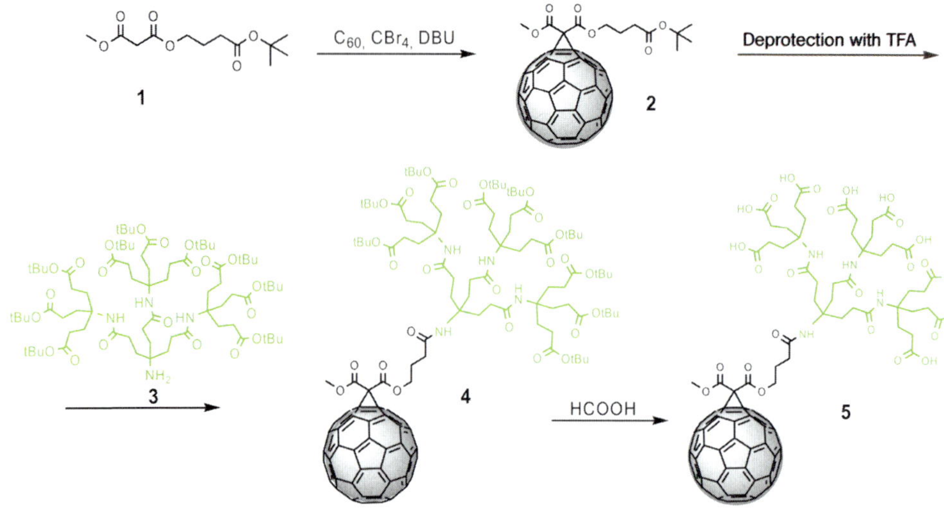

Fig. 4.33 Synthesis of water-soluble mono-malonic ester adducts **5** of fullerene (according to Hirsch *et al.*)

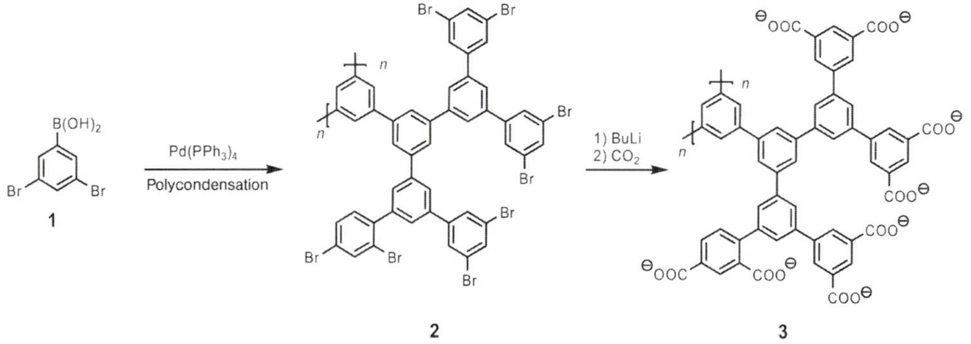

Fig. 4.34 Hyperbranched "polyphenylenes" with terminal carboxylate

4.1.8.2 **Polyanionic dendrimers**

Polycationic dendritic molecules can generally be prepared in the following ways:

- Introduction of positively charged centres in the course of synthesis
- Transformation of neutral groups of the dendrimer into cationic ones
- Incorporation of transition metals into the dendrimer

An example of the introduction of positively charged units during synthesis is provided by water-soluble phosphorus-based dendrimers. They were examined for their suitability as transfection agents for the luciferase gene present in 3T3 cells [63]. This gene from fireflies causes bacteria to produce the enzyme luciferase [64], which, as an oxidoreductase, oxidises luciferin with a high quantum yield to oxyluciferin giving strong (bio)luminescence. On use of this methodolo-

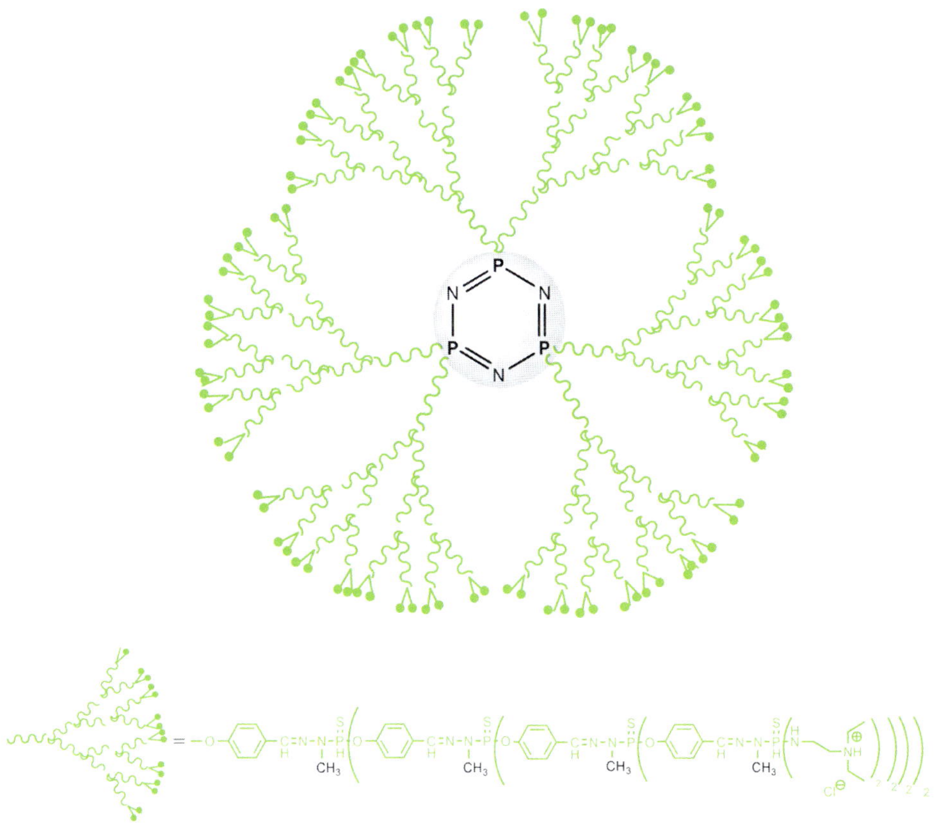

Fig. 4.35 Polycationic phosphorus-based dendrimers (according to Majoral, Caminade *et al.*)

gy in medical applications, it is conceivable that certain bacteria present in low concentration could be visualised by their fluorescence and thus located with the aid of such genes.

The ability to transport nucleic acid into cells depends upon the size of the dendrimer. Thus phosphorus-based dendrimers of the third, fourth, and fifth generation are more efficient than those of lower generations (Fig. 4.35). The nature of the quaternary ammonium end groups also has an influence on transfection ability: A dendrimer that is methylated at the terminal nitrogens is less effective than a protonated dendrimer.

Stoddart *et al.* developed a convergent route to polycationic dendrimers based on mesitylene units [65]. Positively charged dendrons obtained by the Menschutkin reaction are attached as branching units around the 1,3,5-tris(diethylaminomethyl)benzene core unit (**3**) (Fig. 4.36). (Anion) exchange of bromide with hexafluorophosphate ions provides better solubility.

Chemoselective polyalkylation of phosphorus-based dendrimers offers an opportunity to produce cationic dendrimers from neutral ones [66].

Formation of core unit:

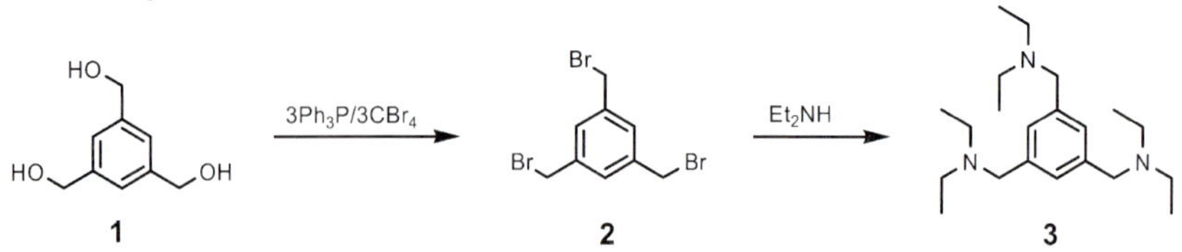

Formation of "monomer":

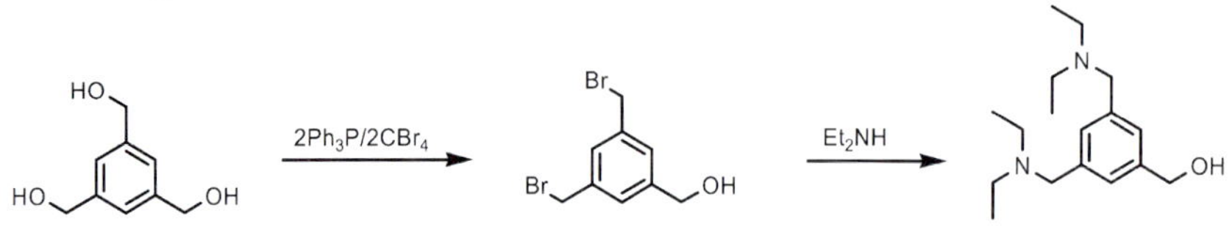

Formation of dendron/Anion exchange:

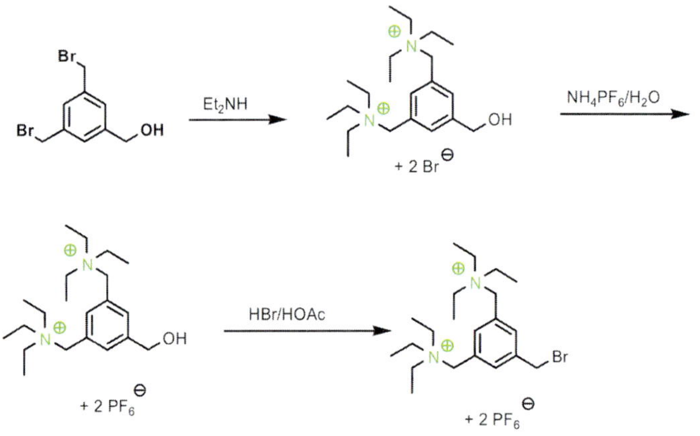

Formation of dendrimer:

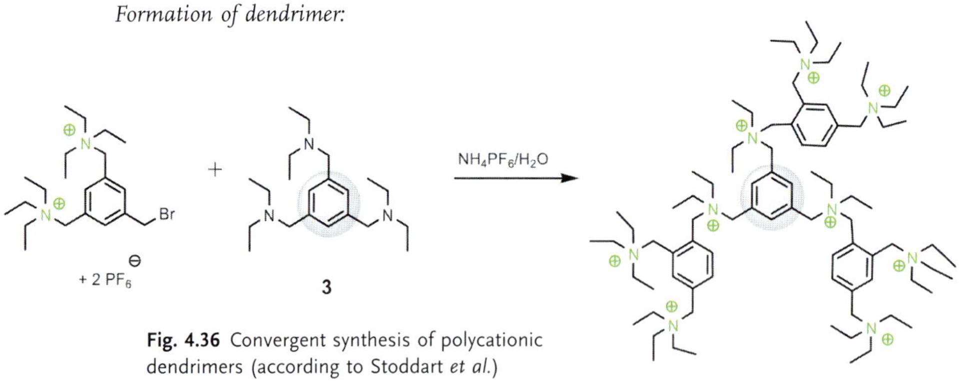

Fig. 4.36 Convergent synthesis of polycationic dendrimers (according to Stoddart *et al.*)

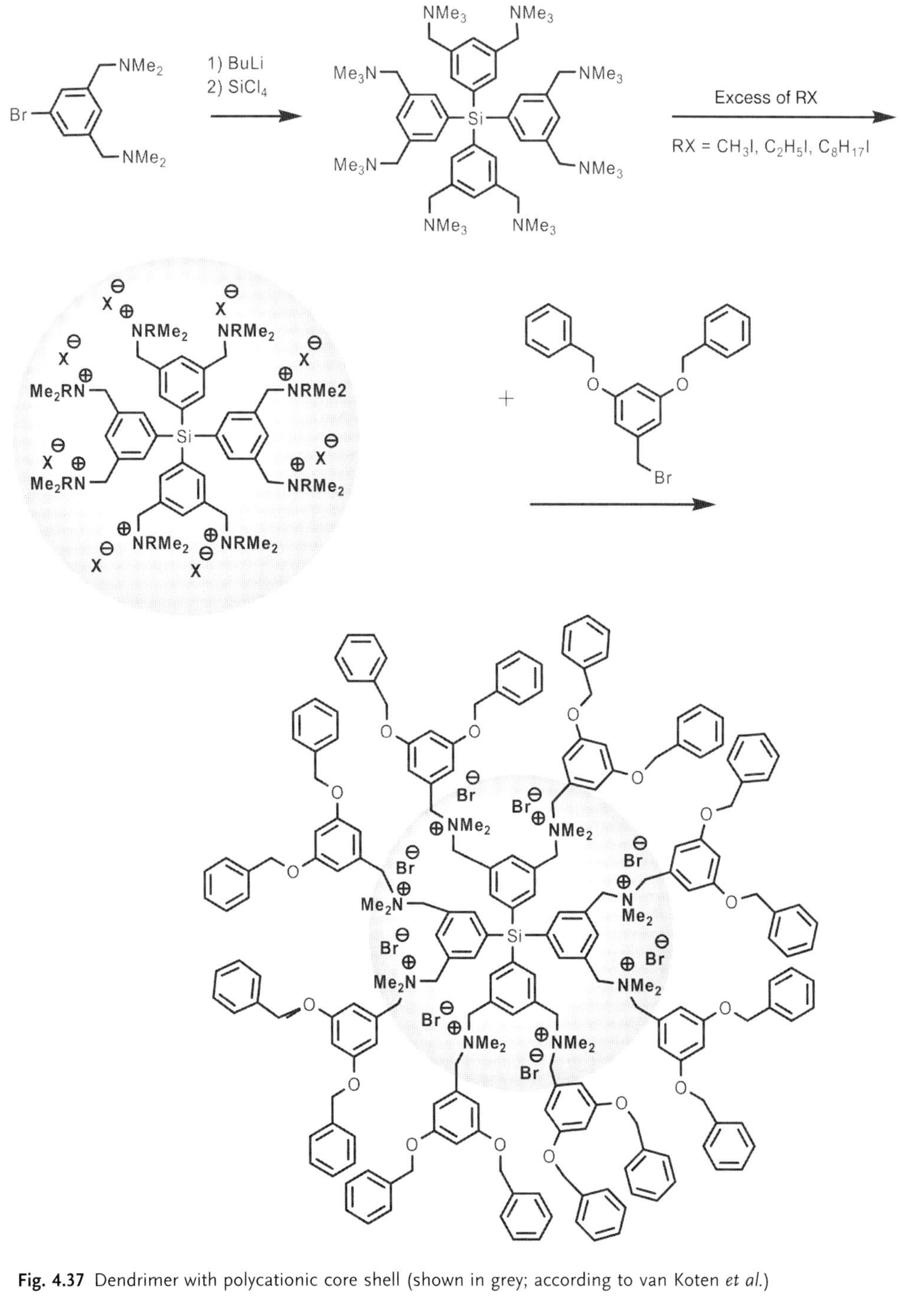

Fig. 4.37 Dendrimer with polycationic core shell (shown in grey; according to van Koten *et al.*)

Van Koten *et al.* successfully prepared silicon-based dendrimers with a polycationic branching shell. The dendrimer type shown in Fig. 4.37 consists of a non-polar core (tetraarylsilane) bearing an ionic layer which, in turn, is surrounded by a less polar polyether "shell" [67]. Such dendrimers are predestined for potential inclusion of anionic guests – with anion exchange – in the positively charged host scaffold.

4.1.9
Silicon-based dendrimers

Siloxanes, prepared in 1989 as representatives of silicon-based dendritic molecules ("silicodendrimers"), were the first dendrimers to contain heteroatoms other than the usual ones (N, O, S, halogens) [68]. As with the phosphodendrimers (Section 4.1.10), their readily modifiable architecture and their pronounced thermostability hold promise of applications, for example, in the form of carbosilanes as liquid-crystalline materials and catalyst supports. They can be subdivided into a number of basic types and their properties are presented below with the aid of characteristic representatives:

- *Polysilanes* with (Si–Si) skeleton
- *Carbosilanes* with (Si–C_n) units
- *Carbosiloxanes* with (Si–O–C) units
- *Siloxanes* with (Si–O–Si) skeleton

4.1.9.1 Silane dendrimers

Independently of each other, Lambert *et al.* [69] and Suzuki *et al.* [70] both gained access to low-generation silane dendrimers (G1, G2) in 1995. The latter prepared a first-generation polysilane dendrimer **3** by a stepped growth polymerisation technique. Coupling of methyl[tris(chlorodimethylsilyl)]silane (**1**) to tris(trimethylsilyl)silyllithium (**2**) led to the first-generation branched dendrimer **3** (Fig. 4.38).

Lambert *et al.* succeeded in developing second-generation dendritic polysilanes [71]. The repetitive steps of this divergent synthesis consist in cleavage of silicon–methyl bonds by trifluorosulphonic acid (CF_3SO_3H) and subsequent formation of silicon–silicon bonds (Fig. 4.39). Starting from tetrasilanes, up to second-generation "starfish-like" dendritic molecules with a molecular mass of 1832.9 g/mol could be prepared in this way [72].

This type of silane dendrimer already displays distortion of the internal silicon scaffold structure in the second generation as a result of steric overcrowding arising from pronounced branching [68]. The advantage of the tendentially spherical dendritic silanes – over low-molecular representatives with less stable Si–Si bonds – lies in the enclosure of these readily cleaved bonds in the interior of the dendrimer. If the exterior shell bears methyl terminal groups, then cleavage reagents (electrophilic or nucleophilic) will hardly be able to access these bonds – nor will acid catalysts be able to effect isomerisation of the scaffold:

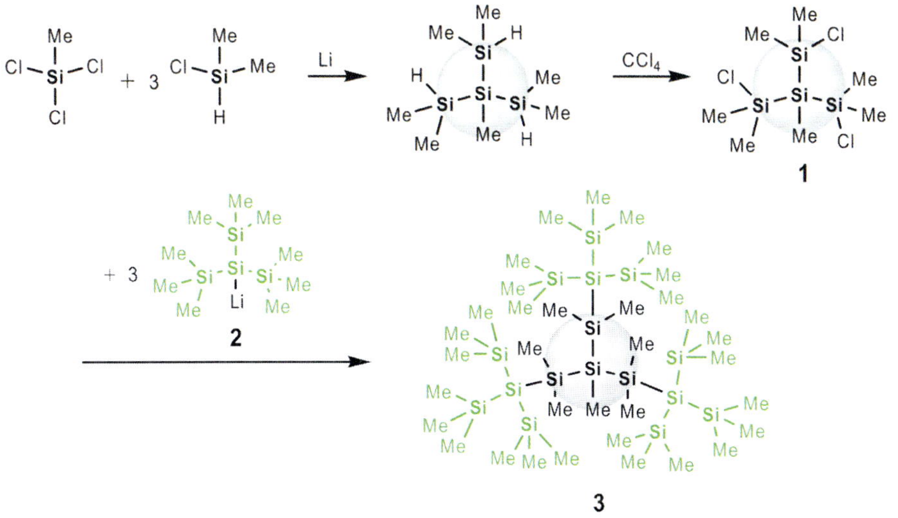

Fig. 4.38 Preparation of a first-generation polysilane dendrimers (according to Suzuki *et al.*)

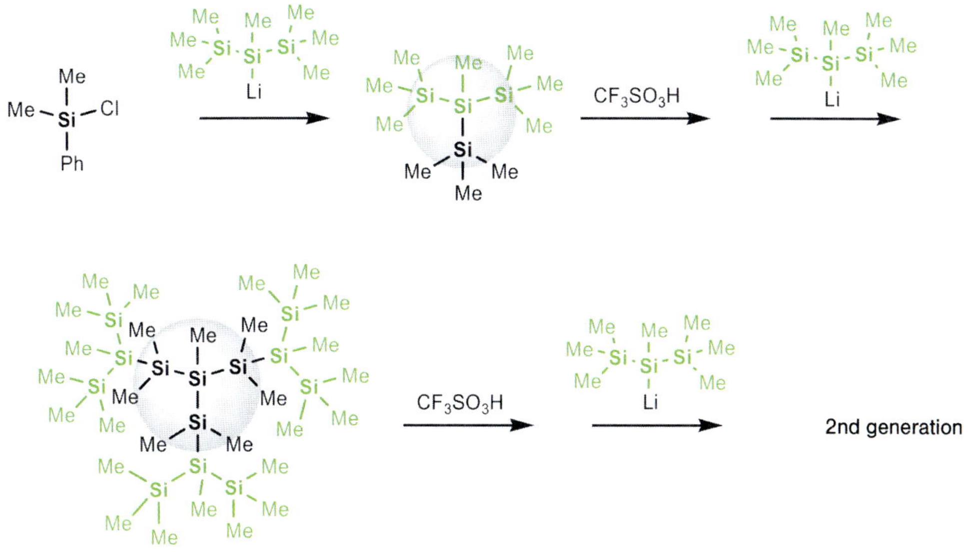

Fig. 4.39 Synthetic route to a second-generation silane dendrimer (according to Sekiguchi *et al.*)

Such a dendritic "protective shell" has also been used to stabilise other metastable core units (see, *e.g.*, Section 4.1.10 and 5.1.2.3).

Silane dendrimers with longer polysilane chains (*e.g.* with thirteen silicon atoms in the longest chain) were prepared by Lambert and Wu *via* a convergent route [73]. These molecules also assume an almost spherical shape.

A more recent synthetic concept is based on the reaction of diethylamino-substituted silyl-lithium compounds with triflate derivatives of silanes or oligosilanes. It leads to stepwise lengthening of Si–Si chains which can be used to assemble polymeric structures [74].

4.1.9.2 Carbosilane dendrimers

Structurally flexible carbosilane dendrimers [75] are characterised by a high kinetic and thermodynamic stability as a consequence of the low dissociation energy of the silicon–carbon bond. As already mentioned, precisely the carbosilanes offer a broad range of possible applications since their structures can be readily modified. Thus the length of the branching units depends upon the length of the alkyl chain of the Grignard reagent, but the multiplicity of branching depends upon the hydrosilylation reagent (Table 4.3).

Reports of carbosilane dendrimers were published from 1990 onwards by the working groups of van der Made, van Leeuwen [76], Roovers [77], Muzafarov [78], and Seyferth [79]. Synthesis was generally accomplished in two steps:

1. Hydrosilylation of a vinyl- or allyl-functionalised silicon compound with a silane
2. ω-Alkenylation of chlorosilanes with a Grignard reagent

Platinum-catalysed hydrosilylation of terminal tetra-allylated silicon core units and subsequent reaction with Grignard reagent showed (Fig. 4.40) that the attainable dendritic generation number depends upon the length of the alkenyl group – which serves as a spacer preventing overcrowding of the branches. Thus a dendrimer of up to the sixth generation could be assembled with decenyl units, whereas only the fourth generation could be attained with vinyl groups and the fifth with allyl units [76 b].

Convergent synthesis [80] permits access to silyl-ethynylated dendrimers of the first and second generations with alternating silicon-ethyne units. The (first-generation) dendrimer shown in Fig. 4.41 can assume an essentially planar conformation owing to its linear triple bonds.

Table 4.3 Scope for variation in the assembly of carbosilanes

Core units	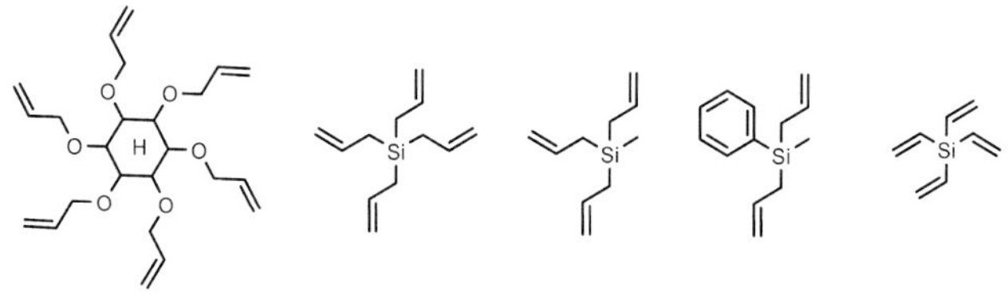		
Hydrosilylation reagent	$SiHCl_3$	$SiHMeCl_2$	$SiHMe_2Cl$
Alkenylation reagent			

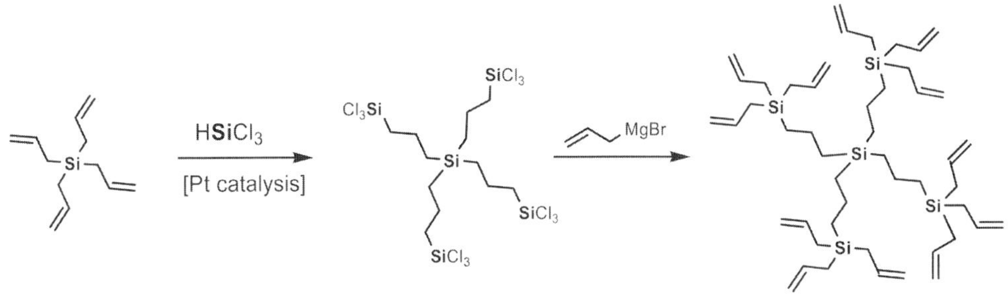

Fig. 4.40 Carbosilane dendrimers (according to van der Made *et al.*)

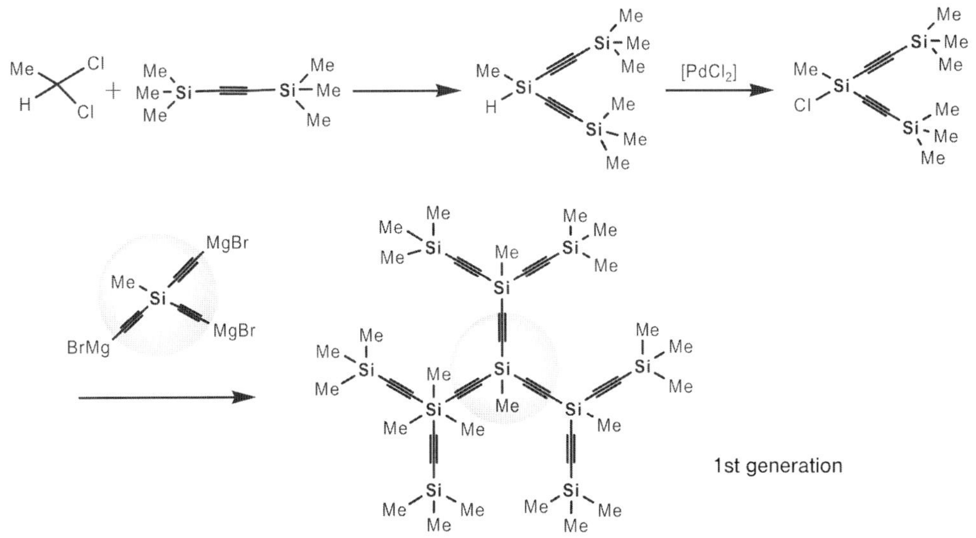

Fig. 4.41 Carbosilane dendrimers (according to Sekiguchi *et al.*)

Both generations exhibit high thermostability (up to 200 °C); however, the silicon–carbon bond is readily cleaved under alkaline or acid conditions. The introduction of phenyl groups at various positions in the carbosilane dendrimer – in the core, at the branching points, or in the periphery – and their acid cleavage with stoichiometric quantities of trifluorosulphonic acid permits nucleophilic substitution at these sites. According to Gade *et al.* this permits targeted insertion of functional groups or catalytically active units at these sites [81].

Hydrosilylation of the protected allyl-glycoside **1** with the carbosilane **2** (by means of Silopren™, a platinum-siloxane complex from Bayer AG) led *via* Si–C bond formation to a glycosidic carbosilane dendrimer (Fig. 4.42) [82].

Very recent work in this area [83] yielded dumbbell-shaped carbosilane dendrimers (Fig. 4.43) [84].

These dendritic dumbbells bearing up to six galabiose units (Gal*α*1-4Gal) in the periphery could serve as artificial inhibitors of the Shiga toxin (Vero toxin).

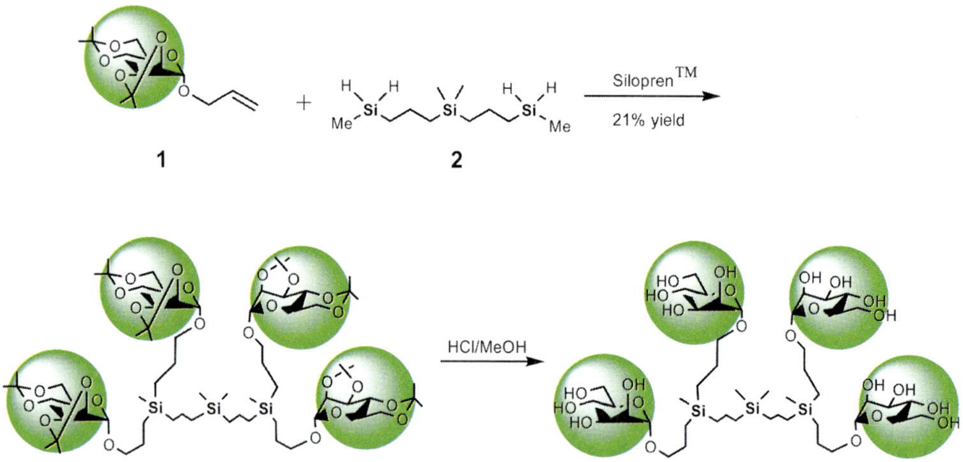

Fig. 4.42 Glycosidic carbosilane dendrimer (according to Lindhorst *et al.*)

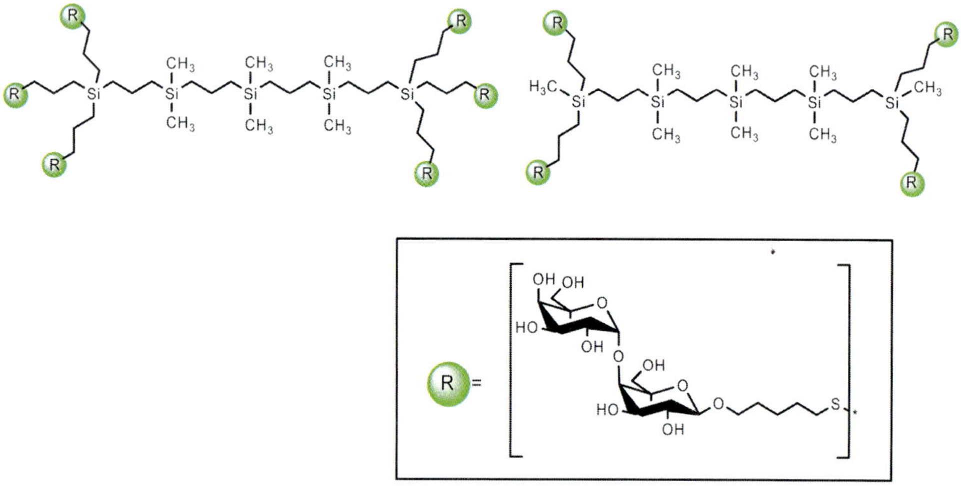

Fig. 4.43 Dendritic carbosilane dumbbells with galabiose branches (according to Terunuma *et al.*)

This toxin produced by *Escherichia coli* bacteria causes diarrhoea and haemolytic uraemia syndrome.

Less usual carbosilane dendrimers were synthesised by reaction of acetyl-protected hydroxyethyl glycosides with chlorosilanes. Introduction of the carbohydrate component proceeds *via* alcoholysis of the chlorosilane with formation of silicon–oxygen bonds; a carbosilane core unit is formed with carbosiloxane side arms [82]. Reverse assembly of carbosilanes with a carbohydrate core is also feasible [54].

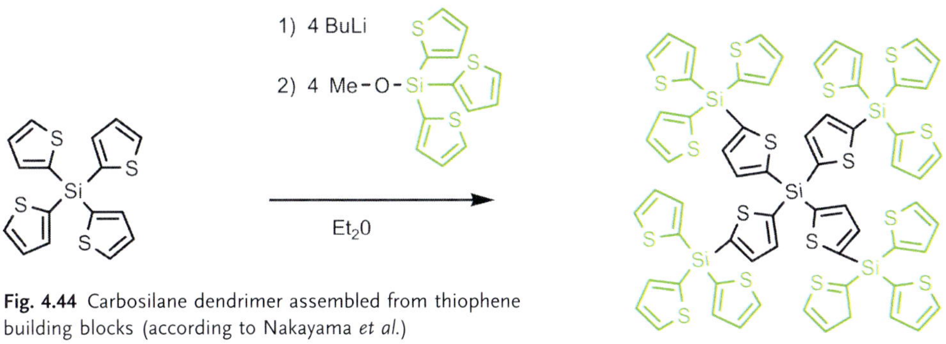

Fig. 4.44 Carbosilane dendrimer assembled from thiophene building blocks (according to Nakayama *et al.*)

A first-generation carbosilane dendrimer consisting of 16 thiophene rings could be isolated in 19% yield, albeit together with various by-products in comparable amounts [85]. The synthetic building block methyl tris(2-thienyl)silyl ether was first prepared from tetramethoxysilane and an excess of thienyllithium (Fig. 4.44).

Carbosilane dendrimers functionalised eight-fold with crown ethers (Fig. 4.45) can accommodate a total of eight potassium ions in their peripheral crown ether units [86].

Terminal palladium-complexed, phosphane-functionalised carbosilane dendrimers have been used as potential catalysts in membrane reactors [87].

4.1.9.3 **Carbosiloxane dendrimers**

The Si–O–C sequence is the distinctive structural unit of carbosilanes. Like that of the carbosilanes, the scaffold of this dendrimer family can be constructed in a targeted manner by the choice of certain core units [88] (Table 4.4). Synthesis of carbosiloxanes generally proceeds by

- hydrosilylation of carbosiloxane core units functionalised with vinyl or allyl groups,
- subsequent alcoholysis.

In order to illustrate the assembly of carbosiloxane dendrimers, the smallest core unit will first be considered. A fourth-generation cauliflower dendrimer (Fig. 4.46) was obtained *via* a repetitive synthetic sequence with the aid of dichloromethylsilane as hydrosilylation and allyl alcohol as alcoholysis reagent [89].

A cyclic core unit can also provide a foundation for carbosiloxane dendrimers (Table 4.4 and Fig. 4.47). The scaffold then shows a somewhat less compact geometry ("cyclic spacing") in the core region – in contrast to the situation with linear spacing.

Kim *et al.* employed a divergent-repetitive synthetic route (see above, carbosilane dendrimers) to obtain mixed carbosilane/carbosiloxane dendrimers

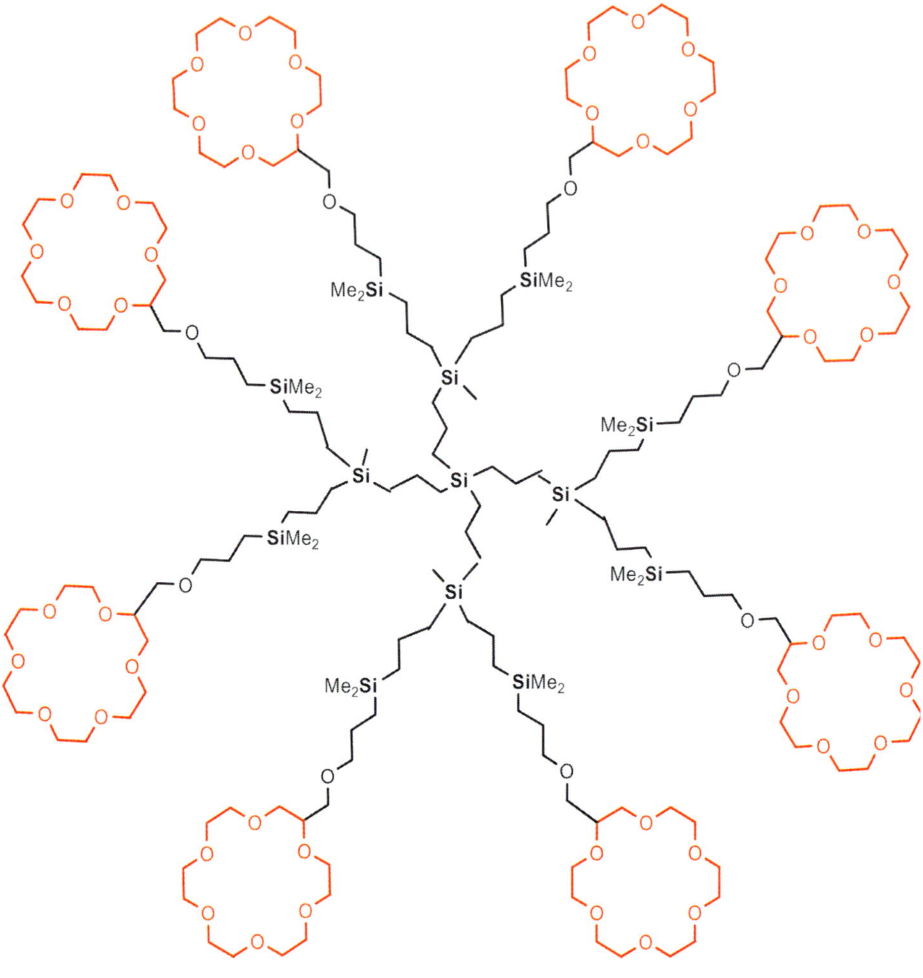

Fig. 4.45 Crown ether-decorated carbosilane (according to Lang *et al.*)

Table 4.4 Representative core units for carbosiloxanes

Core units for assembly of carbosiloxanes				
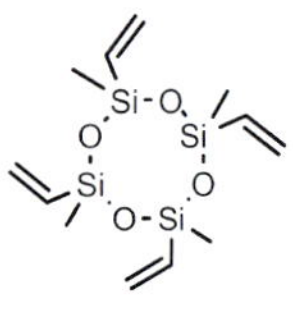	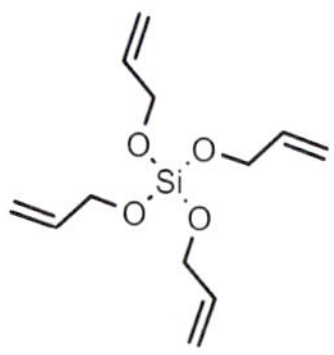	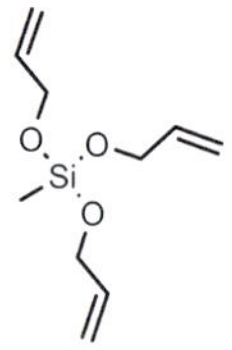	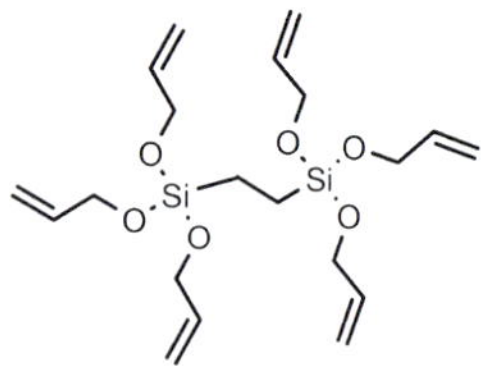	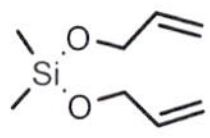

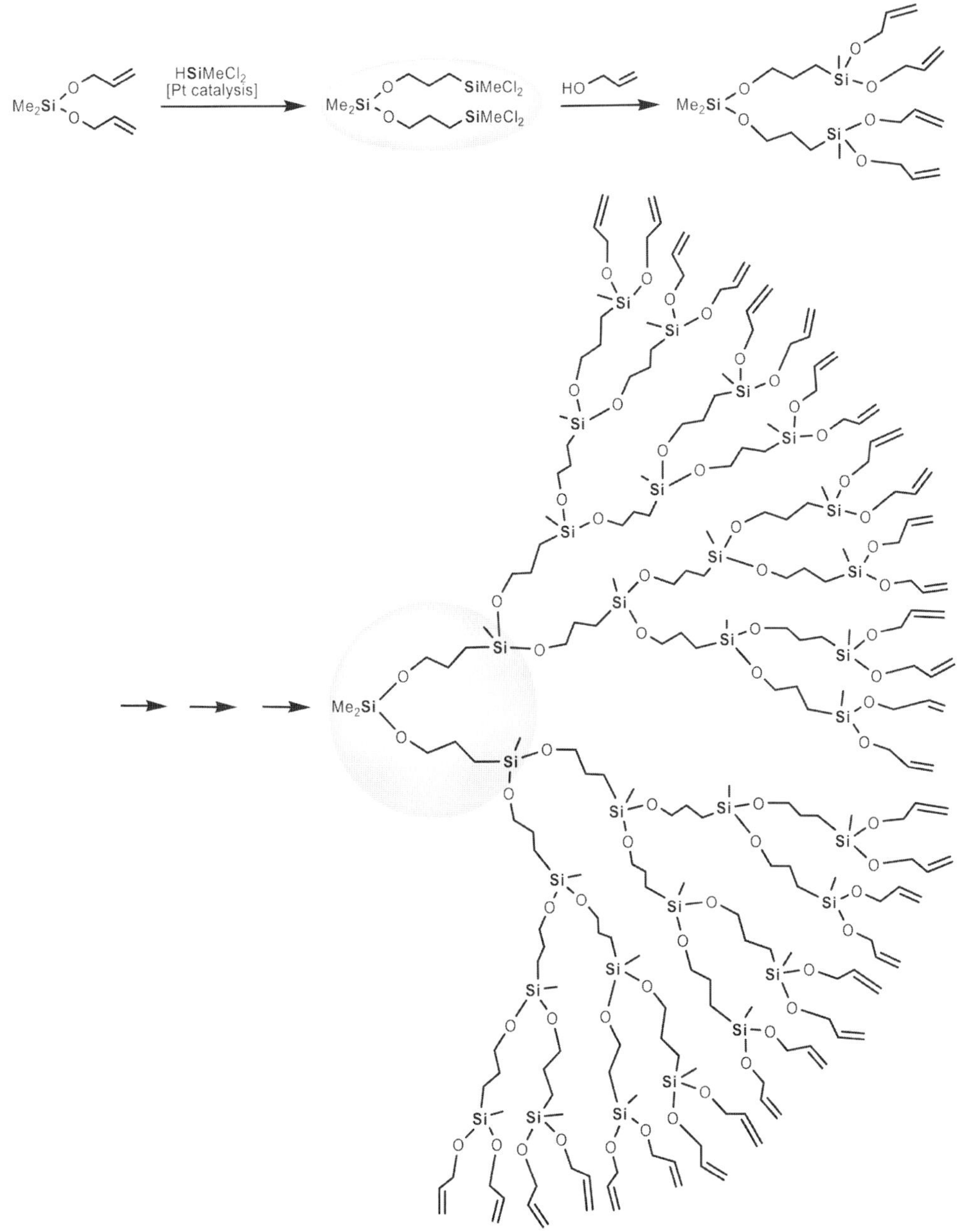

Fig. 4.46 Synthesis of a "cauliflower dendrimer" (according to Lang *et al.*)

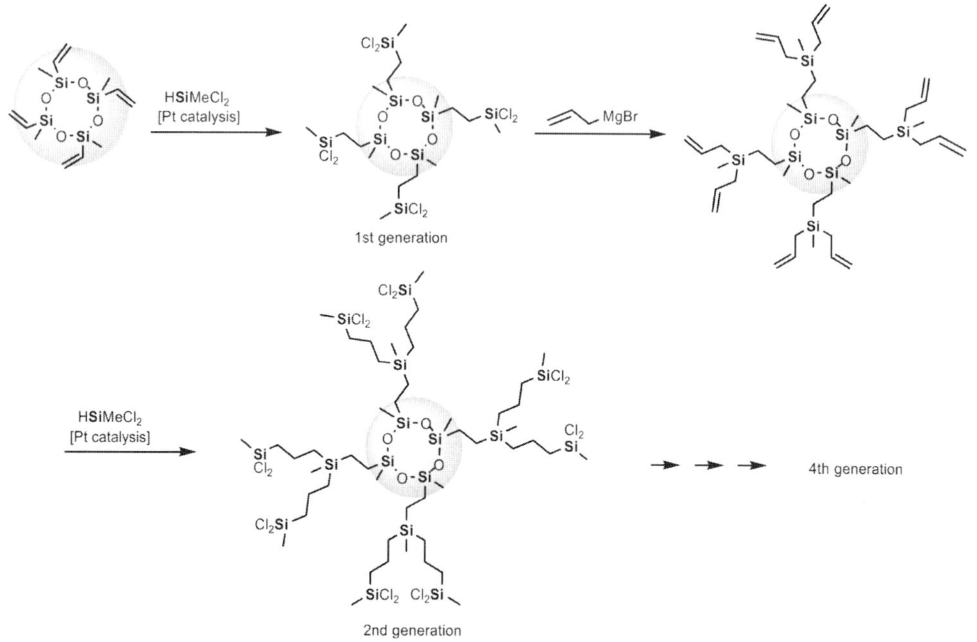

Fig. 4.47 Carbosilane with cyclic core unit (according to Kim *et al.*)

(Fig. 4.47) of up to the fourth generation with terminal chloride functions [90]. These could be transformed into the corresponding terminal functional groups, for example by reaction with *p*-phenylphenol, *p*-bromophenol, lithium phenylacetylide, or allylmagnesium bromide.

Starting from the same core unit, the terminal Si–Cl groups formed on reaction with hydrosilylation reagent could react with a hydroxyterpyridine to form up to third-generation dendrimers. After complexation with $PtCl_2$ and subsequent coordination of 2,2′:6′,2″-terpyridine, terminal SH groups were introduced which served as anchor groups for attachment of the dendrimers to gold surfaces (Au 111) [91].

Carbosiloxane dendrimers with SiH end groups (Fig. 4.48) [92] are generally accessible by hydrosilylation of $MeCOCH_2CH_2CH{=}CH_2$ with chlorosilanes to yield Si–Cl terminated compounds and subsequent reduction with lithium aluminium hydride to form the corresponding alcohol with terminal SiH groups.

Functionalisation of the carbosiloxane dendrimers is just as versatile as the above possibility of using different core units. Thus various metallo-carbosiloxane dendrimers can be constructed, for example with titanocene dichloride end groups [93], ferrocene [94], or dicobalthexacarbonyl units [95].

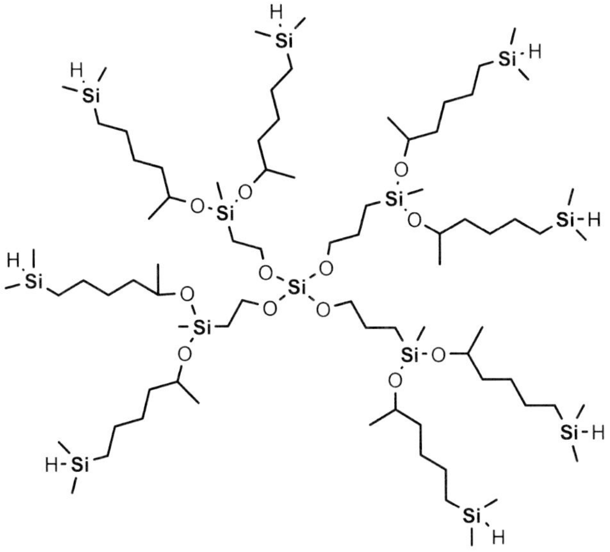

Fig. 4.48 Carbosiloxane dendrimer with terminal Si–H groups (according to Lang *et al.*)

4.1.9.4 **Siloxane dendrimers**

The longest known representatives of dendrimers containing heteroatoms – apart from the usual ones such as N, O, S, halogen – were first constructed in 1989 in the form of siloxane dendrimers [96] of up to the fourth generation [97]. This type of dendrimer is a kind of bridge between inorganic and organic chemistry. Starting from methyltrichlorosilane as trifunctional core unit, the three chlorine atoms were nucleophilically substituted by bis(ethoxy)methylsilane groups in a first (repetitive) step (Fig. 4.49). In the following, likewise repeatable, sequence the resulting ethoxy units were refunctionalised with thionyl chloride ($SOCl_2$) to form terminal chloride units. The fourth-generation dendrimer ultimately obtained contains 48 terminal chlorine substituents.

Masamune *et al.* constructed a polysiloxane dendrimer with an oligosilane core unit [98]. A repetitive synthetic strategy, consisting in catalytic oxidation of the terminal SiH to SiOH groups and substitution of the hydroxyl groups by a spacer, provided access to polysiloxane dendrimers of up to the third generation.

The bromosilane obtained by reaction of the phenylsiloxane **1** with bromine in the presence of triethylamine or sodium siloxide led Kakimoto *et al.* to the siloxane core building block **4**. It contains three phenylsilane-terminated disiloxane branching units, which should minimise steric hindrance on construction of subsequent generations. A sequence of bromination, amination, and alcoholysis ultimately leads to the third-generation polysiloxane **5** (Fig. 4.50) [99].

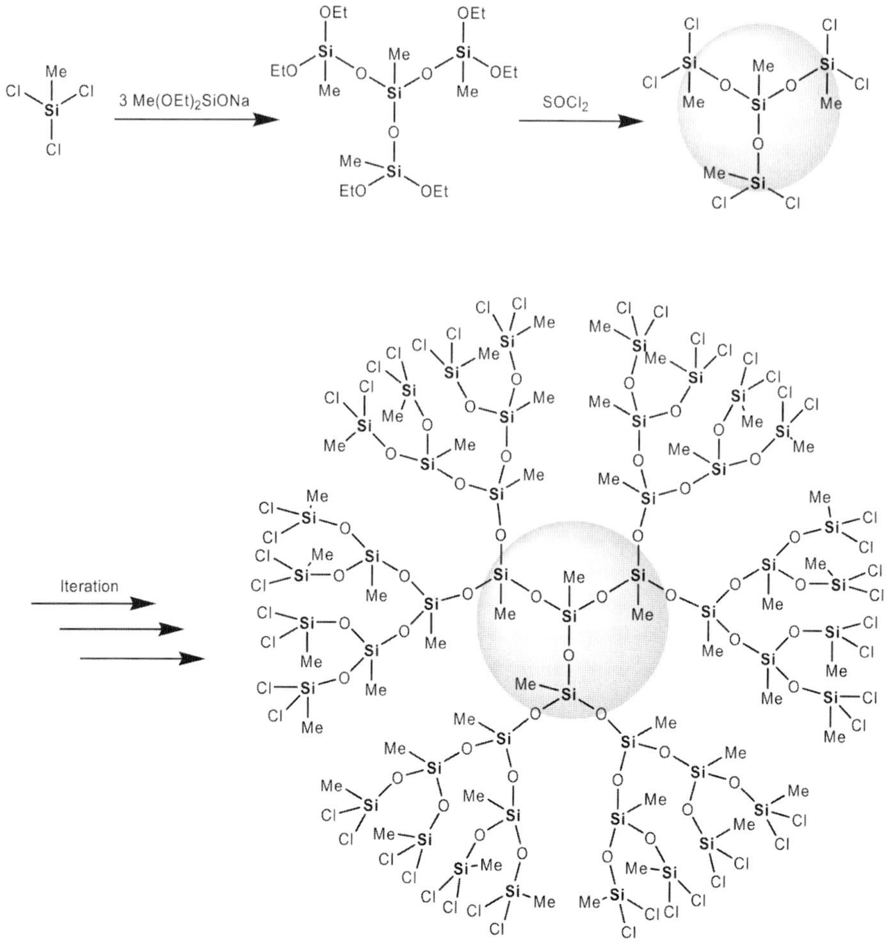

Fig. 4.49 Synthesis of fourth-generation siloxane dendrimers (according to Muzfarov *et al.*)

4.1.9.5 **Hyperbranched silicon-based polymers**

While not so structurally perfect (see Sections 1.3 and 1.4) as their dendritic relatives, these hyperbranched dendritic compounds (see Section 2.7) can be produced faster and at less expense, making them especially interesting for the materials sector. The various commercially available monomers (Fig. 4.51) offer great structural versatility.

Like monodisperse carbosiloxane dendrimers, hyperbranched carbosiloxanes can be produced by hydrosilylation [100]. The first compounds of this kind were synthesised in 1991 by Mathias and Carothers [101] by polymerisation of an allyl-tris(dimethylsiloxy)silane monomer with addition of chloroplatinic acid hydrate (Fig. 4.52). The reaction proceeds relatively fast; however, addition of more catalyst does not increase the molecular masses (19000 g/mol). The NMR spec-

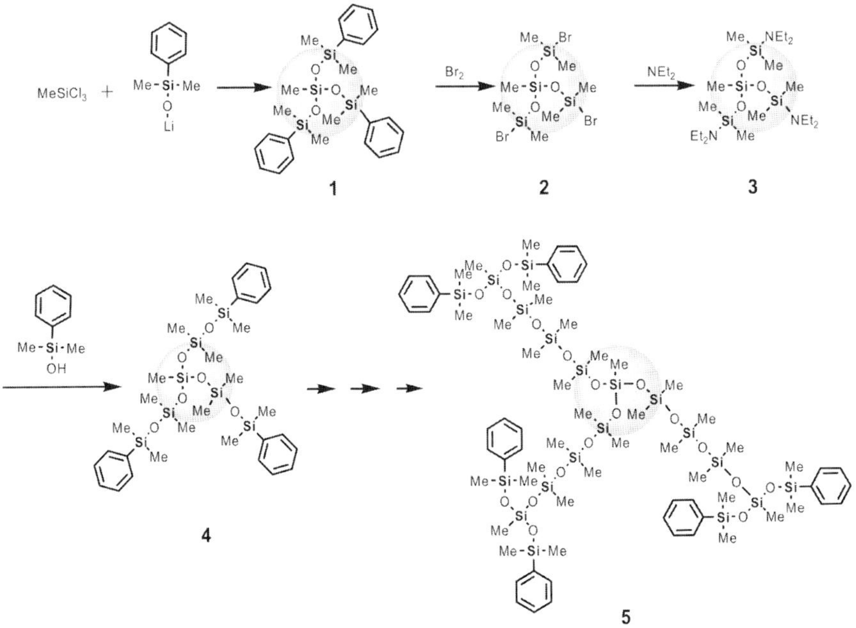

Fig. 4.50 Polysiloxane dendrimers (according to Kakimoto *et al.*)

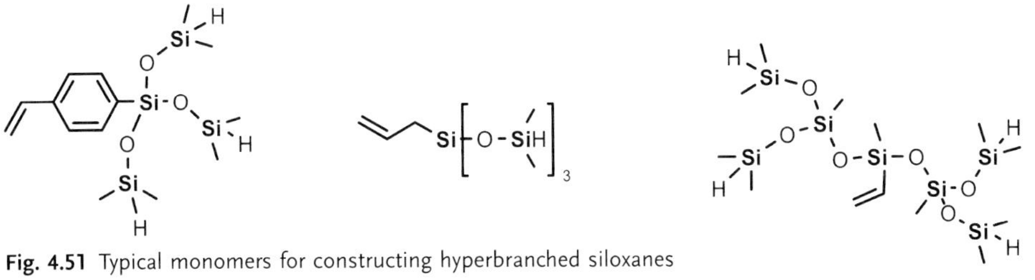

Fig. 4.51 Typical monomers for constructing hyperbranched siloxanes

tra of these hyperbranched compounds show practically no vinyl signals, indicating almost complete polymerisation of the product.

In order to lower the reactivity, the Si–H groups were hydrosilylated with allyl phenyl ethers in later work [102]. Hyperbranched polycarbosilanes (Fig. 4.53) can be accessed by Grignard coupling of chloromethyltrichorosilane [103]. The initialising step of polymerisation is the – almost quantitative – formation of the Grignard compound Cl_3SiCH_2MgCl. The polymeric product contains various structural units (linear, singly branched, doubly branched). Subsequent reduction with lithium aluminium hydride results in a highly branched polymer having an empirical formula of $(SiH_{1.85}CH_2)_n$ determined approximately by elemental analysis, ^{1}H-, ^{13}C-, ^{29}Si-NMR, IR, and GPC and a $PDI = M_w/M_n = 5200/750 = 6.9$ (see Chapter 1).

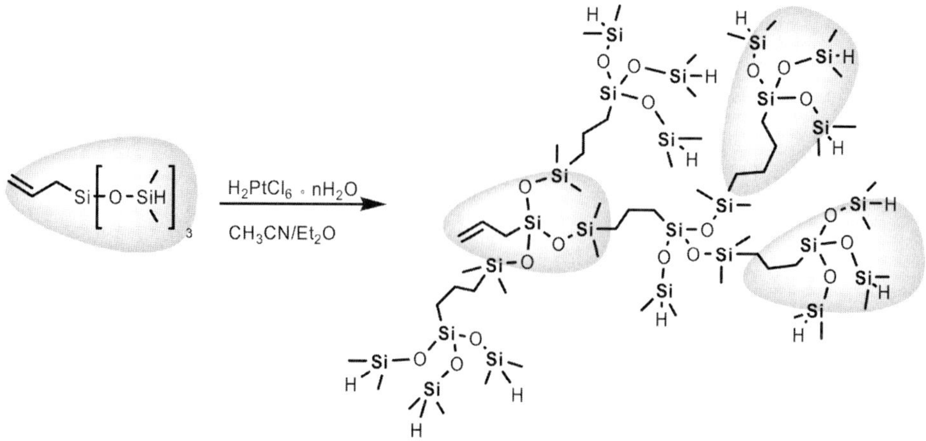

Fig. 4.52 Hyperbranched carbosiloxane (according to Mathias and Carothers). Some of the monomeric building blocks are shown with a grey background for the sake of illustration

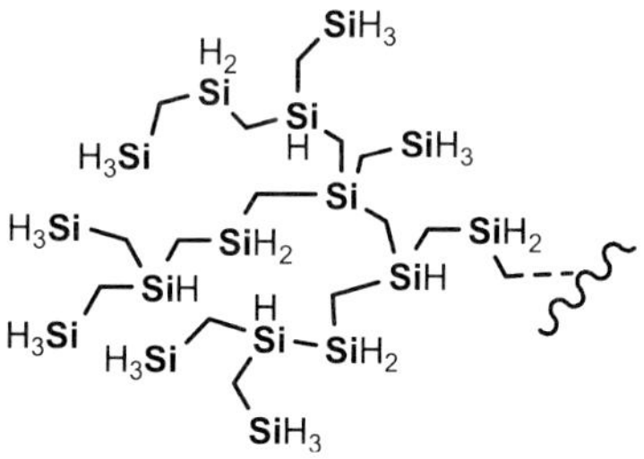

Fig. 4.53 Hyperbranched polycarbosilane (according to Interrante *et al.*)

4.1.10
Phosphorus-based dendrimers

The first neutral phosphorus-containing dendrimers ("phosphodendrimers") were described in 1994 by Majoral *et al.*, who subsequently substantially expanded the range of compounds. The phosphorus was bonded to other heteroatoms such as N and O [104] and also to carbon [105]. However, dendrimers are also known in which phosphorus atoms are located in the core unit or in the periphery, and also both in the core and at the branching points as well as the periphery [106]. An example with pentavalent phosphorus atoms in the core unit and at the branching points is shown in Fig. 4.54.

In an initial assembly step, trichlorothiophosphorus is allowed to react with the sodium salt of 4-hydroxybenzaldehyde. Subsequent activation by addition of a hydrazine derivative yields a first-generation phospho-dendrimer (Fig. 4.54). Iteration of this synthetic sequence ultimately leads to the fourth generation (molar mass 11 269 Da).

The periphery changes its functionality according to the generation (formyl groups or dichlorothiophosphoryl units). The resulting dendrimers are very

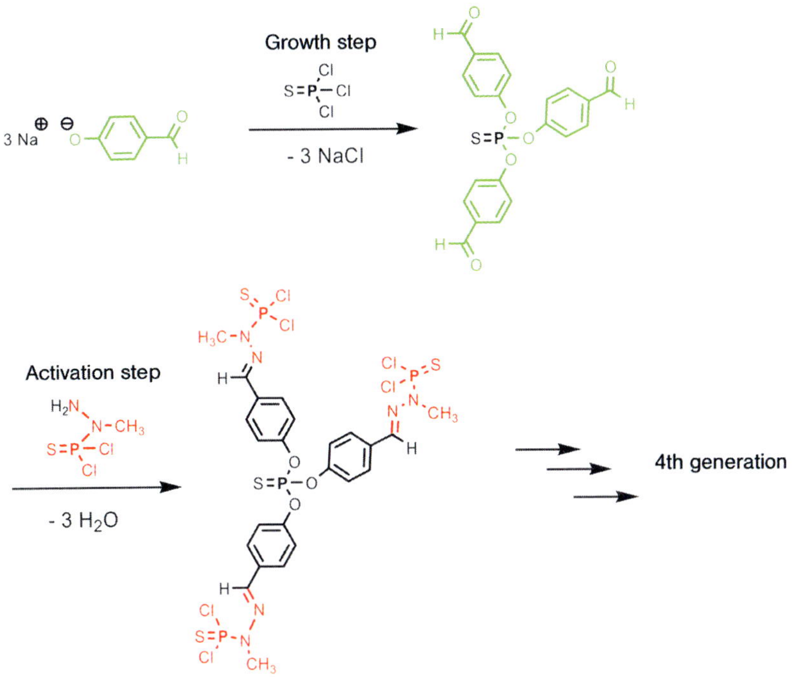

Fig. 4.54 Phosphorus-containing dendrimers (according to Majoral *et al.*)

stable and soluble in most of the common solvents. It is a distinct advantage that the phosphorus atoms of the individual generations can be distinguished on the basis of their different chemical shifts and signal intensities in the ^{31}P-NMR spectrum and precisely assigned to the individual "onion skins" (generations) of the molecule [104]. This permits straightforward assessment of the completeness (perfection) of the preceding reactions (and the purity of the intermediate generations) – at least as far as the nature and number of the participating phosphorus atoms are concerned. (The invisibility of any possible non-P-containing impurities is beneficial because the lack of signal interference simplifies interpretation; however, the drawback is that references to the dendrimer purity only apply to the immediate vicinity of the phosphorus atoms in the molecule.) MALDI-TOF mass spectroscopy was also used in addition to ^{31}P-NMR spectroscopy for characterisation of phosphorus-based dendrimers [107].

Reaction of dichlorothiophosphoryl units with allylamines gave corresponding phosphodendrimers with terminal allyl functions while reaction with propargylamine afforded molecules with terminal alkyne groups. Peripheral tri- and tetrafunctionality could thus be attained [108].

Phosphodendrimers are of relevance as "modifiers" of material surfaces in the development of DNA chips [109]. Organic phosphodendrimers also have potential applications as new gelators; in small quantities they permit gelling of organic and organometallic substances in water (hydrogels) under mild conditions [110].

Being soluble in water, such dendrimers also become attractive for medical applications because phosphorus-containing dendrimers show low toxicity. Majoral *et al.* successfully synthesised a two-photon tracer for *in vivo* imaging (similar to the contrast agent mentioned in Chapter 8). In order to avoid aggregation, a stilbene-type two-photon active lipophilic chromophore unit was incorporated into the core of a dumbbell-shaped dendrimer whose periphery is decorated with cationic groups (Fig. 4.55). In addition to endowing water solubility, the cationic shell forms a kind of protective coating for the central chromophore against external influences [111]. Other ionic phosphorus-based dendrimers can be found in Section 4.1.8.2.

Phosphodendrimers bearing peripheral λ^3-phosphorus in the form of diphenylphosphine groups can also act a ligands in palladium, platinum, and rhodium complexes [112]. The metals do not have to be located exclusively on the periphery, but can also serve as core unit, as in the case of phosphodendrimers with ferrocene as core. Electron-withdrawing P=N–P=S entities were coupled to such dendrimers – up to the fifth generation – which strongly affected the electrochemical properties of the ferrocene [113]. Phosphorus-based dendrimers containing ferrocene moieties in the core, the branching units, and the periph-

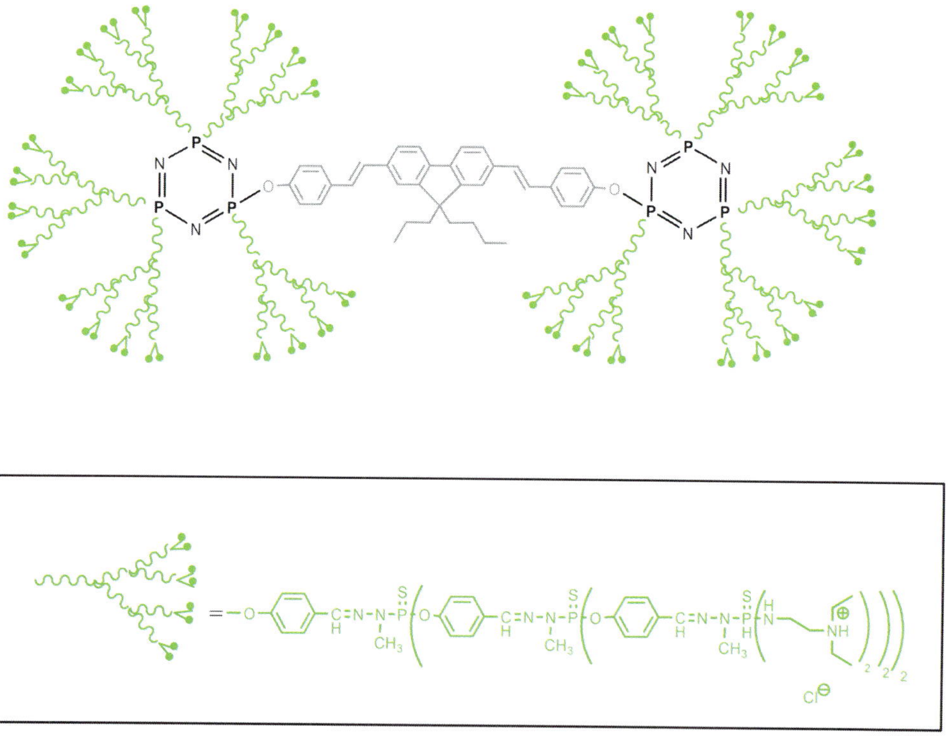

Fig. 4.55 Phosphodendrimers with stilbene units as two-photon tracers (according to Majoral *et al.*)

ery can also be prepared [114]. Polycationic phosphodendrimers should be mentioned here for the sake of completeness; more details on this topic will be found in Section 4.1.5 (ionic dendrimers) [115].

4.1.11
Metallodendrimers (and Newkome dendrimers)

On account of their physical, photo-physical, or catalytic properties, metallodendrimers have become a widespread class of compounds. Combination of the characteristics of dendrimers with those of transition metals can, for example, produce light-harvesting effects (see Section 5.2) and energy-transfer gradients.

Metallodendrimers can be of supramolecular construction [116] (see Section 2.6), formed by metal/ligand complexation (coordination chemistry), or also produced by covalent bonding.

As already described in Chapter 3, metallodendrimers can be classified as follows according to the position occupied by the metal in the dendritic molecule (Fig. 4.56):

A. *Dendrimers with metal core*
B. *Dendrimers with peripherally complexed metal*
C. *Dendrimers with metal in the dendritic branches*
D. *Dendrimers with metals at the branching points.*

A. *Dendrimers with metal core*

Synthesis of dendritic structures with a metal complex as core can proceed in two different ways:

Covalent coupling: A pre-existing metal complex is coupled covalently with dendrons and is thus surrounded by a dendritic shell. Appropriate examples are the redox-active metallodendrimers of Kaifer *et al.* [117] or the dendritic porphyrin/metal complexes prepared and characterised by Diederich *et al.* [118] and Aida *et al.* [119].

Supramolecular or coordinative [120] *complexation:* A method of preparing dendrimers with central metal complex entity accomplished by Kawa and Fréchet consists in the complexation of a metal cation with suitably functionalised dendron ligands. Self-assembly of three Fréchet-type dendrons, each with a carboxy-

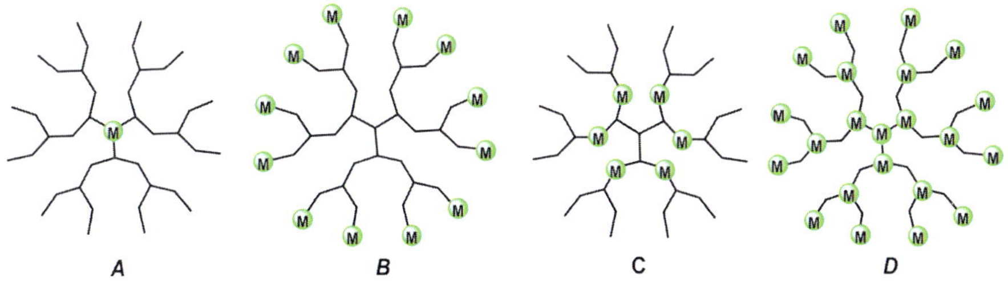

Fig. 4.56 Classification of metallodendrimers

late group at the focal point, around a trivalent lanthanide ion [erbium(III), europium(III), or terbium(III)] as core unit (Chapter 2, Fig. 2.9) yielded dendritic complexes with remarkable photo-physical properties [121]. Thus these metallodendrimers exhibit significantly stronger luminescence than the isolated (non-dendritic) lanthanide complex as a consequence of energy transfer from the aryl benzyl ether groups to the metal, and of the shielding of the metal ions from one another (see Chapter 5).

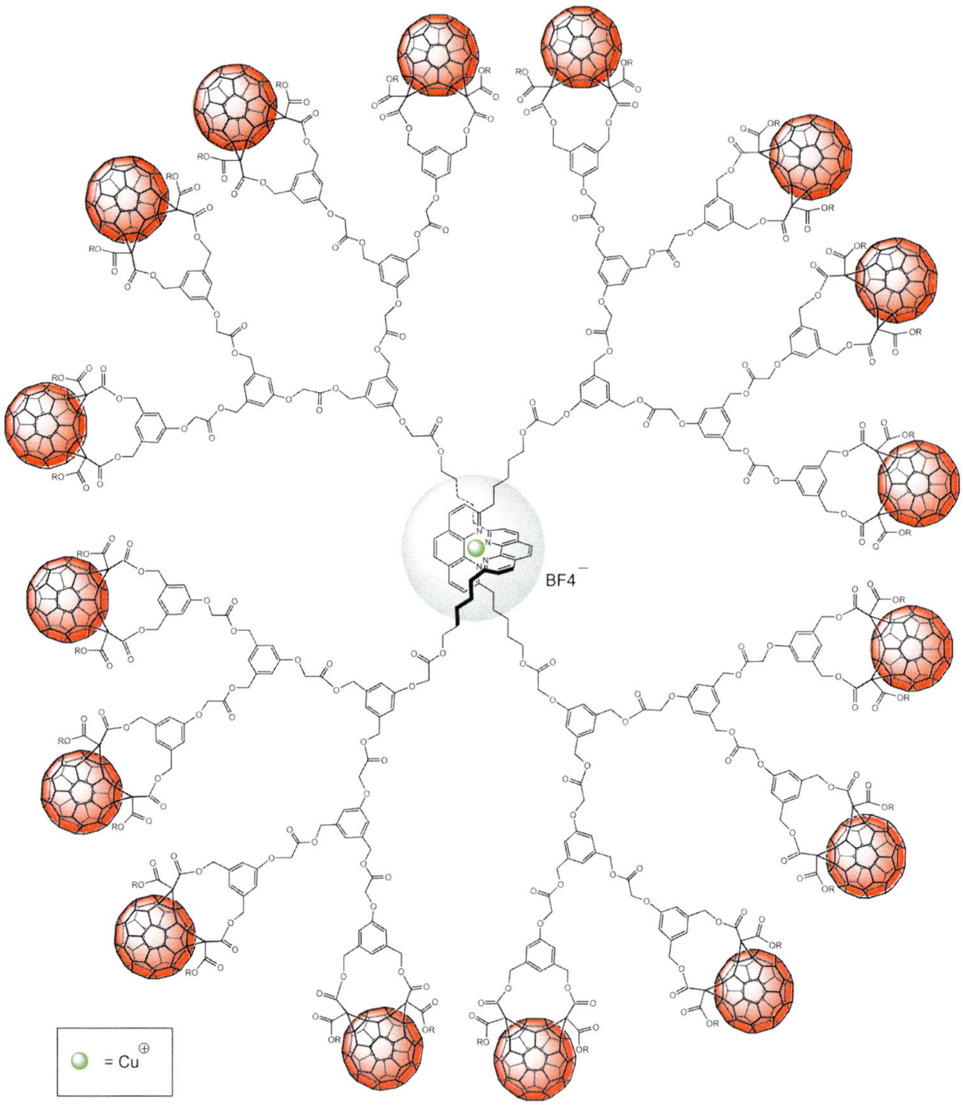

Fig. 4.57 Metallodendrimer with Cu(I)-phenanthroline core and terminal fullerenes (according to Nierengarten *et al.*)

A metallodendrimer with a copper(I) core unit coordinatively bound to phenanthroline ligands and whose dendrons bear terminal fullerene units was synthesised by Nierengarten *et al.* (Fig. 4.57) [122].

B. Dendrimers with peripherally complexed metal

Metallocenes have frequently been used as terminal moieties in dendrimer chemistry – as already demonstrated in previous sections. They are of interest primarily because of their potential application in catalysis [123]. An unusual metallodendrimer with peripheral ferrocene entities and optically active ferrocenyldiphosphine ligands (josiphos ligands) was prepared by Togni *et al.* (Fig. 4.58) [124]. Adamantanetetracarboxlic acid was one of the core units employed.

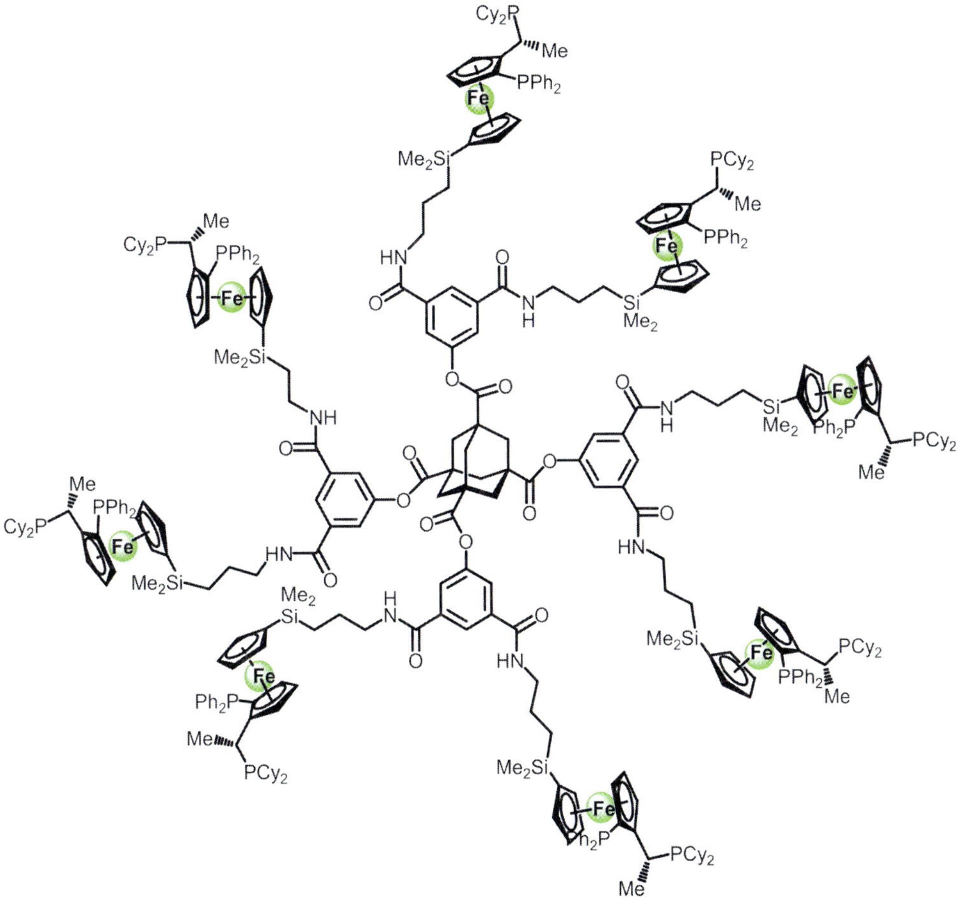

Fig. 4.58 Metallodendrimer with optically active ferrocene entities at its periphery (according to Togni *et al.*)

Integration of chiral entities, especially of already proven complex ligands in – specifically designed – dendritic molecules (see Section 4.2), extends the range of applications because asymmetrically catalysed reactions can now also be undertaken with dendrimers.

According to Reetz *et al.*, dendrimers with diaminobutyl units (DAB dendrimers) can be furnished with terminal diphosphanyl groups – for example, by double phosphanation; these end groups are predestined for complexation [125]. The bidentate dendritic N-$(CH_2PPh_2)_2$-groups acting as ligands open the way – according to van Koten *et al.* – to peripheral palladium-, nickel-, iridium-, and rhodium transition metal complexes with applications in diverse catalytic reactions (*e.g.* Heck coupling), and also as sensors for sulphur dioxide and other gases (see Chapter 8) [126].

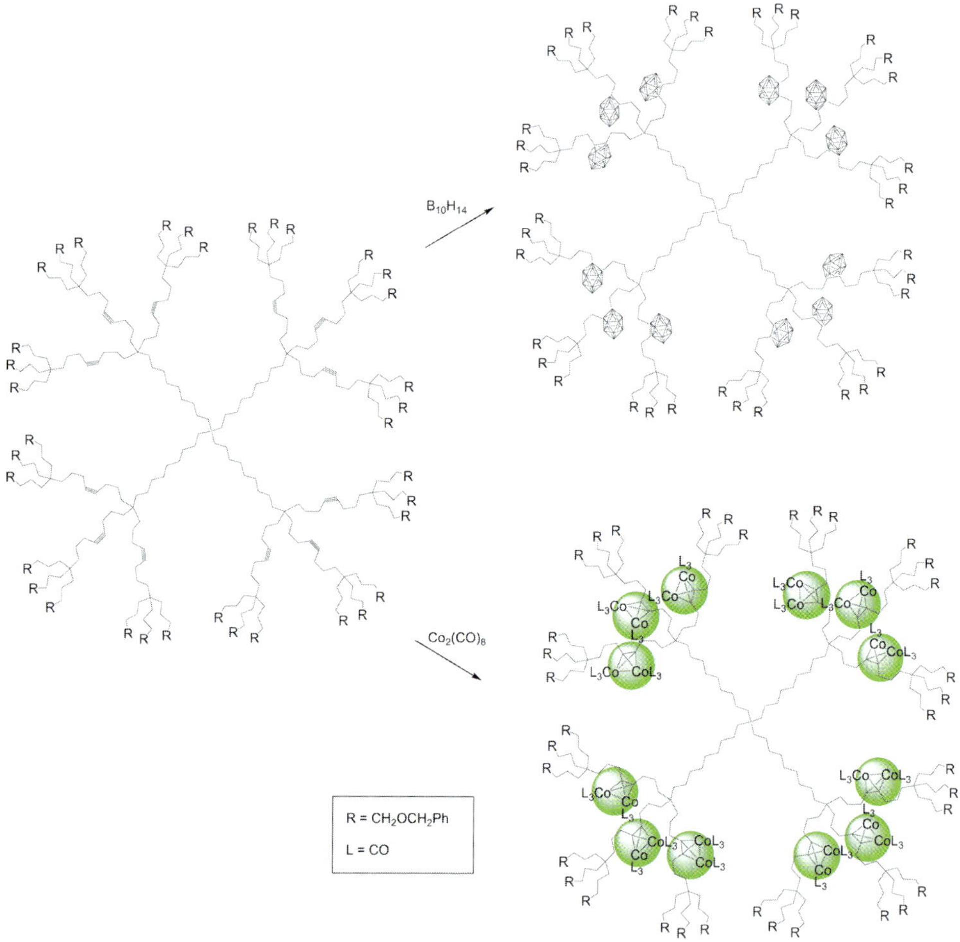

Fig. 4.59 Dendrimers with precisely inserted – interior – coordination sites (according to Newkome *et al.*)

C. Dendrimers with metal in the dendritic branches

Dendrimers with metal complex moieties in their branches require the prior incorporation of specific coordination sites into the dendrimer scaffold. Newkome *et al.* used such a dendrimer with twelve alkyne units for spot-on introduction of 1,2-dicarba-*closo*-dodecaborane groups (Fig. 4.59, above right) [127]. Moreover, on-target coordination with dicobalt-octacarbonyl to form a metallodendrimer with twelve dicobalt-hexacarbonyl units was also accomplished. These units can serve as protective groups on the one hand [128], and as catalysts on the other [129].

D. Dendrimers with metals at the branching points

A dendrimer with metal complexes both in the core and in the branches was described by Balzani *et al.* The luminescent, heteroleptic (having different ligands), dendritic polypyridine-ruthenium or polypyridine-osmium complex can be prepared both divergently and convergently [130] (cf. Section 2.5.2).

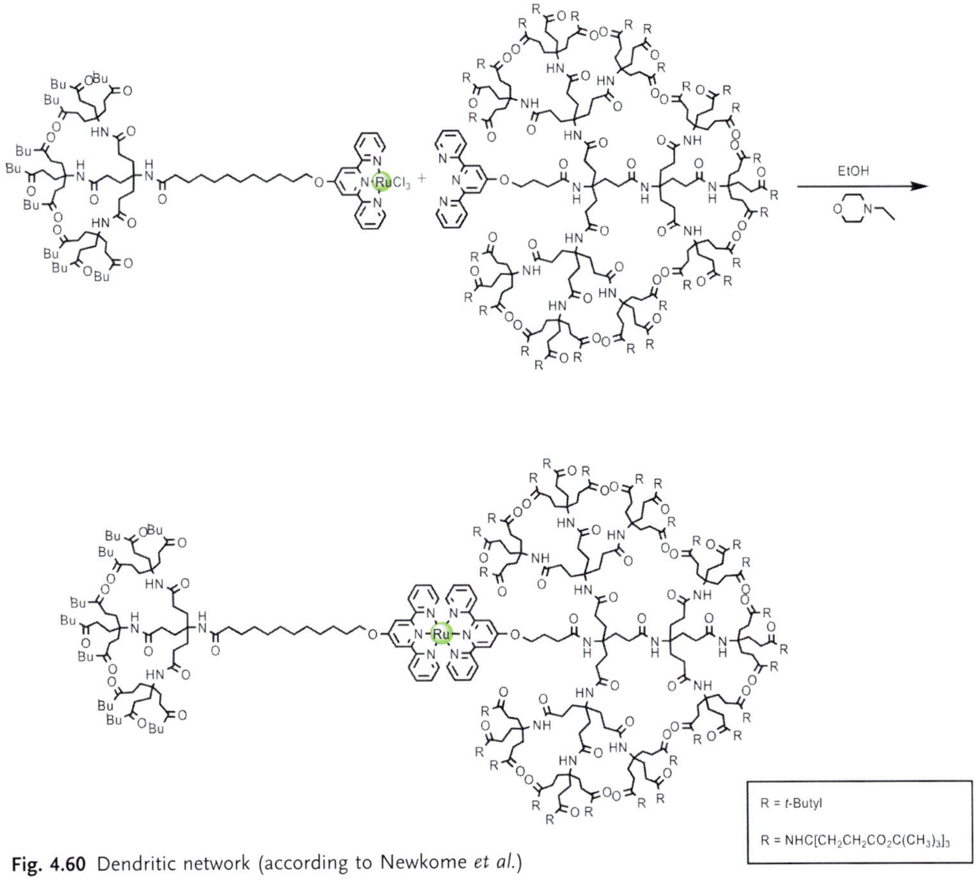

Fig. 4.60 Dendritic network (according to Newkome *et al.*)

Newkome *et al.* constructed metallodendrimers by coordination chemistry through ligand/ruthenium/ligand coupling. This synthetic concept permits linkage of preconstructed dendrons (Fig. 4.60) to form a dendritic network. The corresponding dendrons are arranged around the ruthenium by the two terpyridine ligands, thus conferring greater stability on the dendrimer – owing to the dendritic protective shell [131].

Similar dendritic architectures were formed by fixing polyether dendrons to a terpyridine [132] core unit by means of iron(II) ions [133].

Newkome-type dendrimers and dendrons with 1 → 3- and 1 → 4-branching recently became commercially available under the name "NTreon" [134a].

The present Chapter has focussed on the principal types of dendrimers and their syntheses. Further examples will be found in the cited literature [134b].

Bibliography and Notes for Section 4.1 "Achiral dendrimers"

Review articles are indicated by the words "Review(s)" or "Book(s)" in bold-faced type.

1 Astramol®; company: DSM, Eindhoven, the Netherlands.

2 J. C. Hummelen, J. L. J. van Dongen, E. W. Meijer, *Chem. Eur. J.* **1997**, *3*, 1489–1493.

3 Data at the following Internet sites: "www.dsm.com and www.dendritech.com"

4 E. Buhleier, W. Wehner, F. Vögtle, *Synthesis* **1978**, 155–158.

5 R. Moors, F. Vögtle, *Chem. Ber.* **1993**, *126*, 2133–2135.

6 M. de Brabander-van den Berg, E. W. Meijer, *Angew. Chem.* **1993**, *105*, 1370–1372; *Angew. Chem. Int. Ed.* **1993**, 32, 1308–1310; C. Wörner, R. Mülhaupt, *Angew. Chem.* **1993**, *105*, 1367–1370; *Angew. Chem. Int. Ed.* **1993**, *32*, 1306–1308.

7 a) F. Vögtle, S. Gestermann, C. Kauffmann, P. Ceroni, V. Vicinelli, B. Balzani, *J. Am. Chem. Soc.* **2000**, *122*, 10398–10404; cf. D. A. Tomalia, *Progr. Polym. Sci.* **2005**, *30*, 294–324; b) J. Aumanen, T. Kesti, V. Sundström, G. Theobaldi, F. Zerbetto, N. Werner, G. Richardt, J. van Heyst, F. Vögtle, J. Korppi-Tomolla, **2008**, in print.

8 B. González, C. M. Casado, B. Alonso, I. Cuadrado, M. Moran, Y. Wang. A. E. Kaifer, *Chem. Commun.* **1998**, 2569–2570.

9 a) B. González, B. Alonso, J. Losada, M. P. Garcia-Armada, C. M. Casado, *Organometallics* **2006**, *25*, 3558–3561; b) V. Vicinelli, P. Ceroni, M. Maestri, M. Lazzari, V. Balzani, S.-K. Lee, J. van Heyst, F. Vögtle, *Org. Biomol. Chem.* **2004**, *2*, 2207–2213.

10 Distributed by Aldrich; proprietary rights: Dendritech®, Michigan, USA.

11 D. A. Tomalia, A. M. Naylor, W. A. Goddard III, *Angew. Chem.* **1990**, *102*, 119–238; *Angew. Chem. Int. Ed.* **1990**, *29*, 138–175.

12 J. D. Eichmann, A.. U. Bielinska, J. F. Kukowska-Latallo, J. R. Baker, *Pharm. Sci. Technol. Today* **2000**, *3*, 232–245.

13 A. Archut, G. C. Azzellini, V. Balzani, L. De Cola, F. Vögtle, *J. Am. Chem. Soc.* **1998**, *120*, 12187–12191; M. W. P. L. Baars, R. Kleppinger, M. H. J. Koch, S.-L. Yeu, E. W. Meijer, *Angew. Chem.* **2000**, *112*, 1341–1344; *Angew. Chem. Int. Ed.* **2000**, *39*, 1285–1288.

14 D. A. Tomalia, H. Baker, J. R. Dewald, M. Hall, G. Kallos, S. Martin, J. Roeck, J. Ryder, P. Smith, *Polym. J.* **1985**, *17*, 117–132; D. A. Tomalia, H. Baker, J. R. Dewald, M. Hall, G. Kallos, S. Martin, J. Roeck, J. Ryder, P. Smith, *Macromolecules* **1986**, *16*, 2466–2468.

15 J. W. Lee, B.-K. Kim, H. J. Kim, S. C. Han, W. S. Shin, S.-H. Jin, *Macromolecules* **2006**, *39*, 2418–2422.

16 Y. Haba, C. Kojima, A. Harada, K. Kono, *Angew. Chem.* **2007**, *119*, 238–241; *Angew. Chem. Int. Ed.* **2007**, *46*, 234–237.

17 I. J. Majoros, C. B. Metha, J. R. Baker Jr., Mathematical Description of Dendrimer Structure; http://www.forsight.org/conferences/MNT8/abstractIMajoros/index.html
18 I. J. Majoros, L. Piechler, D. A. Tomalia, J. R. Baker Jr., Synthesis and Characterisation of Novel POPAM-PAMAM (POMAM) Hybrid Dendrimers as a Building Block in Nanotechnology. http://www.forsight.org/conferences/MNT8/abstract/Baker/index.html
19 J. Friedhofen, *PhD Thesis*, Kekulé Institute of the University of Bonn, **2006**: "http://hss.ulb.uni-bonn.de/diss_online"
20 B. Qualmann, M. M. Kessels, H.-J. Musiol, W. D. Sierralta, P. W. Jungblut, L. Moroder, *Angew. Chem.* **1996**, *108*, 970–973; *Angew. Chem. Int. Ed.* **1996**, *35*, 909–911.
21 R. G. Denkewalter, J. F. Kolc, W. J. Lukasavage, *US Pat. 4360646*, **1979**; R. G. Denkewalter, J. F. Kolc, W. J. Lukasavage, *US Pat.* 4289872, **1981**; R. G. Denkewalter, J. F. Kolc, W. J. Lukasavage, *US Pat. 4410688*, **1983**.
22 a) V. Balzani, P. Ceroni, S. Gestermann, M. Gorka, C. Kauffmann, F. Vögtle, *J. Chem. Soc., Dalton Trans.* **2000**, 3765–3771; D. Parker, *Coord. Chem. Rev.* **2000**, *205*, 109. V. Vicinelli, P. Ceroni, M. Maestri, V. Balzani, M. Gorka, F. Vögtle, *J. Am. Chem. Soc.* **2002**, *124*, 6461–6468; b) F. Vögtle, M. Gorka, V. Vicinelli, P. Ceroni, M. Maestri, V. Balzani, *Chem. Phys. Chem.* **2001**, *12*, 769–773; c) C. Giasante, P. Ceroni, V. Balzani, F. Vögtle, *Angew. Chem.* **2008**, *120*, 5502–5505; *Angew. Chem. Int. Ed.* **2008**, *47*, 5422–5425; d) M. Claeys-Bruno, O. Vandenabeele-Trambouze, M. Sergent, M. Geffart, D. Bodet, M. Dobrijevic, A. Commeyras, R. Phan Tan Luu, *Chemometrics and Intelligent Laboratory Systems* **2006**, *80*, 186–197; e) Commercially available Dendrigraft polylysines: COLCOM, Cap Alpha, Avenue de l'Europe-Clapiers, 34940 Montpellier Cedex 9, France; cf. I. Tsogas, Th. Theodossiou, Z. Sideratou, C.M. Paleos, H. Collet, J.C. Rossi, B. Romestand, A. Commeyras, *Biomacromolecules* **2007**, *8*, 3263–3270.
23 ***Review***: H. Hart, *Pure Appl. Chem.* **1993**, *65*, 27–34; H. Hart, *J. Org. Chem.* **1991**, *56*, 6905–6912; K. Shahlai, H. Hart, *J. Am. Chem. Soc.* **1990**, *112*, 3687–3688.
24 ***Book***: J. K. Young, J. S. Moore, (Eds. P. J. Stang, F. Diederich), *Modern Acetylene Chemistry*, VCH, Weinheim, New York **1995**; Z. Xu, J. S. Moore, *Angew. Chem.* **1993**, *105*, 261–264; *Angew. Chem. Int. Ed.* **1993**, *32*, 1354–1357; rigid macrocycles, see also: W. Zhang, J. S. Moore, *Angew. Chem.* **2006**, *118*, 4524–4548; *Angew. Chem. Int. Ed.* **2006**, *45*, 4416–4439; S. Höger, *Angew. Chem.* **2005**, *117*, 3872–3875; *Angew. Chem. Int. Ed.* **2005**, *44*, 3806–3808; G.-B. Pan, X.-H. Cheng, S. Höger, W. Freyland, *J. Am. Chem. Soc.* **2006**, *128*, 4218–4219; S. Klyatskaya, N. Dingenouts, C. Rosenauer, B. Müller, S. Höger, *J. Am. Chem. Soc.* **2006**, *128*, 3150–3151.
25 Z. Xu, J. S. Moore, *Acta Polymer* **1994**, *45*, 83–87.
26 J. H. Friedhofen, F. Vögtle, *New J. Chem.* **2006**, *30*, 32–43.
27 V. Percec, J. G. Rudick, M. Peterca, S. R. Staley, M. Wagner, M. Obata, C. M. Mitchell, W.-D. Cho, V. S. K. Balagurusamy, J. N. Lowe, M. Glodde, O. Weichold, K. J. Chung, N. Ghionni, S. N. Magonov, P. A. Heiney, *Chem. Eur. J.* **2006**, *12*, 5731–5746; V. Percec, A. E. Dulcey, M. Peterca, M. Ilies, S. Nummelin, M. J. Sienkowska, P. A. Heiney, *Proc. Nat. Acad. Sci.* **2006**, *103*, 2518–2523.
28 S. A. Soomro, R. Benmouna, R. Berger, H. Meier, *Eur. J. Org. Chem.* **2005**, 3586–3593.
29 T. M. Miller, T. X. Neenan, *Chem Mater.* **1990**, *2*, 346–349. T. M. Miller, T. X. Neenan, R. Zayas, H. E. Bair, *J. Am. Chem. Soc.* **1992**, *114*, 1018–1025. We use the family name polybenzene dendrimers (generally polyarene dendrimers) instead of "polyphenylene dendrimers" because the rules of nomenclature restrict phenylene to $-C_6H_6-$.
30 ***Review***: U.-M. Wiesler, T. Weil, K. Müllen, *Top. Curr. Chem.* (Volume Ed. F. Vögtle) **2000**, *212*, 1–41.
31 F. Morgenroth, K. Müllen, *Tetrahedron* **1997**, *53*, 15349–15366.
32 U.-M. Wiesler, A. J. Berresheim, F. Morgenroth, G. Lieser, K. Müllen, *Macromolecules* **2001**, *34*, 187–199.
33 E. V. Andreitchenko, C. G. Clark, Jr., R. E. Bauer, G. Lieser, K. Müllen, *Angew.*

Chem. **2005**, *117*, 6506–6512; *Angew. Chem. Int. Ed.* **2005**, *44*, 6348–6354.

34 a) U.-M. Wiesler, K. Müllen, *Chem. Commun.* **1999**, *22*, 2293–2294; b) U.-M. Wiesler, *PhD Thesis*, Johannes Gutenberg University, Mainz **2001**, C. Hampel, *PhD Thesis*, Johannes Gutenberg University, Mainz **2001**.

35 F. Morgenroth, E. Reuther K. Müllen, *Angew. Chem.* **1997**, *109*, 647–649; *Angew. Chem. Int. Ed.* **1997**, *36*, 631–634; J. Schmidt, W. Pisula, D. Sebastiani, K. Müllen, *J. Am. Chem. Soc.* **2006**, *128*, 9526–9534; R. E. Bauer, V. Enkelmann, U. M. Wiesler, A. J. Berresheim, K. Müllen, *Chem. Eur. J.* **2002**, *8*, 3858–3864.

36 A. Herrmann, G. Mihov, G. W. M. Vandermeulen, H.-A. Klok, K. Müllen, *Tetrahedron* **2003**, *59*, 3925–3935.

37 H. Meier, M. Lehmann, *Angew. Chem.* **1998**, *110*, 666–669; *Angew. Chem. Int. Ed.* **1998**, *37*, 643–645; S. A. Soomro, R. Benmouna, R. Berger, H. Meier, *Eur. J. Org. Chem.* **2005**, 3586–3593, M. Lehmann, C. Köhn, H. Meier, S. Renker, A. Oehlhof, *J. Mater. Chem.* **2006**, *16*, 441–451. ***Review***: H. Meier, M. Lehmann, in H. S. Nalwa: *Encyclopedia of Nanoscience and Nanotechnology* **2004**, *10*, 95–105.

38 ***Review***: V. Vicinelli, P. Ceroni, M. Maestri, M. Lazzari, V. Balzani, S.-K. Lee, J. van Heyst, F. Vögtle, *Org. Biomol. Chem.* **2004**, 2, 2207–2213.

39 Y. H. Kim, O. W. Webster, *J. Am. Chem. Soc.* **1990**, 112, 4592–4593.

40 J. K. Stille, F. W. Harris, R. O. Rakutis, H. Mukamal, *J. Polym. Sci., Part B* **1966**, *4*, 791–793; H. Mukamal, F. W. Harris, J. K. Stille, *J. Polym. Sci., Part A* **1967**, *5*, 2721.

41 T. Horn, S. Wegner, K. Müllen, *Macromol. Chem. Phys.* **1995**, *196*, 2463–2474; A. Müller, R. Stadler, *Macromol. Chem. Phys.* **1996**, *197*, 1373–1385; A. D. Schlüter, *Polym. Preprints* **1995**, *36*, 592.

42 C. Hawker, J. M. J. Fréchet, *J. Chem. Soc., Chem. Commun.* **1990**, 1010–1013; C. J. Hawker, J. M. J. Fréchet, *J Am. Chem. Soc.* **1990**, *112*, 7638–7647.

43 K. L. Wooley, C. J. Hawker, J. M. J. Fréchet, *J. Am. Chem. Soc.* **1991**, *113*, 4252–4261.

44 A. Klaikherd, B. S. Sandanaraj, D. R. Vutukuri, S. Thayumanavan, *J. Am. Chem. Soc.* **2006**, 128, 9231–9237; C. Akpo, E. Weber, J. Reiche, *New J. Chem.* **2006**, *30*, 1820–1833.

45 ***Review***: S. Nummelin, M. Skrifvars, K. Rissanen, *Top. Curr. Chem.* (Volume Ed. F. Vögtle) **2000**, *210*, 8–38.

46 E. Gilles, J. M. J. Fréchet, *J. Am. Chem. Soc.* **2002**, 124, 14137–14146.

47 ***Review***: N. Röckendorf, T. K. Lindhorst, *Top. Curr. Chem.* (Volume Eds. F. Vögtle, C. A. Schalley) **2002**, *217*, 201–241.

48 P. R. Ashton, V. Balzani, M. Clemente-Leon, B. Colonna, A. Credi, N. Jayaraman, F. M. Raymo, J. F. Stoddart, M. Venturi, *Chemistry* **2002**, *8*, 673–684.

49 R. Roy, D. Zanini, J. Meunier, A. Romanowska, *J. Chem. Soc., Chem. Commun.* **1993**, 1869–1890; R. Roy, *Polymer News* **1996**, *21*, 226–232.

50 F. Bambino, R. T. Brownlee, F. C. Chiu, *Tetrahedron Lett.* **1994**, *35*, 4619–4622.

51 P. R. Ashton, S. E. Boyd, C. L. Brown, N. Jayaraman, S. A. Nepogodiev, J. F. Stoddart, *Chem. Eur. J.* **1996**, *2*, 1115–1128.

52 K. Sadalapure, T. K. Lindhorst, Angew. *Chem.* **2000**, *112*, 2066–2069; *Angew. Chem. Int. Ed.* **2000**, *29*, 2010–2013.

53 L. V. Backinowsky, P. I. Abronina, A. S. Shashkov, A. A. Grachev, N. K. Kocetkov, S. A. Nepogodiev, J. F. Stoddart, *Chem. Eur. J.* **2002**, *8*, 4412–4423.

54 M. M. K. Boysen, T. K. Lindhorst, *Org. Lett.* **1999**, 1925–1927.

55 M. Dubber, T. K. Lindhorst, *Chem. Commun.* **1998**, 1265–1266.

56 K. I. Sugiura, H. Tanaka, T. Matsumoto, T. Kawai, Y. Sakata, *Chem. Lett.* **1999**, 1193–1194.

57 H. F. Chow, T. K.-K. Mong, M. F. Nongrum, C.-W. Wan, *Tetrahedron* **1998**, *54*, 8543–8660.

58 T. Dwars, E. Paetzold, G. Oehme, *Angew. Chem.* **2005**, *117*, 7338–7364; *Angew. Chem. Int. Ed.* **2005**, *44*, 7174–7199.

59 G. R. Newkome, C. N. Moorefield, G. R. Baker, M. J. Saunders, S. H. Grossman, *Angew. Chem.* **1991**, *103*, 1207–1209; *Angew. Chem. Int. Ed.* **1991**, *30*, 1178–1180.

60 C. J. Hawker, K. L. Wooley, J. M. J. Fréchet, *J. Chem. Soc. Perkin Trans 1* **1993**, 1287–1297.

61 M. Braun, S. Atalick, D. M. Guldi, H. Lang, M. Brettreich, S. Burghardt, M.

Hatzimarinaki, E. Ravanelli, M. Prato, R. van Eldik, A. Hirsch, *Chem. Eur. J.* **2003**, *9*, 3867–3875.

62 Y. H. Kim, O. W. Webster, *J. Am. Chem. Soc.* **1990**, *112*, 4592–4593.

63 C. Loup, M.-A. Zanta, A.-M. Caminade, J.-P. Majoral, B. Meunier, *Chem. Eur. J.* **1999**, *5*, 3644–3650.

64 "Hatschepsut", *Geo Magazin* **2002**, 07.

65 J. F. Stoddart, P. R. Ashton, K. Shibata, A. N. Shipway, *Angew. Chem.* **1997**, *109*, 2902–2905; *Angew. Chem. Int. Ed.* **1997**, *36*, 2781–2783.

66 C. Larre, A.-M. Caminade, J.-P. Majoral, *Angew. Chem.* **1997**, *109*, 614–617; *Angew. Chem. Int. Ed.* **1997**, *36*, 596–599.

67 A. W. Kleij, R. van de Coevering, R. J. M. Klein Gebbink, A.-M. Noordman, A. L. Spek, G. van Koten, *Chem. Eur. J.* **2001**, *7*, 181–192.

68 ***Book/Reviews***: D. Y. San in *"The Chemistry of Organic Silicon Compounds"*, Z. Rappoport, Y. Apeloig (Eds.), Vol. 3, Chapter 13, 745–803, Wiley, New York **2001**; J.-M. Majoral, A.-M. Caminade, *Chem. Rev.* **1999**, *99*, 845–880.

69 J. B. Lambert, J. L. Pflug, C. L. Stern, *Angew. Chem.* **1995**, *107*, 106–108; *Angew. Chem. Int. Ed.* **1995**, *34*, 98–99.

70 H. Suzuki, Y. Kimata, S. Satoh, A. Kuriyama, *Chem. Lett.* **1995**, *24*, 293–294.

71 J. B. Lambert, J. L. Pflug, J. M. Denari, *Organometallics* **1996**, *15*, 615–625.

72 ***Review***: M. Nanjo, A. Sekiguchi, C. Kuboto, H. Sakurai, *The Sendai International Symposium on the Frontiers of Organosilicon Chemistry*, Sendai, Japan **1994**; A. Sekiguchi, M. Nanjo, C. Kubuto, H. Sakurai, *J. Am. Chem. Soc.* **1995**, *117*, 4195–4196; D. Gudat, *Angew. Chem.* **1997**, *109*, 2039–2043; *Angew. Chem. Int. Ed.* **1997**, *36*, 1951–1955.

73 J. B. Lambert, H. Wu, *Organometallics* **1998**, *17*, 4904–4909.

74 W. Uhlig, *Z. Naturforsch.* **2003**, *58b*, 183–190.

75 C. Schlenk, H. Frey, *Monatsh. Chem.* **1999**, *130*, 3–14.

76 a) A. W. van der Made, P. W. N. M. van Leeuwen, *J. Chem. Soc., Chem. Commun.* **1992**, 1400–1401; b) A. W. van der Made, P. W. N. M. van Leeuwen, J. C. de Wilde, R. A. C. Brandes, *Adv. Mater.* **1993**, *5*, 466–468.

77 L.-L. Zhou, J. Roovers, *Macromolecules* **1993**, *26*, 963–968.

78 A. M. Muzafarov, E. A. Rebrov, V. S. Papkov, *Usp. Kim.* **1991**, *60*, 1596–1612; G. M. Ignat'eva, E. A. Rebrov, V. D. Myakushev, T. B. Chenskaya, A. M. Muzafarov, *Polym. Sci. Ser. A* **1997**, *39*, 843.

79 D. Seyferth, D. Y. Son, A. L. Rheingold, R. L. Ostrander, *Organometallics* **1994**, *13*, 2682–2690.

80 T. Matsuo, K. Uchida, A. Sekiguchi, *Chem. Commun.* **1999**, 1799–1800.

81 A. Tuchbreiter, H. Wemer, L. Gade, *Dalton Trans.* **2005**, 8, 1394–1402.

82 M. M. K. Boysen, T. K. Lindhorst, *Tetrahedron* **2003**, 59, 3895–3898.

83 Initial work on saccharide-substituted carbosilane dendrimers: K. Matsuoka, M. Terabatake, Y. Esumi, D. Terunuma, H. Kuzuhara, *Tetrahedron Lett.* **1999**, *40*, 7839–7842; K. Matsuoka, H. Kurosawa, Y. Esumi, D. Terunuma, H. Kuzuhara, *Carbohydr. Res.* **2000**, *329*, 765–772; K. Matsuoka, H. Oka, T. Koyma, Y. Esumi, D. Terunuma, *Tetrahedron Lett.* **2001**, *42*, 3327–3330; T. Mori, K. Hatano, K. Matsuoka Y. Esumi, E. J. Toone, D. Terunuma, *Tetrahedron* **2005**, *61*, 2751–2760.

84 A. Yamada, K. Hatano, K. Matsuoka, T. Koyama, Y. Esumi, H. Koshino, K. Hino, K. Nishikawa, Y. Natori, D. Terunuma, *Tetrahedron* **2006**, *62*, 5074–5083.

85 J. Nakayama, J. S. Lin, *Tetrahedron Lett.* **1997**, *38*, 6043–6046.

86 R. Buschbeck, H. Lang, *Organomet. Chem.* **2005**, *690*, 696–703.

87 D. de Groot, J. N. H. Reek, P. C. J. Kamer, P. W. N. M. van Leeuwen, *Eur. J. Org. Chem.* **2002**, 1085–1095.

88 H. Lang, B. Lühmann, *Adv. Mater.* **2001**, *13*, 1523–1540.

89 R. Buschbeck, K. Brüning, H. Lang, *Synthesis* **2001**, *15*, 2289–2298.

90 C. Kim, E. Park, *J. Korean Chem. Soc.* **1998**, *42*, 277.

91 K.-H. Jung, H.-K. Shin, C. Kim, Y.-S. Kwon, *Materials Science & Engineering*, C: *Biomimetic and* Supramolecular *Systems* **2004**, *C24(1–2)*, 177–180; C. Kim, S. Son, *J. Organomet. Chem.* **2000**, *599*, 123–127.

92 R. Buschbeck, H. Sachse, H. Lang, *Organomet. Chem.* **2005**, *690*, 751–763.

93 R. Buschbeck, H. Lang, *Organomet. Chem.* **2005**, *690*, 1198–1204.

94 *Book*: A. Mutluay, P. Jutzi, in N. Auner, J. Weis (Eds.), *Organosilicon Chemistry. From Molecules* to *Materials*, Wiley-VCH, 1st Edit., Weinheim **2000**, Vol. 4, 531.

95 K. Brüning, B. Lühmann, H. Lang, *Z. Naturforsch.* **1999**, *54b*, 751–756.

96 R. Bischoff, S. E. Cray, *Progr. Polym. Sci.* **1999**, *24*, 185–219. ***Reviews***: J.-M. Majoral, A.-M. Caminade, *Chem. Rev.* **1999**, *99*, 845–880; H. Lang, B. Lühmann, *Adv. Mater.* **2001**, *13*, 1523–1540.

97 E. A. Rebrov, A. M. Muzafarov, V. S. Papkov, A. A. Zhdanov, *Dokl. Akad. Nauk. SSSR* **1989**, *309*, 376–380.

98 H. Uchida, Y. Kabe, K. Yoshino, A. Kawamata, T. Tsumuraya, S. Masamune, *J. Am. Chem. Soc.* **1990**, *112*, 7077–7079.

99 A. Morikawa, M. Kakimoto, Y. Imai, *Macromolecules* **1991**, *24*, 3469–3474.

100 L. J. Mathias, T. W. Carothers, *J. Am. Chem. Soc.* **1991**, *113*, 4043–4044.

101 ***Review***: H. Frey, C. Schlenk, in *Top. Curr. Chem.* (Volume Ed. F. Vögtle) **2000**, *210*, 69–129.

102 L. J. Mathias, T. W. Carothers, R. M. Bozen, *Polym. Prepr. Am. Chem. Soc. Div. Polym. Chem.* **1991**, *32*, 633–639.

103 C. K. Whitmarsh, L. V. Interrante, *Organometallics* **1991**, *10*, 1336–1344.

104 N. Launay, A.-M Caminade, R. Lahana, J.-P. Majoral, *Angew. Chem.* **1994**, *106*, 1682–1684; *Angew. Chem. Int. Ed.* **1994**, *33*, 1589–1592.

105 A. Miedaner, C. J. Curtis, R. M. Barkley, D. L. DuBois, *Inorg. Chem.* **1994**, *33*, 5482–5490.

106 ***Reviews***: J.-M. Majoral, A.-M. Caminade, *Chem. Rev.* **1999**, *99*, 845–880; J.-P. Majoral, A.-M. Caminade, *Top. Curr. Chem.* (Volume Ed. F. Vögtle) **1998**, *197*, 79–124; J.-M. Majoral, A.-M. Caminade, R. Laurent, P. Sutra, *Heteroatom Chemistry* **2002**, *13*, 474–485.

107 J.-C. Blais, C.-O. Turrin, A.-M. Caminade, J.-P. Majoral, *Anal. Chem.* **2000**, *72*, 5097–5105.

108 M.-L. Lartigue, M. Slany, A.-M. Caminade, J.-P. Majoral, *Chem. Eur. J.* **1996**, *2*, 1417–1426.

109 ***Review***: A.-M. Caminade, J.-P. Majoral, *Acc. Chem. Res.* **2004**, *37*, 341–348; J.-M. Majoral, C.-O. Turrin, R. Laurent, A.-M. Caminade, *Macromolecular Symposia*, **2005**, *229*, 1–7.

110 C. Marmillon, F. Gauffre, T. Gulik-Krzywicki, C. Loup, A.-M. Caminade, J.-P. Majoral, J.-P. Vors, E. Rump, *Angew. Chem.* **2001**, *113*, 2696–2699; *Angew. Chem. Int. Ed.* **2001**, *40*, 2626–2629.

111 T. R. Krishna, M. Parent, M. H. V. Werts, L. Moreaux, S. Gmouh, S. Chapak, A.-M. Caminade, J.-P. Majoral, M. Blanchard-Desce, *Angew. Chem.* **2006**, *118*, 4761–4764; *Angew. Chem. Int. Ed.* **2006**, *45*, 4645–4648.

112 C.-O. Turrin, B. Donnadieu, A.-M. Caminade, J.-P. Majoral, *Z. Anorg. All. Chem.* **2005**, *631*, 2881–2887; C.-O. Turrin, J. Chiffre, D. de Montauzon, G. Balavoine, E. Manoury, A.-M. Caminade, J.-P. Majoral, *Organometallics* **2002**, *21*, 1891–1897.

113 M. Bardaji, M. Kustos, A.-M. Caminade, J.-P. Majoral, B. Chaudret, *Organometallics* **1997**, *16*, 403–410.

114 C.-O. Turrin, J. Chiffre, D. de Montauzon, J.-C. Daran, A.-M. Caminade, E. Manoury, G. Balavoine, J.-P. Majoral, *Macromolecules* **2000**, *33*, 7328–7336.

115 C. Loup, M.-A. Zanta, A.-M. Caminade, J.-P. Majoral, B. Meunier, *Chem. Eur. J.* **1999**, *5*, 3644–3650.

116 ***Review***: H.-J. van Manen, F. C. J. M. van Veggel, D. N. Reinhoudt, *Top. Curr. Chem.* (Volume Eds. F. Vögtle, C. A. Schalley) **2002**, *217*, 121–162.

117 C. M. Cardona, A. E. Kaifer, *J. Am. Chem. Soc.* **1998**, *120*, 4023–4024.

118 a) P. J. Dandliker, F. Diederich, M. Gross, C. B. Knobler, A. Louti, E. M. Sanford, *Angew. Chem.* **1994**, *106*, 1821–1824; *Angew. Chem. Int. Ed.* **1994**, *33*, 1739–1742; b) P. J. Dandliker, F. Diederich, J. P. Gisselbrecht, A. Louati, M. Gross, *Angew. Chem.* **1995**, *107*, 2906–2909; *Angew. Chem. Int. Ed.* **1995**, 34, 2725–2728.

119 Y. Tomoyose, D. L. Jiang, R. H. Jin, T. Aida, T. Yamashiti, K. Horie, E. Yashima, Y. Okamoto, *Macromolecules* **1996**, *29*, 5236–5238.

120 See note [30a] in Chapter 2.

121 M. Kawa, J. M. J. Fréchet, *Chem. Mater.* **1998**, *10*, 286–296.

122 J.-F. Nierengarten, D. Felder, J.-F. Nicoud, *Tetrahedron Lett.* **1999**, *40*, 273–276.

123 E. Alonso, D. Astruc, *J. Am. Chem. Soc.* **2000**, *122*, 3222–3223; G. E. Oosterom, J.

N. H. Reek, P. C. J. Kamer, P. W. N. M. van Leeuwen, *Angew. Chem.* **2001**, *113*, 1878–1901; *Angew. Chem. Int. Ed.* **2001**, *40*, 1828–1849; B. González, C. M. Casado, B. Alonso, I. Cuadrado, M. Morán, Y. Wang, A. E. Kaifer, *J. Chem. Soc., Chem. Commun.* **1998**, 2569–2570.
124 C. Köller, B. Pugin, A. Togni, *J. Am. Chem. Soc.* **1998**, *120*, 10274–10275.
125 M.T. Reetz, G. Lohmer, R. Schwickardi, *Angew. Chem.* **1997**, *109*, 1559–1562; *Angew. Chem. Int. Ed.* **1997**, *36*, 1526–1529.
126 M. Albrecht, R. A. Gossage, A. L. Spek, G. van Koten, *J. Chem. Soc., Chem. Commun.* **1998**, 1003–1004.
127 G. R. Newkome, C. N. Moorefield, G. R. Baker, A. L. Johnson, R. K. Behera, *Angew. Chem.* **1991**, *103*, 1205–1207; *Angew. Chem. Int. Ed.* **1991**, *30*, 1176–1178.
128 K. M. Nicholas, R. Pettit, *Tetrahedron Lett.* **1971**, *12*, 3475–3478.
129 C. Exon, P. Magnus, *J. Am. Chem. Soc.* **1983**, *105*, 2477–2478; M. J. Knudsen, N. E. Schore, *J. Org. Chem.* **1984**, *49*, 5025–5026; P. Magnus, R. T. Lewis, J. C. Huffman, *J. Am. Chem. Soc.* **1988**, *110*, 6921–6923.
130 G. Denti, S. Serroni, S. Campagna, V. Ricevuto, V. Balzani, *Inorg. Chim. Acta* **1991**, *182*, 127–129; S. Campagna, S. Serroni, A. Juris, V. Balzani, *Inorg. Chem.* **1992**, *31*, 2982–2984; ***Reviews***: S. Campagna, S. Serroni, V. Balzani, G. Denti, A. Juris, M. Venturi, *Acc. Chem. Res.* **1998**, *31*, 29–34; S. Campagna, S. Serroni, V. Balzani, A. Juris, M. Venturi, *Chem. Rev.* **1996**, *96*, 756–833; Y.-H. Liao, J. R. Moos, *J. Chem. Soc., Chem. Commun.* **1993**, 1774–1777.
131 G. R. Newkome, C. N. Moorefield, G. R. Baker, A. L. Johnson, R. K. Behera, *Angew. Chem.* **1991**, *103*, 1205–1207; *Angew. Chem. Int. Ed.* **1991**, *30*, 1176–1178; G. R. Newkome, R. Güther, C. N. Moorefield, F. Cardullo, L. Echegoyen, E. Pérez-Cordero, H. Luftmann, *Angew. Chem.* **1995**, *107*, 2159–2162; *Angew. Chem. Int. Ed.* **1995**, *34*, 2023–2026; G. R. Newkome, X. Lin, *Macromolecules* **1991**, *24*, 1443–1444; G. R. Newkome, C. N. Moorefield, *Polym. Prepr. Am. Chem. Soc. Div., Polym. Chem.* **1993**, *34*, 75–76.
132 ***Book***: U. S. Schubert, H. Hofmeier, G. R. Newkome, *Modern Terpyridine Chemistry*, Wiley-VCH, New York, Weinheim **2006**.
133 H.-F. Chow, I. Y.-K. Chan, D. T. W. Chan, R. W. M. Kwok, *Chem. Eur. J.* **1996**, *2*, 1085–1091; U. S. Schubert, C. Eschbaumer, *Angew. Chem.* **2002**, *114*, 3016–3050; *Angew. Chem. Int. Ed.* **2002**, *41*, 2892–2926.
134 a) Frontier Scientific, Inc.: http://www.frontiersci.com; b) e.g. Calixaren-based dendrimers: N. Cheriaa, M. Mahouachi, A. Ben Othman, L. Baklouti, Y. Kim, R. Abidi, J. Vicens, *Supramol. Chemistry* **2006**, *18*, 265–271; Oligothiophene dendrimers: C.-Q. Ma, E. Mena-Osteritz, T. Debaerdemaeker, M. M. Wienk, R. A. Jansen, P. Bäuerle, *Angew. Chem.* **2007**, *119*, 1709–1713; *Angew. Chem. Int. Ed.* **2007**, *46*, 1689–1683.

4.2 Chiral dendrimers

4.2.1 Classification of chiral dendrimers

Mention of chirality in dendritic architectures can be traced back to patents of Denkewalter *et al.*, which describe the construction of peptide-like dendritic structures from L-lysine units [1]. In spite of the demanding nature of some of these syntheses, numerous chiral dendritic structures have meanwhile been prepared and characterised [2]. This cannot be explained solely by the somewhat academic interest in the effect of chiral monomeric building blocks on the chirality of the overall molecule. The prospect of using chiral dendrimers as model compounds for biopolymers in molecular recognition processes, the development of new dendritic materials for sensor technology and asymmetric catalysis also continue to be driving forces for research in this area.

In the classification of chiral dendrimers, the way in which chirality was introduced into the molecule, or the site at which chiral modules are implanted, can be employed as distinguishing criterion. The chiral dendrimers so far described in the literature can accordingly be assigned to one of the following groups (Fig. 4.61) [2a, c, 3]:

- *(A) Dendrimers with chiral core and achiral branching scaffold (see Section 4.2.3)*
- *(B) Dendrimers with chiral end groups (see Section 4.2.5)*
- *(C) Dendrimers with chiral spacers or chiral branching units (see Section 4.2.4)*
- *(D) Dendrimers with achiral core unit and at least three constitutionally different dendrons*
- *(E) Dendrimers with both chiral core as well as chiral branching units and chiral end groups.*

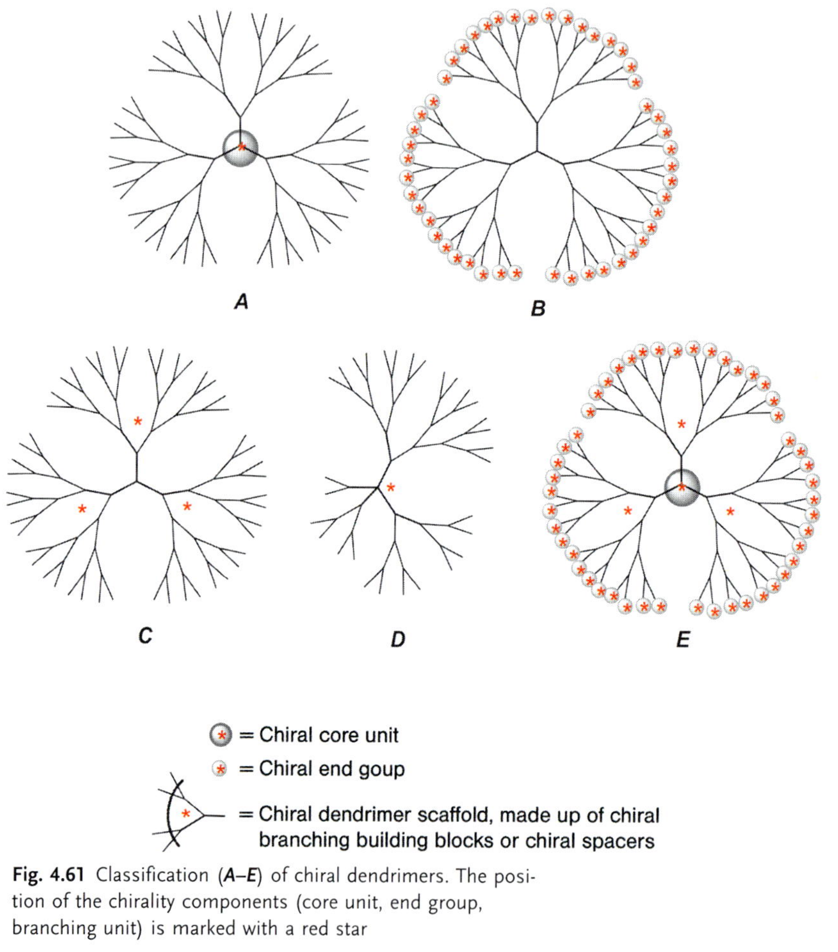

Fig. 4.61 Classification (*A–E*) of chiral dendrimers. The position of the chirality components (core unit, end group, branching unit) is marked with a red star

The last two dendrimer types are less well known, because in the case of ***D*** the chirality is low or in the case of ***E*** the interpretation of chiral effects probably proves difficult (see Section 4.2.7).

In most cases chirality results from the presence of stereogenic centres (stereocentres). Thus optically active natural products, such as amino acids [4], carbohydrates [5], or nucleic acids [6] are frequently used as building blocks or end groups to impart chirality to dendrons or dendrimers. In addition, chiral dendrimers are also produced with synthetic chiral building blocks and are used for systematic studies of the influence of chiral building blocks on dendrimer conformation. These also include dendritic structures whose chirality originates from planar chiral [7], axially chiral [8], topologically chiral [9], or cycloenantiomeric [7b] units. Dendrimers with chiral core (case ***A*** in Fig. 4.61) are most frequently encountered.

4.2.2 Studies on the chirality of dendritic molecules

4.2.2.1 Chiroptical studies

Studies on the chirality of dendritic molecules are generally based on the interpretation of analytical data such as optical rotation or circular dichroism. These chiroptical parameters are sensitive to conformational changes, which is why chiroptical methods are often employed for structural elucidation in solution [10]. Thus information about the effect of chiral building blocks on conformation can usually be deduced from comparison of the chiroptical properties of dendrimers with those of the corresponding chiral monomers or suitable model compounds [11]. In the case of dendrimers containing more than one monomeric chiral building block, an additive relationship between the chiroptical parameters of the individual monomers and the overall dendrimer structure indicates that the conformational equilibrium of the monomeric building blocks is not perturbed by incorporation into the dendrimer framework. In contrast, if deviations from this additivity are observed, for example in the case of dendrimers with chiral peripheral groups, then packing effects could be operative and the chiral monomeric building blocks of the dendrimer may be present in conformations that are different from those in the free monomers. Depending upon the type of dendrimer, the former monomers are located in different parts of the dendrimer scaffold, each with a different local environment. In such cases, deviation from additivity can also be explained in terms of differences in composition [11]. When conducting comparative studies on dendrimers of a given series it should be noted that large differences in mass generally exist between dendrimers of different generations and that the specific rotations may vary with the concentration of the measured solution. It therefore often makes more sense to consider the molar rotation $[\Phi]_D$, since this parameter gives the specific rotation as a function of molecular weight and is thus based on the same number of molecules.

The influence of the three-dimensional geometry of chiral elements on the macroscopic chiroptical properties of dendrimers can be studied by placing suitably selected chiral building blocks in predetermined positions within the dendrimer scaffold with the aid of convergent or divergent synthetic methods.

The principal goal is to define those factors which lead to the macroscopic chirality of the dendrimer. Despite numerous studies on this topic, the relation between the molecular chirality of the dendritic building blocks and the macroscopic chirality of molecules has still not been completely elucidated [12]. Yet an understanding of this relation is important for the development of new materials, including polymers, whose properties and function depend upon their macroscopic chirality [13].

4.2.2.2 Possible applications of chiral dendrimers

The ability of chiral dendrimers to undergo stereoselective host/guest interaction and to form clathrates is of interest with a view to potential applications. Such properties could lead to uses in sensor technology [14] and to the development of new dendrimer-based chiral materials for chromatography (chiral stationary phases, CSP). Use of chiral dendritic structures as catalysts or ligands in asymmetric synthesis is also expected [15] (see Section 8.2.3). Ideally having a monodisperse and asymmetric, globular structure, higher-generation dendrimers made up of optically active natural products resemble naturally occurring macromolecular systems. Interest is therefore focussed primarily on the possible role of such chiral dendrimers in biological systems [16] and less on the study of their chiroptical properties. Biocompatible dendrimers of this kind are of interest mainly as possible enzyme mimics [17]. In addition, their suitability as carriers for active substances in medicine is under investigation.

The following sections address the various types (***A–E*** in Fig. 4.61) of chiral dendrimers. Individual publications which have shed light on the relation between the molecular chirality of the dendritic building blocks and the macroscopic chirality of the molecules as well as applications-relevant properties of chiral dendrimers are presented.

4.2.3
Dendrimers with chiral core and achiral branching scaffold

Dendrimers with chirality residing in the core are obtained relatively simply in a convergent synthesis by linkage of achiral dendrons to a chiral core unit. Numerous Fréchet-type poly(aryl ether) dendrimers with chiral core unit have already been prepared in this way (Williamson ether coupling). Thanks to their structural purity, convergently synthesised dendrimers prove particularly useful in studying the following questions relating to the chirality of synthetic macromolecules:

- Is the optical activity of the chiral core unit retained on linkage to a branching scaffold made up of achiral building blocks?
- Can the chirality of the core unit be transferred to a distant part of the molecule and induce a chiral conformational order?

- Does the chiral core unit permit enantioselective (dendrimer) host/(substrate) guest complexation in the interior of the dendrimer scaffold?

4.2.3.1 **Chiroptical studies on dendrimers with chiral cores**

The first dendrimers with chiral cores for studies on the influence of the stereogenic centres of a core unit on the chiroptical properties of the overall molecule were presented by Seebach's group [18]. These workers first synthesised dendrimers based on a chiral tris(hydroxymethyl)methane core unit. To these were attached zeroth- to second-generation Fréchet dendrons, either directly or separated from the core by an aliphatic (*n*-propyl) or an aromatic spacer (*p*-xylylene) (Fig. 4.62). Remarkably, the dendrimers with aliphatic spacer showed no significant optical activity. This "loss" of chiral information was attributed to a "dilution effect", resulting from linkage of the achiral dendron to the chiral core unit,

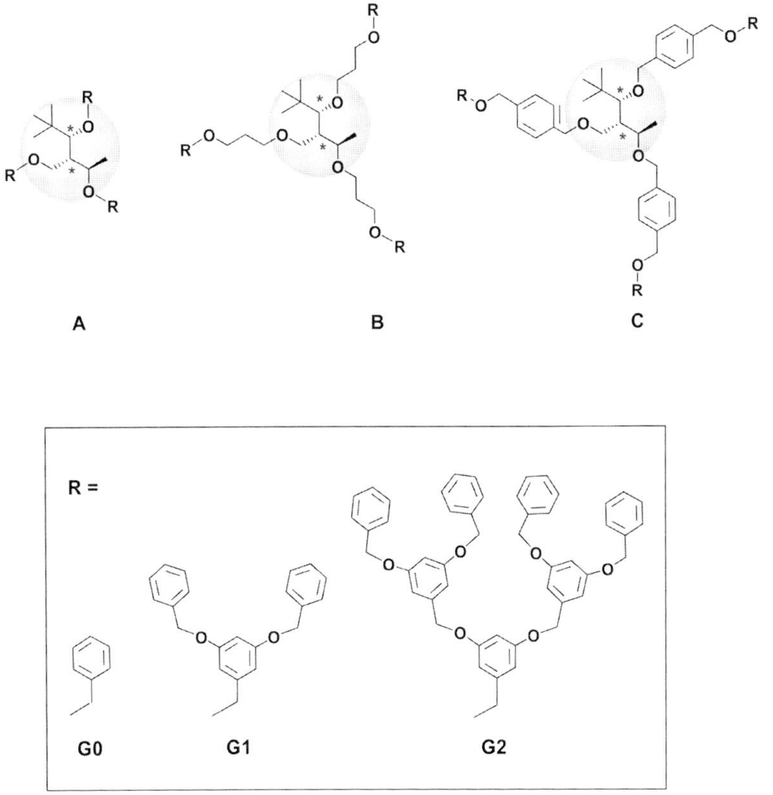

Fig. 4.62 Chiral-core dendrimers of various generations (according to Seebach *et al.*), consisting of a chiral tris(hydroxymethyl)methane core unit and poly(aryl ether) dendrons (G0, G1, or G2), which are linked to the core either directly (A), *via* an aliphatic spacer (B), or *via* an aromatic spacer (C)

and to the enhanced flexibility of the dendron, resulting from the conformationally mobile aliphatic spacer. In contrast, almost constant molar rotations were measured for the dendrimer series with aromatic spacer (*p*-xylylene) regardless of the generation number ($[\Phi]_D^{RT}$ G0: +98; G1: +101; G2: +103). This indicates that the chirality of the core unit is "retained", yet fails to induce any significant chiral substructures in the dendrimer scaffold. Among the dendrimers without spacers between core unit and dendrons, a slight decrease of the molar rotation values was initially noted on transition from the first to the second generation ($[\Phi]_D^{RT}$ G0: +48; GI: +46) whereas the values were almost doubled on going to the second generation ($[\Phi]_D^{RT}$ G1: +46; G2: +87). This was interpreted as indicating that the chiral core unit is able to induce chiral conformational order in the direction of the dendrimer periphery within the larger dendrimer owing to the generally more rigid overall structure.

Overall, studies on poly(benzyl ether) dendrimers with various chiral core units show that the influence of the achiral dendrimer scaffold on the chiroptical properties of the core unit depend primarily upon the origin of their chirality.

Thus the molar rotations of enantiomerically pure dendritic analogues of axially chiral (*S*)-1,1′-bi-2-naphthol (Fig. 4.63) become increasingly negative on transition from the second to the fifth dendrimer generation. Quantitative analysis of the CD data revealed that this could be attributed to enlargement of the torsional angle between the two naphthyl groups as a result of increasing steric hindrance between the dendritic substituents [8 c].

Further studies have shown that the torsional angle of the binaphthyl unit depends not only upon the steric demands of the two dendritic substituents, but

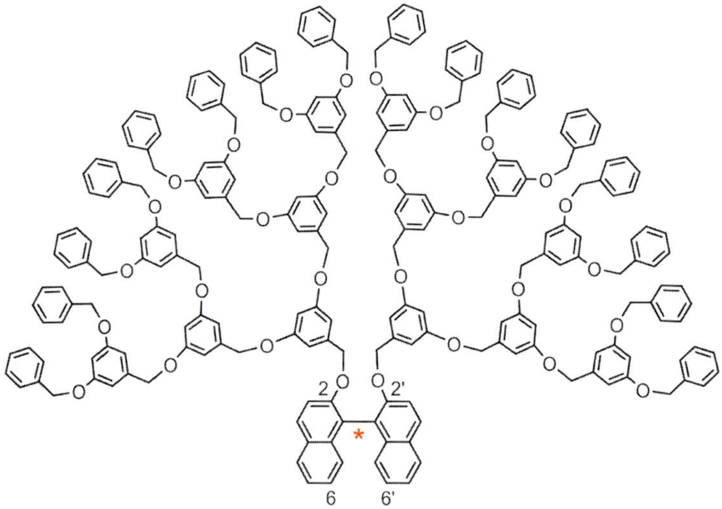

Fig. 4.63 Third-generation poly(benzyl ether) dendrimer with axially chiral binaphthyl core unit

also upon the position at the core unit. Thus attachment of Fréchet dendrons to positions 6 and 6′ has a significantly less pronounced effect on the torsional angle of the binaphthyl core unit than attachment to positions 2 and 2′; the molar optical rotation values remain almost constant regardless of the generation number.

Less common representatives of chiral core dendrimers include those with planar chiral [2.2]paracyclophanes, topologically chiral [2]catenanes, and molecular knots [9] or cycloenantiomeric [2] rotaxanes [7b] as core unit. CD-spectroscopic studies on a series of planar chiral "dendrophanes" [7b, 19] (Fig. 4.64) and topologically chiral "dendroknots" [9a, b] (Fig. 4.65) with Fréchet type poly-(benzyl ether) dendrimer scaffold revealed an amplification of the CD signal with increasing size of the dendritic substituents. This dendritic effect (see Section 6.3) indicates that the chiral core induces circular dichroism in the "inherently" achiral poly(benzyl ether) dendrimer scaffold by forcing the dendrimer branches into a chiral conformational geometry through the influence of increasing steric hindrance (see Section 4.2.7).

The local chirality of a core unit is not necessarily manifested in a macroscopic chirality of the dendrimer. This is clearly apparent from the example of the dendrimers constructed by Meijer *et al.* from a glycerol unit as chiral core unit and four Fréchet-type dendrons of different generations (Fig. 4.66). No optical activity could be measured with available instrumentation for the two enantiomeric forms (*S*)-**1** and (*R*)-**1** [2a].

Such enantiomerically pure chiral compounds, which are optically inactive in the generally investigated UV/Vis spectral range from 200 to 800 nm, can be designated as "cryptochiral" according to Mislow *et al.*, since the chiral information is hidden in the molecule (*i.e.* present in cryptic form) [12a, 20]. Cryptochirality of the dendrimer is explained on the one hand by the pronounced conformational mobility of the dendrimer branches and on the other by the slight electronic difference between the dendritic substituents.

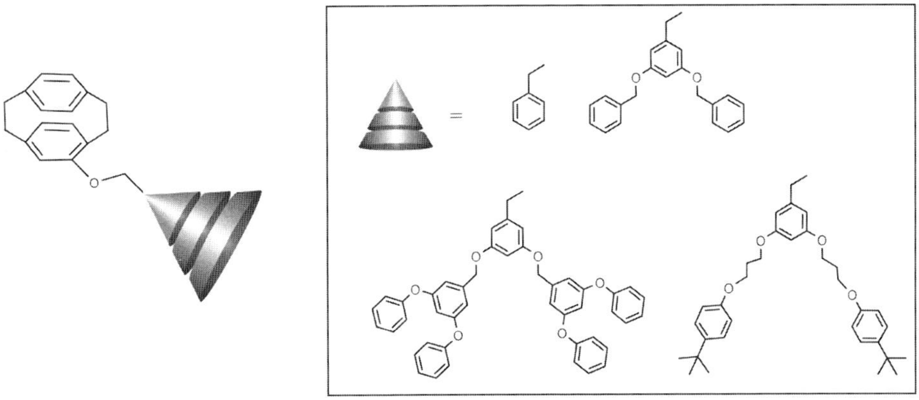

Fig. 4.64 Dendryl substituents on a [2.2]paracyclophane core: dendrophanes

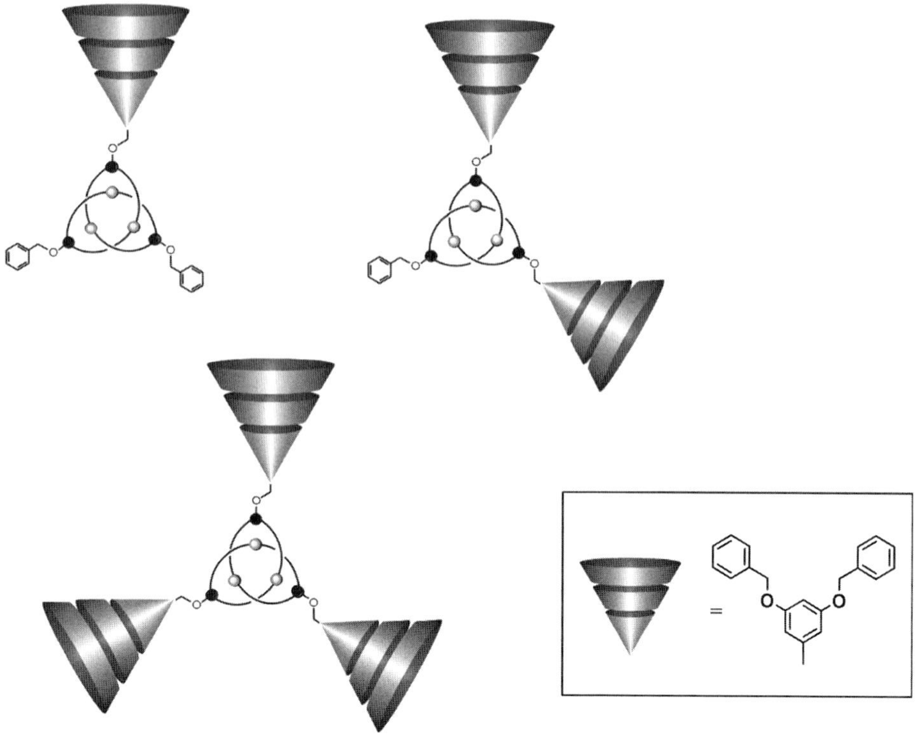

Fig. 4.65 Dendron-substituted knots (schematic): dendroknots

In contrast, attachment of poly(benzyl ether) dendrons with more sterically demanding branching units (2,6-substitution pattern on the benzene ring; cf. (*S*)-**2** in Fig. 4.66) to an analogous (*S*)-(+)-configured core afforded a dendrimer which surprisingly exhibited optical activity – at low temperatures [21]. However, the optical activity of the dendrimer vanishes again at 30 °C, explicably as a consequence of the greater flexibility of the dendrimer scaffold at this temperature. Hence the molecular scaffold of such dendrimers must possess a certain degree of structural rigidity in order that the core can induce a chiral conformation in the dendrimer scaffold and the local chirality of the core is manifested in a macroscopic chirality characteristic of the overall molecule. This conclusion is in accord with the results of chiroptical studies on other dendrimers with chiral core unit.

Information about "transmission" of chirality in dendrimer molecules also came from various generations of dendrimers derived from optically active alkaloids. Starting from atropine or quinine, corresponding ammonium salts **1** or **2**, respectively, of up to the third generation were synthesised by quaternisation at the aliphatic nitrogen with Fréchet-type dendritic benzyl bromides (Fig. 4.67) [22].

Enantiomeric resolution of the corresponding racemates was accomplished by HPLC on chiral stationary phases (CSP) [23]. Chiroptical studies by means of circular dichroism indicated that the chiral dendrimer core exerted some influence

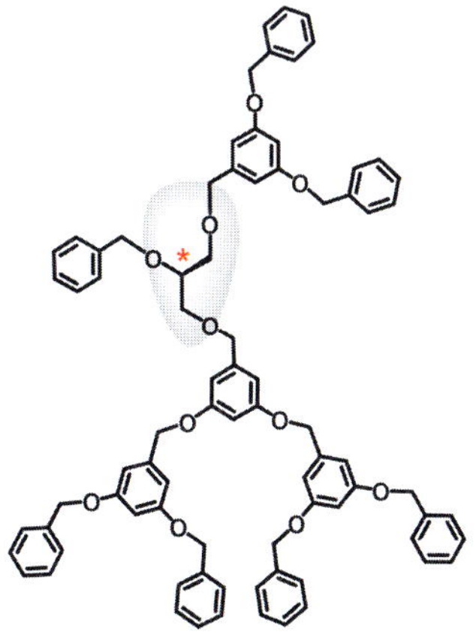

(*S*)-**1**

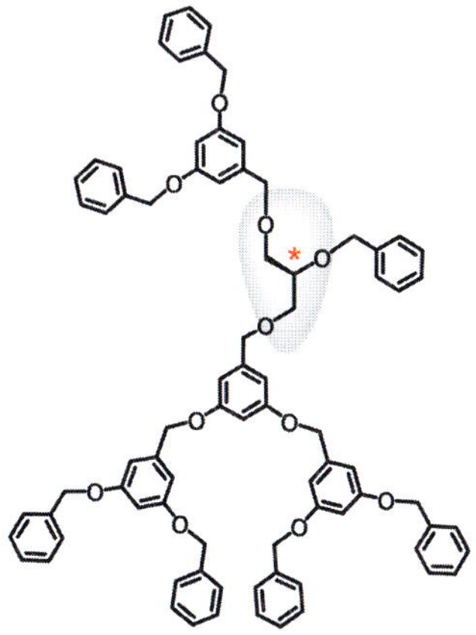

(*R*)-**1**

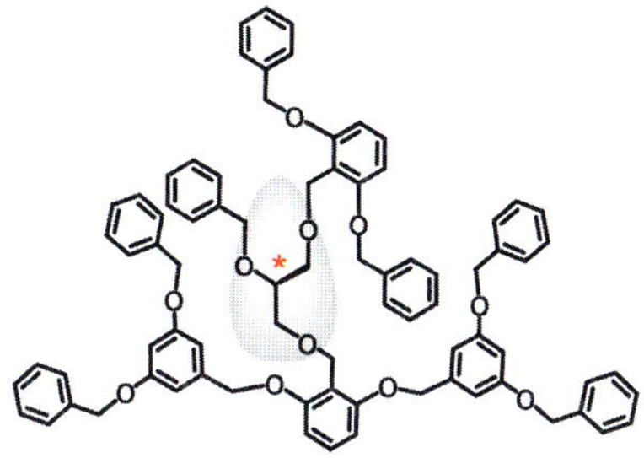

(*S*)-**2**

Fig. 4.66 While no optical activity could be measured for the enantiomers of the "cryptochiral" dendrimer **1** (with 3,5-substituted benzene rings), the constitutionally isomeric (2,6-substituted) dendrimer **2** showed a slight, yet measurable optical activity

on the (achiral) dendritic part of the molecule. The chirality proved to be dependent upon the nature of the chirality elements and the generation number of the formally achiral dendritic branches attached to the chiral core: "dendritic effect" (see Section 6.3). The approach adopted and the conclusions drawn will now be presented in somewhat greater detail for the purpose of illustration [22]:

Figure 4.68 depicts the circular dichrograms of the zeroth to second generations. The almost mirror-image appearance of the Cotton effect of the enantiomers (with maximum at around 220 nm) is immediately obvious in the left-hand diagram. The intensities of molar circular dichroism increase with increasing dendrimer generation (dendritic effect). It was therefore assumed that, for such an amplified dichroism effect, a CD signal is induced in the achiral branches if they are placed in a chiral environment. Since the number of chromophoric groups (benzene and dimethoxybenzene units) increases at higher generations, a higher total extinction coefficient was also obtained for the enantiomers. The somewhat shifted maximum for the higher generations at 220 nm can be explained by overlapping of the carbonyl absorption of the atropine unit

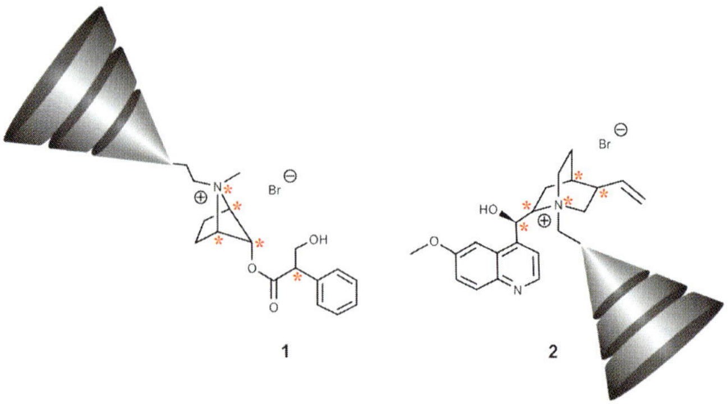

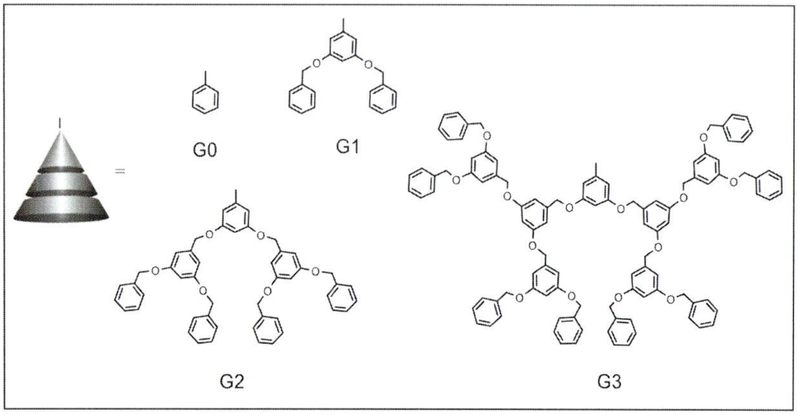

Fig. 4.67 Dendritically substituted alkaloid derivatives (according to Vögtle *et al.*)

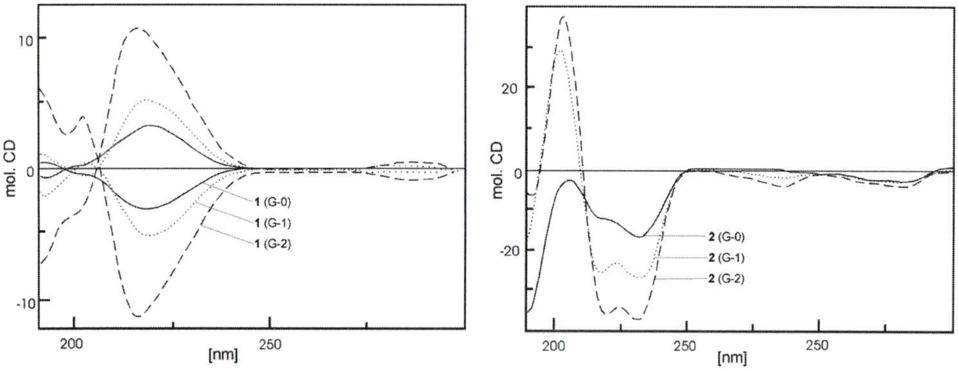

Fig. 4.68 Circular dichrograms of dendritically substituted atropines (**1**) and quinines (**2**); each of zeroth to second generation (according to Vögtle *et al.*)

and the intense absorption of the greater number of chromophores of the poly(-benzyl ether) dendron. Stronger Cotton effects were also observed on increasing the size of the dendritic branches attached to the chiral core.

It could also be demonstrated by vibrational circular dichroism (VCD) that the observed dendritic effect can be attributed more to the benzene and COC groups than to the CH_2 subunits of the dendritic branches. Stronger CD signals were also observed with increasing generation number in VCD; this effect was ascribed to the increasing number of benzene and dimethoxybenzene moieties (chromophores) at higher generations.

Overall, this increase of the (positive) dendritic effect with increasing generation number is reminiscent of the situation in dendrophanes and dendroknots (see Figs. 4.64 and 4.65). Compared to those chiral architectures (with planar chiral or topologically chiral elements), the dendron-substituted alkaloids described here possess relatively simple centrochiral units, with both ammonium salts exhibiting five stereocentres. This phenomon of induced circular dichroism can be expected to occur comparatively frequently; a possible explanation is that a chirally induced propeller-like arrangement of the benzene rings exists in the dendritic "blades". Chirality of a particular sense (clockwise or anticlockwise) could be preferred in such "microdomains" and lead to enhanced Cotton intensities. In the future, theoretical calculations could help to provide a more detailed interpretation of such findings. At present, however, despite considerable advances in the interpretation of CD spectra, exact calculations are still difficult in the case of such flexible and large molecules which can adopt numerous energetically similar conformations.

Studies performed on chiral core dendrimers have provided valuable information about the influence of the achiral dendrimer scaffold on the chiroptical properties of the core unit. Yet they also show that prediction of the chiroptical properties of the dendrimer is difficult, since the chiral relationship between the local chirality of the core unit and the nanoscopic conformation of the overall dendrimer structure is influenced by numerous structural factors. Further studies will be required to attain a fuller understanding of how an individual chiral building block can induce chirality in the entire dendrimer architecture (see Section 4.2.7).

In conclusion, mention should be made of dendronised chiral salen ligands and their Co^{2+} and Ni^{2+} complexes, which were prepared for use as Jacobsen-type catalysts by diaza-Cope rearrangement [24]. However, instead of the *meso* compounds produced, the enantiomers will have to be synthesised directly in enantioselective manner (Fig. 4.69).

4.2.3.2 **Possible applications of chiral-core dendrimers**

The first example of the use of a chiral-core dendrimer as enantioselective receptor in molecular recognition processes was reported by Diederich (see Fig. 4.70). This dendrimer, designated as a "dendrocleft", acts as enantioselective receptor for monosaccharides. It bears a central axially chiral 9,9′-spirobi[9*H*-fluorene] unit which is linked *via* two 2,6-bis(carbonylamino)pyridine spacer groups, each in 2,2′ position, with triethylene glycol monomethyl ether dendrons [14b]. Both

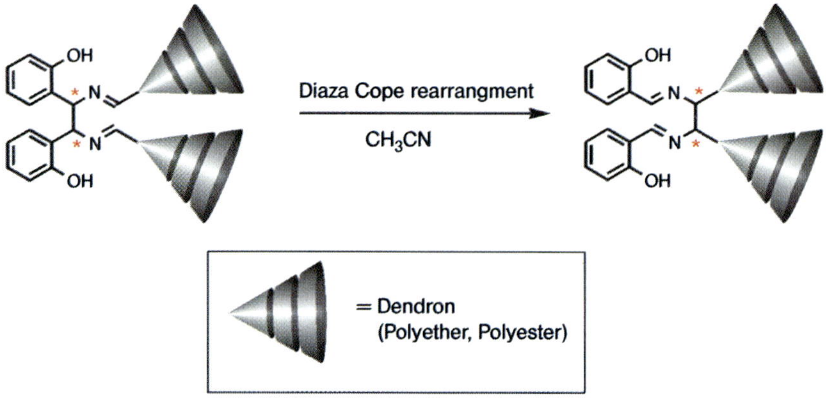

Fig. 4.69 Chiral salen ligands (according to Vögtle, Portner)

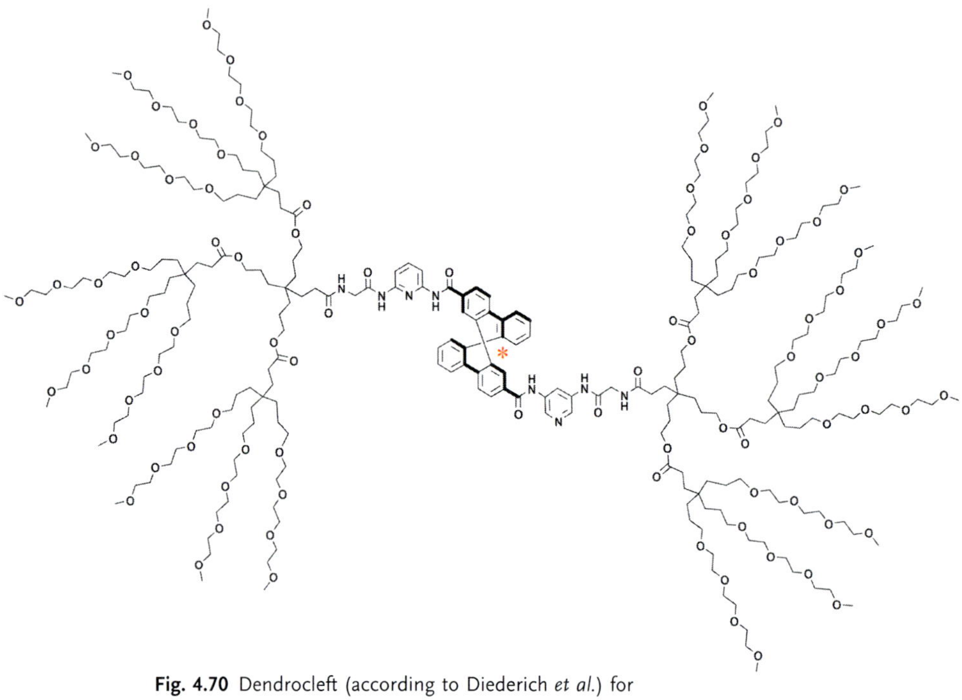

Fig. 4.70 Dendrocleft (according to Diederich *et al.*) for stereoselective host/guest interaction depending upon the size of the dendrimer scaffold surrounding the chiral core (marked with a red asterisk)

^{1}H-NMR studies and analysis of sensor responses in the CD spectrum – after addition of sugar as guest to the dendrocleft host – show that the ability of this dendritic host to enantioselectively recognise monosaccharides decreases with increasing generation number of the dendron surrounding the chiral core unit.

In contrast, the diastereoselectivity of the dendritic host increases. This indicates that on shielding of the chiral core unit with sterically more demanding dendrons in higher-generation dendroclefts the monosaccharide guests are no longer bound in the immediate vicinity of the chiral core unit; instead, a less specific host/guest interaction takes place with the dendrimer scaffold.

The prototype of a first fluorescence sensor for enantioselective recognition of chiral amino alcohols was synthesised by Pu *et al.* by linking phenylacetylene dendrons to an axially chiral binaphthol core unit (see Fig. 4.71) [14c, d, 25].

Such chiral luminescent dendrimers, whose fluorescence is quenched significantly more strongly by one enantiomer of a chiral guest molecule than by its mirror-image counterpart, would be of interest for rapid determination of the

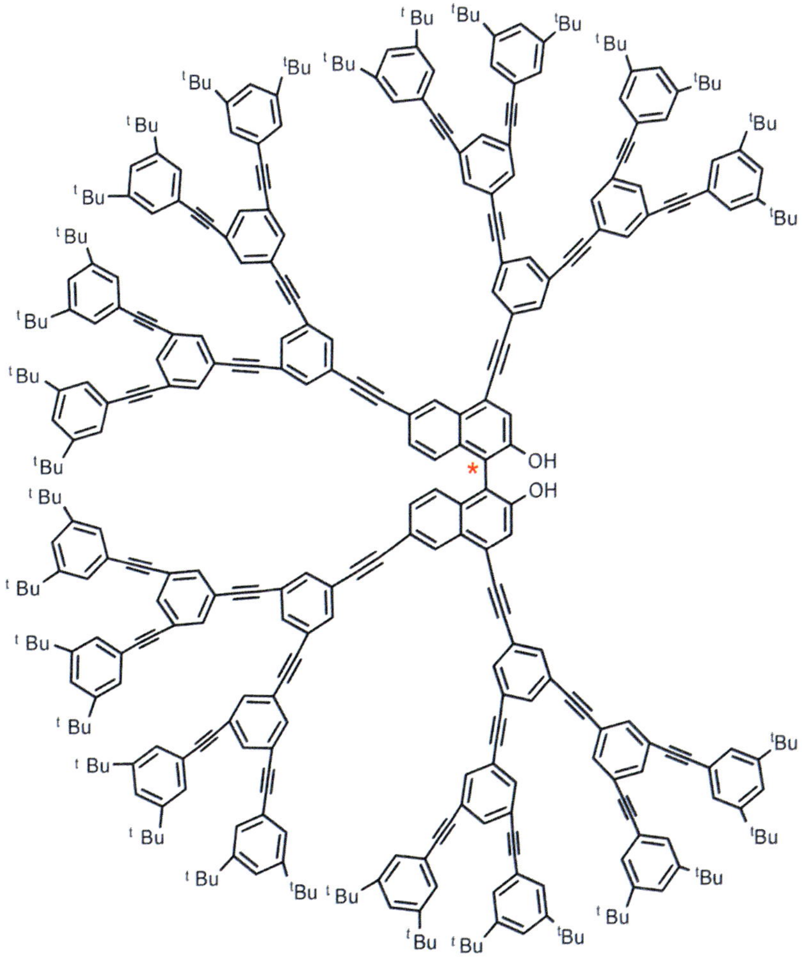

Fig. 4.71 Axially chiral dendritic fluorescence sensor for enantioselective recognition of chiral amino alcohols (according to Pu *et al.*). The chiral core is marked with a red asterisk

enantiomeric composition of chiral compounds and could find application in the combinatorial search for chiral catalysts.

4.2.4
Dendrimers with chiral building blocks as spacers or branching units

4.2.4.1 Chiroptical studies on dendrimers with chiral dendrimer scaffold

Sharpless *et al.* noted an increase of the specific rotation with increasing generation number in the case of dendrimers having trimesic acid as achiral core unit and a completely chiral polyether dendrimer skeleton based on 1,2-diol branching units (Fig. 4.72) [27]. The observed molar optical rotations corresponded to

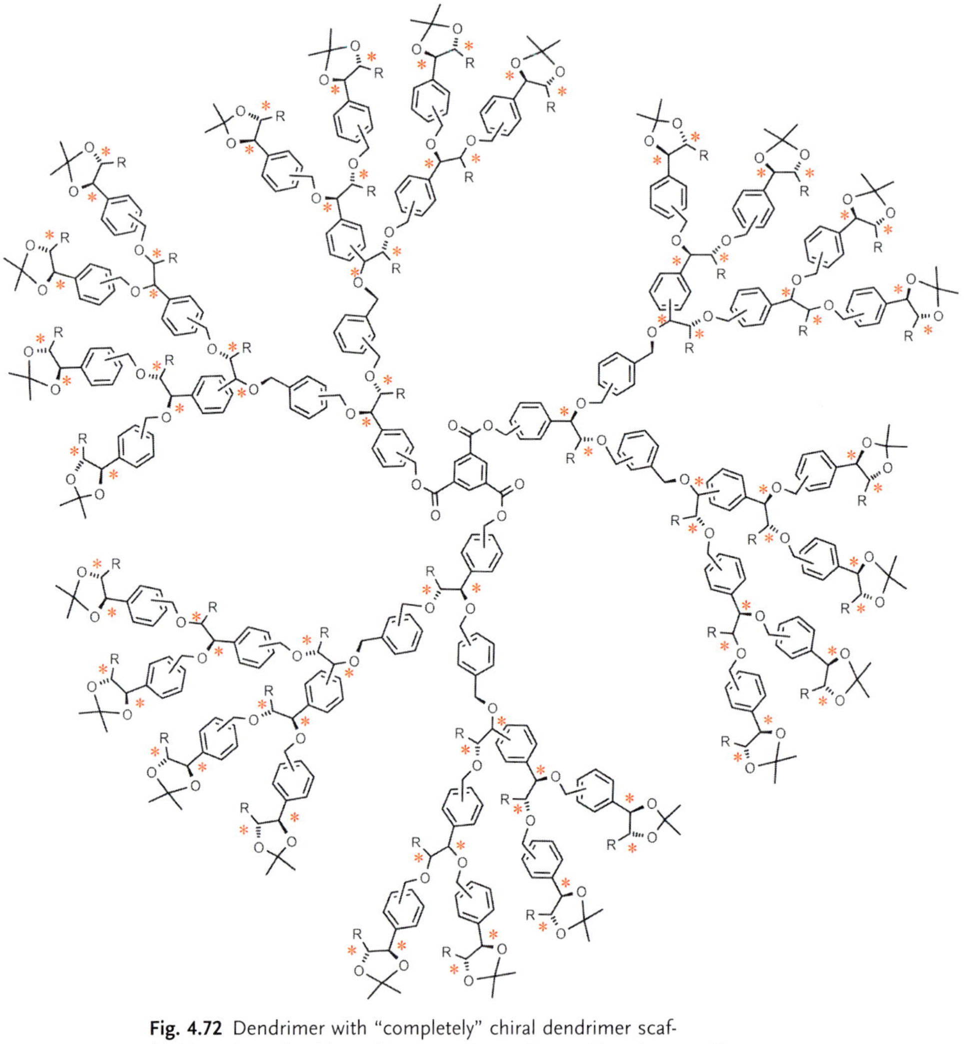

Fig. 4.72 Dendrimer with "completely" chiral dendrimer scaffold based on chiral branching units (according to Sharpless *et al.*)

the sum of the rotations of the chiral monomeric building blocks. This indicates that the optical activity is due to the chiral units and not to a chiral conformational geometry (orientation) of the dendrimer scaffold (see also Section 4.2.7).

Chow and Mak came to a similar conclusion on investigating the chiroptical properties of dendrimers containing enantiomerically pure threitol building blocks obtained from tartaric acid as spacers between the achiral phloroglucin branching units (see Fig. 4.73) [27]. They found that the chiral spacers in the dendrimer scaffold do not influence one another and contribute additively to the overall rotation. Moreover, they also observed that on introduction of both enantiomers one (*R,R*)-threitol unit precisely compensated the rotational contribution of one (*S,S*)-threitol unit, provided that the enantiomeric building blocks were located at equivalent positions within the dendrimer scaffold. However, CD-spectroscopic data revealed that the contribution of the exterior threitol units to the total rotation must be slightly different from that of the interior units.

Starting from fifth- and third-generation phosphorus-containing dendrimers with peripheral planar chiral ferrocene groups, Majoral *et al.* constructed higher dendrimer generations [28]. In this way, they were able to specifically position the optically active groups within the molecular scaffold at a given generation level. On comparison of the chiroptical properties, a degree of agreement was found between the behaviour of higher-generation dendrimers and classical polymers. Thus the chiroptical properties are influenced only by the number of chiral groups and their chemical environment, but not by their position within the scaffold of the molecule.

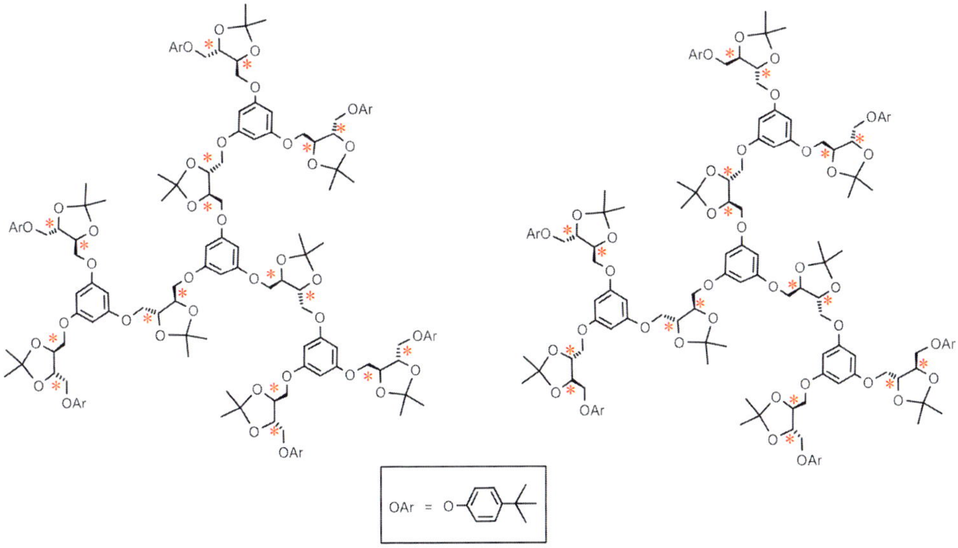

Fig. 4.73 Targeted incorporation of the two enantiomers (*RR*, *SS*) of a chiral spacer (threitol) compensates the rotation in these dendrimers (according to Chow and Mak; stereocentres are marked with a red asterisk)

4.2.4.2 Possible applications of dendrimers with chiral branching scaffold

Owing to their size and the feasibility of controlling the dendrimer structure and globular morphology *via* their synthesis, dendrimers with chiral branching scaffold are of interest as potential protein mimics. In addition, the introduction of chiral branching units or spacers into the molecular scaffold should lead to the development of non-symmetrical macromolecular conformations and provide chiral cavities for asymmetrical catalysis or chiral recognition processes.

4.2.5 Chirality in the periphery

One of the most commonly adopted approaches to the production of chiral dendrimers consists in the attachment of chiral units to the peripheral functionalities of a divergently constructed dendrimer.

Alternatively, dendrimers with chirality in the molecular periphery can also be constructed by the convergent approach starting from chiral monomeric units.

4.2.5.1 Chiroptical studies on dendrimers with peripheral chiral units

The first example of a dendrimer with peripheral chiral units was reported by Newkome *et al.* in 1991 [29].

They functionalised up to second-generation polyether amide dendrimers (arborols; see Chapter 1) with enantiomerically pure tryptophan units (Fig. 4.74). The measured optical activities per chiral terminal group were roughly constant for all molecules of this dendrimer series.

In contrast, the rotation per chiral end group of first- to fifth-generation POPAM dendrimers with sterically demanding Boc-protected amino acids as end groups (Fig. 4.75) decreased significantly with increasing generation number and approached the value zero [30].

Meijer *et al.* interpreted this finding by assuming that individual conformations are "frozen in", as a result of increasing steric hindrance, and thus cancel each other out. This idea is supported by the observation that the rotations per chiral group again become of roughly equal magnitude from the first to the fifth generation on introduction of a flexible alkyl chain between the L-phenylalanine units and the POPAM dendrimer scaffold. The aliphatic spacer increases the radius of the dendrimers and thus assures a greater distance between the chiral end groups, which thus exert much less influence upon one another [2a]. A remarkable indication of chirality within the so-called "dendritic box" formed by 64 Boc-protected L-phenylalanine units comes from the occurrence of induced circular dichroism on inclusion of certain achiral dye molecules [31]: Both Newkome *et al.* and Meijer *et al.* found that an excessively dense packing of chiral end groups linked together *via* hydrogen bonding will destroy the optical activity of the dendrimers [29, 30a].

Studies by Newkome, Meijer, and other research groups [29, 30a, 32] revealed that, in dendrimers with chiral end groups and otherwise achiral molecular

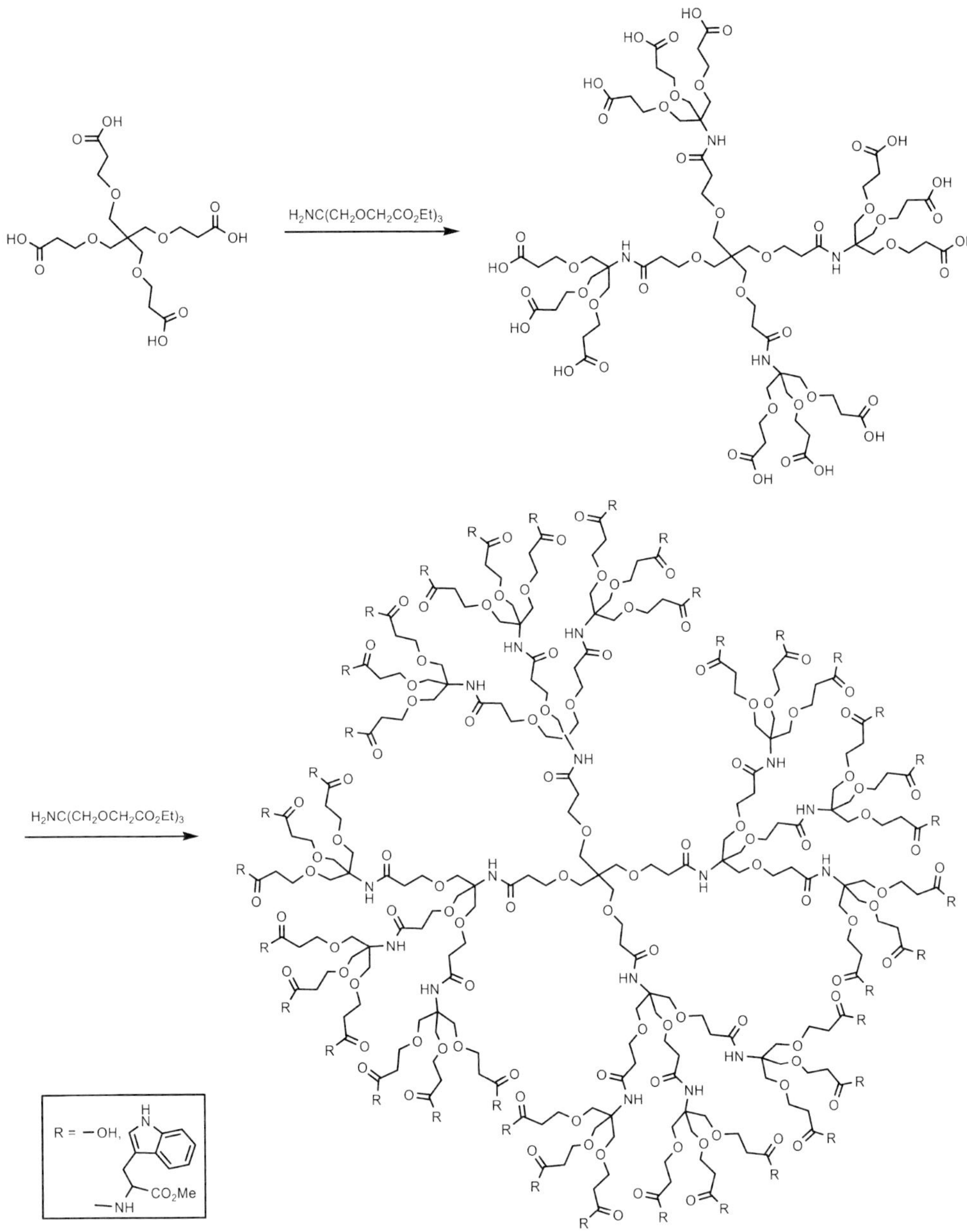

Fig. 4.74 Second-generation peripherally tryptophan-substituted arborol (according to Newkome *et al.*)

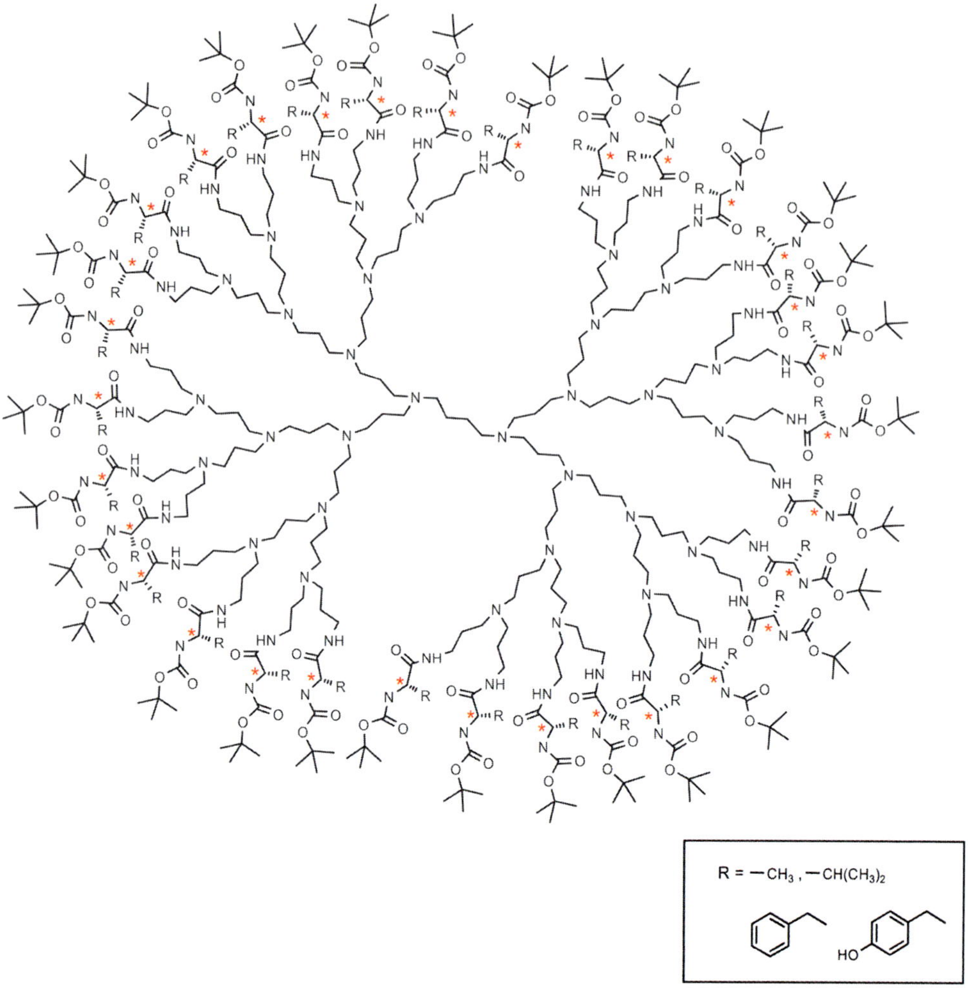

Fig. 4.75 POPAM dendrimer with chiral Boc-protected amino acids in the molecular periphery (according to Meijer *et al.*)

scaffold, the optical activity per end group generally remains approximately constant for the various generations, provided that the chiral groups are not sensitive to differences in packing.

4.2.5.2 **Possible applications of dendrimers with peripheral chiral units**

Dendrimers with readily accessible chiral end groups have been synthesised largely for use in asymmetric catalysis or for molecular recognition processes.

In the hope of obtaining biocompatible dendrimers suitable for medical use in the prevention of infection, PAMAM dendrimers [33], polyether amide dendrimers, arborols [29, 34] (cf. Fig. 4.74), and POPAM dendrimers [30, 35] were furnished with peripheral amino acid or sugar units.

4.2.6
Chiral dendrimers for asymmetric catalysis

Much effort and money have been, and continue to be, invested in the development and improvement of catalysts for asymmetric catalysis. The lack of any means of recovery means that the – large scale – use of soluble chiral metal complexes remains uneconomical. Recent years have seen the publication of numerous studies concerned with the possibility of recovering and recycling such expensive asymmetric catalysts. One conceivable solution involves the incorporation of catalytically active units such as chiral metal ligands into dendrimers. In order to perform asymmetric catalysis with dendrimers, the catalytically active units can be introduced into different sites of the dendrimer molecule, for example as chiral units in the periphery or the interior of the dendrimer. And dendrimers containing achiral catalytically active units surrounded by chiral branches in the interior of the dendrimer are conceivably also of use in asymmetric catalysis. The regular structure of the dendrimer scaffold ensures that every catalytically active unit has an almost identical chiral environment.

Chiral dendrimers with catalytically active peripheral groups were investigated by Meijer *et al.*, who functionalised POPAM dendrimers of various generations with enantiomerically pure amino alcohols [2a, 34, 36]. On use of these surface-modified dendrimers for enantioselective addition of diethylzinc to benzaldehyde, the catalytic performance was found to decrease with increasing size of the dendrimer. Both the yields and the selectivity declined and reached a minimum at a fifth-generation dendrimer with 64 peripheral units. This phenomenon was ascribed to increasing steric hindrance in the dendrimer periphery and the concomitant impairment of conformational freedom of the catalytically active groups. More information about dendritic effects in catalysis will be found in Section 6.3.2.

4.2.7
Interpretation of the chirality of dendritic molecules

Previous sections concerned with chiral dendrimers have often mentioned a "dendritic chirality effect" in the form of "chiral amplification" or "chiral attenuation" on going to higher generations (positive or negative dendritic effect, respectively). This Section concludes the current Chapter by presenting possible explanations for these important findings, whose importance extends well beyond the bounds of dendrimer chemistry. This is not only of general theoretical interest (for an understanding of chirality), but also relevant to applications (*e.g.* catalysis; see previous Section 4.2.6, Section 6.3.2, and Section 8.2). Moreover, findings relating to dendrimers may possibly permit conclusions to be drawn about corresponding effects in polymers.

General conclusions concerning circular dichroism and chirality of dendrimers are presented in Section 7.9.

Chiral amplification and attenuation: Chiroselective host/guest recognition and chiral induction – in chemical reactions – have become increasingly important in recent years. Ever-increasing demands are placed upon separation performance in the chromatographic resolution of racemates on chiral column materials (CSP: chiral stationary phases) [37c]: Since chiral amplifications are meanwhile ascribed considerable importance in stereoselective reactions and especially in asymmetric catalysis (see also Section 8.2.3), a number of explanatory approaches are considered here in detail. Recent findings in polymer chemistry prove to be of value in stereoselective polymerisations, in the course of which centrochiral substituted alkene monomers react – intermolecularly – to form helical polymer strands, thus generating a chiral secondary structure. In certain cases, their overall chirality is greater than would result from the stereocentres themselves. The formation of chiral clusters of achiral molecules on surfaces leading to two-dimensional chiral objects – three-dimensional on considering the surface on just one side, and observations of corresponding dynamic racemisation effects on surfaces are likewise valuable for interpretations [38].

Figure 4.76 depicts an intramolecular dynamic equilibrium of a dendrimer with four blade dendrons, in the course of which the (helical) propeller conformation is inverted, *i.e.* the sense of the chirality changes [37a]. In such cases it was observed that all four dendrons undergo correlated motion, rather than just one dendron changing its conformation without the other three dendrons rotat-

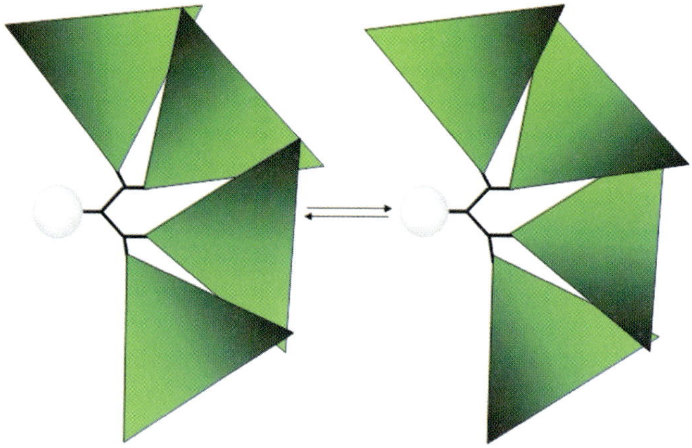

Fig. 4.76 A dendrimer with flat dendrons with a propeller-like arrangement, in which correlated intramolecular conformational motion leads to inversion of the sense of chirality: Owing to a partial steric intermeshing, when one of the dendrons is inverted the other three are also constrained to undergo inversion (cf. Fig. 4.72). In this way, conformational information (chirality in this case) can be transmitted from one position in a dendrimer/polymer to another distant one (schematic, according to Parquette *et al.*)

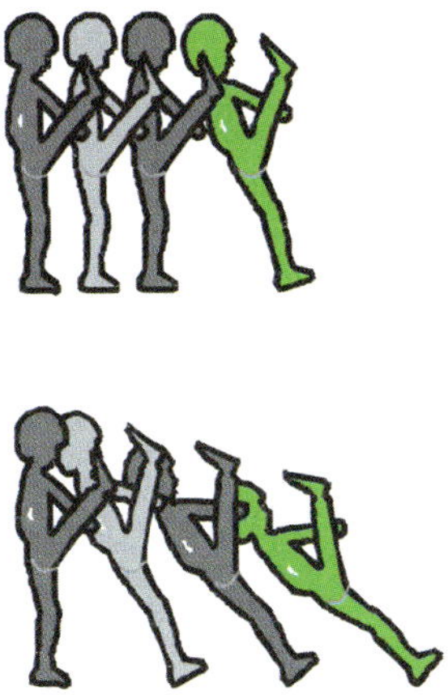

Fig. 4.77 "*Kick-line analogy*" (schematic; according to Parquette *et al.*) for "allosteric transmission" of local perturbations starting from a certain molecule (or part of a molecule) to another: Through specific interactions, a line of dancers forms a defined dance formation (kick line), shown here as a two-dimensional chiral arrangement (three dimensional, if the underlying plane of the paper is taken into consideration – reminiscent of two-dimensional chiral arrangements of clusters on metal surfaces) [38]. A small perturbation ("little green man" falling backwards on the right-hand side of the Figure) at a certain site of the polymerising row of monomers (intermolecular) or at a particular site of a polymer/dendrimer molecule (intramolecular) leads to all other persons (molecules or parts thereof) in contact therewith likewise falling over in the same way in the same direction, with the sense of chirality being retained owing to the close conformational connection. Conformational cooperativity of the monomer units along the backbone the of the molecule increases the energy difference between the two helical polymer configurations. If one of the dancers kicking in synchronised fashion happens to kick in the wrong direction, *i.e.* to kick in the opposite direction to the majority of the others, then the penalty is that the other dancers, whose movements are correlated within the line, are destabilised. This model, taken from polymer chemistry, was quantified by statistical thermodynamics; it can serve as a model for the amplification of chiral information also in other systems – such as dendrimers

ing. This correlated conformational mobility clearly depends upon the size, shape, and rigidity of the dendrons. In any case, this shows that even distant (nanometres apart) groups in a dendrimer are interdependent and "communicate with one another" with regard to their mobility and also with regard to their chirality. Such processes could be detected with the aid of temperature-dependent NMR spectroscopy (cf. Section 7.3.4).

Figure 4.77 suggests how such interaction can take place over large distances: According to the so-called "kick-line analogy" for "allosteric transmission" of local perturbations, a whole series of components of a molecule – shown here as a group of dancers or a chorus line – join in the same movement. If a part of a molecule on the outside right (shown here in green) falls over, then all other components coupled therewith will fall over in a similar manner, permitting transmission of a perturbation even to the most distant parts. This means that mutually adjacent groups transmit a perturbation ("message", "chirality information") from one site of the dendrimer to another, with the sense of the chirality (left or right helix) also being influenced in this way.

That such conformational and configurational effects do actually occur in the transmission of chirality in molecules is demonstrated by corresponding findings and conceptual models from the field of polymer chemistry: The *sergeant-and-soldier principle* and the *majority rules model* [38].

In the sergeant-and-soldier effect, a small number of molecules or parts thereof (*i.e.* sergeants) influence the (conformational) properties of a large number of achiral substrate molecules or parts of molecules (*i.e.* soldiers). Specifically, a low concentration of optically active monomer in a largely achiral monomer leads to predominance of a single helical chirality sense in the resulting polymer (polyisocyanate). The chiral monomer (in green) illustrated in Fig. 4.78 b, which is capable of inversion under the prevailing conditions, is constrained by the existing kick-line **a)** to kick in the same direction (Fig. 4.78 c left), thus

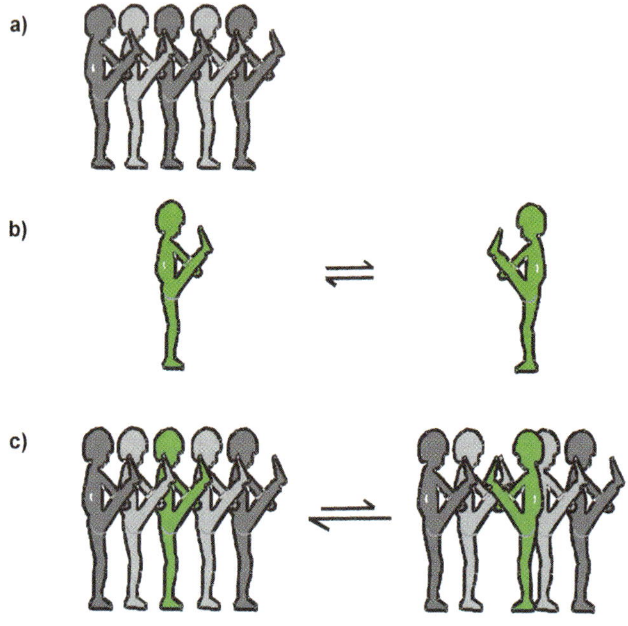

Fig. 4.78 "Majority rules principle" as model for chirality amplification in (intermolecular) polymerisation of (unsymmetrically substituted) monomeric ethenes to form polymers, which can also convey the concept of (intramolecular) transmission of chiral perturbations (*e.g.* emanating from a stereogenic centre) in dendritic molecules: The perturbation introduced into a system having a prior order (dance group **a**) in the form of an excess of one of the two enantiomers **b**, which are interconvertible under the prevailing conditions, acts owing to the defined contact (interactions) within the dance group in such a way that one of the enantiomeric guests (shown in green) is preferentially (selectively) admitted ("incorporated") into the ranks of the majority and thus taken out of the equilibrium **b**) (schematic; according to Parquette *et al.*)

avoiding a "mismatch destabilisation" (**c**, right-hand side) arising from kicking in the wrong direction. Owing to the stability of a polymer helix once formed, this process proceeds through the entire polymer strand and leads to a high chiral amplification: *Just a few (green) sergeants control the spatial orientation of many soldiers.*

According to the majority rules model the chirality of the enantiomer of the chiral monomer present in excess is amplified during polymerisation [38] (Fig. 4.78). Even if just a small enantiomeric excess is present, the resulting polymer shows a helical inequality identical to that of the corresponding chiral homopolymer (see the caption of Fig. 4.78 for details).

Both effects (sergeant/soldier and majority rules) require helical conformations of the polymer strands undergoing formation (with mobile helical points of inversion).

We have gone into considerable detail in the above account since such conformational synchronisation effects are also responsible for many unusual properties of materials in nature (proteins), such as in non-linear responses to external stimuli. Small perturbations of a given parameter can effect drastic changes in the structure and function of proteins in this way. Correlated movements also facilitate the allosteric transfer of local structural perturbations throughout the entire structure and have important consequences for the thermodynamics of folding, complexation, and catalysis of and with proteins. This approach also permits small energy differences to be correspondingly amplified in synthetic functional materials, which is of considerable interest in applications, such as in sensor technology: global response to a local molecular event.

Bibliography and Notes for Section 4.2 "Chiral Dendrimers"

Review articles are indicated by the words "Review(s)" or "Book(s)" in bold-faced type.

1 a) R. G. Denkewalter, J. F. Kolc, W. J. Lukasavage, *US Pat.* 4410688, **1979**; 4289872, **1981**; 4360646, **1982**.

2 ***Reviews***: a) H. W. I. Peerlings, E. W. Meijer, *Chem. Eur. J.* **1997**, *3*, 1563–1579; b) C. W. Thomas, Y. Tor, *Chirality* **1998**, *10*, 53–59; c) D. Seebach, P. B. Rheiner, G. Greiveldinger, T. Butz, H. Sellner, *Top. Curr .Chem.* (Volume Ed. F. Vögtle) **1998**, *197*, 125–164; B. Romagnoli, W. Hayes, *J. Mater. Chem.* **2002**, *12*, 767–799.

3 Classification of chiral dendritic molecules is treated in the ***Review***: C. W. Thomas, Y. Tor, *Chirality* **1998**, *10*, 53–59.

4 a) J. F. G. A. Jansen, H. W. I. Peerlings, E. M. M. de Brabander-van den Berg, E. W. Meijer, *Angew. Chem.* **1995**, *107*, 1321–1324; *Angew. Chem. Int. Ed.* **1995**, *34*, 1206–1209; b) M. Niggemann, H. Ritter. *Acta Polymer* **1996**, *47*, 351–356; c) S. J. E. Mulders, A. J. Brouwer, R. M. J. Liskamp, *Tetrahedron Lett.* **1997**, *38*, 3085–3088; d) A. J. Brouwer, S. J. E. Mulders, R. M. J. Liskamp, *Eur. J. Org. Chem.* **2001**, 1903–1915.

5 ***Reviews***: N. Röckendorf, T. K. Lindhorst, *Top. Curr. Chem.* (Volume Eds. F. Vögtle, C. Schalley), **2001**, *217*, 201–236; M. Dubber, T. K. Lindhorst, *Chem. Commun.* **1998**, 1265–1266.

6 R. H. E. Hudson, M. J. Damha, *J. Am. Chem. Soc.* **1993**, *115*, 2119–2124; L. J. Twymann, A. E. Beezer, J. C. Mitchell, *Tetrahedron Lett.* **1994**, *15*, 4423–4424.

7 a) J. Issberner, M. Böhme, S. Grimme, M. Nieger, W. Paulus, F. Vögtle, *Tetrahedron: Asymmetry* **1996**, *7*, 2223–2232; b) C. Reuter, G. Pawlitzki, U. Wörsdörfer, M. Plevoets, A. Mohry, T. Kubota, Y. Okamoto, F. Vögtle, *Eur. J. Org. Chem.* **2000**, 3059–3067.

8 All known compounds of this kind are derived from 1,1′-binaphthyl derivatives: a) H. W. Peerlings, E. W. Meijer, *Eur. J. Org. Chem.* **1998**, *63*, 573–577; b) S. Yamango, M. Furukawa, A. Azuma, J.-I. Yoshida, *Tetrahedron Lett.* **1998**, *39*, 3783–3786; c) C. Rosini, S. Superchi, H. W. I. Peerlings, E. W. Meijer, *Eur. J. Org. Chem.* **2000**, 61–71; d) A. Bahr, B. Felber, K. Schneider, F. Diederich, *Helv. Chim. Acta* **2000**, *83*, 1346–1376; e) P. Ganghi, B. Huang, J.C. Gallucci, J. R. Parquette, *Org. Lett.* **2001**, *3*, 3129–3132; f) P. Rajakumar, K. Ganesan, *Tetrahedron: Asymmetry* **2005**, *16*, 2295–2298.

9 a) J. Recker, W. M. Müller, U. Müller, T. Kubota, Y. Okamoto, M. Nieger, F. Vögtle, *Chem. Eur. J.* **2002**, *8*, 4434–4442; b) O. Lukin, T. Kubota, Y. Okamoto, A. Kaufmann, F. Vögtle, *Chem. Eur. J.* **2004**, *10*, 2804–2810; O. Lukin, F. Vögtle, *Angew. Chem.* **2005**, *117*, 2–23; *Angew. Chem. Int. Ed.* **2005**, *44*, 1456–1477.

10 a) A. Ritzen, T. Frejd, *Eur. J. Org. Chem.* **2000**, *65*, 3771–3782; b) ***Books***: K. Nakanishi, N. Berova, R. W. Woody, *Circular Dichroism*, VCH, New Volk **1994**; c) P. Schreier, A. Bernreuther, M. Huffer, *Analysis of Chiral Organic Molecules*, de Gruyter, Berlin **1995**.

11 J. R. McElhanon, D. V. McGrath, *J. Am. Chem. Soc.* **1998**, *120*, 1647–1656.

12 Terms such as *macroscopic, nanoscopic,* and *mesoscopic chirality* are known from studies by *Mislow* [a) A. B. Buda, T. Auf der Heyde, K. Mislow, *Angew. Chem.* **1992**, *104*, 1012–1031; *Angew. Chem. Int. Ed.* **1992**, *32*, 989–1007] and *Avnir* [b) O. Katzenelson, H. Z. Hel-Or, D. Avnir, *Chem. Eur. J.* **1996**, *2*, 174–181] and serve for definition of the chirality of large supramolecular and macromolecular systems (*e.g.* chiral clusters, aggregates, polymers, or dendrimers).

13 ***Review***: O. W. Matthews, A. N. Shipway, J. F. Stoddart, *Prog. Polym. Sci.* **1998**, *23*, 1–56.

14 a) Q.-S. Hu, V. Pugh, M. Sabat, L. Pu, *J. Org. Chem.* **1999**, *64*, 7528–7536; b) D. K. Smith, A. Zingg, F. Diederich, *Helv. Chim. Acta* **1999**, *82*, 1225–1241; c) V. Pugh, Q.-S. Hu, L. Pu, *Angew. Chem.* **2000**, *112*, 3784–3787; *Angew. Chem. Int. Ed.* **2000**, *39*, 3638–3641; d) V. J. Pugh, Q.-S. Hu, X. Zuo, F. D. Lewis, L. Pu, *J. Org. Chem.* **2001**, *66*, 6136–6140.

15 a) ***Reviews*** of dendritic catalysts: D. Astruc, F. Chardac, *Chem. Rev.* **2001**, *101*, 2991–3023; b) R. Kreiter, A. W. Kleij, R. J. M. Klein Gebbink, G. van Koten, in *Top. Curr. Chem.* (Volume Eds. F. Vögtle, C. Schalley) **2001**, *217*, 163–199; c) R. van Heerbeek, P. C. J. Kamer, P. W. N. M. van Leeuwen, J. N. H. Reek, *Chem. Rev.* **2002**, *102*, 3717–3756. Special articles: a) A. R. Schmitzer, S. Franceschi, E. Perez, I. Rico-Lattes, A.Lattes, L. Thion, M. Erard, C. Vidal, *J. Am. Chem. Soc.* **2001**, *123*, 5956–5961; c) G. D. Engel, L. H. Gade, *Chem. Eur. J.* **2002**, *8*, 4319–4329; Y. Ribourdouille, G. D. Engel, M. Richard-Plouet, L. H. Gade, *Chem. Commun.* **2003**, *11*, 1228–1229; d) Y.-C. Chen, T.-F. Wu, J.-G. Deng, H. Liu, X. Cui, J. Zhu, Y.-Z. Jiang, M. C. K. Choi, A. S. C. Chan, *J. Org. Chem.* **2002**, *67*, 5301–5306; e) Y.-C. Chen, T.-F. Wu, L. Jiang, J.-G. Deng, H. Liu, J. Zhu, Y.-Z. Jiang, *J. Org. Chem.* **2005**, *70*, 1006–1010; general: T. Ikariya, K. Murata, R. Notori, *Org. Biomol. Chem.* **2006**, 4, 393–406.

16 a) A. Pessi, E. Bianchi, F. Bonelli, L. Chiappinelli, *J. Chem. Soc., Chem. Commun.* **1990**, 8–9; b) T. D. Pallin, J. P. Tam, *Chem. Commun.* **1996**, 1345–1346; c) M. S. Shchepinov, I. A. Udalova, A. J. Bridgman, E. M. Couthern, *Nucleic Acids Res.* **1997**, *25*, 4447–4454; d) D. Ranganathan, S. Kurur, *Tetrahedron Lett.* **1997**, *38*, 1265–1268.

17 Peptidic dendrimers as protein mimics: a) C. Douat-Casassus, T. Darbre, J.-L. Reymond, *J. Am. Chem. Soc.* **2004**, *126*, 7817–7826; b) G. Sanclimens, L. Crespo, E. Giralt, F. Albericio, M. Royo, *J. Org. Chem.* **2005**, *70*, 6274–6281.

18 a) D. Seebach, J.-M. Lapierre, G. Greiveldinger, K. Skobridis, *Helv. Chim. Acta* **1994**, *77*, 1673–1688; b) D. Seebach, J.-M. Lapierre, K. Skobridis, G. Greivel-

dinger, *Angew. Chem.* **1994**, *106*, 457–458; *Angew. Chem. Int. Ed.* **1994**, *33*, 440–442; c) P. K. Murer, D. Seebach, *Angew. Chem.* **1995**, *107*, 2297–2300; *Angew. Chem. Int. Ed.* **1995**, *34*, 2116–2119.

19 The term *dendrophane* was coined by Diederich: a) S. Mattei, P. Seiler, F. Diederich, *Helv. Chim. Acta* **1995**, *78*, 1904–1912; b) B. Kenda, F. Diederich, *Angew. Chem.* **1998**, *110*, 3357–3361; *Angew. Chem. Int. Ed.*. **1998**, *37*, 3154–3158.

20 K. Mislow, P. Bickart, *Isr. J. Chem.* **1977**, *15*, 1–6.

21 H. W. I. Peerlings, D. C. Trimbach, E. W. Meijer, *Chem. Commun.* **1998**, 497–498.

22 U. Hahn, A. Kaufmann, M. Nieger, O. Julinek, M. Urbanova, F. Vögtle, *Eur. J. Org. Chem.* **2006**, 1237–1244.

23 ***Review***: *Top. Curr. Chem.* (Volume Eds. K. Sakai, N. Hirayama, R. Tamura) **2007**, Vol. 296.

24 K. Portner, F. Vögtle, M. Nieger, *Synlett* **2004**, 1167–1170; H. Kim, S. M. So, J. Chin, B. M. Kim, *Aldrichimica Acta* **2008**, *41*, 77–88.

25 L.-Z. Gong, Q.-S. Hu, L. Pu, *J. Org. Chem.* **2001**, *66*, 2358–2367.

26 H.-T. Chang, C.-T. Chen, T. Kondo, G. Siuzdak, K. B. Sharpless, *Angew. Chem.* **1996**, *108*, 202–206; *Angew. Chem. Int. Ed.* **1996**, *35*, 182–186.

27 ***Review***: H.-F. Chow, C. C. Mak, *Pure Appl. Chem.* **1997**, *69*, 483–488.

28 C.-O. Turrin, J. Chiffre, D. de Montauzon, G. Balavoine, E. Manoury, A.-M. Caminade, J. P. Majoral, *Organometallics* **2002**, *21*, 1891–1897.

29 G. R. Newkome, X. Lin, C. D. Weis, *Tetrahedron: Asymmetry* **1991**, *2*, 957–960.

30 a) J. F. G. A. Jansen, H. W. I. Peerlings, E. M. M. de Brabander-van den Berg, E. W. Meijer, *Angew. Chem.* **1995**, *107*, 1321–1324; *Angew. Chem. Int. Ed.* **1995**, *34*, 1206–1209; b) J. F. G. A. Jansen, E. M. M. de Brabander-van den Berg, E. W. Meijer, *Science* **1994**, *266*, 1226–1229.

31 J. F. G. A. Jansen, E. M. M. de Brabander-van den Berg, E. W. Meijer, *Recl. Trav. Chim. Pays-Bas* **1995**, *114*, 225–230.

32 a) H.-F. Chow, C. C. Mak, *Tetrahedron Lett.* **1996**, *37*, 5935–5938; b) M. L. Lartigue, A. M. Caminade, J. P. Majoral, *Tetrahedron: Asymmetry* **1997**, *8*, 2697–2708; c) C. O. Turrin, J. Chiffre, J. C. Daran, D. de Mountauzon, A. M. Caminade, E. Manoury, G. Balavoine, J. P. Majoral, *Tetrahedron* **2001**, 57, 2521–2536; d) J. Issberner, M. Böhme, S. Grimme, M. Nieger, W. Paulus, F. Vögtle, *Tetrahedron: Asymmetry* **1996**, *7*, 2223–2232.

33 a) K. Aoi, K. Itoh, M. Okada, *Macromolecules* **1995**, *28*, 5391–5393; b) T. K. Lindhorst, C. Kieburg, *Angew. Chem.* **1996**, *108*, 2083–2086; *Angew. Chem. Int. Ed.* **1996**, *35*, 1953–1956.

34 G. R. Newkome, Z.-Q. Yao, G. R. Baker, V. K. Gupta, *J. Org. Chem.* **1985**, *50*, 2003–2004.

35 P. R. Ashton, S. E. Boyed, C. L. Brown, S. A. Nepogodiev, E. W. Meijer, H. W. I. Peerlings, J. F. Stoddart, *Chem. Eur. J.* **1997**, *3*, 974–984.

36 M. S. T. H. Sanders-Hovens, J. F. G. A. Jansen, J. A. J. M. Vakemans, E. W. Meijer, *Polym. Mater. Sci. Eng.* **1995**, *73*, 338–339.

37 ***Reviews***: a) J. W. Lockman, N. M. Paul, J. R. Parquette, *Progr. Polym. Sci.* **2005**, *30*, 423–452; and further literature cited therein; b) M. Crego-Calama, D. N. Reinhoudt (Volume Eds.), *Supramolecular Chirality, Top. Curr. Chem.* **2006**, 265; c) see also: E. Francotte, W. Lindner (Eds.), Chirality in Drug Research, in: *Methods and Principles* in *Medicinal Chemistry*, Vol. 33, VCH-Wiley, Weinheim-New York **2007**; d) V. V. Borovkov, Y. Inoue, *Top. Curr. Chem.* (Volume Eds. M. Crego-Calama, D. N. Reinhoudt) **2006**, *265*, 89–146.

38 S. Weigelt, C. Busse, L. Petersen, E. Rauls, B. Hammer, K. V. Gothelf, F. Besenbacher, T. R. Linderoth, *Nature Materials* **2006**, *5*, 11; ***Review***: K.-H. Ernst, *Top. Curr. Chem.* (Volume Eds. M. Crego-Calama, D. N. Reinhoudt) **2006**, *265*, 209–252; cf. also: K. Toyofuku, A. Alam, A. Tsuda, N. Fujita, S. Sakamoto, K. Yamaguchi, T. Aida, *Angew. Chem.* **2007**, *119*, 6596–6600; *Angew. Chem. Int. Ed.* **2007**, *46*, 6476–6480; ***Books***: K. Soai (ed.), *Amplification of Chirality, Top. Curr. Chem.* **2008**, vol. *284*; D. B. Amabilino (ed.), *Chirality at the Nanoscale,* Wiley-VCH, Weinheim **2009**.

5
Photophysical properties of dendritic molecules

Dendrimers attract avid interest on account of their physical properties, whether due to their viscosity behaviour, to photoinduced energy transfer and electron transfer, or to light-harvesting effects [1]. In particular, the photophysical properties have been investigated owing to the challenge of, say, amplifying their luminescence by attachment of many similar – or different, mutually interacting – photoactive molecular building blocks (chromophores, fluorophores, luminophores). In the following section these properties will therefore be presented in more general terms – from the standpoint of basic research – whereas other, primarily applications-oriented properties will be considered in Chapter 8.

5.1
Luminescence and energy transfer

5.1.1
Luminescence

Luminescence is defined as emission of light by materials in the visible, UV, and IR spectral range after input of energy [2]. Emission of light quanta by molecules from a (photo)excited state proceeds *via* transition of an electron to a lower energy level. Depending upon the way in which the excited state is generated, a distinction is made between photoluminescence, chemoluminescence, radioluminescence, and thermoluminescence (Table 5.1). In the case of photo-

Table 5.1 Types of luminescence and their mode of excitation

Types of luminescence	Mode of excitation
Photoluminescence	Light absorption
Chemoluminescence	Chemical reaction
Radioluminescence	Nuclear radiation
Thermoluminescence	Thermally activated ion recombination
Electroluminescence	Electric field

Dendrimer Chemistry. Fritz Vögtle, Gabriele Richardt and Nicole Werner

ISBN: 978-3-527-32066-0

luminescence, the length of time between excitation and emission determines whether the phenomenon is classed as fluorescence or phosphorescence.

Fluorescence: The electron returns from the excited state (lifetime *ca.* 10^{-9} s) to the original state (ground state) (Fig. 5.1 a). Emission proceeds spontaneously after absorption – the time lapse between excitation and emission is negligibly small in the case of fluorescence – at a longer wavelength. The absorption spectrum and the fluorescence spectrum are often almost mirror images of each

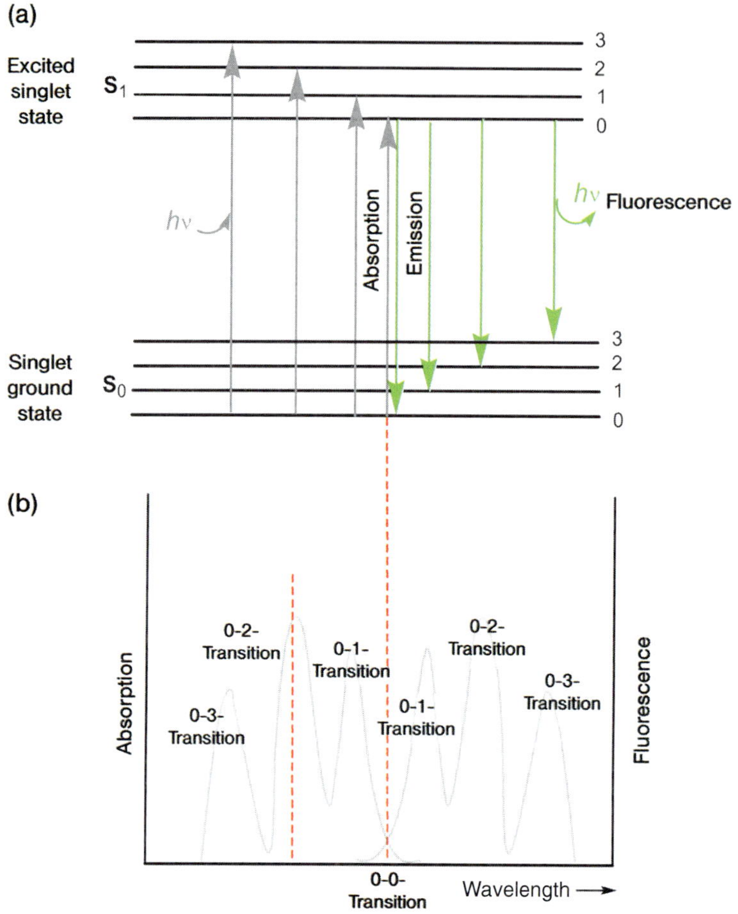

Fig. 5.1 Absorption and emission (fluorescence). a) Overview of electronic and vibronic energy levels (schematic). The absorption spectrum displays a vibrational structure which is characteristic of the photoexcited state (S_1). b) Band spectra with *Stokes* shift (schematic). The fluorescence spectrum closely resembles the mirror image of the absorption spectrum. The vibrational structure of the fluorescence spectrum is characteristic of the ground state (S_0); it is shifted to longer wavelengths. Only the two 0-0 transitions lie at the same place. The Stokes shift provides information about (absorption/emission) shifts of other transitions

other (Fig. 5.1 b). The wavelength difference between the absorption and emissions maxima is also designated as the “Stokes shift”.

The emission of fluorescence light quanta may be suppressed by fluorescence quenching. In such cases, excitation energy

- can be lost by non-radiative deactivation, when molecules collide with other molecules, or
- is emitted in the form of phosphorescence light quanta.

Fluorescence spectroscopy is often used for detection of atoms and molecules. Laser light is increasingly used for excitation (LIF: Laser-induced fluorescence).

The *fluorescence yield* describes the ratio of the number of photons emitted in the excited singlet state (S_1) to the number of photons absorbed (values between 0 and 1).

Phosphorescence: In phosphorescence the electron falls from the excited singlet state by inter-system crossing to the lowest triplet state (T_1; lifetime 10^{-6}–10^{-3} s), where it initially remains because a transition to the singlet state is symmetry-forbidden, and subsequently returns to the ground state (Fig. 5.2). The length of time between excitation and emission is therefore significantly greater than in fluorescence.

The *phosphorescence yield* is defined as the ratio between the number of photons emitted in the triplet state and the number of photons absorbed (values between 0 and 1).

Excimer is an abbreviation of the term “excited dimer” used to describe a short-lived compound made up of two molecules (or atoms), which can only exist in the excited state. In the ground state it decomposes into its components which may also repel each other. All components of an excimer molecule have to be identical. Such molecules are used in laser technology.

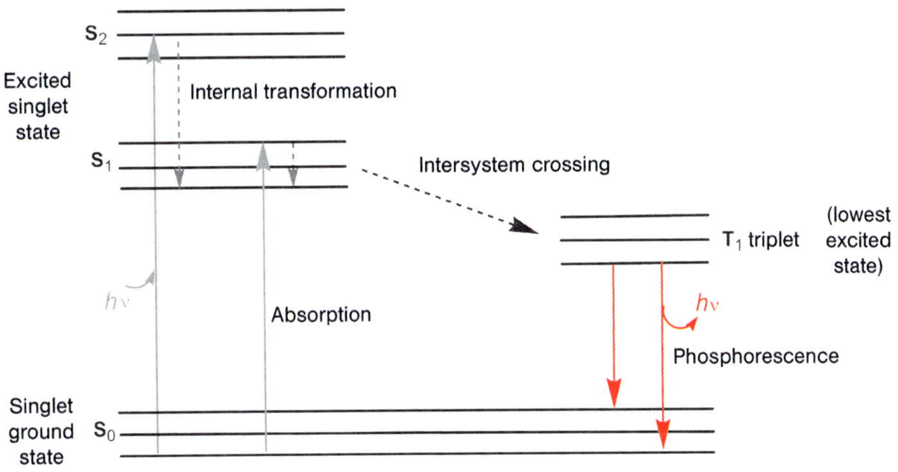

Fig. 5.2 Phosphorescence (schematic)

Exciplex is a combination of the words *excited* and *complex*. Exciplexes are also non-existent in the ground state. In contrast to an excimer, however, the components of a – short-lived – exciplex can be molecules or atoms differing in nature.

5.1.2
Energy transfer

5.1.2.1 Dexter mechanism: Energy transfer by radiative emission

The Dexter mechanism describes electron exchange from the excited state of the donor to the excited state of the acceptor, accompanied by a simultaneous exchange of an electron of the ground state from the acceptor to the donor (Fig. 5.3). This electron exchange necessitates overlapping of the donor- and acceptor orbitals, yet no spectral overlapping is required. This is a short-range interaction (over a distance of less than 10 Å) which decreases exponentially with increasing distance.

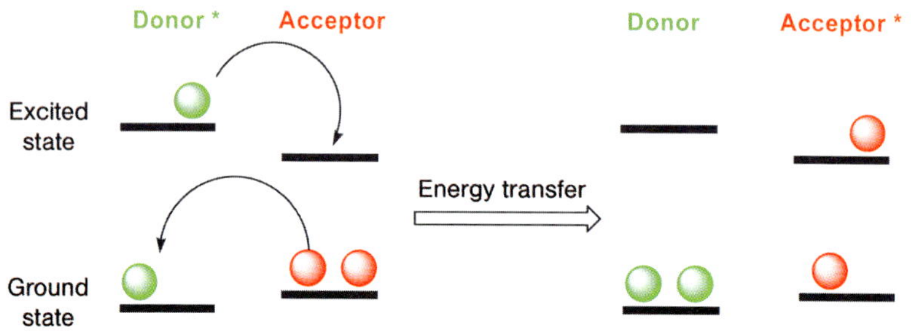

Fig. 5.3 Energy transfer according to the Dexter mechanism (schematic)

5.1.2.2 Förster mechanism: Energy transfer by dipole-dipole interactions

A non-radiative energy transfer solely by dipole-dipole interactions – without recourse to electron exchange – from energy donor to an acceptor (Fig. 5.4) is described by the *Förster* mechanism.

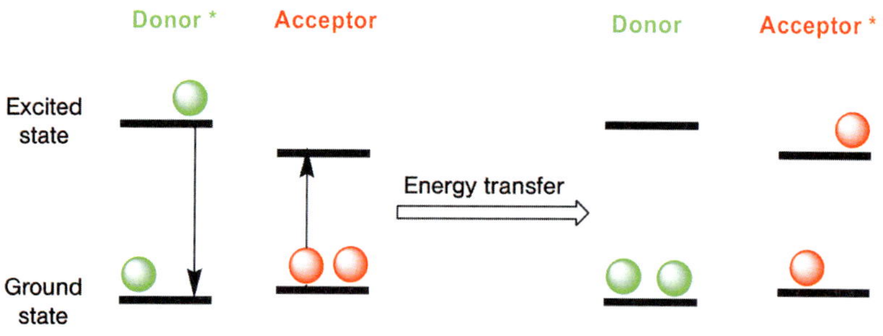

Fig. 5.4 Energy transfer according to the Förster mechanism (schematic)

An important condition for energy transfer in this manner is an overlapping of the fluorescence emission spectrum of the donor with the absorption spectrum of the acceptor. A small distance (less than 10 nm) between donor and acceptor is likewise important because according to Förster the efficiency of energy transfer decreases with the inverse sixth power of the distance between donor and acceptor:

$$E = 1/[1 + (r/R_0)^6] \tag{5.1}$$

r = distance between donor and acceptor
R_0 = Förster radius

Here the chromophores can be farther apart (10–100 Å; long-range interactions) since this mechanism does not require any overlapping of orbitals. The Förster radius designates the distance between donor and acceptor at which the efficiency of energy transfer amounts to exactly 50%. Half of the excited donor molecules are then deactivated by fluorescence resonance energy transfer, and the other 50% by fluorescence or phosphorescence.

5.1.2.3 Examples from the field of dendritic molecules

The photophysical effects and mechanisms described in general terms in the above sections will now be illustrated with the aid of findings and interpretations from the world of dendrimers. Since straightforward luminescence effects are also mentioned elsewhere in this book in the context of syntheses (Chapter 2) and the individual types of compounds (Chapter 4) and are, moreover, considered as part of more complex events in Section 5.2, just a few characteristic examples of fluorescence and phosphorescence of dendrimers will be presented: The adjacency of many groups in a molecule may lead to the occurrence of complex processes, which sometimes make it difficult to draw precise conclusions, but may sometimes also be useful and amenable to optimisation for various applications.

In principle, dendrimers and dendrons can be designed in such a way that luminescent building blocks (fluorophores, luminophores) are present in the dendrimer itself, in appropriate interaction partners (guests), or in both and can be observed during their interaction. This provides a means of analysing supramolecular (host/guest) interactions thermodynamically and kinetically, elucidating sensor effects, and developing diagnostic applications (cf. also ref. [3]).

a) Fluorescent dendrimer

The dendrimer of mixed POPAM/Fréchet type shown in Fig. 5.5 formally bears several types of fluorophores in three “shells”: naphthyl units on the periphery; two layers with dimethoxybenzene character further inside; and finally the strongly fluorescing dansyl groups – adding up to a total of 64 luminophoric moieties. Incident UV radiation interacts first with the naphthyl groups, which

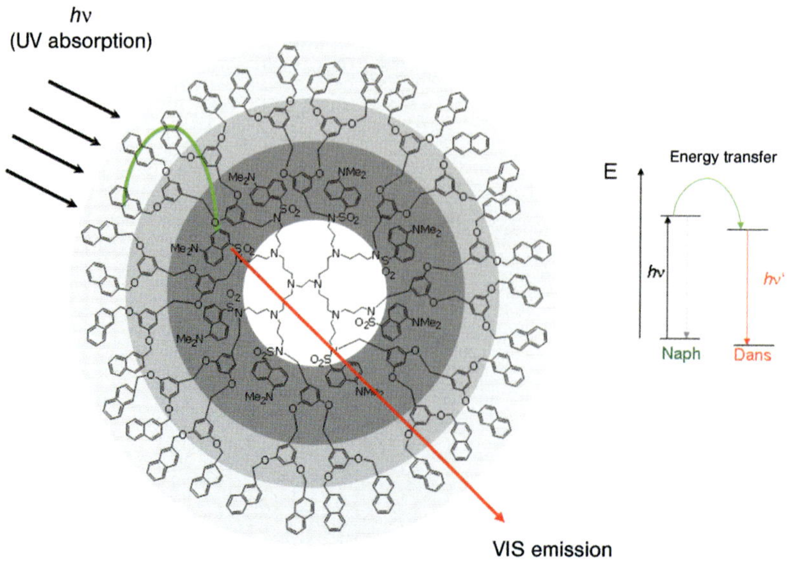

Fig. 5.5 Incident UV light energy is transferred from the 32 peripheral naphthyl chromophores to the dansyl moieties and emitted as visible light (schematic; according to Balzani, Vögtle *et al.*)

then channel the energy *via* energy transfer (ET, *vide supra*) to the intramolecularly adjacent dansyl groups – with lower-lying electronic energy levels (see Fig. 5.5). The dansyl system ultimately emits visible light (λ 514 nm).

Addition to such fluorescent dendrimers of non-fluorescing guest compounds capable of interacting with the hetero atoms of the dendrimer – Cu^{2+} or Co^{2+} ions in the simplest case – leads to (partial) quenching of fluorescence. A specific example of such behaviour will be found in Section 6.2.3.3.

b) Fluorescent guests (in a non-fluorescent dendrimer)

Figure 5.6 depicts the octahedral complex of three non-luminescent, doubly dendron-substituted bipyridines with the photoactive ruthenium cation [4]. The dashed Ru^{2+}-N bonds are so stable that even the peripheral carboxyl groups can be transformed, for example, into the acyl chloride and subsequently into the carboxamide, without detriment to the complex. Balzani *et al.* showed that the lifetime of the photoexcited state, and thus the photostability of the dendritic complexes – whose numerous ether groups ensure water solubility – is greater than for the non-substituted reference compound (tris-bipyridine-ruthenium complex). Moreover, it also increases significantly with increasing generation number – both in the presence and in the absence of atmospheric oxygen: "positive dendritic effect". This property is highly desirable in the case of tris-bipyridine-ruthenium complexes because such species are used in the photocleavage of water (hydrogen technology).

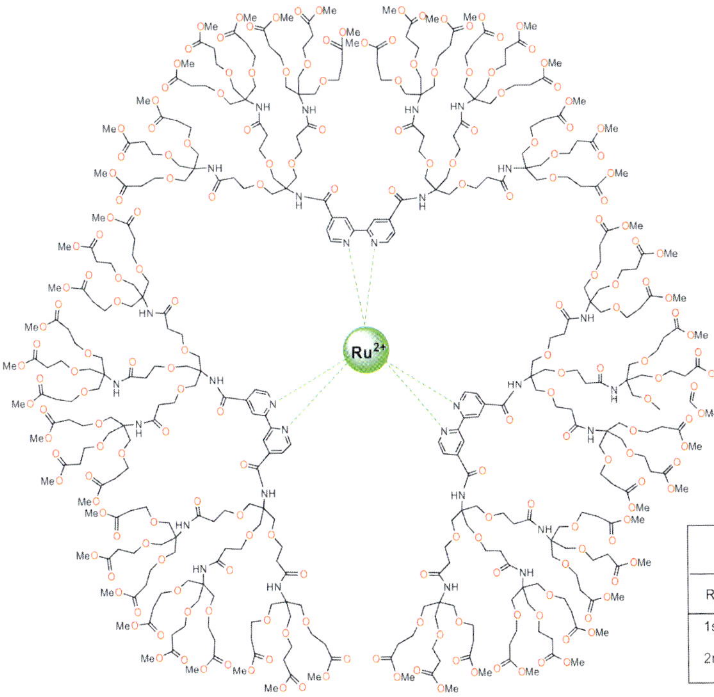

Shortened lifetime through quenching with O_2

	with O_2 [ns]	Without O_2 [ns]
$Ru[(bipy)_3]^{2+}$	172	990
1st generation	760	
2nd generation	1010	1740

Fig. 5.6 Lifetime of the photoexcited state of dendritic Ru^+-bipyridine complexes – in the presence and absence of atmospheric oxygen (according to Balzani, De Cola, Vögtle *et al.*), as compared with unsubstituted ruthenium-tris-bipy

Figure 5.7 compares the luminescence lifetimes of various homo- and heteroleptic $[Ru(bipy)_3]^{2+}$ complexes with acyclic and macrocyclic bipyridines together with the dendritic species just considered [4]. Here too it is apparent that the stability towards photooxidation is again greater for the more sterically shielded homoleptic complexes than for the less strongly shielded heteroleptic species, which each contain another unsubstituted bipyridine ligand. However, the variant with "dendritic protective shield" proves to be the "champion" (dendritic effect).

c) Fluorescent dendrimer and fluorescent guests
This case is illustrated in Fig. 5.8. The fluorescent dendrimer already introduced in Fig. 5.5 was used. On addition of the fluorescing guest molecule eosin and irradiation with UV light, the eosin collects the light energy of all 64 chromophoric groups because of its lower-lying energy levels (light harvesting, see Section 5.2). As a consequence, visible light of the (fluorescence) wavelength of eosin (λ 555 nm) is emitted [5]. Such effects are of interest in view of the possible future artificial imitation of basic photosynthetic processes (Fig. 5.9) [6].

Another example of this type of interaction is given in Section 3.2.1 (Fig. 3.14).

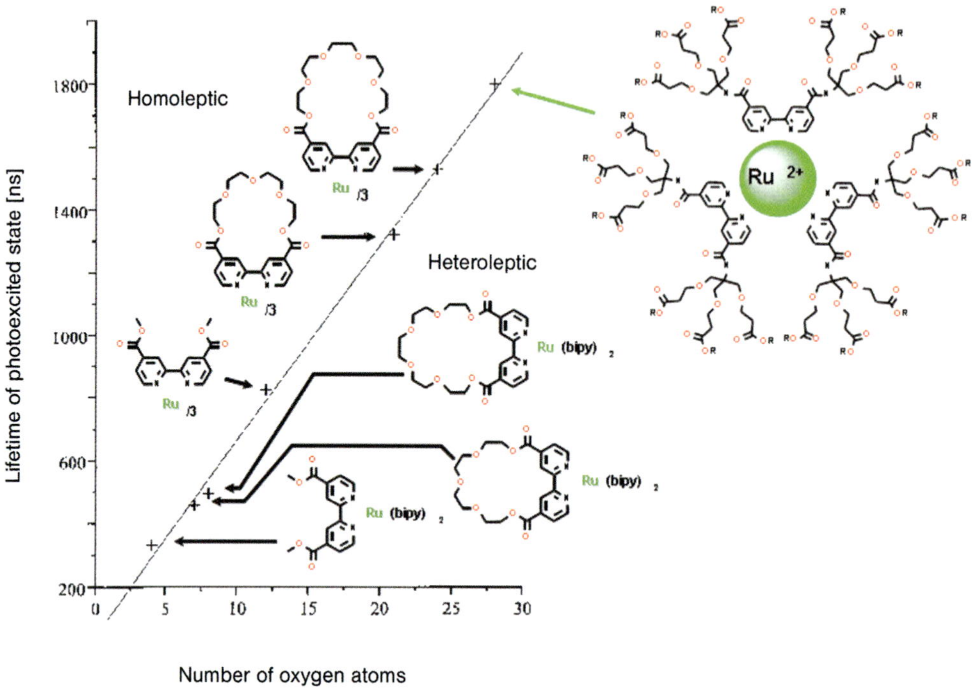

Fig. 5.7 Luminescence of homoleptic and heteroleptic $[Ru(bipy)_3]^{2+}$ complexes as a function of the number of oxygen atoms in the molecule – and hence of the steric shielding of the central ion (according to Balzani, Vögtle *et al.*)

5.2 Antenna effect and photoisomerisation of dendrimers

5.2.1 Antenna effect

The possibility of arranging many functional groups around a core moiety opens the way to purposeful construction of dendritic light-harvesting systems, permitting, for example, energy transfer between periphery and core. Light-harvesting antenna systems generally contain a multitude of mutually interacting light-absorbing chromophore molecules which – in their capacity as donors – transmit the excitation energy to an acceptor unit in the interior of the dendrimer (Fig. 5.5) [3, 7]. If the chromophores do not interact, the energy collected will be dissipated and the excited states return to the ground state without any energy transfer.

The principle of the antenna effect is based on long-range interactions (see Section 5.11) between periphery and core. These make it possible to influence

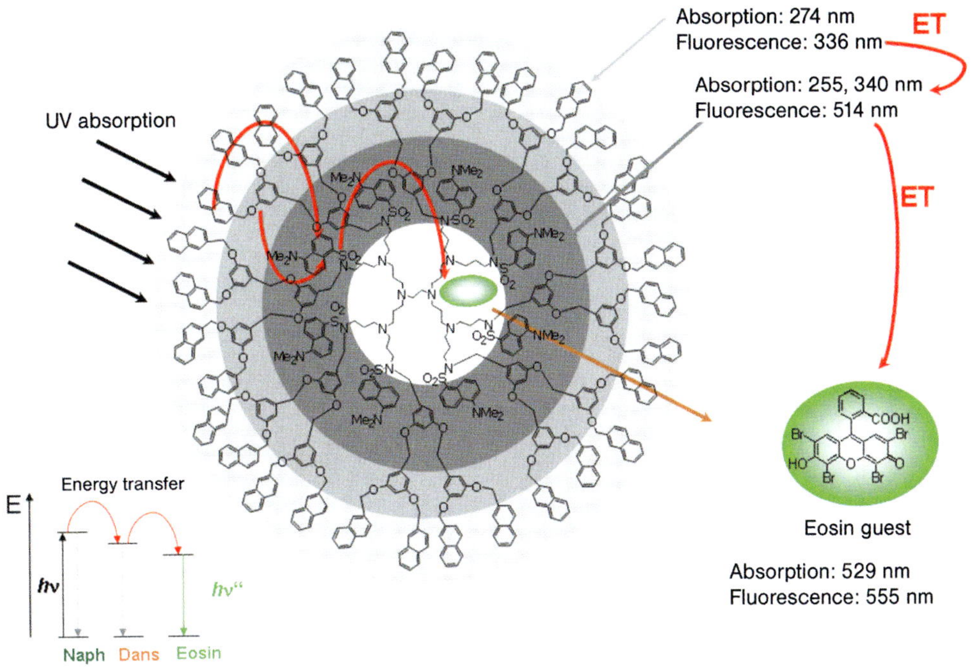

Fig. 5.8 Quenching of the fluorescence of the dendrimer host in the presence of eosin (as guest) by energy transfer starting from the photoactive dansyl groups. The guest molecule eosin collects the light energy of all 64 chromophoric groups (naphthalene, dialkoxyphenyl, dansyl) (schematic; according to Vögtle, Balzani, Ceroni, *et al.*)

the core instead of just the periphery, as would be the case in the absence of these interactions. Dendritic light-harvesting systems (antennae) [8] could be constructed by targeted end-group functionalisation of a dendrimer with appropriate mutually interacting chromophore units – in order to attain a large absorption cross section. A convergent synthetic approach afforded dye-bearing dendrimers (Fig. 5.10) in which incident light is directed *via* the chromophores on the periphery to the fluorescent core.

Energy transfer does not proceed *via* a multistage hopping process; instead the energy is transported in targeted fashion to the focal point, that is the core. To this end, cumarin 2 served as – multiple – terminal donor chromophore and cumarin 343 as acceptor dye in the core of the dendritic molecule [9]. Energy transfer probably proceeds *via* dipole/dipole interactions (Förster mechanism, Section 5.1), because the selected chromophoric system possesses strong dipole moments. Energy transfer *via* the (covalent) dendrimer scaffold can be excluded on the basis of the absorption maxima. Energy transfer was seen to proceed with high efficiency in such dendrimers of up to the fourth generation.

Owing to their spheroidal molecular shape, fullerenes represent attractive functional core units for light-harvesting dendritic systems. In fullerene dendrimers of

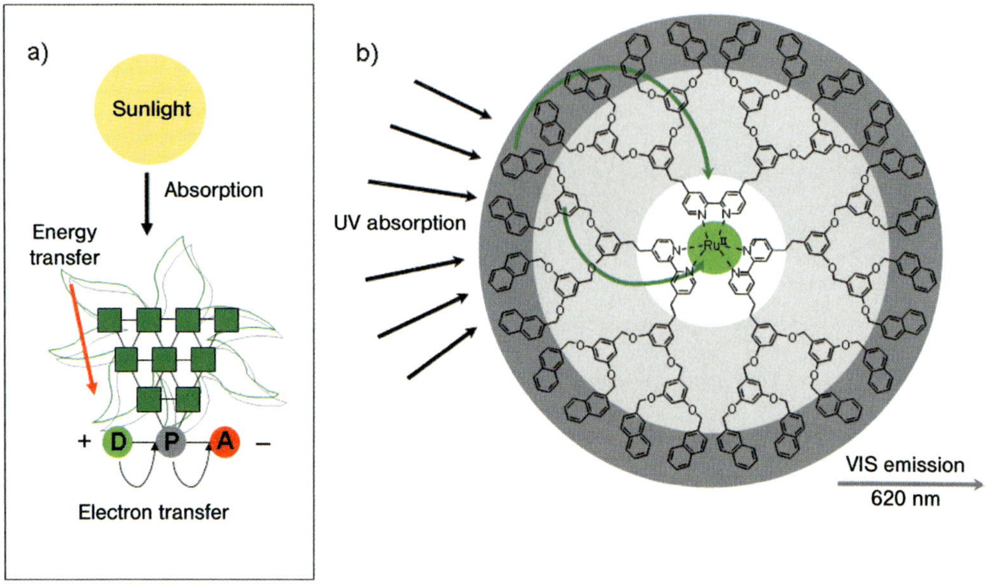

Fig. 5.9 a) Light energy is harvested with the aid of chlorophyll and is used by plants for photosynthesis. b) Example of a simple dendritic light-harvesting system: Nanometre-scale complex made up of photoactive dendrons and photoactive metal ion. The UV light (λ 270 nm) absorbed by the naphthyl groups is transferred to the ruthenium which emits visible light of 620 nm wavelength (according to Balzani, De Cola, Vögtle *et al.*)

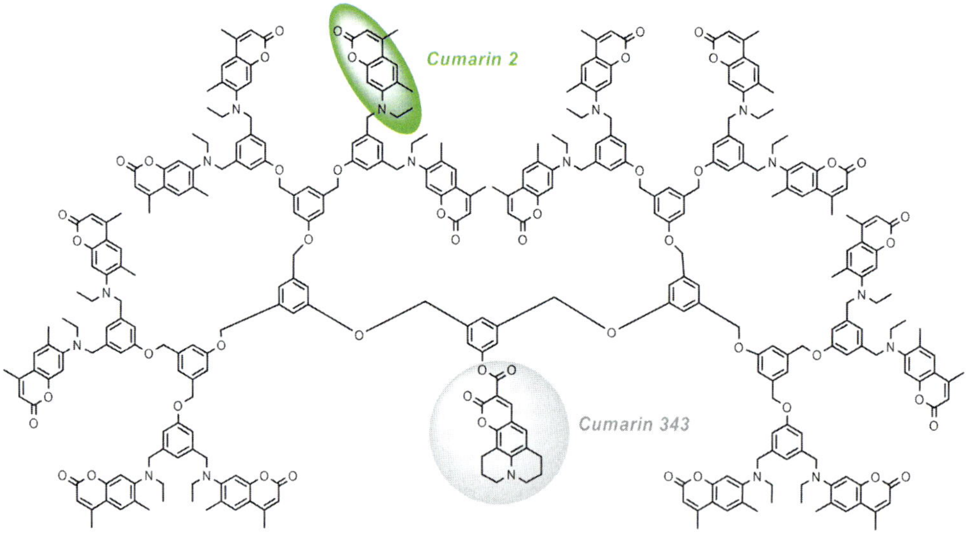

Fig. 5.10 Light-harvesting dendritic antenna (according to Fréchet *et al.*)

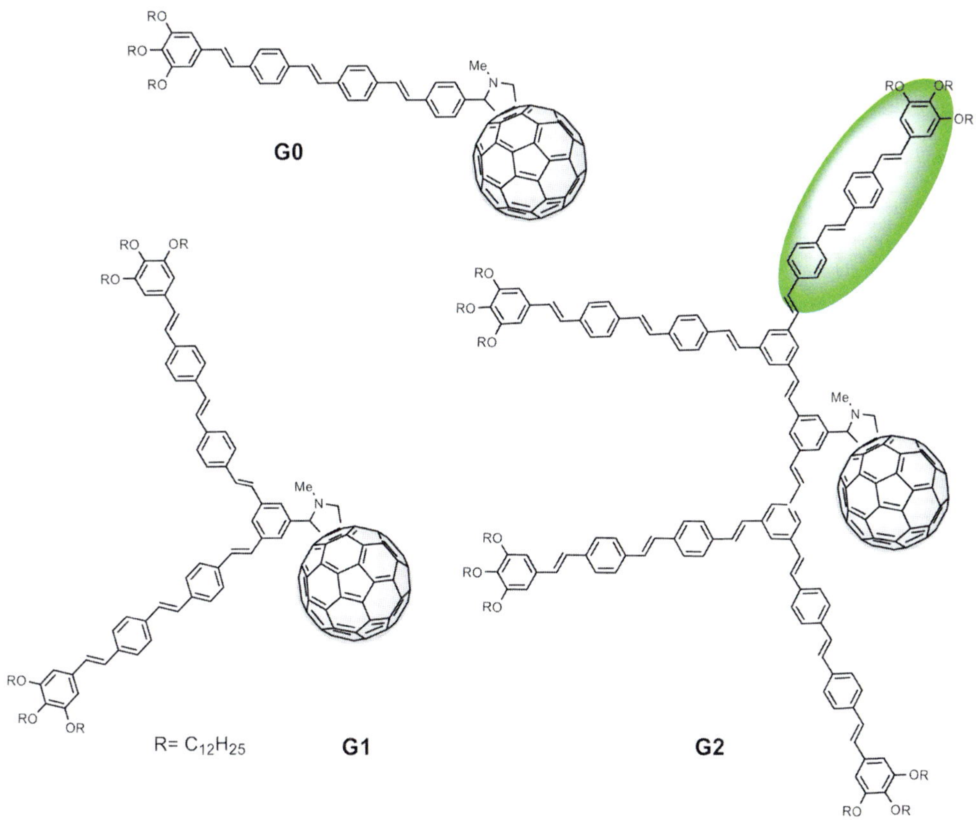

Fig. 5.11 Zeroth- to second-generation fullerene dendrimers (according to Nierengarten *et al.*)

the type shown in Fig. 5.11 the fluorescence of the terminal oligophenylenevinylene groups arising on excitation at the absorption maximum is quenched.

This constitutes evidence for a singlet-singlet energy transfer of oligophenylenevinylene groups to the fullerene core. At 394 nm – corresponding to the oligophenylenevinylene absorption maximum – the molar absorption of the fullerene dendrimer increases with increasing generation number (Gn). The light-harvesting properties become more pronounced with increasing polyene groups [10].

Corresponding studies of the dynamics of energy transfer from the periphery to the core in two different types of light-harvesting dendrimers of different generation numbers have shown that interactions over shorter distances are of decisive importance owing to overlapping of wave functions between periphery and core. The high speed of energy transfer is explained in terms of the rapid relaxation from the Frank-Condon state in the periphery [11].

Moore *et al.* used a convergent approach to construct first- to sixth-generation dendrons based on arylacetylene units. These were functionalised at the focal point with a – fluorescent – perylene unit and investigated with regard to their flu-

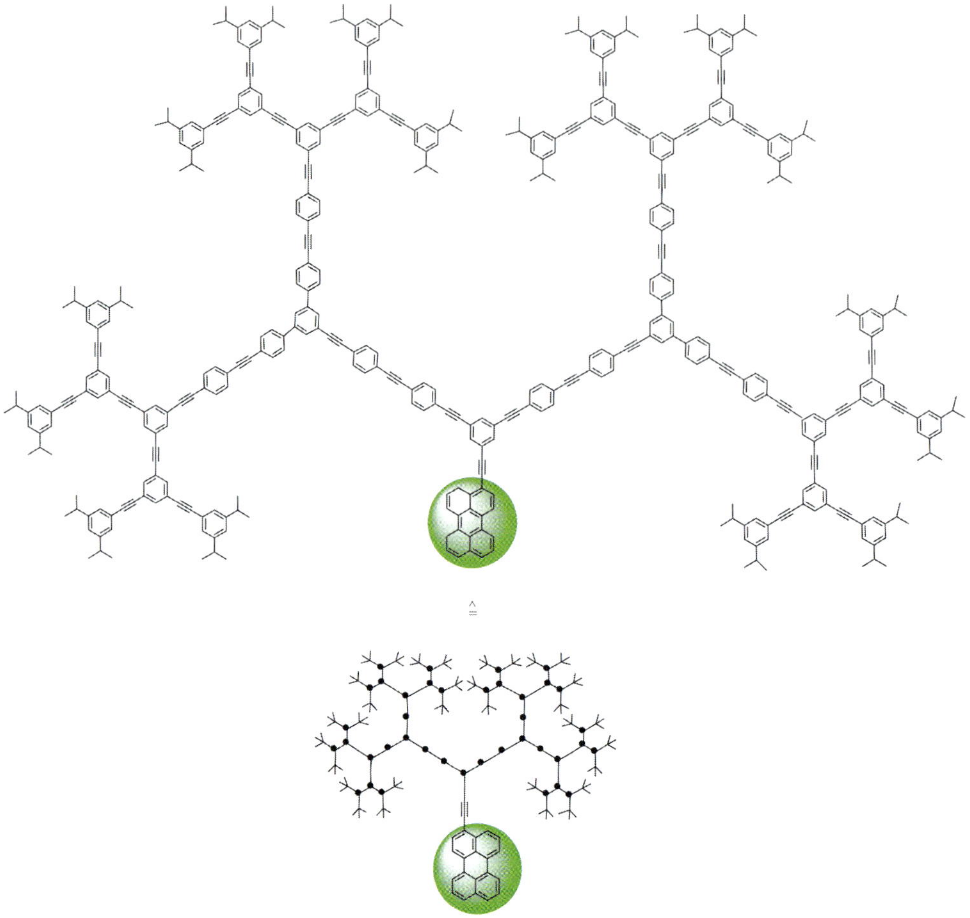

Fig. 5.12 Dendritic energy funnel (according to Moore *et al.*)

orescence properties [12, 13]. An increase of fluorescence intensity was observed with increasing generation number, which is attributed to a growing molar extinction coefficient (positive dendritic effect). Accordingly, a light harvesting-effect is again observed here. In contrast, the efficiency of energy transfer in the dendritic molecule from the arylacetylene units to the perylene group – as acceptor – decreases with increasing dendrimer generation number (negative dendritic effect) [13]. Acceleration of energy transfer from the periphery to the focal point along a gradient was accomplished by extension of the electron system with additional tolane units (Fig. 5.12), with the number of arylacetylene spacers being increased one at a time from the outside inwards with increasing degree of branching. The electron falls to ever lower energy levels on passing through the individual branching units of the molecule. Excitation reaches just the chromophores in the scaffold, and is then transferred to perylene at the focal point and appears in the form of fluorescence. This dendritic system therefore acts as a kind of energy funnel [14].

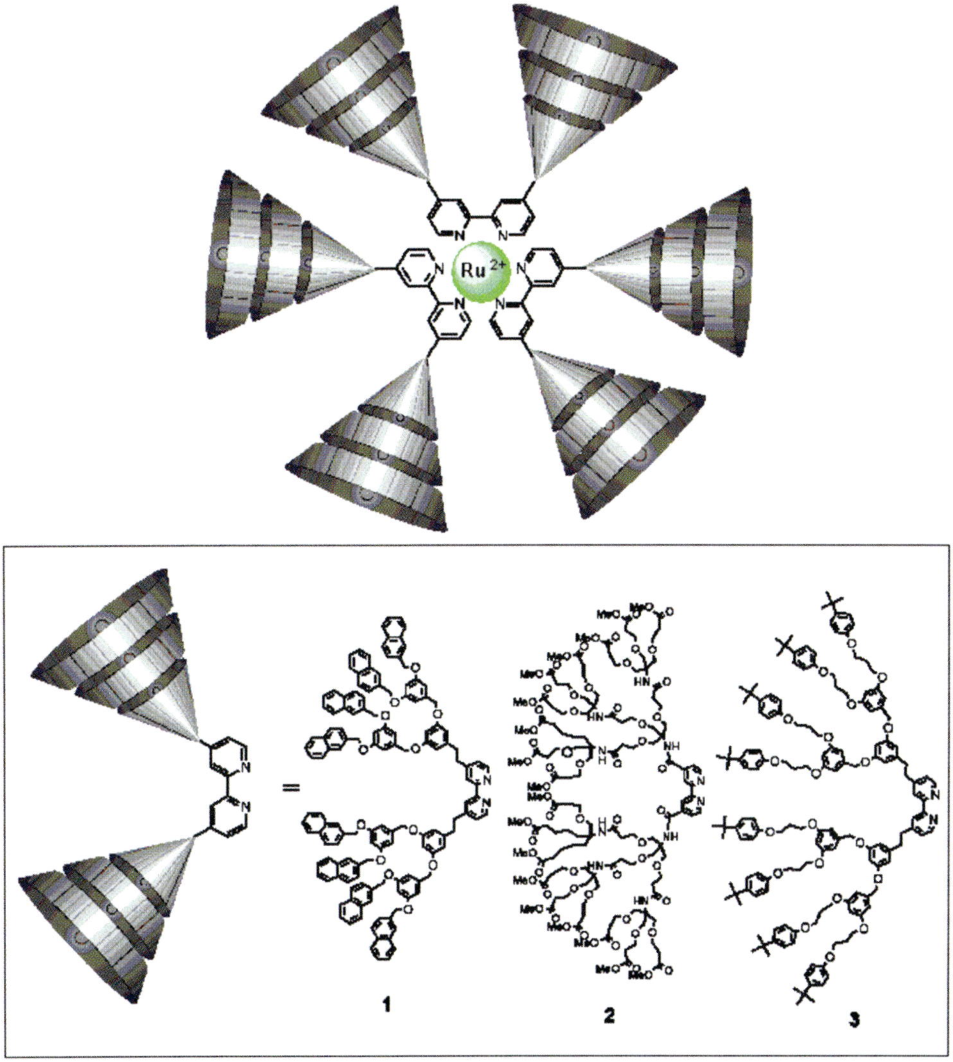

Fig. 5.13 Dendritic tris(bipyridine)-ruthenium(II) complexes (according to Balzani, De Cola, Vögtle *et al.*)

Vögtle *et al.* [4] synthesised the first dendritically substituted 2,2′-bipyridines, which likewise display an antenna effect in light harvesting with amplification of luminescence. Starting from these ligands, corresponding dendritic tris(bipyridine)-ruthenium(II) complexes were prepared (Fig. 5.13) [15]. The antenna effect exerted by the outer naphthalene units in **1** is manifested in amplified luminescence, which is attributable to efficient energy transfer from light-absorbing naphthalene groups to the light-emitting ruthenium cation [16]. The quan-

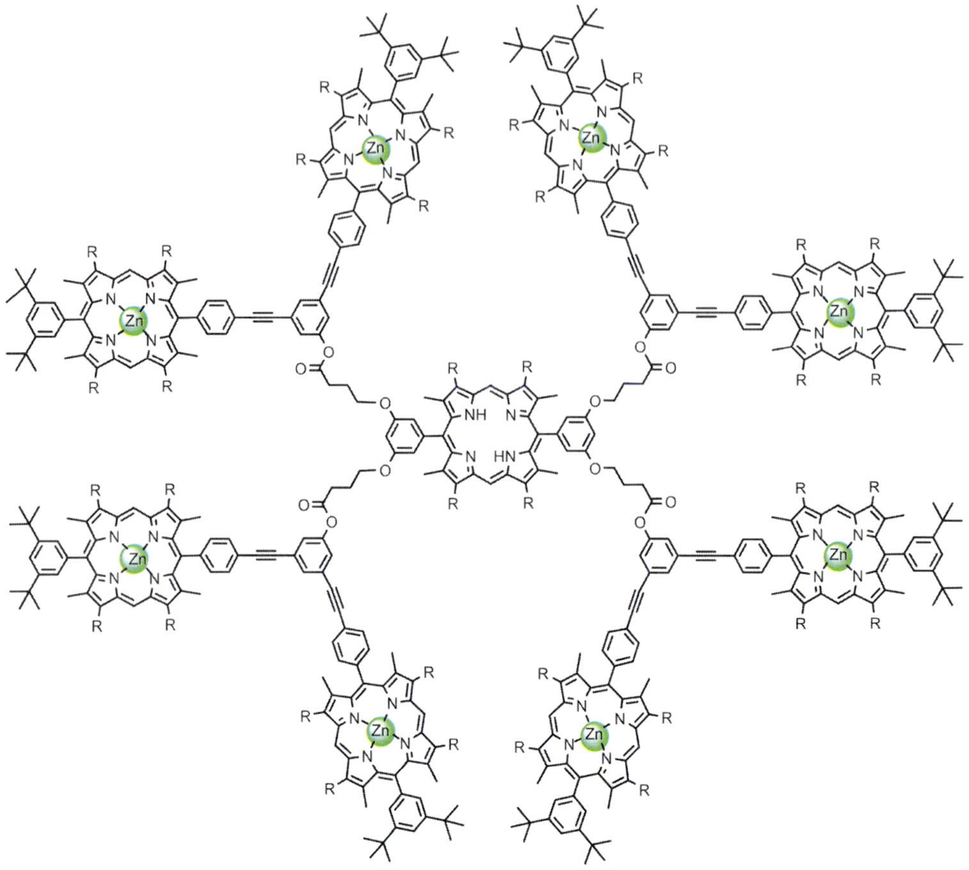

Fig. 5.14 Dendritic multiporphyrin system (according to Aida *et al.*)

tum yield and the lifetime of the photoexcited state increase with increasing generation number in the absence, but also – albeit less strongly – in the presence of molecular oxygen (cf. Section 5.1.2.3 b).

This desired stabilisation effect – also with regard to photooxidation – of the excited state is effected by the steric shielding offered by the dendrimer shell, which protects the "luminophoric" ruthenium-bipyridine centre from quenchers such as oxygen molecules. Also of value in applications, water solubility was achieved with the synthesis of the dendritic tris(bipyridine)-ruthenium complex **2** having a polyetheramide shell (cf. Fig. 5.6). Compared to non-dendritic complexes, this metallodendrimer has the advantage that the lifetime of the photoexcited state can be prolonged up to 50-fold relative to the non-dendritic reference substance. The shielding effect is accordingly not limited to dendrons of Fréchet type. The dendritic polyetheramide shell, synthesised divergently using the method of Newkome *et al.* (generations 1 to 3), likewise effects a reduction

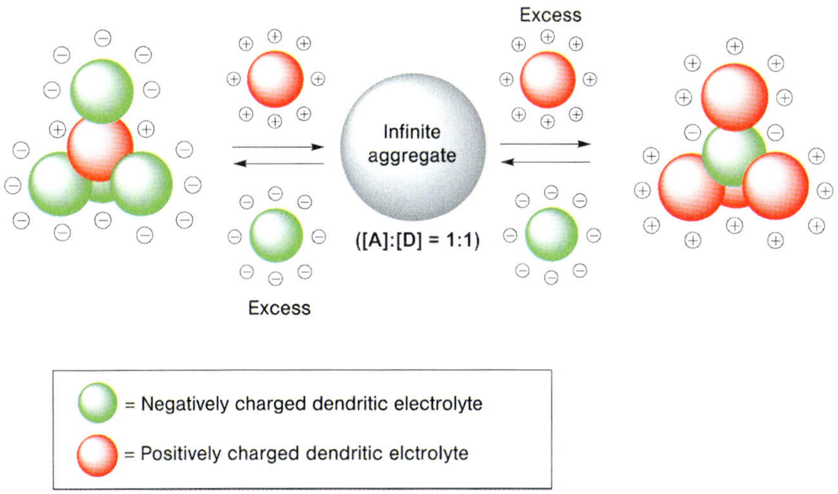

Fig. 5.15 Electrostatic interaction of two oppositely charged dendritic electrolytes (A: acceptor; D: donor; schematic; according to Aida *et al.*)

of fluorescence quenching by molecular oxygen with increasing size of the dendritic groups [4]. Balzani, De Cola, and Vögtle *et al.* also investigated the influence of further dendritic substituents of the photoactive and electroactive core units of this supramolecular dendrimer and other types of dendrimers [17].

The design of dendritic multiporphyrin systems [18] permits energy transfer over longer distances. The outer shell of the dendrimer shown in Fig. 5.14 is made up of eight porphyrin-zinc complexes as energy donor units. Excitation of the units of the outer shell leads to fluorescence emission of the metal-free porphyrin core as a result of energy transfer from the periphery to the energy acceptor [19].

Aida *et al.* also observed an electrostatic association between negatively charged third generation polycarboxylate-functionalised porphyrin dendrimers (as donor) and positively charged ones bearing terminal ammonium (as acceptor) in protic solvents. Depending upon the molar ratio, differently organised arrangements of oppositely charged dendritic electrolytes were found (Fig. 5.15) [20]. In the resulting nanoscale associates, an intermolecular singlet energy transfer can take place from one dendrimer to a neighbouring one. This process depends upon the fine structure (topography) of the dendrimer cluster, which in turn is determined by the molar ratio.

The scaffold of the two porphyrin dendrimers used is identical (Fig. 5.16); they differ solely in the functionalisation of their periphery. The dendrimer shown schematically in red in Fig. 5.15 corresponds to this basic scaffold, for example with $R = CONH(CH_2)_2N^+Me_3$; the dendrimer shown in green bears the group $R = COO^-K^+$.

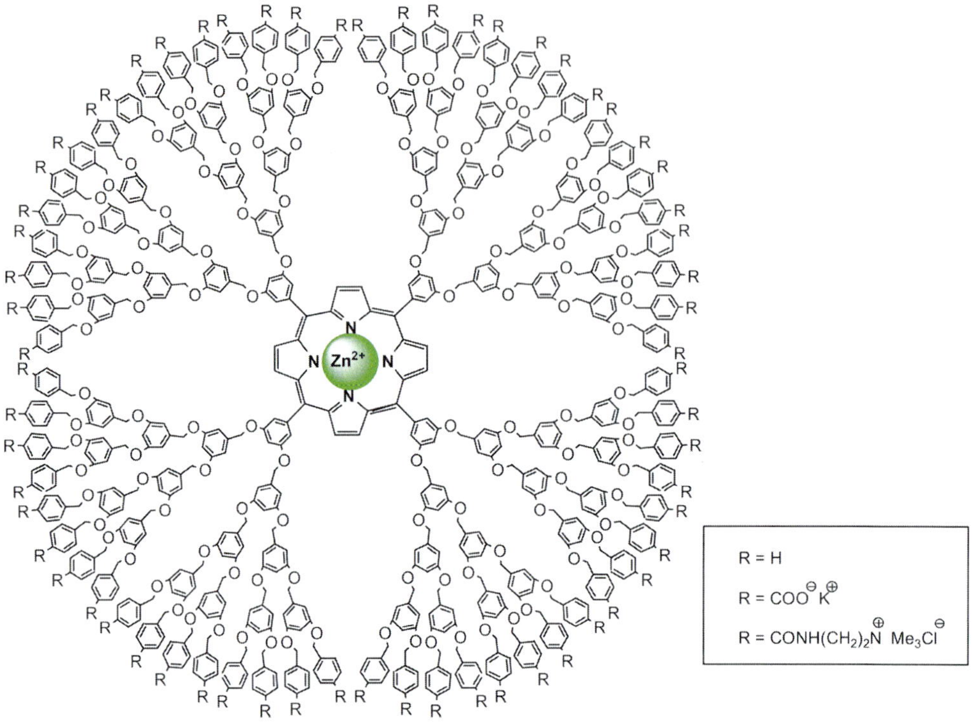

Fig. 5.16 Ionic porphyrin dendrimers (according to Fréchet and Aida *et al.*)

Dendrimers with cyclam core unit (1,4,8,11-tetraazacyclotetradecane) and Fréchet type dendrons, decorated with eight or 16 naphthyl units (Fig. 5.17) were examined for changes of their luminescence and absorption spectra during protonation [21]. The core unit itself is photoinactive, but can interact with photoactive groups in the dendritic branches, influencing the emission properties of the chromophore units and giving rise to the possibility of new emission bands.

In acetonitrile/dichloromethane solution this type of dendrimer shows three kinds of emission bands having their origin in the excited state localised within the naphthyl group, a naphthyl excimer, and a naphthyl/amine exciplex. Titration with trifluoroacetic acid revealed that, in spite of formally possessing four nitrogen atoms, the cyclam core undergoes only two successive protonation steps which significantly affect the luminescence properties.

The protonated dendrimer exhibits a far more intense excimer band than corresponding Fréchet dendrons without a cyclam core. One possible reason for this behaviour is that excimer formation is facilitated by folding of the – flexible – benzyl ether framework. Whereas a change in emission intensity is observed in the course of protonation of the cyclam dendrimer, a reference substance containing no cyclam and benzene moieties (Fig. 5.17) shows a linear increase

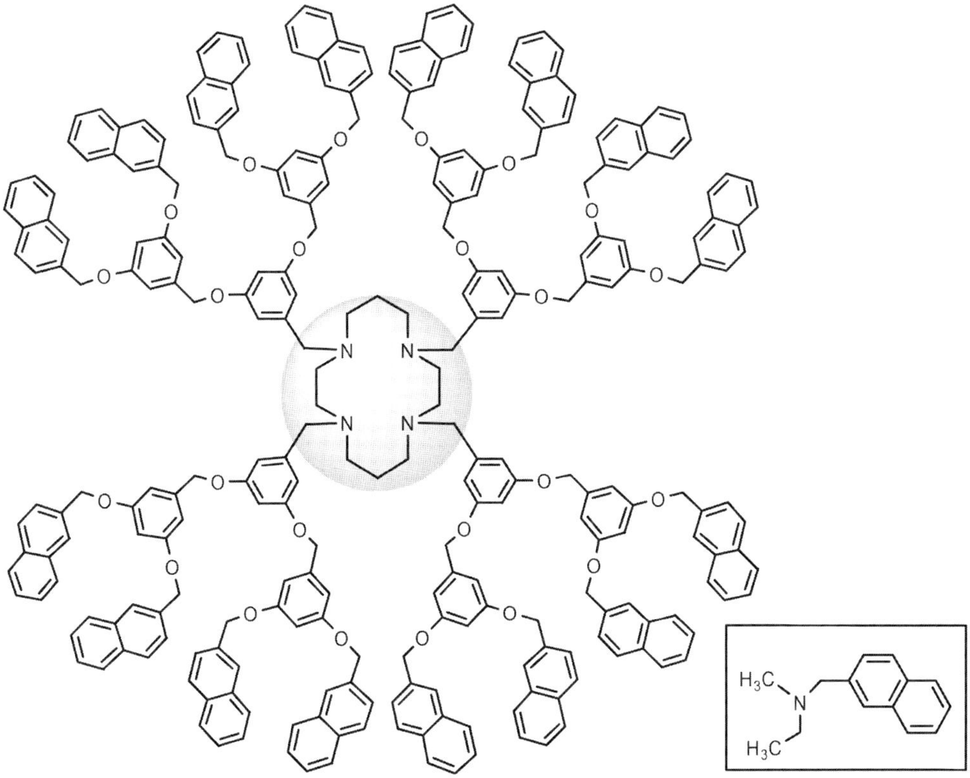

Fig. 5.17 Naphthyl-decorated dendrimer with cyclam core unit (according to Balzani, Vögtle *et al.*); reference substance without dendritic branching shown in box

of the naphthyl-localised band but a decrease in intensity of the exciplex in accord with the above interpretation.

The non-linear spectral changes of the naphthyl-localised exciplex and excimer bands occur after addition of the first two equivalents of trifluoroacetic acid. It is thus unnecessary to protonate all of the nitrogen atoms in order to suppress exciplex formation since – expressed figuratively – the nitrogen atoms share the protons. Moreover, protonation not only provides protection against exciplex formation but also leads to conformational changes in the cyclam unit itself, which in turn affects excimer formation between peripheral naphthyl units of the dendrimer.

Within their scaffolds, POPAM dendrimers also possess amine groups which can undergo protonation or coordination with metal ions. The spectroscopic and photochemical properties of the first to the fourth POPAM generation with peripheral fluorescing naphthylsulfonamide groups were compared with those of reference compounds **A** (*N*-methylnaphthalenesulfonamide) and **B** (*N*-(3-dimethylaminopropyl)-2-naphthalene-l-sulfonamide) [22] (Fig. 5.19).

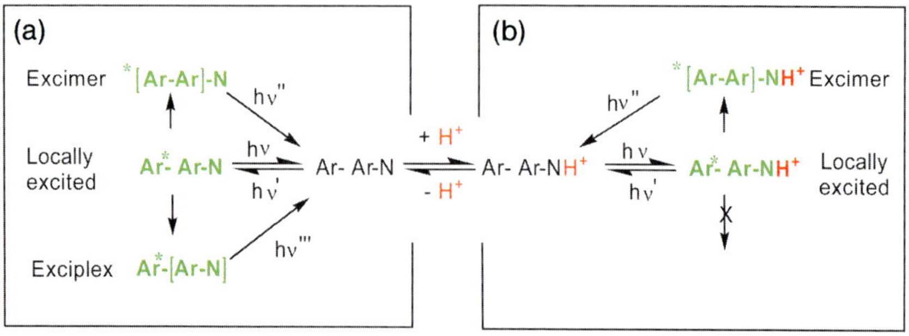

Fig. 5.18 Photoexcited states of the cyclam dendrimer shown in Fig. 5.17 which are responsible for the observed emission; a) before, b) after protonation (schematic). Ar stands for an arene, such as a naphthalene unit, N for the N atom of an amine; Ar-Ar-N symbolises a dendrimer, in which interactions are possible between two arene units or between an arene and an amine unit. The asterisk indicates localisation of the electronic excitation

Both protonation and metal coordination influence the absorption/emission properties of the fluorophoric groups – and also their interactions. The absorption spectra of the dendrimers deviate decisively from those of the reference substances. Additionally, the intensity of the fluorescence band of the dendrimers at λ_{max} 343 nm decreases rapidly with increasing generation number (dendritic effect, see also Section 6.3) and is accompanied by the appearance of a weakened broad emission of low energy. Protonation of the amine groups of the dendrimer with CF_3SO_3H in acetonitrile causes a pronounced increase of luminescence intensity of the naphthylsulfonamide entities and a modification of the emission bands. Titration with $Zn(CF_3SO_3)_2$ shows similar effects only at higher concentrations. In contrast, use of $Co(NO_3)_2 \cdot 6\ H_2O$ leads only to a slight increase in fluorescence intensity of the naphthylsulfonamide entities in the dendrimer.

These results indicate that in the individual dendrimers, interactions take place both in the ground state and in the excited state between the naphthyl units and also between naphthyl and amine units of the dendrimer branches; which result in dimer/excimer and charge-transfer/exciplex excited states.

5.2.2
Photoisomerisation

Reversible photoisomerisation [23] numbers among the readiest reactions in photochemistry. The presence of a photoswitchable moiety is a prerequisite for this reaction. Since photoisomerisation commands comparable interest in physics and chemistry – as well as biology – it will be considered here at the conclusion of the photophysical chapter and before the beginning of the following chapter on chemical reactions.

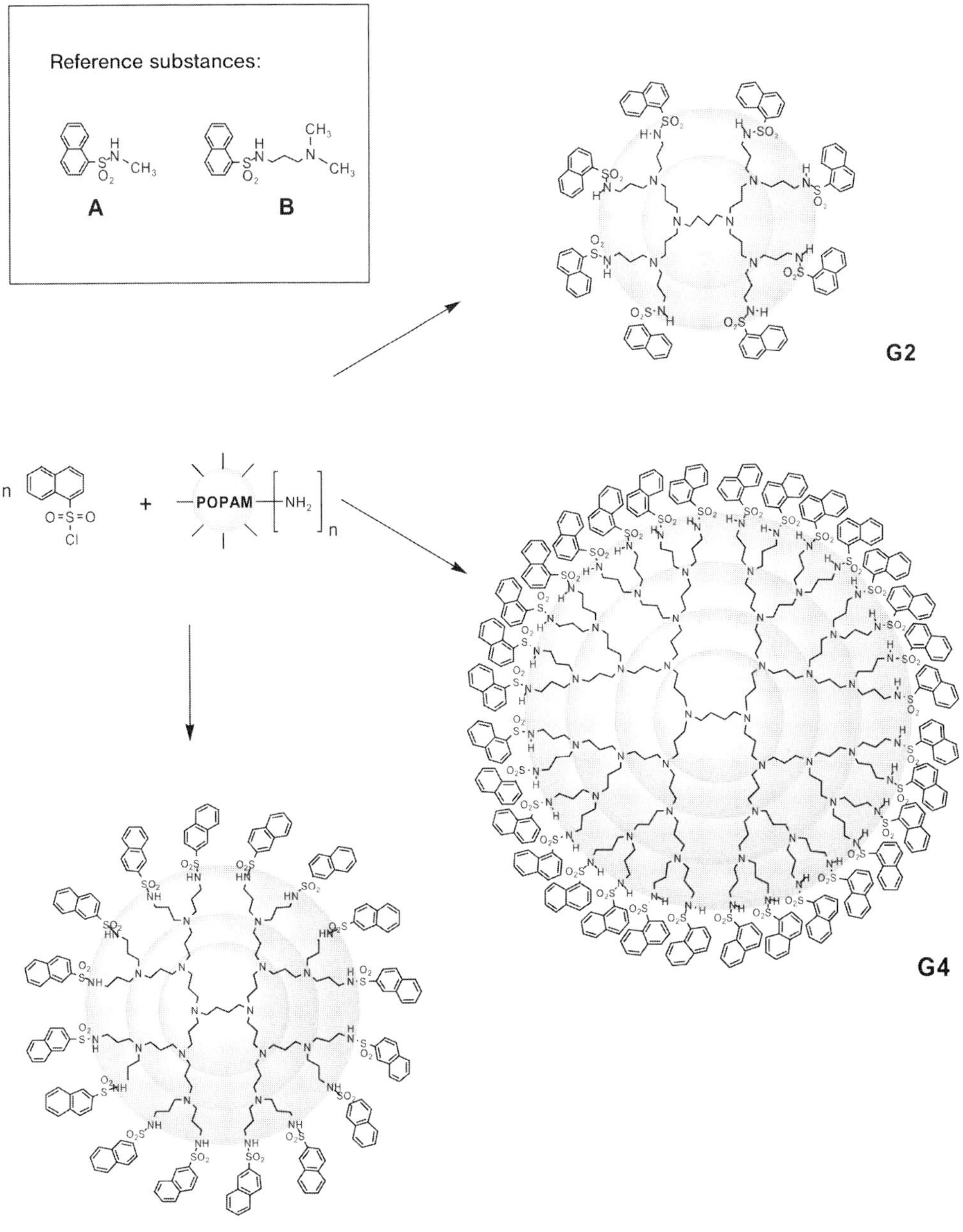

Fig. 5.19 Sulfonamide-substituted G2–G4 POPAM dendrimers; corresponding monosulfonamides **A** and **B** as reference substances (according to Pina, Balzani, Vögtle *et al.*)

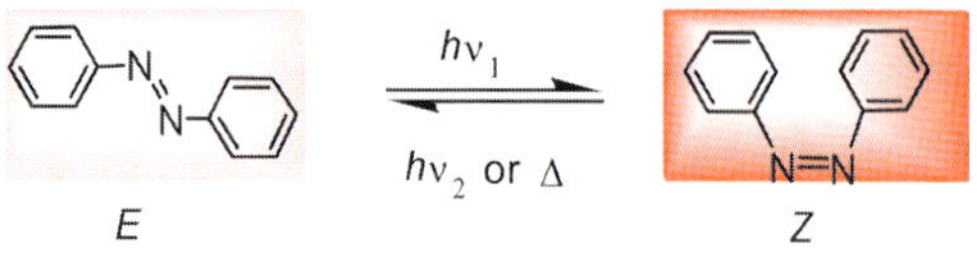

Fig. 5.20 *E/Z* photoisomerisation of azobenzene

Photoisomerisations of azobenzene units have been found to proceed particularly clearly and to be well suited for experimental studies. The thermodynamically more stable *E*-isomer can be transformed photophysically into the *Z*-form, which can return to the *E*-form on photoexcitation or thermal treatment (Fig. 5.20).

The isomers exhibit different absorption spectra: The *E*-form displays an intense $\pi\pi^*$ band in the near UV range and a broad, symmetry-forbidden $n\pi^*$ band in the visible region. On isomerisation to the *Z*-form, the $\pi\pi^*$ band is shifted to shorter wavelengths, with a concomitant increase of the $n\pi^*$ band. The $E \rightarrow Z$ isomerisation engenders a pronounced structural change of the molecule, which is manifested in an increase of dipole moment (see also Section 8.8.1).

The efficiency of the *E/Z* photoreaction – and especially the speed of the thermal *Z/E* reverse reaction – is strongly dependent upon the electronic and steric influence of the substituents on the arene units. π-Donors in *para* position accelerate it; π-acceptors inhibit it; push-pull substituents are particularly effective in lowering the barrier.

If the periphery of a POPAM dendrimer bears azobenzene units then dye molecules can be included as guests in the dendrimer scaffold (see also host/guest chemistry in Section 6.2.3) [23]. Here the *E*- and *Z*-isomers (or their enriched versions) differ in their capacity for accommodating guest molecules. In principle, guest molecule inclusion can be controlled (switched) in this way (Fig. 5.21).

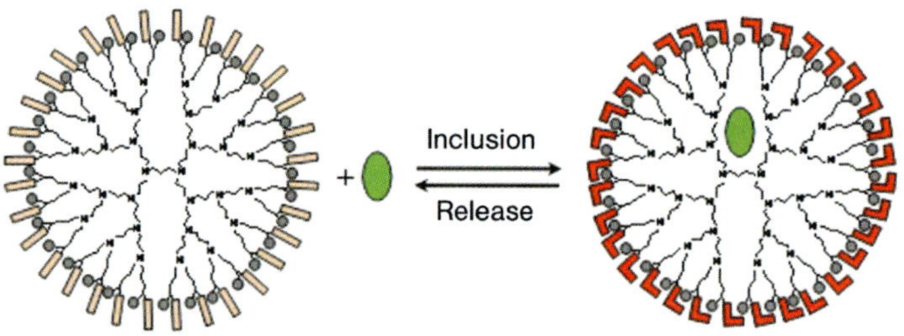

Fig. 5.21 A POPAM dendrimer with peripheral azobenzene units, on the left in the *E*-configuration, on the right in the *Z*-configuration (schematic). The *E*- and *Z*-isomers interact differently with eosin as guest

Studies on the supramolecular interaction of these azobenzene-decorated dendritic host compounds with eosin as guest substance revealed that the *Z*-form accommodates more eosin than the *E*-form. This permits inclusion/liberation of guest molecules by appropriate irradiation. This is possibly relevant for medical applications relating to controlled release of drugs at certain sites in the organism through the action of light. The rate of thermal $E \rightarrow Z$ isomerisation is of the order of $3 \times 10^6\ s^{-1}$ [24–26].

Functionalisation of POPAM-dendrimers with the push/pull azobenzene derivative methyl orange (cf. Fig. 5.21) [27] offers further photoswitching opportunities. The change in colour of this pH indicator from red (pH 3.1) to yellow (pH 4.4) can be explained by protonation of the azo function to form the mesomerically stabilised azonium ion (Fig. 5.22).

Methyl orange can also be subjected to reversible potoisomerisation [28]. Corresponding decoration of the outer shell of a POPAM dendrimer accordingly affords two control mechanisms – conformational and pH control. As with azobenzene-functionalised POPAM dendrimers, the photophysical properties are independent of the generation number, with the exception of the molar extinc-

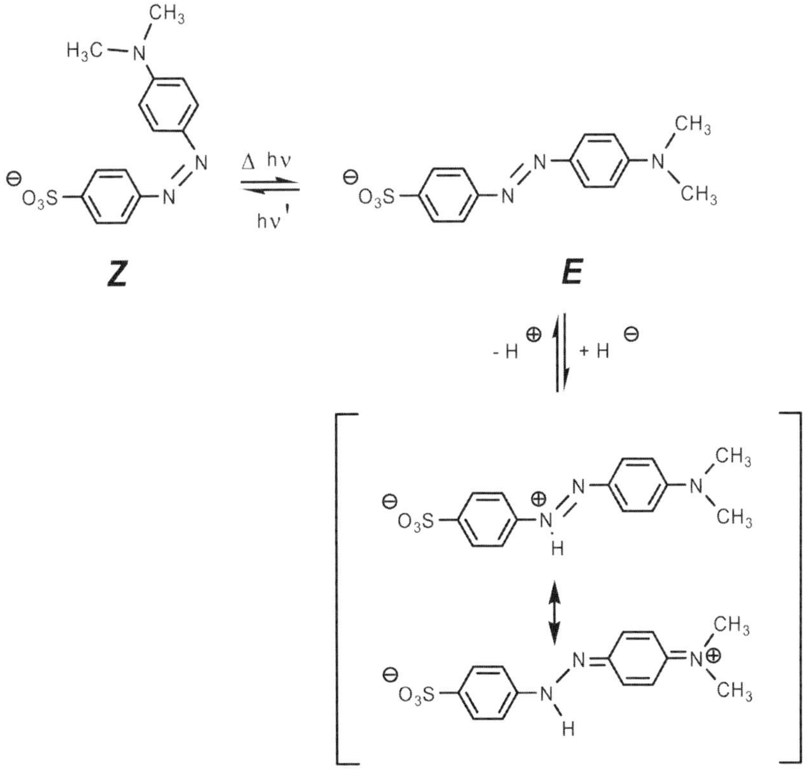

Fig. 5.22 Reason for the change of colour of the pH indicator methyl orange

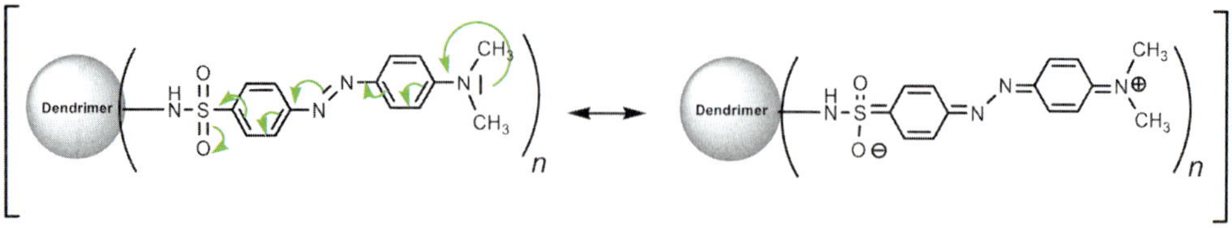

Fig. 5.23 Dendrimer with push/pull π-azobenzene system (schematic)

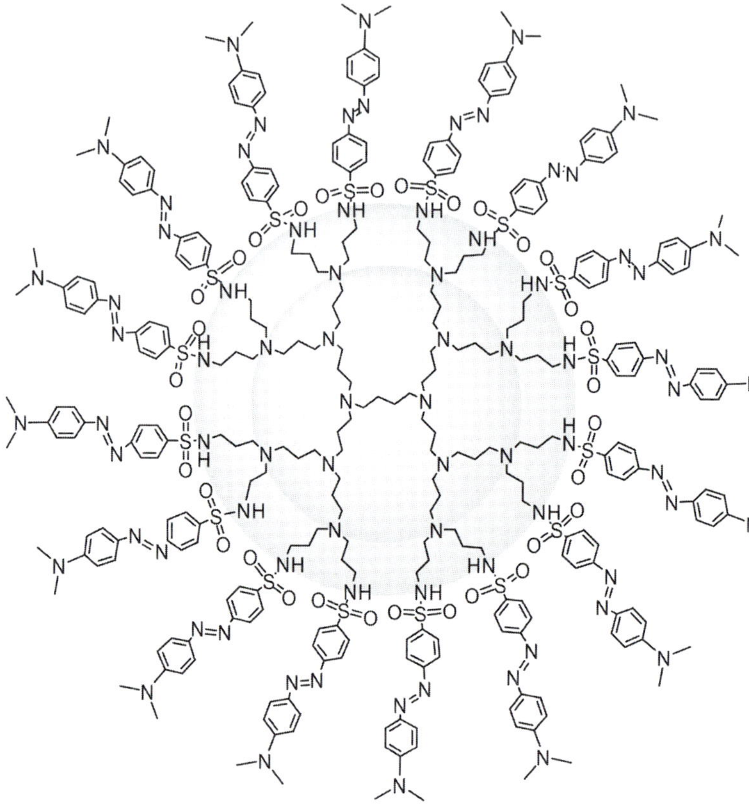

Fig. 5.24 Methyl orange-functionalised third-generation POPAM dendrimer (according to De Cola, Vögtle, Pina, *et al.*)

tion coefficient (Lambert-Beer law), which deviates considerably from the expected value in the case of the fourth and fifth generation due to the exponential increase in the number of methyl orange functions. The *E/Z*- and *Z/E* isomerisation rates are higher than for normal azobenzenes. The reason is to be found in the push-pull system (Fig. 5.23) formed by the dimethylamino group as π-electron donor and the SO_2 group as π-electron acceptor, which lowers the energy barrier for the isomerisation process.

The pH-dependence of methyl orange-functionalised dendrimers (Fig. 5.24) is on the one hand generation-dependent (dendritic effect, see Section 6.3) and on the other hand strongly influenced by the tertiary amino groups in the core. As the most basic component in the dendrimer scaffold the latter are protonated first, which also influences the protonation of the methyl orange moieties. At higher generations, the colour change is not so fast and unequivocal as in the case of monomeric methyl orange and only attainable by addition of an excess of acid [29].

A family of photoswitchable dendrimers with azobenzene core units was described by Aida *et al.* Fréchet-type dendrons – of up to the fifth generation – were attached to the four *meta* positions of azobenzene (Fig. 5.25) [30]. Isomerisation with infrared light – from the *Z* to the *E* conformer – proceeds 260 times faster at 21 °C than without irradiation. However, photoisomerisation of the azobenzene core requires a higher energy than can normally be supplied by infrared light quanta. An explanation of this unexpected phenomenon is attempted in terms of the antenna effect exerted by numerous aryl ether units in the dendrons (Section 5.2.1). The peripheral units absorb and harvest the energy and transfer it to the focal point. Since the core of the dendrimer molecule is isolated by the bulky outer dendrons, the energy apparently cannot be dissipated. Poly(benzylether) dendrimers with anthracene core exhibit antenna effects and can undergo intramolecular [$4\pi+4\pi$] cycloaddition reactions (optical switching) [31].

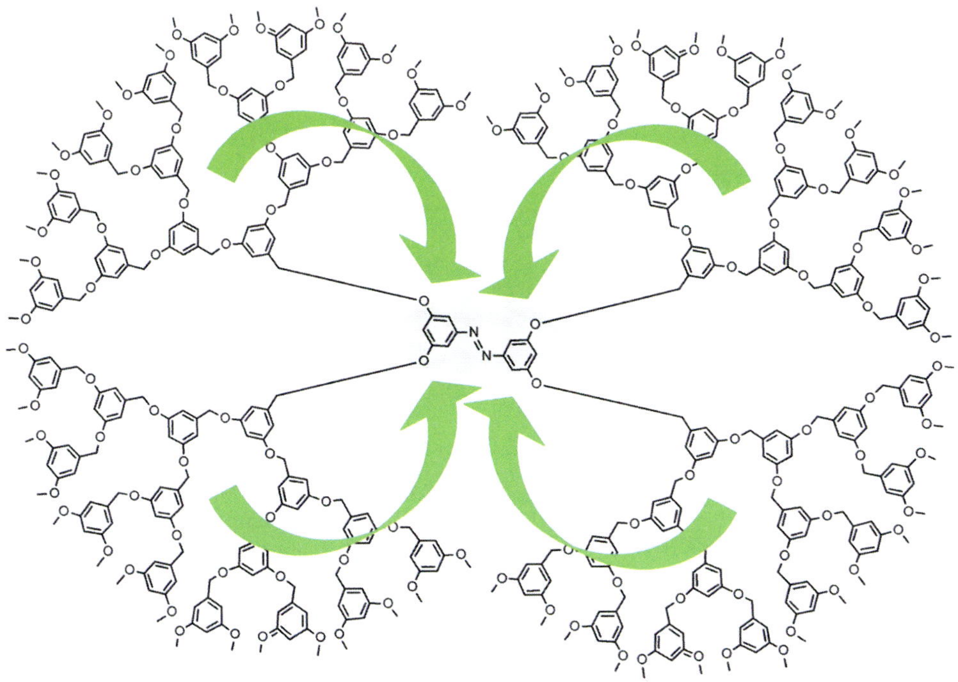

Fig. 5.25 Energy transfer through the dendritic shell to the azobenzene core unit (according to Aida *et al.*)

Bibliography and Notes for Chapter 5 "Photophysical properties of dendritic molecules"

Review articles are indicated by the words "Review(s)" or "Book(s)" in bold-faced type.

1 ***Review***: A. Adronov, J. M. J. Fréchet, *Chem. Commun.* **2000**, 1701–1710. ***Book***: V. Balzani, P. Piotrowiak, M. A. J. Rodgers, J. Mattay, D. Astruc, H. B. Gray, J. Winkler, S. Shunichi, T. E. Mallouk, Y. Haas, A. P. de Silva, I. Gould (Eds.), *Electron Transfer in Chemistry*, Wiley-VCH, Weinheim **2001**; G. Bergamini, P. Ceroni, M. Maestri, V. Balzani, S. K. Lee, F. Vögtle, *Photochem. Photobiol. Sci.* **2004**, *3*, 898–905; P. Ceroni, V. Vicinelli, M. Maestri, V. Balzani, S. K. Lee, J. van Heyst, M. Gorka, F. Vögtle, *J. Organomet.* Chem. **2004**, *689*, 4375–4383.

2 ***Books***: P. W. Atkins *Kurzlehrbuch Physikalische Chemie*, 3rd Edit., Wiley-VCH, Weinheim **2001**; B. Valeur, Molecular Fluorescence – Principles and Application, Wiley-VCH, Weinheim **2002**; see also H. A. Staab, *Einführung in die Theoretische Organische Chemie*, Verlag Chemie, Weinheim **1962**; W. Foerst (Ed.), *Optische Anregung Organischer Systeme*, Verlag Chemie, Weinheim 1966.

3 V. Balzani, P. Ceroni, C. Giansante, V. Vicinelli, F.-G. Klärner, C. Verhaelen, F. Vögtle, U. Hahn, Angew. Chem. **2005**, *117*, 4650–4654; *Angew. Chem. Int. Ed.* **2005**, *44*, 4574–4578; M. Z. Yin, J. Shen, R. Gropeanu, G. O. Pflugfelder, T. Weil, K. Müllen, *Small* **2008**, *4*, 894–898.

4 J. Issberner, F. Vögtle, L. De Cola, V. Balzani, *Chem. Eur. J.* **1997**, *3*, 706–712.

5 U. Hahn, M. Gorka, F. Vögtle, V. Vicinelli, P. Ceroni, M. Maestri, V. Balzani, *Angew. Chem.* **2002**, *114*, 3747–3750; *Angew. Chem. Int. Ed.* **2002**, *41*, 3595–3598; G. Teobaldi, F. Zerbetto, *J. Am. Chem. Soc.* **2003**, *125*, 7388–7393.

6 M. Plevoets, F. Vögtle, L. De Cola, V. Balzani, *New J. Chem.* **1999**, *23*, 63–69; N. Armaroli, V. Balzani, *Angew. Chem.* **2007**, *119*, 52–67; *Angew. Chem. Int. Ed.* **2007**, *46*, 52–66.

7 a) A. Dirksen, U. Hahn, F. Schwanke, M. Nieger, J. N. H. Reek, F. Vögtle, L. De Cola, *Chem. Eur. J.* **2004**, *10*, 2036–2047; b) C. Giasante, P. Ceroni, V. Balzani, M. Maestri, S.-K. Lee, F. Vögtle, *New J. Chemistry* **2007**, *31*, 1250–1258; V. Vicinelli, G. Bergamini, P. Ceroni, V. Balzani, F. Vögtle, O. Lukin, *J. Phys. Chem. B* **2007**, *111*, 6620–6627..

8 L. Jullien, J. Canceill, B. Valeur, E. Bardez, J.-M. Lehn, *Angew. Chem.* **1994**, *106*, 2582–2584; *Angew. Chem. Int. Ed.* **1994**, *33*, 2438–2439; L. Jullien, J. Canceill, B. Valeur, E. Bardez, J.-P. Lefèvre, J.-M. Lehn, V. Marchi-Artzner, R. Pansu, *J. Am. Chem. Soc.* **1996**, *118*, 5432–5442. J. Seth, V. Palaniappan, T. E. Johnson, S. Prathapan, J. S. Lindsey, D. F. Bocian, *J. Am. Chem. Soc.* **1996**, *118*, 11194–11207; M. S. Vollmer, F. Würthner, F. Effenberger, P. Emele, D. U. Meyer, T. Stümpfig, H. Port, H. C. Wolf, *Chem. Eur. J.* **1998**, *4*, 260–269.

9 S. L. Gilat, A. Adronov, J. M. J. Fréchet, *Angew. Chem.* **1999**, *111*, 1519–1524; *Angew. Chem. Int. Ed.* **1999**, *38*, 1422–1427.

10 G. Accorsi, N. Armaroli, J.-F. Eckert, J.-F. Nierengarten, *Tetrahedron Lett.* **2002**, *43*, 65–68. ***Review***: J.-F. Nierengarten, *Top. Curr. Chem.* (Volume Eds. C. A. Schalley, F. Vögtle), **1998**, *228*, 87–110; cf. also: J. Iehl, R. P. de Freitas, B. Delavaux-Nicot, J.-F. Nierengarten, *J. Chem. Soc., Chem. Comm.* **2008**, 2450–2452; for photophysical properties of fullerene-rich dendrimers see: U. Hahn, J.-F. Nierengarten, F. Vögtle, A. Listorti, F. Monti, N. Armaroli, *New J. Chem.* **2008**, in print.

11 I. Akai, H. Nakao, K. Kanemoto, T. Karasawa, H. Hashimoto, M. Kimura, *J. Luminesc.* **2005**, *112*, 449–453.

12 C. Devadoss, P. Bharathi, J. S. Moore, *Angew. Chem.* **1997**, *109*, 1706–1709; *Angew. Chem. Int. Ed.* **1997**, *36*, 1709–1711.

13 C. Devadoss, P. Bharathi, J. S. Moore, *J. Am. Chem. Soc.* **1996**, *118*, 9635–9644,

14 Z. Xu, J. S. Moore, *Acta Polymer* **1994**, *45*, 83–87; R. Kopelman, M. Shortreed, Z.-Y. Shi, W. Tan, Z. Xu, J. S. Moore, A. B. Haim, J. Klafter, *Phys. Rev. Lett.* **1997**, *78*, 1239–1242.

15 M. Plevoets, F. Vögtle, L. De Cola, V. Balzani, *New J. Chem.* **1999**, 63–69.
16 Related dendrons and their energy-harvesting and energy-transfer properties: G. M. Stewart, M. A. Fox, *J. Am. Chem. Soc.* **1996**, *118*, 4354–4360.
17 M. Plevoets, F. Vögtle, M. Nieger, G. C. Azzellini, A. Credi, L. De Cola, V. de Marchis, M. Venturi, V. Balzani, *J. Am. Chem. Soc.* **1999**, *121*, 6290–6298.
18 D. L. Officer, A. K. Burrell, D. C. W. Reid, *Chem. Commun.* **1996**, 1657–1658. C. C. Mak, N. Bampos, J. K. M. Sanders, *Angew. Chem.* **1998**, *110*, 3169–3172; *Angew. Chem. Int. Ed.* **1998**, *37*, 3020–3023.
19 M.-S. Choi, T. Yamazaki, I. Yamazaki, T. Aida, *Angew. Chem.* **2003**, *116*, 152–160; *Angew. Chem. Int. Ed.* **2003**, *42*, 150–158. M.-S. Choi, T. Aida, T. Yamazaki, I. Yamazaki, *Chem. Eur. J.* **2002**, *8*, 2667–2678; cf.: W.-S. Li, K. K. Kim, D.-L. Jiang, H. Tanaka, T. Kawai, J. H. Kwon, D. Kim, T. Aida, *J. Am. Chem. Soc.* **2006**, *128*, 10527–10532.
20 N. Tomioka, D. Takasu, T. Takahashi, T. Aida, *Angew. Chem.* **1998**, *110*, 1611–1614; *Angew. Chem. Int. Ed.* **1998**, *37*, 1531–1534.
21 C. Saudan, V. Balzani, P. Ceroni, M. Gorka, M. Maestri, V. Vicinelli, F. Vögtle, *Tetrahedron* **2003**, *59*, 3845–3852; G. Bergamini, P. Ceroni, V. Balzani, L. Cornelissen, J. van Heyst, S.-K. Lee, F. Vögtle, *J. Mater. Chem.* **2005**, *15*, 2959–2964.
22 F. Pina, P. Passaniti M. Maestri, V. Balzani, F. Vögtle, M. Gorka, S.-K. Lee, J. van Heyst, H. Fakhrnabavi, *ChemPhysChem* **2004**, *5*, 473–480.
23 ***Books*** and ***Reviews*** about – non-dendritic – azobenzene photoisomerisation and other photoswitches: V. Balzani, F. Scandola, *Supramolecular Photochemistry*, Ellis Horwood, New York **1991**; N. N. P. Moonen, A. H. Flood, J. M. Fernández, J. F. Stoddart, *Top. Curr. Chem.* **2005**, *262*, 99–132; B. L. Feringa (Ed.), *Molecular Switches*, Wiley-VCH, Weinheim **2001**; B. L. Feringa, N. Koumura, R. A. van Delden, M. K. J. ter Wiel, *Appl. Phys. A* **2002**, *75*, 301–308; C. Dugave (Ed.), *cis-trans Isomerization in Biochemistry*, Wiley-VCH, Weinheim, New York **2006**.
24 A. Archut, G. C. Azzellini, V. Balzani, L. De Cola, F. Vögtle, *J. Am. Chem. Soc.* **1998**, *120*, 12187–12191; F. Puntoriero, P. Ceroni, V. Balzani, G. Bergamini, F. Vögtle, *J. Am. Chem. Soc.* **2007**, *129*, 10714–10719.
25 A. Archut, F. Vögtle, L. De Cola, G. C. Azzellini, V. Balzani, P. S. Ramanucham, R. H. Berg, *Chem. Eur. J.* **1998**, *4*, 699–706.
26 P. Sierocki, H. Maas, P. Dragut, G. Richardt, F. Vögtle, L. De Cola, F. (A. M.) Brouwer, J. I. Zink, *J Phys. Chem. B.* **2006**, *110*, 24390–24398.
27 A. Dirksen, E. Zuidema, R. M. Williams, L. De Cola, C. Kauffmann, F. Vögtle, A. Roque, F. Pina, *Macromolecules* **2002**, *35*, 2743–2747.
28 D. M. Junge, D. V. McGrath, *J. Am. Chem. Soc.* **1999**, *121*, 4912–4913.
29 A. Dirksen, L. De Cola, *C. R. Chimie* **2003**, *6*, 873–882.
30 D. L. Jiang, T. Aida, *Nature* **1997**, *388*, 455–456.
31 D. Cao, S. Dobis, C. Gao, S. Hillmann, H. Meier, *Chemistry Eur. J.* **2007**, 9317–9323.

6
(Special) chemical reactions of dendritic molecules

6.1
Covalent chemical reactions

The versatility of dendritic molecules is manifested not only in their physical properties, but also in their ability to undergo chemical reactions. Steering of the chemical properties with regard to core, branches, and periphery can already be attempted at the design stage prior to the start of any synthesis. Often readily accessible to reagents, the dendrimer surface with its well-defined number of end groups offers broad scope for functionalisation. A suitable balance of hydrophilicity/lipophilicity properties and appropriate choice of the terminal groups on the periphery and of the conformational flexibility of the branching units permits a certain degree of adaptation to the reaction partners and parameters. The molecular structure also determines the density distribution within the dendrimer and thus also permanent or dynamic niches formed in the course of interactions which are, in turn, important for host/guest chemistry.

Since the covalent reactions on the periphery of dendrimers (functionalisation and transfunctionalisation) which immediately spring to mind have already been described in Chapter 3, a number of further, more specialised kinds of reactions will be considered here. These also include reactions in which bonds are formed or broken between core and periphery in the interior of the dendrimer molecules. So far, however, there are more supramolecular (non-covalent reversible) examples in this field (see Section 6.2) than covalent ones, which although somewhat specialised are also highly original and will now be considered in the following Section 6.1.1.

6.1.1
Metathesis

Fréchet and Liang synthesised a fifth-generation benzyl ether dendrimer with 22 internal allyl groups. Subsequent olefin metathesis [1] led on the one hand to intramolecular linking, and on the other hand to creation of a covalent linkage site for olefin-functionalised guests (Fig. 6.1) [2]. In this way, intermolecular cross-linking – between G5 dendrimer molecules – could also be attained.

Dendrimer Chemistry. Fritz Vögtle, Gabriele Richardt and Nicole Werner

ISBN: 978-3-527-32066-0

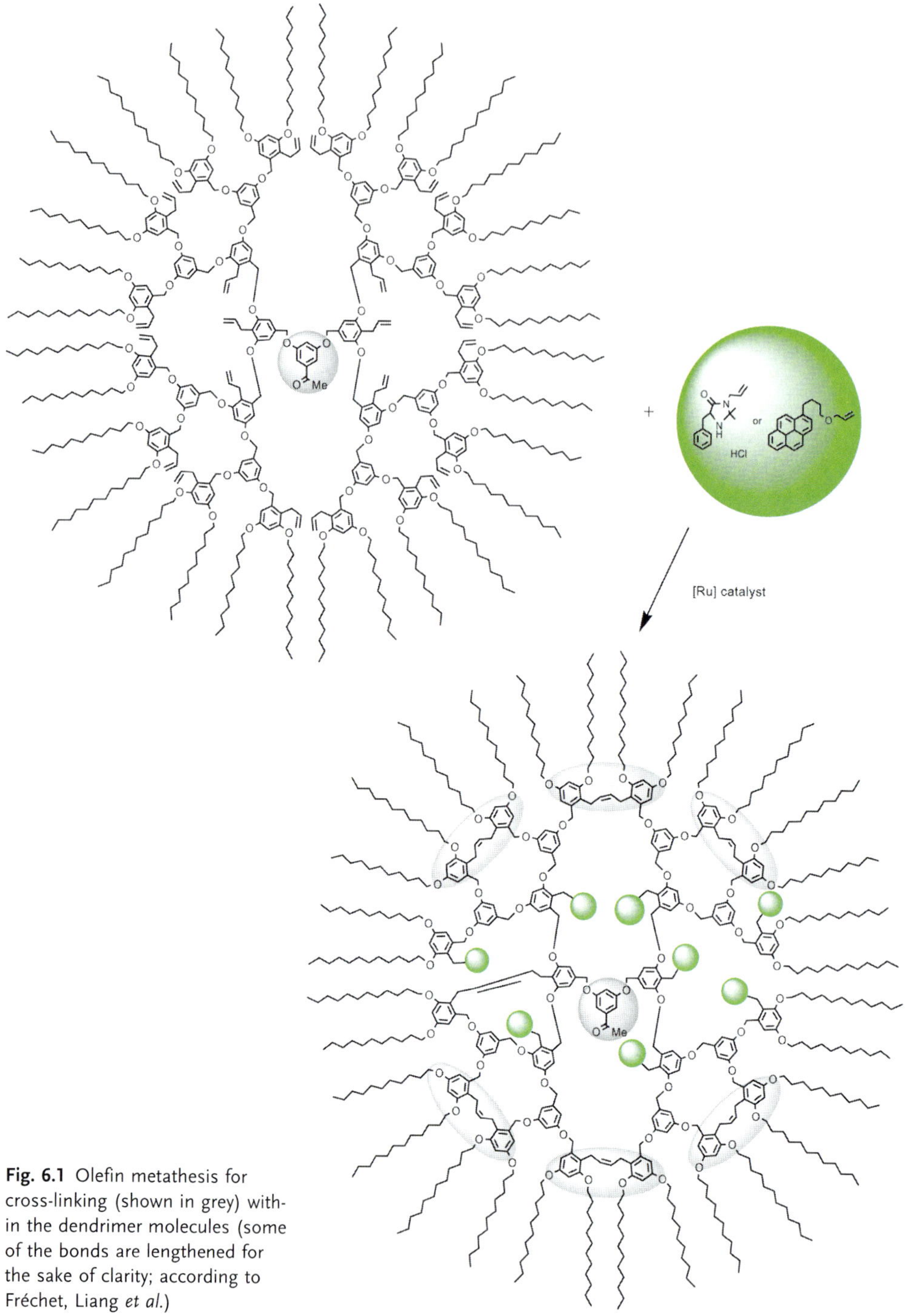

Fig. 6.1 Olefin metathesis for cross-linking (shown in grey) within the dendrimer molecules (some of the bonds are lengthened for the sake of clarity; according to Fréchet, Liang *et al.*)

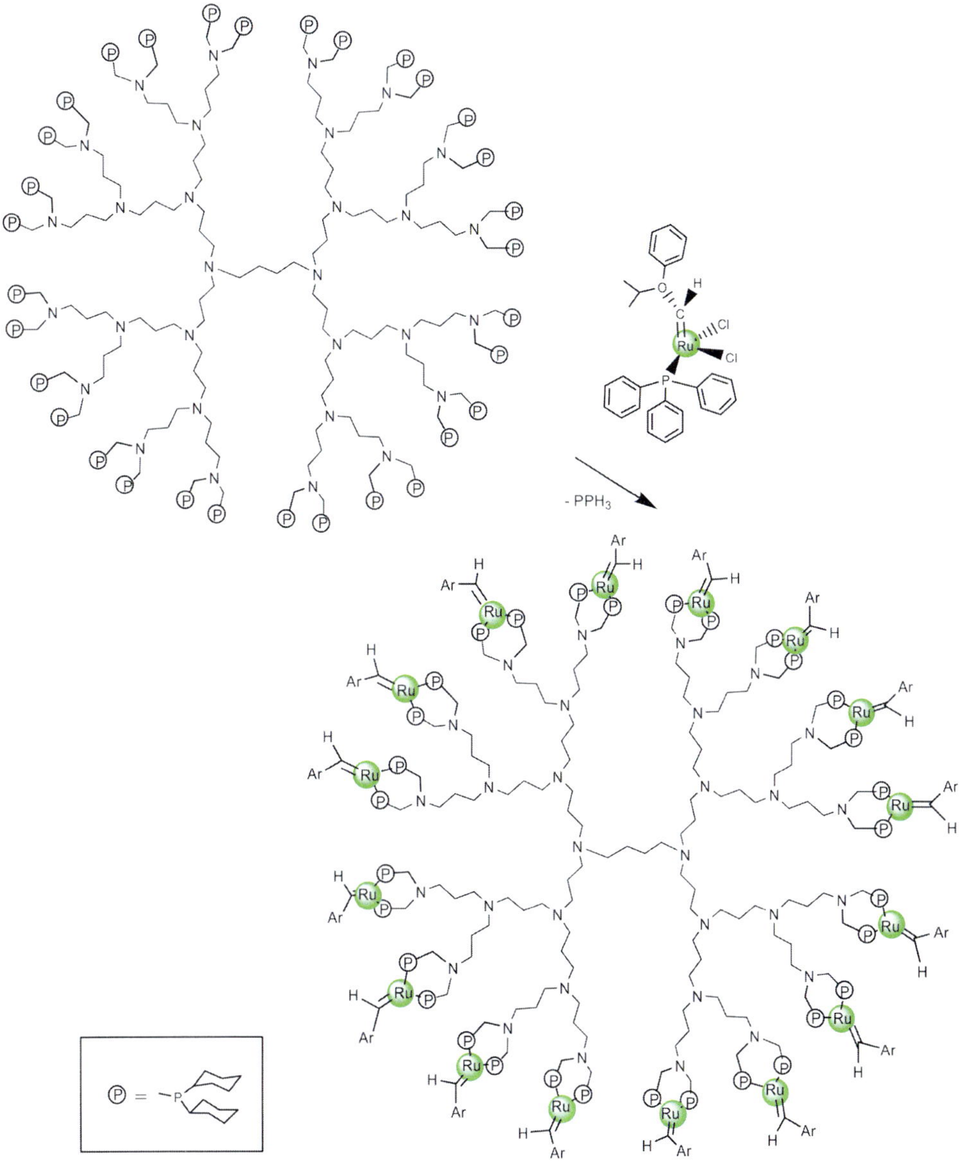

1

Fig. 6.2 Ruthenium-carbene complexed dendrimer (according to Astruc *et al.*; the two cyclohexyl groups on each of the P atoms are omitted for the sake of clarity)

1

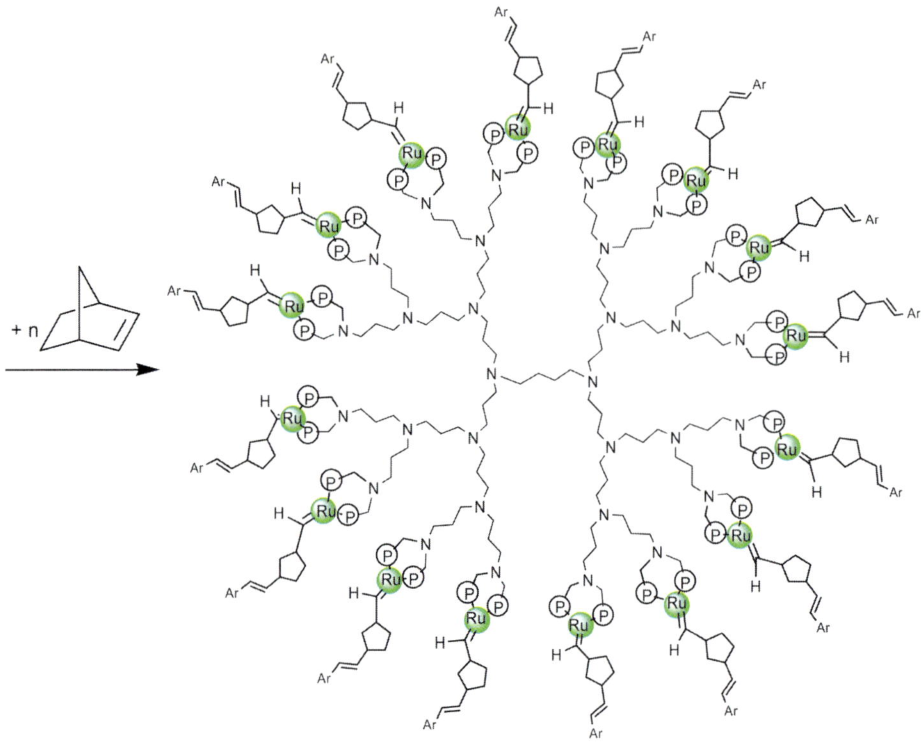

2

Fig. 6.3 Ring-opening metathesis polymerisation (ROMP) of norbornene (the two cyclohexyl groups on each of the P atoms are omitted for the sake of clarity; according to Astruc *et al.*)

Up to third-generation ruthenium-carbene complexed dendrimers (Fig. 6.2) prepared by Astruc *et al.* contain a chelating diphosphane, which is sufficiently stable for construction of the dendritic architecture while also sufficiently reactive to permit synthesis of the dendrimer depicted in Fig. 6.3 by ring-opening metathesis polymerisation (ROMP) [3].

This reaction is performed with a Hoveyda catalyst [Ru(=CH-*o*-O-i-Pr-C_6H_4) · $Cl_2(PPh_3)$] modified by introduction of an isopropyloxy substituent in the *ortho* position of the benzylidene ligand, replacing one of the phosphane ligands by the hemilabile chelating ether ligand [4]. The ruthenium-carbene complexed dendrimer thus prepared was subsequently used for ring-opening metathesis polymerisation of norbornene (Fig. 6.3).

Branched metallodendrimer catalysts have also been described by van Koten, Hoveyda, and Verdonck [5].

Catalytic crossed olefin metathesis can also be used to stereoselectively mono- or bifunctionalise polyolefin dendrimers [6]. Grubbs *et al.* were already able to show that, with aid of second-generation Ru catalysts, a terminal alkene in the presence of another one (with electron-withdrawing substituents) can be subjected to a metathesis reaction, which proceeds stereoselectively to give high yields of crossed *E*-configured olefins [7].

When this approach is applied to the covalent chemistry of water-soluble dendrimers, an oligomethylbenzene such as mesitylene or hexamethylbenzene is activated in an initial step by a cyclopentadienyl complex of iron (**1** in Fig. 6.4). Subsequent reaction with KOH and allyl bromide (or 1-iodoundecane) in tetrahydrofuran and followed by exposure to visible light (in MeCN) affords the dendritic core **2** after decomplexation. Since terminal polyolefins tend to undergo ring closing metathesis, the core unit is "lengthened" by hydrosilylation with $HSiMe_2CH_2Cl$ to prevent such reaction. Reaction with the allyl compound $HOC_6H_4C(CH_2CH{=}CH_2)_3$ and renewed lengthening and reaction with the long-chain alkene $HOC_6H_4O(CH_2)_9CH{=}CH_2$ transforms the first-generation 27-fold allyl dendrimer **3** into the 27-fold olefin dendrimer **4**. Concluding reaction with Grubbs catalyst and an alkene with electron-withdrawing groups leads to the desired polyolefin **5** with the corresponding groups (R = H or methyl).

Polycarboxylic acid dendrimers obtained in this way were reacted with ferrocene-substituted tertiary amines $N[(CH_2)_4\text{-}Fc]_3$ (Fc = ferrocenyl), leading to ammonium salts containing a total of 243 terminal ferrocene units [6].

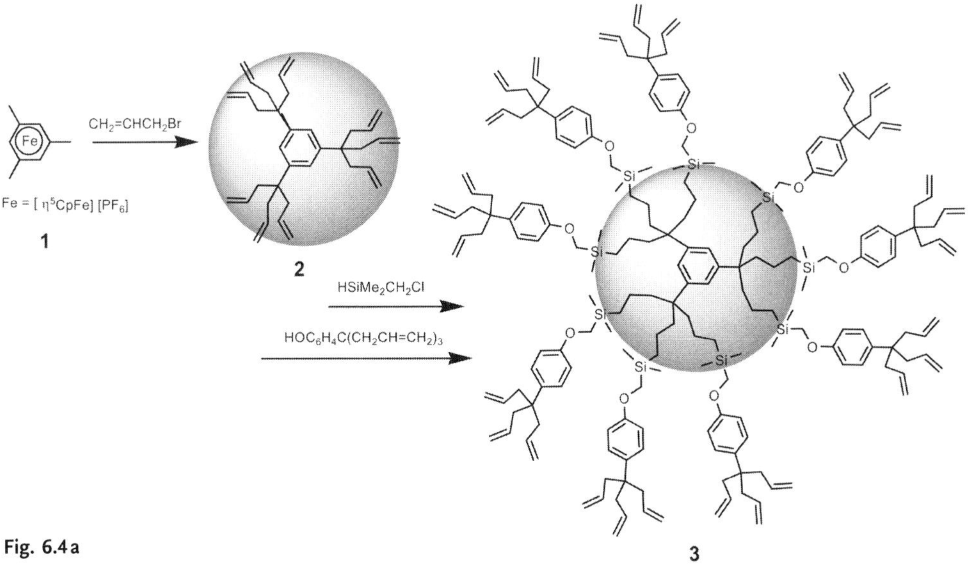

Fig. 6.4 a

3

Fig. 6.4 a, b Olefin metathesis on dendrimers (according to Astruc *et al.*)

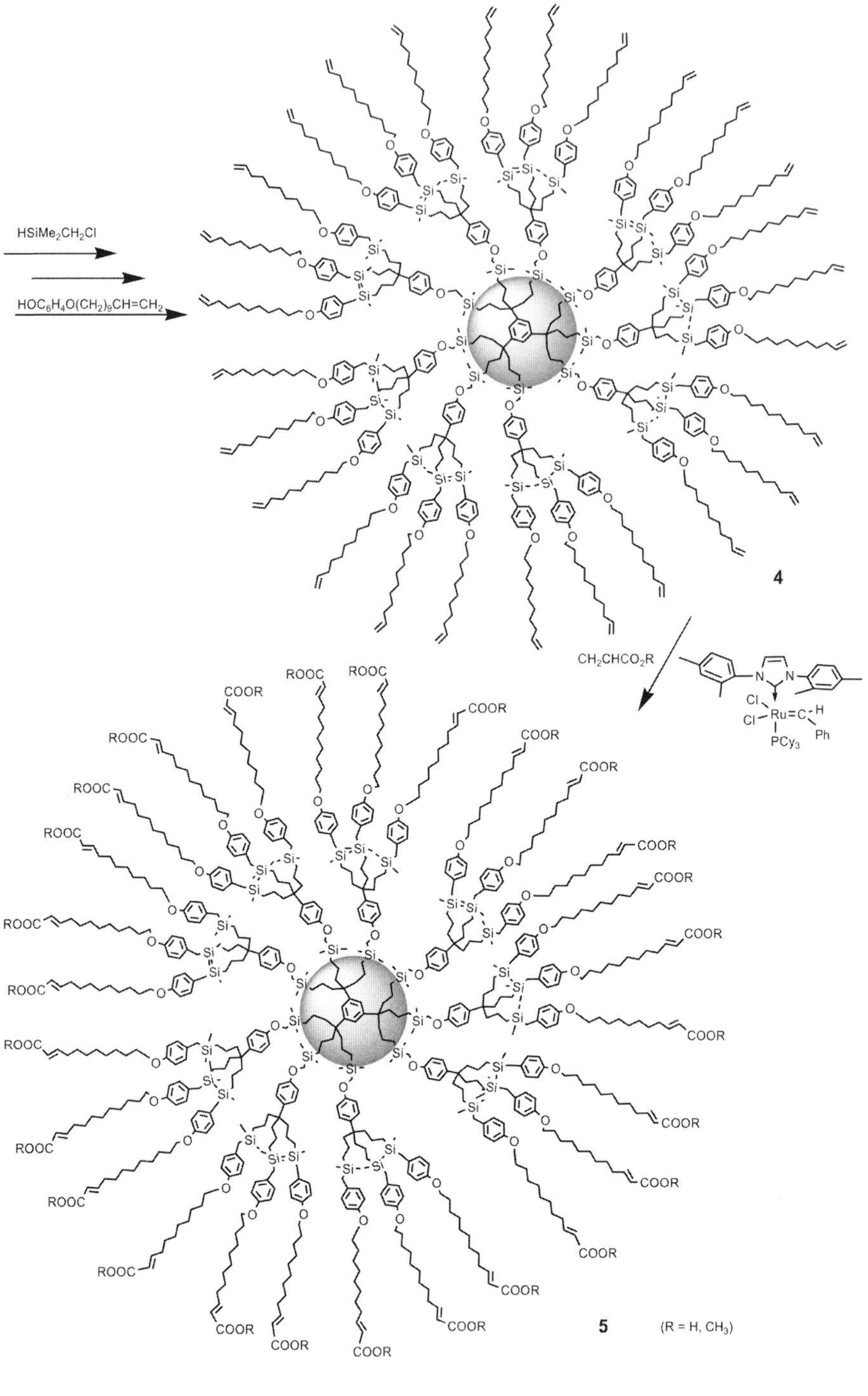

Fig. 6.4 b

6.1.2 Molecular imprinting

Molecular imprinting [8] also involves construction of a polymeric network. In contrast to conventional host/guest chemistry, the process does not use low-molecular-weight host substances, but a polymeric network (matrix) with pre-formed cavities (pores). These are prepared and used for selective guest inclusion. To this end, monomers (*e.g.* styrene) are reversibly multiply bound around a convex template – the desired guest at a later stage – and then polymerised with the aid of a cross-linker (*e.g.* ethylene dimethacrylate). The formerly monomeric building blocks then enclose the template [9] as a molecularly imprinted polymer (MIP). The template itself is not involved in polymerisation, but serves merely as place holder for the later (concave nano)pores. After removal of the template by washing/extraction, there remain well-defined cavities ("imprints"), whose shape and polarity properties are complementary to those of the template, rather like a "plaster cast" (Fig. 6.5).

Attachment of the alkene monomers to the template (Fig. 6.5), which must be reversible – readily formed and readily broken – to permit removal of the template after polymerisation, can generally be accomplished in two ways: covalently or non-covalently. While the latter interactions (ionic, hydrophobic, π–π, hydrogen bonding) can easily be reversed, there is less scope for reversible covalent linkages. One of these is the formation of boronic esters – from boronic acid units of the monomers and OH groups of sugar templates.

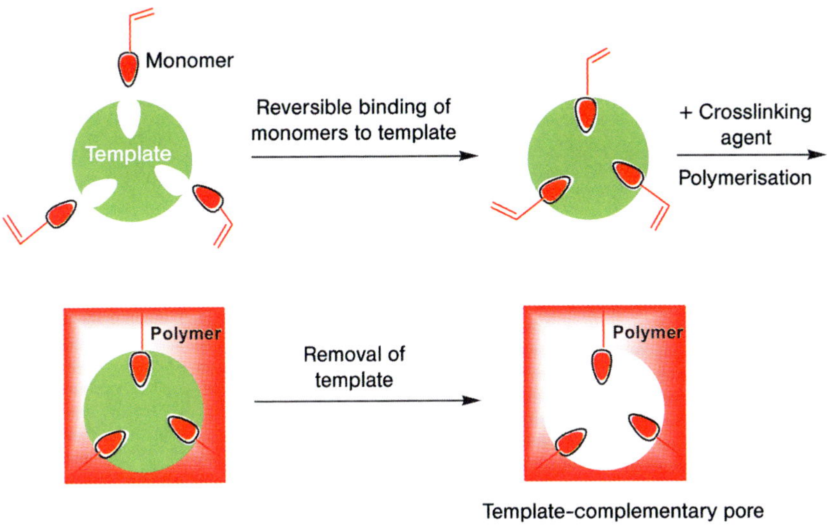

Fig. 6.5 Principle of molecular imprinting (schematic, according to Wulff *et al.*)

Molecular imprinting with dendrimers

Zimmerman *et al.* adopted a convergent route to synthesise dendrimers with a benzene-1,3,5-tricarboxlic acid core surrounded by 3,5-dihydroxybenzy1 ether dendrons with terminal homoallyl ether functions. The dendrimer is closed at the periphery by ring-closing metathesis (RCM) with the aid of a Grubbs catalyst in highly diluted solution (Fig. 6.6). The trimesic acid serving as covalent template is removed by subsequent hydrolysis, leaving behind just a well-defined cavity [10, 11]. It is necessary to perform the reaction in dilute solution to avoid intermolecular reactions such as dimer formation. Undesired side reactions occur if a concentration of 10^{-5} mol/L is exceeded.

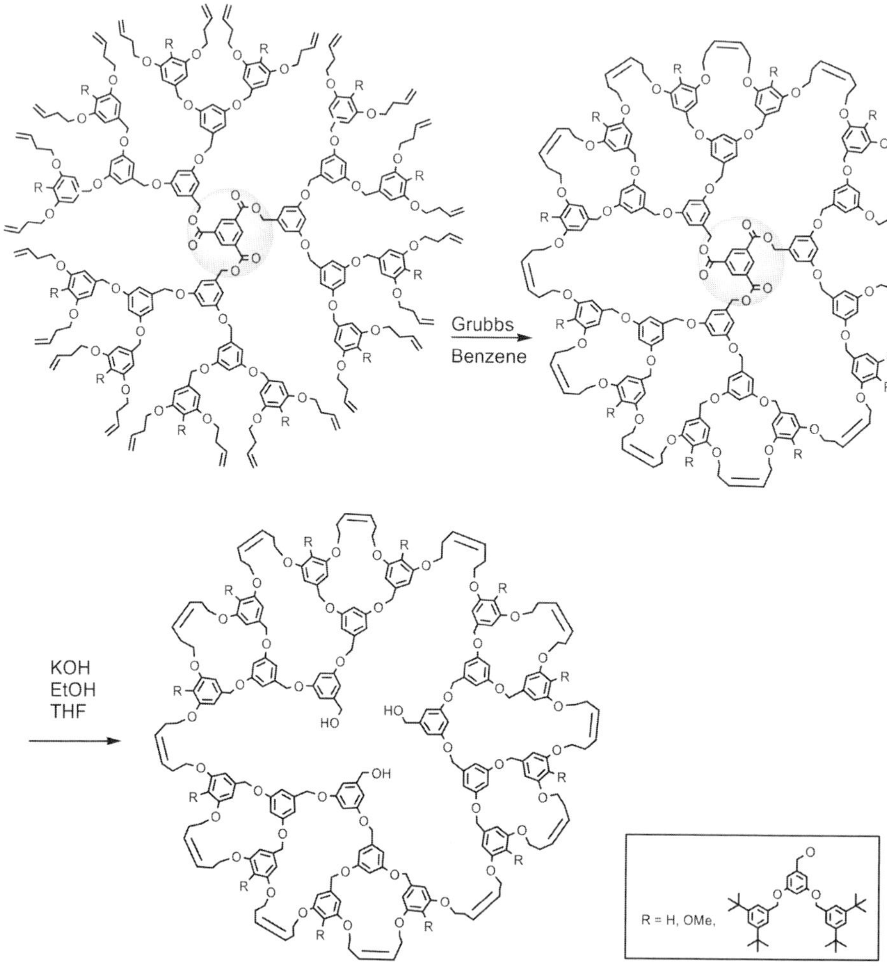

Fig. 6.6 Molecular imprinting with dendrimers (according to Zimmerman *et al.*)

In order to suppress these reactions the periphery of the reactant dendrimer was modified by introduction of space-filling benzyl ether entities between the homoallyl groups (R in Fig. 6.6) while retaining the architecture of the scaffold. Subsequent execution of ring-closing metathesis showed that reaction is also feasible at higher concentrations (10^{-3} mol/L) without any significant intermolecular cross-linking [11]. The molecular imprinting process was also accomplished starting from dendrimer scaffolds with 5,10,15,20-tetrakis(4-hydroxyphenyl)porphyrin as well as 5,10,15,20-tetrakis(3,5-dihydroxyphenyl)porphyrin cores as templates [12]. Molecular imprinting also afforded nanotubes. Moreover, molecularly imprinted surfaces have also proved valuable as recognition layers for acoustic and optical sensors as well as for the design of "MIP arrays" and biochips. Further information will be found in Chapter 8 ("Special properties and potential applications") [13].

6.1.3
Covalent introduction of functionalities in the interior of dendritic molecules

As already described in Chapter 2, dendrimers can be synthesised convergently or divergently. The convergent synthetic strategy permits controlled placement of functionalities within a dendron scaffold or at the focal point (Fig. 6.7). The availability of suitable dendrons is therefore important. Appropriate modification of dendrimer scaffolds and of function-bearing sites is essential for subsequent introduction of reactive moieties within the branching units, for steric shielding of any substituents present ("collar formation"), and for solubility tuning by lipophilic or hydrophilic dendrons. This approach can be used, for example, to prevent or make difficult (partial) luminescence quenching by colli-

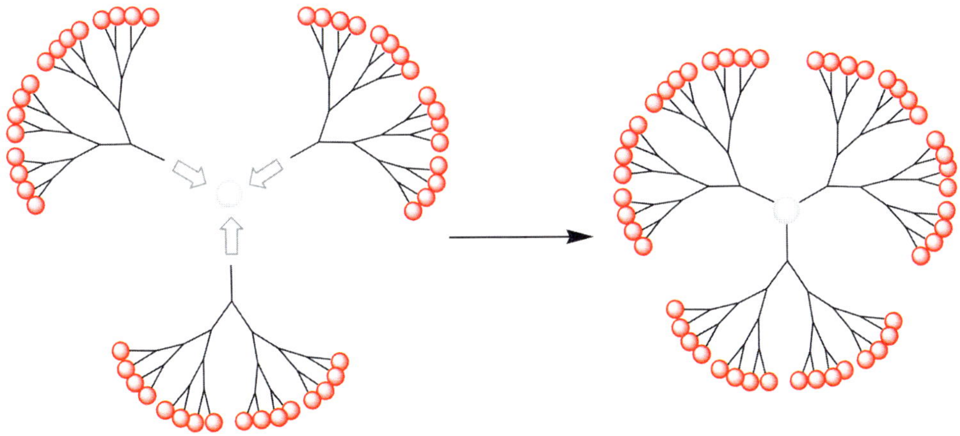

Fig. 6.7 Convergent covalent attachment of – inert or functional – dendrons during dendrimer construction.

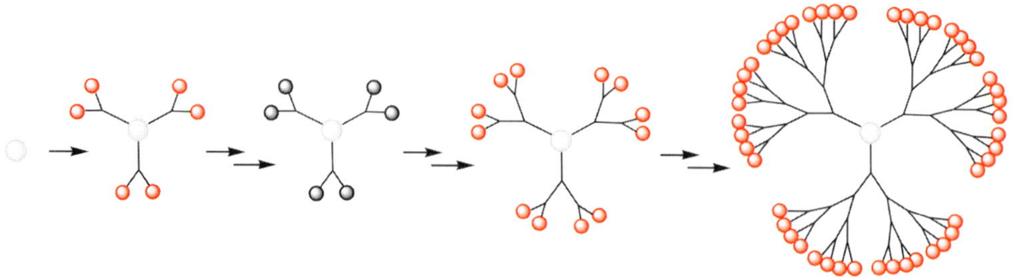

Fig. 6.8 Divergent covalent linkage of – inert or functional – building blocks during dendrimer construction.

sions between luminescent dendrimer arms that approach one another too closely [14].

However, the divergent synthetic strategy also permits covalent construction of a dendritic shell around a function-bearing core unit (Fig. 6.8; further examples are given in Section 4.1).

Steric envelopment of the core, the branches, or the end groups (Fig. 6.8) can limit access of reactants to reactive sites. This permits the creation of local micro-environments exerting an influence on the chemical, electrical, or "photonic" [15] properties of the enveloped functional groups. Compared to non-dendritic reference compounds, the electron density, polarity, or hydrophilicity/lipophilicity prevailing in the newly created micro-environment can alter the solvatochromism and redox behaviour of the reactive site, to name just a few of the properties possibly affected [16]. The concept of enveloping catalytically active entities by surrounding them with dendrons can provide a means of influencing (selective) access of substrates to the catalyst and thus of creating or improving substrate selectivity, regioselectivity, or enantioselectivity [17]. Majoral's dendrimers permit particularly versatile subsequent covalent bonding in the dendrimer interior by exploiting the possibilities of phosphorus chemistry (details are given in Section 4.1.10).

6.2 Supramolecular (host/guest) interactions

6.2.1 Non-covalent modification of a dendrimer periphery

While the previous sections have largely addressed variation of the dendrimer periphery by covalent transfunctionalisation, an alternative concept is based on modification of the dendrimer surface by non-covalent interactions [18]. Selective interactions of guest molecules with dendritic hosts depend upon the nature of both the dendrimer core and the dendrimer shell.

Kim *et al.* discovered a method of reversibly fixing pseudorotaxane moieties to the outer shell by non-covalent interactions (Fig. 6.9). The prefix *pseudo* de-

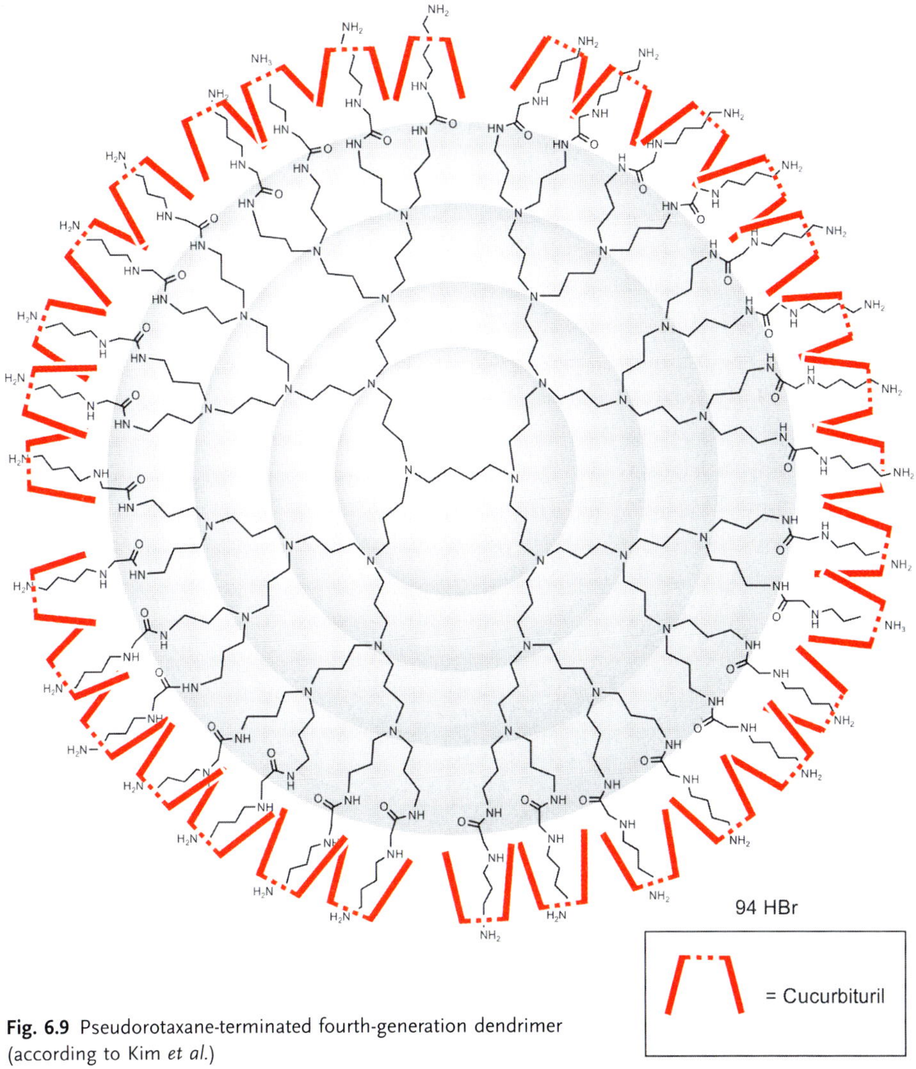

Fig. 6.9 Pseudorotaxane-terminated fourth-generation dendrimer (according to Kim *et al.*)

scribes rotaxanes which can readily dethread because their stopper groups are small or absent. Such terminal groups form a relatively rigid dendrimer shell which can be used, for example, for (reversible) inclusion of guests in the interior of the dendrimer – depending upon the degree of dethreading of the rotaxane wheel under certain reaction conditions. After release of the guests as a result of dethreading of the wheel the "rigid surface" regains a degree of flexibility.

Meijer *et al.* developed a generally applicable supramolecular method for modifying the periphery of POPAM dendrimers. Covalently bonded urea end groups

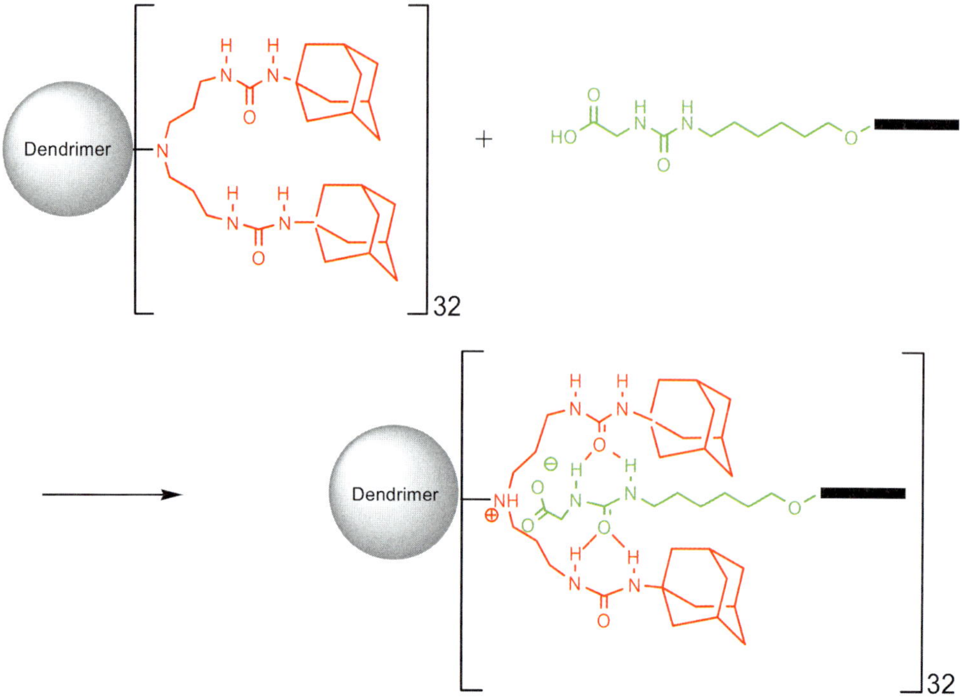

Fig. 6.10 Glycinylurea guest molecules (green) dock onto the urea-functionalised, niche-like periphery (red) of a POPAM dendrimer (shown schematically as grey sphere, with 32 terminal groups) owing to "tailor-made" supramolecular interactions (red bonds) (according to Meijer *et al.*)

were attached to a scaffold consisting of DAB-dendr-(NHCONH-Ad)$_n$, with $n = 4$, 8, 16, 32, and 64, where DAB-dendr stands for the dendrimer scaffold with a 1,4-diaminobutane core, and Ad for adamantyl. This permitted relatively strong yet reversible bonding of glycinylurea guest molecules by means of directional multiple interactions (Fig. 6.10), namely ionic and hydrogen bonding interactions within the niche-like two-dimensional planes of the peripheral bis-adamantylurea entities. The dendritic supramolecules formed by attachment of the guests exhibit a significant increase of dimensional rigidity at the periphery while conformational flexibility is retained at the far end of the selectively bound guest molecules [19].

Further examples of non-covalent modifications of dendrimer exterior shells will be found in the literature [20].

6.2.2
Self-assembly of dendrimers

Tomalia *et al.* used ionic interactions to construct "core/shell tectodendrimers". They originally used the term to designate clusters made up of dendritic molecules with a PAMAM dendrimer as core surrounded by covalently attached lower-generation dendrimers. In this case, the core building block and the dendritic shell unit added in excess undergo (ionic) self-assembly [21 a] with charge neutralisation to form a supramolecular core/shell architecture. Subsequent formation of covalent bonds with the aid of an amine-terminated dendritic core as limiting reagent and addition of an excess of carboxylic acid-terminated dendrimer shell reagent leads to tectodendrimers [21 b]. One advantage of these nanoscale entities is that different functionalities can be assigned to each part-dendrimer.

Complete self-assembly of discrete supramolecular dendrimers was achieved on use of a homotritopic (*i.e.* with three identical host units) Hamilton receptor **1** (cf. Section 6.2.3.1) as core unit with host character. The branching unit is an AB_2 element **2**, made up – covalently – of two Hamilton receptors and a complementary cyanuric acid substrate. One set each of $3 \cdot 2^n - 3$ (n = generation number) molecules of the AB_2 unit **2**, a core (**1**), and the barbiturate or cyanuric end groups **3** go to form the supramolecular dendrimer **4**. In the ideal case that all the hydrogen bonds between the building blocks **1** and **3** are used, an n-th generation supramolecular dendrimer can be expected to form under thermodynamic control on mixing of the three reactants in the ratio $1{:}(3 \cdot 2^n - 3){:}(3 \cdot 2^n)$ (core:branching unit:end group) (Fig. 6.11) [22]. This concept was also applied to the construction of chiral depsipeptide dendrimers [23]. Other supramolecular dendrimers will be found in Sections 2.6 and 4.1.

Control of self-assembly by switchable redox processes in complexes between dendrimers with π-donor and π-acceptor units permitted specific prediction of the space occupied by dendritic aggregates. Use of cucurbit[8]uril (see Fig. 6.9) as donor-host compound for viologen guests (dendronised 4,4′-bipyridines as acceptors) promotes formation of such charge-transfer complexes [24].

6.2.3
Inclusion of guest species in dendritic host molecules

6.2.3.1 Dendrimers with multiple receptor units

The periphery of dendrimers is predestined for attachment of numerous receptor units. In principle, this is also possible in the core or in the branching units [25].

One of the best known receptor units is that introduced by Hamilton *et al.* [26]. Often designated briefly as "Hamilton receptor", this barbiturate receptor niche (Fig. 6.12) consists of an isophthaloyl spacer flanked by two acylated 2,6-diaminopyridine moieties. Macrocyclisation with a rigid diphenylmethane unit generates a non-collapsing host cavity suitable for selective complexation of barbiturate guest molecules *via* formation of six hydrogen bonds.

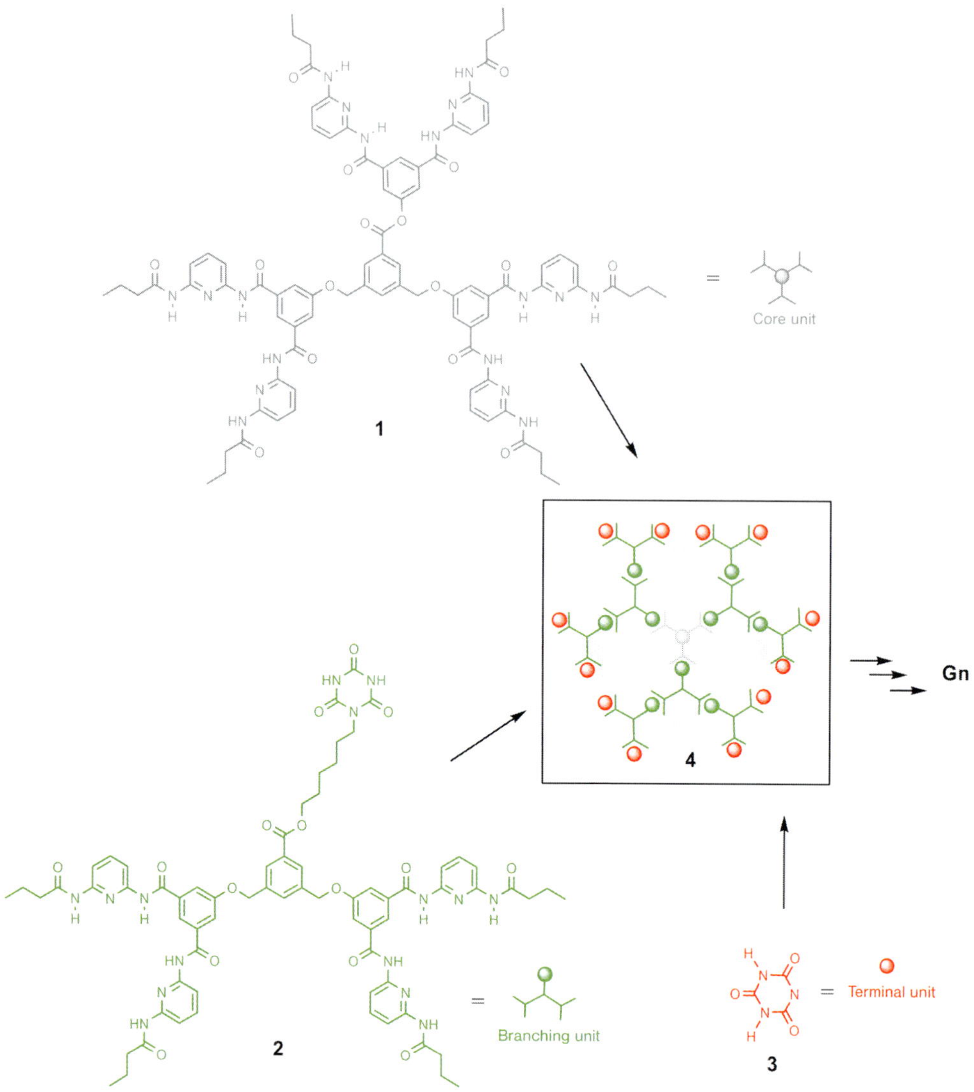

Fig. 6.11 Supramolecular self-assembly of supramolecular dendrimers (according to Hirsch *et al.*)

This proven host system was transferred to dendrimers by functionalising the periphery of POPAM dendrimers with the acyclic Hamilton variant (Fig. 6.13). After monofunctionalisation with an amino group on the isophthaloyl system, it could be multiply covalently bonded to the dendrimer. In this way, up to fourth-generation POPAM dendrimers could be peripherally "embellished" with barbiturate receptors. The mode of action of the individual dendrimers as multivalent

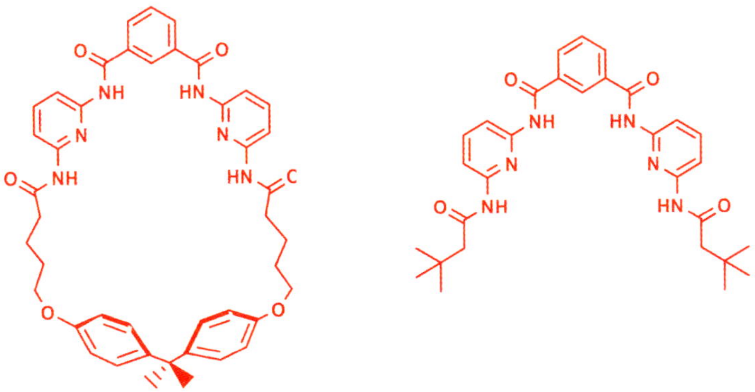

Fig. 6.12 Macrocyclic Hamilton receptor for barbiturate guest molecules (left-hand side). A non-cyclic Hamilton host compound with significantly weaker bonding to corresponding guest molecules is shown (right-hand side) for the sake of comparison

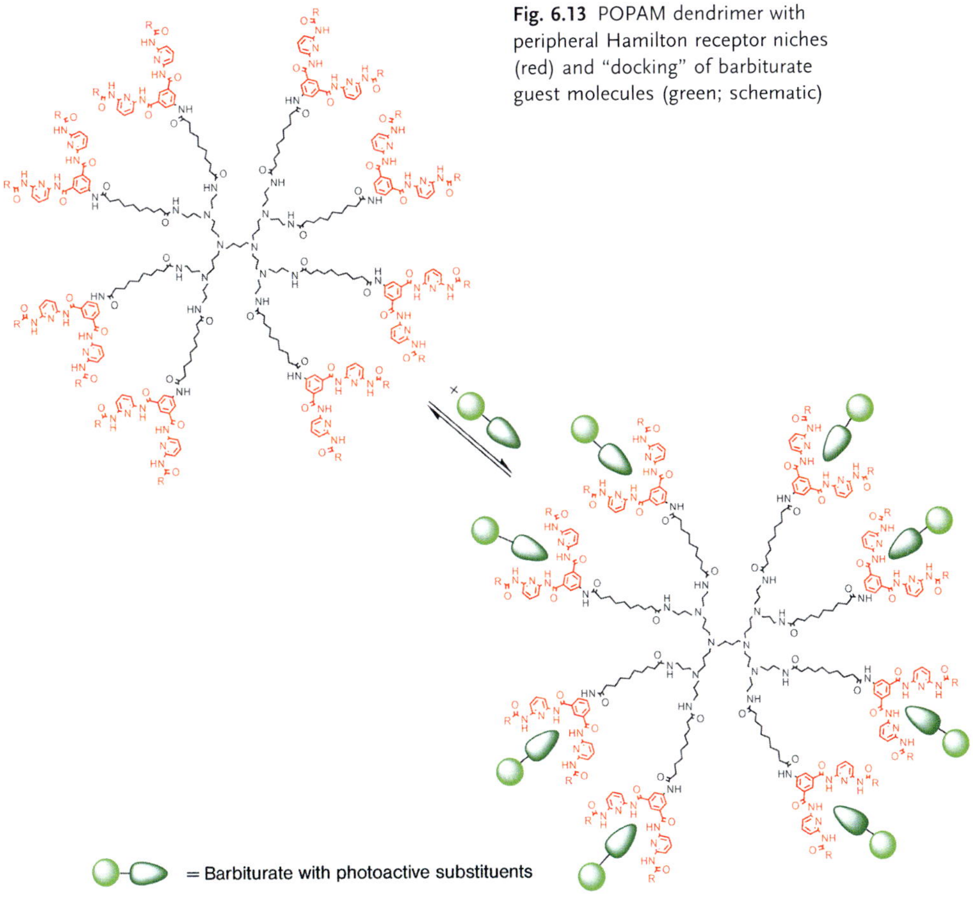

Fig. 6.13 POPAM dendrimer with peripheral Hamilton receptor niches (red) and "docking" of barbiturate guest molecules (green; schematic)

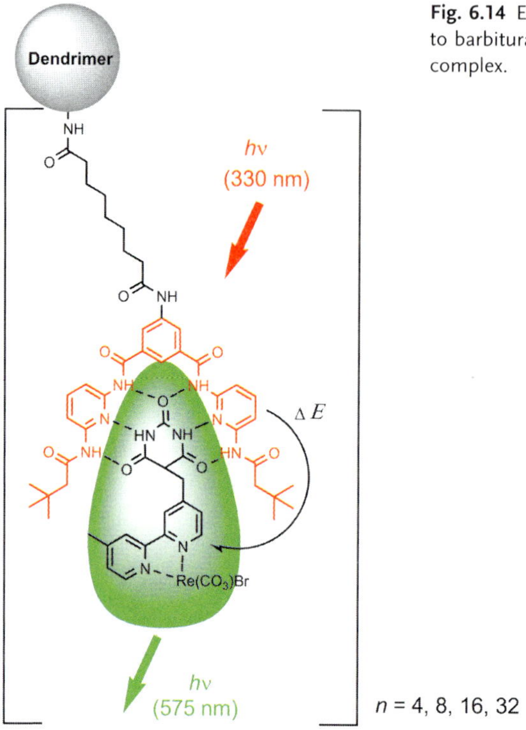

Fig. 6.14 Energy transfer from Hamilton host to barbiturate guest with emitting rhenium complex.

host compounds could be confirmed by ^{1}H-NMR spectroscopy and photophysical studies, which also revealed dendritic effects (Section 6.3).

An increase in the quantum yield of the emission of "Hamilton receptor dendrimers" (Fig. 6.13) with increasing generation number was noted. The reason for this increase was ascribed to the increasing conformational rigidity of the Hamilton receptors with increasing generation number – arising from bonding of the guest molecules – and the associated greater steric hindrance at the periphery of the host/guest complex. Bonding of the rhenium-complex-substituted barbiturate shown in Fig. 6.14, which exhibits a photoexcited state of lower energy than the Hamilton moiety, leads to a host → guest energy transfer ($3.6 \times 10^{10}\ s^{-1}$).

Summarising, it can be said that stable host/guest systems with a pronounced capacity for guest inclusion can be obtained by functionalisation of the periphery of POPAM dendrimers [27].

POPAM dendrimers bearing up to 32 photoswitchable azobenzene groups at the periphery were used as host compounds for *eosin Y* (2′,4′,5′,7′-tetrabromofluorescein dianion) (see also Section 5.1.2). One reason for the choice of this dye was that it shows strong fluorescence which should be influenced by being enveloped by the dendrimer. The other reason was that the energy of its lowest triplet state is higher than that of the lowest triplet state of azobenzene, leading

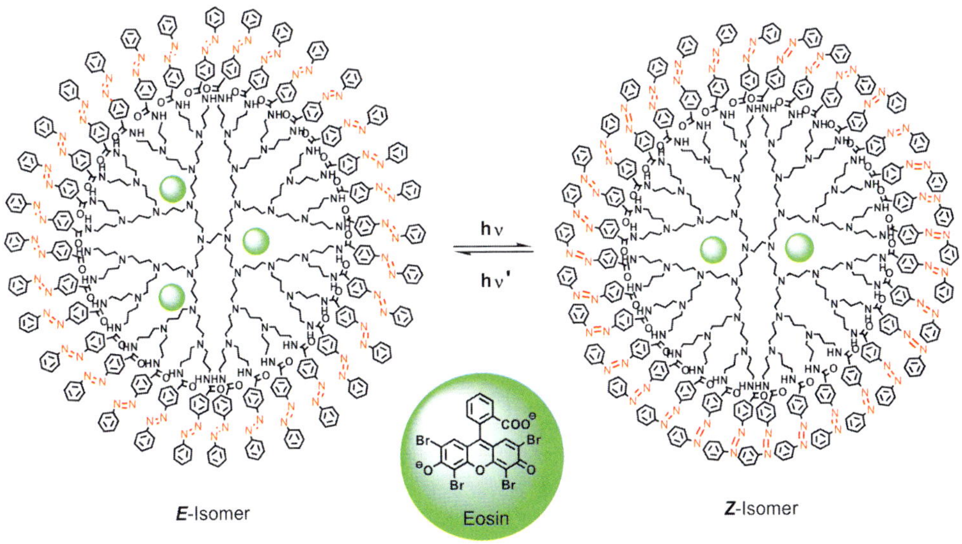

Fig. 6.15 Different forms of inclusion of eosin guest molecules in (isomeric) photoswitchable POPAM host compounds. The number and position of the included eosin molecules vary according to the conditions (schematic; only a few of the guest molecules are shown here for the sake of simplicity; cf. Section 5.2.2 (Fig. 5.21) according to Balzani, De Cola, Vögtle *et al.*)

to sensitisation of photoisomerisation of the peripheral units of the POPAM dendrimer. The *E* form of the azobenzene dendrimer can be – reversibly – switched to the *Z* form by photoexcitation (cf. Section 5.2.2). Both dendritic oligo(azobenzene) isomers quench the fluorescence of eosins, probably as the result of an electron-transfer reaction between the excited singlet state of eosins and the tertiary amine groups of the POPAM core. Quenching is more efficient for the *Z* than for the *E* form. *E*→Z and *Z*→*E* photoisomerisation of the azobenzene units of the dendrimer is sensitised by eosin *via* triplet-triplet energy transfer. The latter proceeds according to the Förster mechanism (see Section 5.1.2), intramolecularly within the dendrimer itself and "intra-supramolecularly" from the dendrimer to the eosin dye. The results support the assumption that the eosin guest penetrates into the interior of the dendrimer (Fig. 6.15) [28].

A light-controlled "dendritic box" was investigated in analogous fourth-generation POPAM dendrimers whose periphery alternately bore azobenzene and naphthyl moieties. Measurements of photoisomerisation at a wavelength of 365 nm in dichloromethane showed that not all azobenzene units in the *E* form are transformed into the *Z* form, but that, for example, four units remain in the *E* form and some 28 units isomerise to the *Z* form. At pH 7, six eosin molecules per host molecule were taken up by this mixture of isomers. On the other hand, the dendrimer having all azobenzene groups in the *E* form takes up eight eosin guest molecules under the same conditions. This can be explained by assuming that the periphery is not so densely packed in the *E* form (cf. Figs. 6.15 and 5.21) and the guest molecules can thus more readily gain access to the niches formed dynamically at changing sites. In the "*almost-Z* form" – adopted by the majority of the azobenzenes in the more voluminous *Z* form – the terminal groups are crowded closer together, making it more difficult for the guest molecules to enter the interior of the dendrimer [29].

6.2.3.2 Guest inclusion by steric compression

Thanks to their conformational flexibility, POPAM dendrimers can serve as host molecules for small guest molecules. The non-covalent reversible inclusion of these guests is attributable to electrostatic, hydrophobic, H-bonding, and acid/base effects in the interior of the dendrimer.

However, irreversible inclusion can be accomplished through formation of a dendritic box by steric compression of the periphery (Fig. 6.16) – after covalent coupling to a sufficient number of large terminal groups. For the inclusion of dye molecules in such dendrimers, see also Section 8.3.6.

An example of non-covalent encapsulation of guests – by host molecules of supramolecular construction – is provided by water-soluble POPAM dendrimers modified with 3,4,5-tris(tetraethylenoxy)benzoyl units, which function as "container molecules" [30] (Fig. 6.17). The interior dendrimer scaffold is basic owing to the tertiary amino groups, and the periphery is hydrophilic. The inclusion of guest molecules was accomplished in buffered aqueous medium at pH 7 with the aid of two anionic water-soluble xanthene dyes. Specifically, Bengal rose and 4,5,6,7-tetrachlorofluorescein were titrated against the host dendrimer. In both cases, a bathochromic shift of the absorption maximum occurred, indicating an interaction between the dendrimer host and the dye guest molecules [31].

The possible formation of niches can also be exploited for the catalysis of chemical reactions. Palladium nanoparticles of 1.7-nm diameter included in a

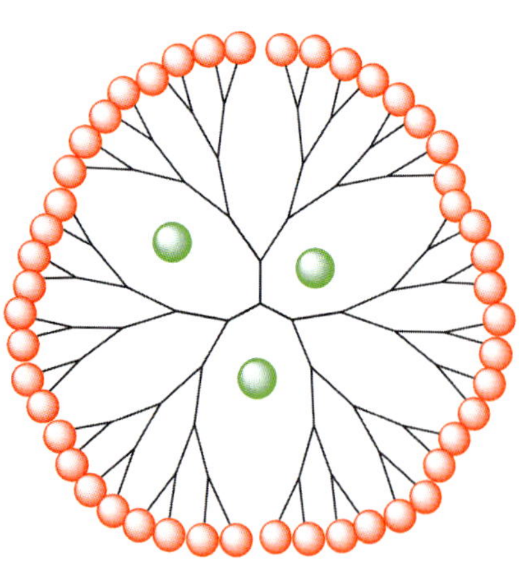

Fig. 6.16 Irreversible encapsulation of guests (green) by compression of the surface to form a dendritic box (schematic, two-dimensional)

Fig. 6.17 POPAM dendrimer with peripheral oligo-ethylene glycol ether units (according to Meijer *et al.*)

fourth-generation PAMAM dendrimer served as catalysts for coupling of haloarenes with organostannanes, which proceeds under mild conditions *via* a Stille reaction. Such coupling reactions usually require high temperatures and organic solvents. Moreover, the conventionally used catalyst has to be produced from a palladium(II) salt by reduction to Pd(0) and stabilised with phosphine ligands. The Pd nanoparticles enclosed in the PAMAM dendrimer have no need of such treatment. Such "dendritically" catalysed Stille coupling proceeds in aqueous medium at room temperature with minimal amounts of Pd (0.1 atom-%) very much in the sense of "green chemistry". The small diameter of the nanoparticles assures a high surface-to-volume ratio, which is important for a high catalytic activity [32].

6.2.3.3 Guest inclusion by dynamic processes (diffusion)

Non-covalent bonding of guest molecules in the interior of a dendrimer molecule can also proceed dynamically. In micelle-like amphiphilic dendrimers, guests can diffuse into and out of the interior of the dendrimer scaffold *via* hydrophilic and hydrophobic interactions (Fig. 6.18).

Balzani, Vögtle *et al.* followed the kinetics of complexation of metal ions (Co^{2+}, Cu^{2+}) for the example of dansyl-bearing POPAM dendrimers (see Chapter 4, Fig. 4.3) [33]. In the first step the cations interacted with the dimethylamino nitrogen of the dansyl units, as can be recognised from the quenching of their fluorescence – which is complete at high ion concentrations (Fig. 6.19a).

If such solutions are allowed to stand for several hours, their fluorescence surprisingly reappears. This is interpreted by assuming that the metal ions slowly penetrate into the interior of the dendrimer to the N atoms of the POPAM-core, again freeing up the fluorescing dansyl units (Fig. 6.19b).

It goes without saying that a non-dendritic monodansyl amino compound as reference substance will be unable to show analogous behaviour because it cannot form intramolecular niches for inclusion of cations. Thus dendritic architectures lead to specific properties which are unattainable or attainable only with difficulty in the absence of branching units.

Newkome *et al.* were the first to synthesise symmetrical, "quater-directional" cascade molecules with a carbon scaffold bearing 36 terminal carboxyl groups – all at an equal distance from the neopentyl core (Fig. 6.20a). The carboxyls were converted into the corresponding ammonium and tetramethylammonium carboxylates. Synthesis of these dendritic "unimolecular micelles" with hydrophobic core and hydrophilic shell was accomplished up to the fourth generation by coupling of a dendritic "hypercore" (constructed from 4,4-bis(4′-hydroxyphenyl)-pentanol monomer) and PEG mesylate (PEG = polyethylene glycol). Dyes such

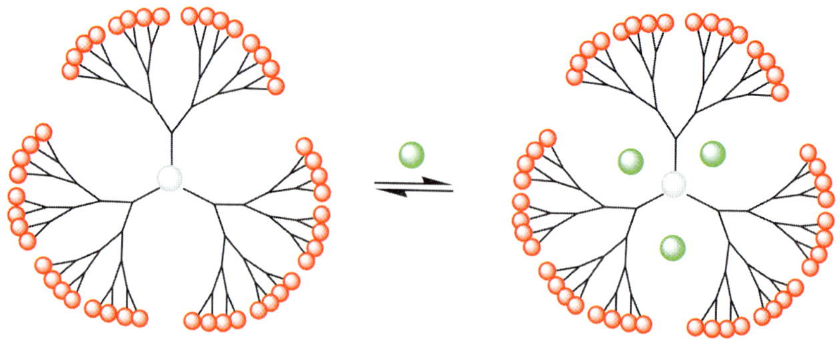

Fig. 6.18 Inclusion and exclusion of guests in dendrimers by diffusion processes

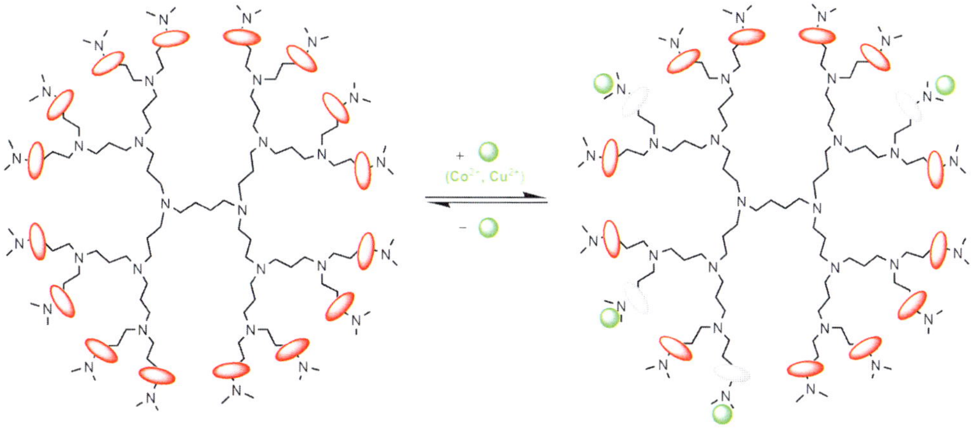

Fig. 6.19 a Interaction of metal ions with the dimethylamino groups of the peripheral dansyl units (shown as red ellipses) of a POPAM dendrimer: The original dansyl fluorescence of the dendrimer is quenched – partially in this case (light grey ellipses)

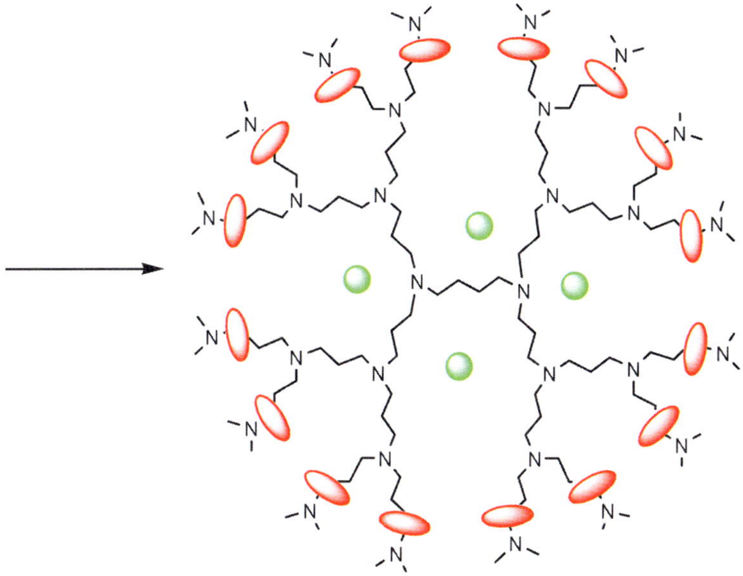

Fig. 6.19 b After a few hours the cations have migrated from the exterior (Fig. 6.19 a) into the interior of the dendrimer: The fluorescence of the dansyl units returns "by itself" (according to Balzani, Ceroni, Vögtle *et al.*)

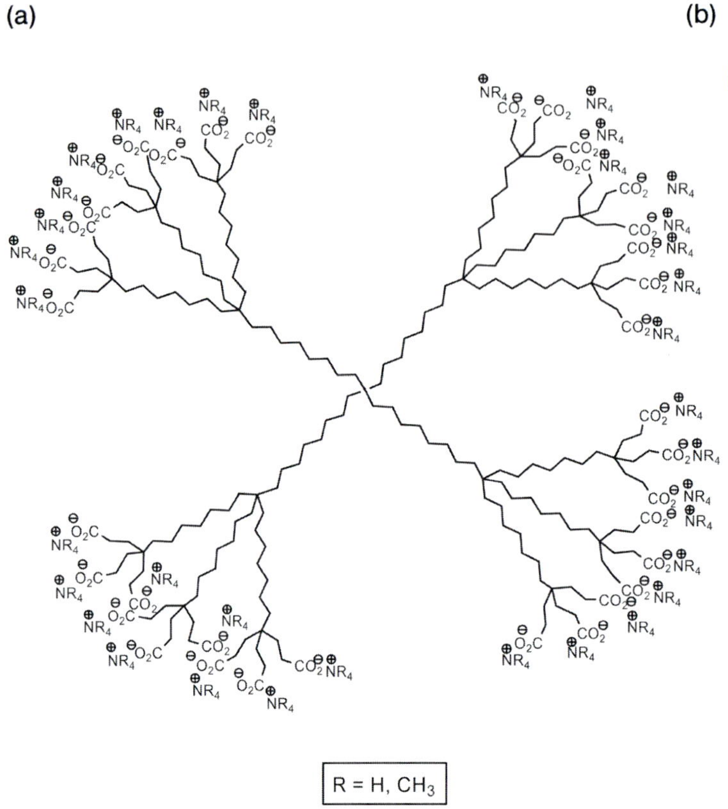

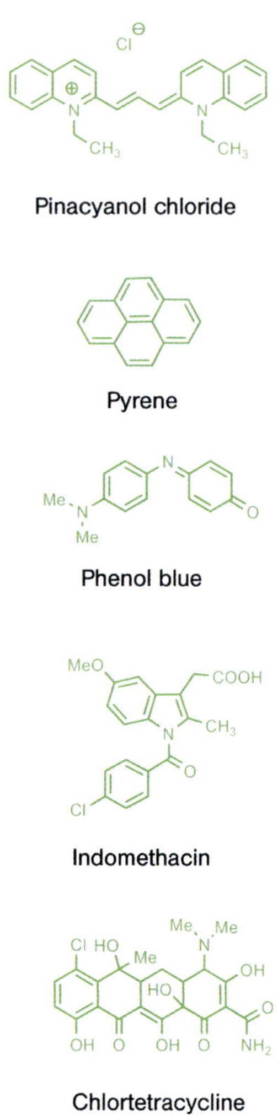

Fig. 6.20 **a** Quater-directional hydrocarbon cascade scaffold with ionic periphery and unimolecular micellar character; **b** guest molecules (according to Newkome, Moorefield *et al.*)

as chlortetracycline, phenol blue, and pinacyanol chloride were included as guests (Fig. 6.20 b). Chlortetracycline was added to the tetramethylammonium cascade compound in aqueous solution. Absorption with subsequent fluorescence at 520 nm was established by fluorescence microscopy and photometry. Neither component – host nor guest – shows fluorescence on its own. On mixing, however, the fluorescence intensity increases linearly with increasing concentration of "micellanoates" (anionic cascade molecules having micellar character). Since chlortetracycline fluoresces only in a lipophilic environment, a host/guest interaction must have taken place in the interior of the cascade scaffold [34].

The host properties are also impressively demonstrated by inclusion of pyrene in aqueous solution (Fig. 6.20 b). It also proved possible to introduce the anti-inflam-

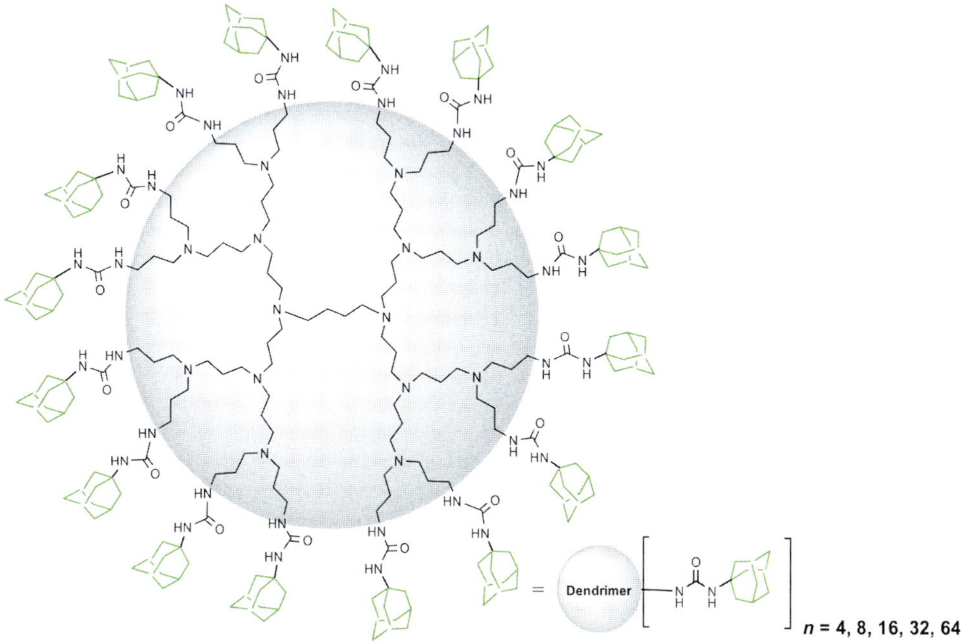

Fig. 6.21 POPAM dendrimers with terminal adamantyl guest units for inclusion in the host cavities of β-cyclodextrin (according to Meijer and Reinhoudt *et al.*)

matory drug indomethacin [35]. Spontaneous self-assembly of fatty acid molecules at the periphery of fourth-generation amino-terminated PAMAM dendrimers leads to transformation of their originally hydrophilic outer shell into a hydrophobic surface – owing to ion-pair formation between the carboxyl groups of the fatty acids and the terminal amino groups of the dendrimer. This ultimately provides access to inverse micellar structures. Such compounds are extractable from aqueous solution into a non-polar medium and can also entrain hydrophilic guest molecules during such transfer (phase transfer process). Apart from the transfer of dyes, these systems are tailor-made for catalytic applications because catalytically active metal nanoparticles thus become soluble in non-polar solvents. The reversible nature of the dendrimer/fatty acid interactions also permits recovery of the catalyst from aqueous solution at pH 2 [36]. Further literature about "dendritic micelles" dye transport and binding of acetylcholine is cited under [37a–c].

6.2.4 Dendrimers as guest molecules

First to fifth-generation POPAM dendrimers with terminal adamantyl groups (Fig. 6.21) were dissolved in water in the presence of β-cyclodextrin. The hydrophobic effect forces the terminal, lipophilic adamantyl groups as guests into the endo-lipophilic host cavities of β-cyclodextrin which offer a complementary fit.

These dendrimers show their best solubility in aqueous cyclodextrin solution at a pH value of 2, because protonation of the tertiary amino groups in the interior of the dendrimer is then complete and the dendrimer assumes a stretched spheroidal shape owing to the mutual repulsion of the positive charges. The dendrimers remain in solution up to a pH value of ≤ 7; they are precipitated under basic conditions, with the exception of the first-generation dendrimers. In the case of the fifth-generation POPAM dendrimer it was found that lack of space on the outer shell due to steric overcrowding precluded supramolecular envelopment of all of its adamantyl groups by cyclodextrin molecules ("supramolecular starburst" situation).

It could also be proved with the aid of the fluorescent probe 8-anilinonaphthalene-l-sulphonate (ANS) that such highly lipophilic substituted dendrimers can themselves act as host compounds for ANS in water. Binding of this fluorescent probe as guest is primarily electrostatic in nature and increases with increasing generation number.

Incomplete coverage of the surface of such a fifth-generation POPAM dendrimer exposes hydrophobic areas of adamantyl units remaining uncomplexed by cyclodextrins on the dendritic outer shell [38]. Pyrenes were used as neutral fluorescence probes to examine whether this might lead to aggregation in water driven by the hydrophobic effect [39 a]. Their inclusion in the dendrimer/cyclodextrin aggregate leads to changes in fluorescence intensity and in the vibrational fine structure. Formation of excimers was also observed.

Covalent bonding of carbohydrate dendrons – by formation of amide bonds – to a ferrocene core affects properties such as water solubility and redox behaviour. For example, if only β-D-glucopyranosyl groups are linked to the ferrocene core, then the hydrophobic redox-active ferrocene units are bound to β-cyclodextrin in water. This leads to a complex consisting of a carbohydrate-based host and a carbohydrate-containing (ferrocenyl) guest. In such complexes the unsubstituted cyclopentadienyl ring of the ferrocene unit is oriented towards the interior of the host cavity (Fig. 6.22 top right). In contrast, if both of the cyclopentadienyl five-membered rings of the ferrocene core bear β-D-glucopyranosyl-containing groups it is understandable that complexation can no longer take place. – For the bioconjugation of biotinylated PAMAM dendrons see lit. [39 b].

Quantitative electrochemical studies afforded information about the protective action of the sugar moieties on the complex. The number of substituents on the amide nitrogen (one or two) and the number of carbohydrate arms present in the substituent (one or three) are of importance for interactions with the solvent [40].

Klärner *et al.* undertook detailed studies of the host/guest relationships between molecular tweezers formed from condensed hydrocarbon arenes and symmetrically and asymmetrically dendryl-substituted viologens – with a 4,4′-bipyridinium core acting as -electron acceptor and a Fréchet-type dendron as π-electron donor (Fig. 6.23). The strong fluorescence of the 1,3-dimethylenoxybenzene units of the Fréchet dendron is quenched as a result of the donor/acceptor interaction. Accordingly, in dichloromethane solution the 4,4′-bipyridinium core

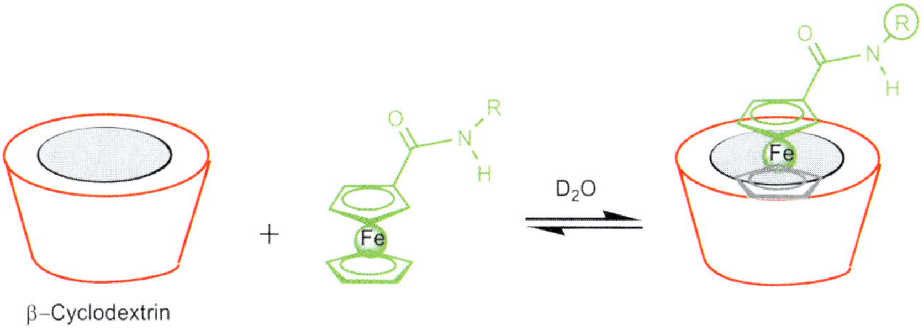

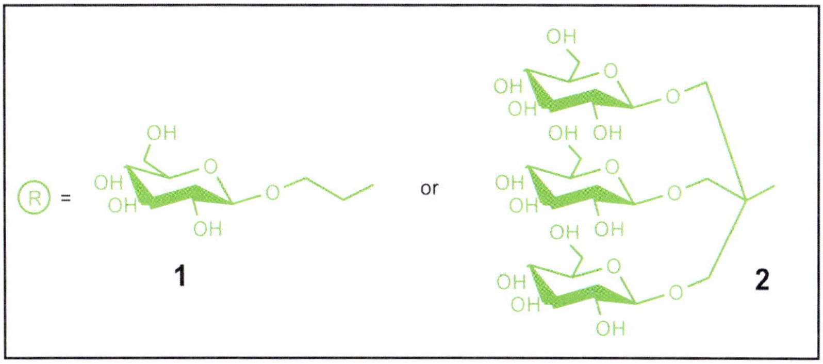

Fig. 6.22 Inclusion of one- or three-armed carbohydrate dendrons **1** or **2** (green), respectively, with ferrocene core unit as guest in the cavity of *β*-cyclodextrin (red; according to Credi, Balzani, Raymo, Stoddart *et al.*; schematic)

is included as guest in the molecular tweezers shown in Fig. 6.23. "Quasi-rotaxanes" with a non-macrocyclic wheel are formed as a result.

Fluorescence measurements and ^{1}H-NMR titrations revealed that the association constant (order of magnitude 10^4 M^{-1}) decreases with increasing generation number of the dendrons owing to increasing steric interactions (dendritic effect, see also Section 6.3). However, the binding constants for the asymmetrically substituted dendrimers are higher than for the corresponding symmetrically substituted ones of the same generation. NMR absorptions of host- and guest protons confirm that the bipyridinium core is completely enclosed in the cavity of the tweezers. Moreover, the tweezers swing from one pyridinium ring of the viologen guest molecule to the other – and back again. In the case of mono-substituted viologens the less-substituted pyridinium ring is preferentially taken up by the tweezers in apolar solvents. Complexation of the bipyridinium core proceeds in various ways, depending upon the size of the cavity (or slot) of the tweezers and of the dendrons: By clipping or tweezering in the case of sym-

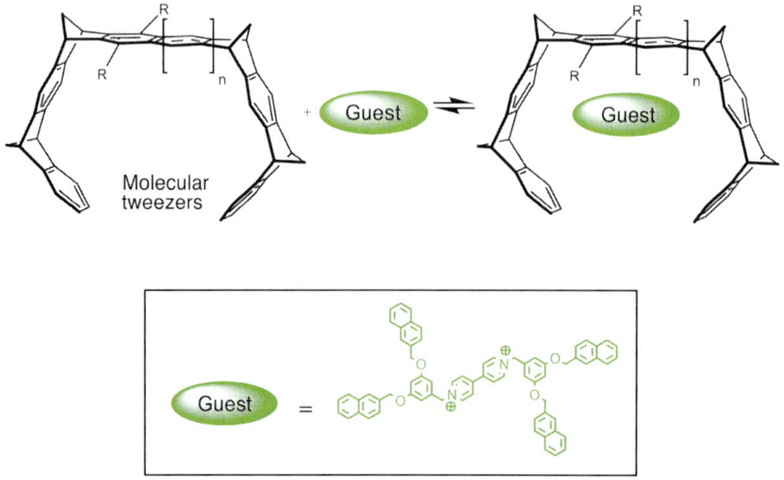

Fig. 6.23 Uptake of a dendritic viologen-type guest molecule in the cavity of molecular tweezers (schematic; according to Klärner *et al.*)

metrical viologens with bulky dendron substituents, but *via* threading for asymmetrically substituted viologens with at least one sufficiently small dendron. Biologically relevant cationic guests can also be selectively bound with tailor-made tweezers of this kind [41, 42].

6.2.5 Dendritic stopper groups (in rotaxanes)

As already implied at the end of the previous section, the use of dendrons as stoppers on rotaxane axes belongs within the area of reversibility in supramolecular chemistry. Because the volume of the dendritic scaffold can be adjusted almost at will, spatial requirements can be tailored such that dethreading of a wheel is hindered to a greater or lesser extent (Fig. 6.24). Classical space-filling entities such as mesityl, 2,6-di-*tert*-butylphenyl, or trityl groups are hardly able to afford similarly bulky, finely tuneable, and also solubility-promoting stoppers. Since dendrons and dendrimers composed of largely aliphatic building blocks are flexible and soft and hence difficult to assess, it was important to measure an effective spatial requirement of dendrons. On use of first-generation (G1) Fréchet dendrons only the free axle but not the rotaxane could be isolated during synthesis: The steric demands of the G1 dendrons were insufficient to prevent dethreading of the wheel (tetralactam ring **1** in Fig. 6.25) from the axle [43]. Rotaxanes with second-generation (G2) and third-generation (G3) Fréchet-type stoppers **3**, each with a facing tritylphenol stopper T1 (**2** in Fig. 6.25), dethread the wheel only on heating, and then *via* the Tl stopper since the bulk of the second and third dendritic stopper generation is greater.

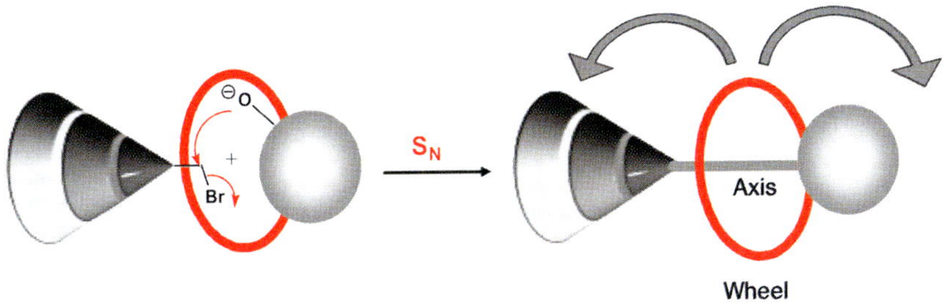

Fig. 6.24 Synthesis of a rotaxane (schematic) with dendritic stopper (cone) and tritylphenol stopper (sphere). S_N = nucleophilic substitution. The grey arrows indicate dethreading of the wheel (red ellipse) – to the left or the right according to choice – from the axle past one of the two stoppers, with the overall less bulky stopper – offering less resistance – being favoured. Specific components are shown in Fig. 6.25 (according to Vögtle *et al.*)

If the steric dimensions of the tritylphenol stopper are further increased by three *tert*-butyl groups (T2 in Fig. 6.25), then dethreading of the tetralactam wheel **1** proceeds over the dendritic Fréchet stopper G2 because the T2 stopper is now spatially more demanding. In further studies only second-generation stoppers were used – one on each side of the axle; here too dethreading occurred and could be followed by NMR (see Section 7.3). On use of different generations (G2, G3) of stoppers on the two sides of one and the same rotaxane dethreading expectedly took place *via* the small second-generation stopper (dendritic effect, see the following Section 6.3).

The results showed the spatial requirements of sterically demanding substituents to increase in the following empirical order:

G1 < T1 < G2 < T2 < G3

6.3 Dendritic effects

The previous section closed by presenting measurable effects of the increase in spatial requirements of dendritic groups with increasing generation number. Dendritic effects have been frequently mentioned elsewhere in this monograph. It is hardly surprising that there have been reports of numerous other consequences for the physical and chemical properties of all kinds of dendrimers – depending upon the generation of the dendrimer or dendrons. These findings will now be reviewed, to our knowledge for the very first time.

The term dendritic effect [44] covers a multitude of diverse chemical and physical effects, whose common feature is that a change of certain properties or phenomena occurs with increasing or deceasing generation number within a dendrimer family. A dendritic effect is described as positive or negative, depending upon whether the observed effect is amplified or attenuated.

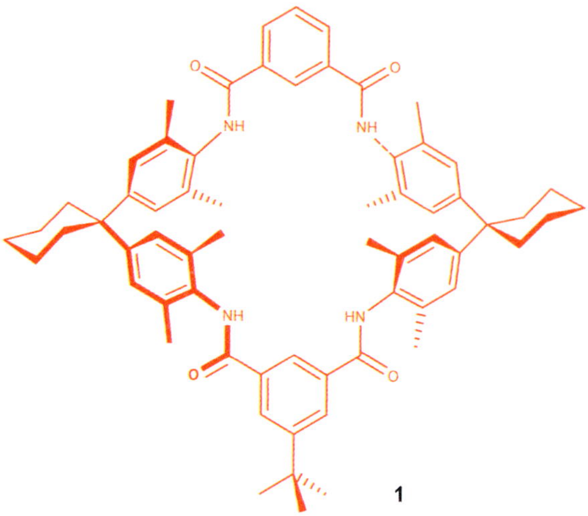

1

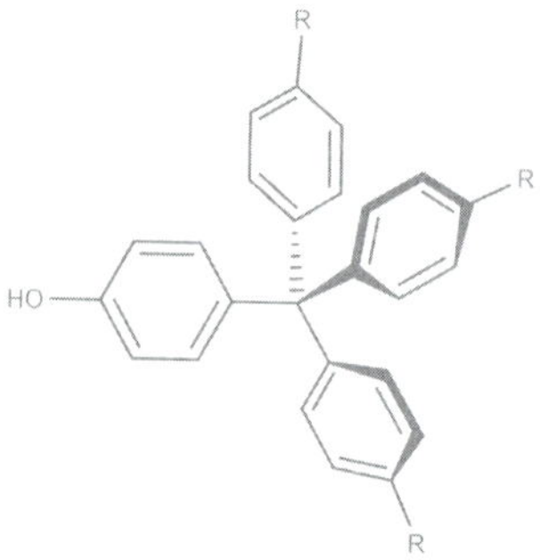

2 R = OH: (T1)
R = tBu: (T2)

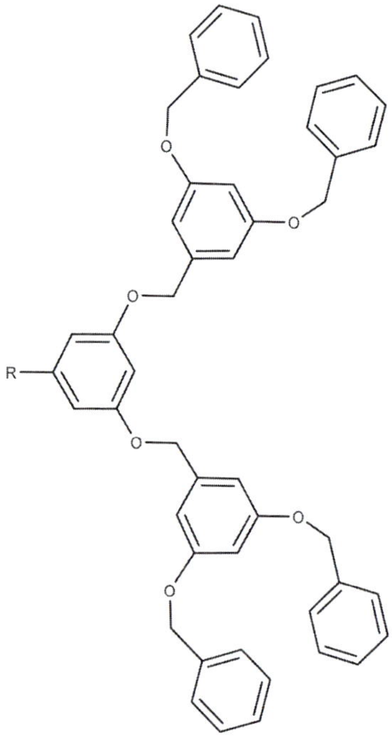

3 R = CH_2Br, OH: (G2)

Fig. 6.25 Components of rotaxanes according to Fig. 6.24: Tetralactam wheel **1**; classical tritylphenol stopper **2** whose size can be varied within certain limits; G2 Fréchet stopper **3** (according to Vögtle *et al.*)

6.3.1
Dendritic effect on inclusion of guests

A dendritic effect was observed in the inclusion of dendritic viologens in molecular tweezers described in Section 6.2.4. Gas-phase studies on this host/guest system revealed the presence of stable 1:1 complexes with the dicationic dendritic viologens (Fig. 6.26). Unstable in isolated form, the dicationic guest undergoes stabilisation through donor/acceptor (charge-transfer) interactions with the electron-rich tweezers. In collision experiments, complexes with dendrons of the zeroth to second generation each show different fragmentation patterns. Dendritic substituents are apparently able to stabilise the viologen core by intramolecular solvation more effectively with increasing generation number: The ability of the larger, conformationally flexible and electron-rich dendrons to fold inwards enables them to enclose the dication. Smaller dendrons, however, are hardly able to enclose the viologen cation and are therefore less effective in stabilising it [42].

PAMAM dendrimers were also found to show generation dependence with respect to permeability towards guests of various sizes. In the fourth generation with 64 terminal groups and a diameter of *ca.* 4.5 nm (for comparison: G1 *ca.* 2 nm, G10 *ca.* 13 nm), the dendrimers assume an almost spheroid shape. In the eighth generation the number of terminal groups is already 1024, whereas the diameter of the dendrimer has only roughly doubled. This leads to increas-

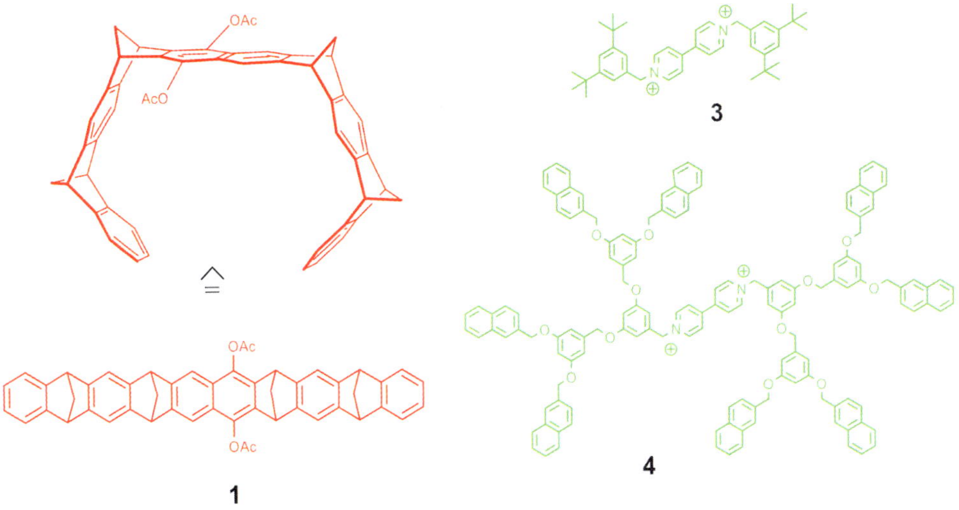

Fig. 6.26 Molecular tweezers **1** as host compound and dicationic, dendritically substituted viologen guest compounds (**3** and **4**) of different generation numbers (according to Balzani, Klärner, Vögtle *et al.*)

ing compression of the dendrimer surface. The space between the individual groups on the periphery thus decreases with increasing generation number. Passage of smaller and larger guest molecules through PAMAM molecules proceeds without problems up to the fourth generation. However, the sixth generation only permits permeability for small guest molecules. A conventional eighth-generation PAMAM dendrimer is ultimately so dense that guest molecules can no longer penetrate into the interior of the molecule [45].

6.3.2
Dendritic effects in catalysis

The expression "dendritic effect" has meanwhile come into general use for characterisation of generation-dependent physical and chemical phenomena in dendrimers. Examples of dendritic effects can also be found in the area of dendritic catalysts [46]. The possibilities offered, for example, by such dendrimers suitably functionalised on the periphery were already discussed in 1994 by Tomalia and Dvornic [47]. Such nanomolecules can combine the advantages of homogeneous and heterogeneous catalytic systems. In addition, their spherical shape generally permits better recovery – of organic catalyst support, metal complexes, or encapsulated catalyst particles – than with soluble polymer-based catalytic systems. It is therefore of interest as to how aspects of importance for catalytic reactions – such as enantioselectivity, catalytic activity, stability of corresponding metal complexes – can be modified through dendritic structures. In some cases a relation has already been found between the above parameters and the generation of the dendrimers. Dendritic catalysts have even exhibited a higher activity than comparable non-dendritic ones in a number of instances (see Section 8.2). In the following, a number of examples of the many dendritic catalysts are compiled and considered in greater detail. Even though some of them have already been described in a different context elsewhere in this book, owing to the outstanding importance of catalysis throughout the entire realm of chemistry the most important ones are again listed with representative formulae strictly in connection with this area of research and development.

6.3.2.1 Metal-containing dendritic catalysts

Access to catalytically active substances with several metal sites was gained with transition metal complexes of N,C,N-chelating ligands and their fixation on the periphery of carbosilane dendrimers. Thus corresponding nickel complex sites were fixed directly on silicon atoms of the dendrimer surface (Fig. 6.27). The preparation of such "nickel dendrimers" proceeded *via* poly-lithiation of a carbosilane ligand and subsequent transmetallation with $[NiCl_2(PEt_3)_2]$; however, not all ligand sites were metallated by this procedure. This catalytic system was investigated with regard to its redox activity in the Kharasch addition of carbon tetrachloride to methyl acrylate by maintaining the nickel concentration constant and varying the size of the dendrimer [48].

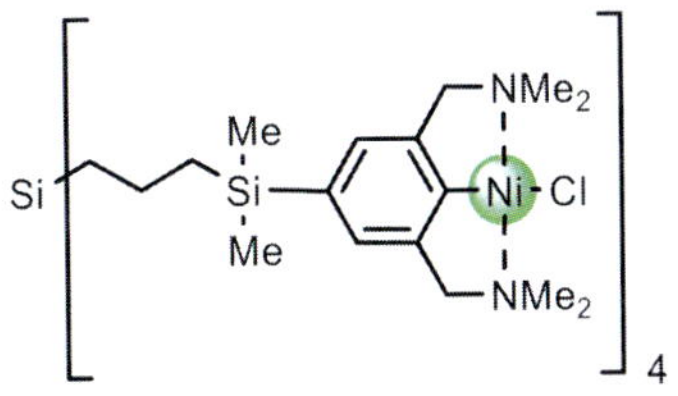

1

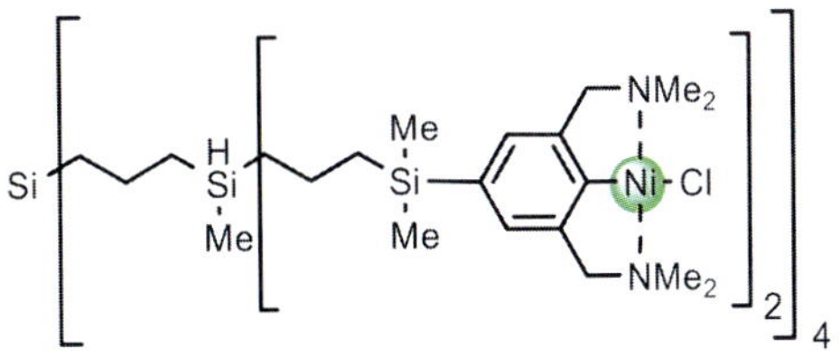

2

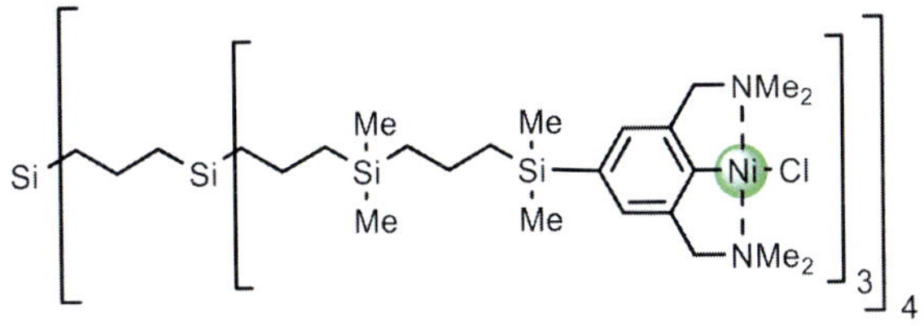

3

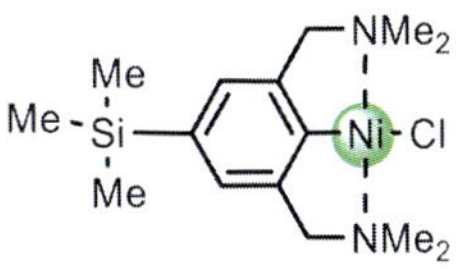

4

Fig. 6.27 Nickel-loaded carbosilane dendrimers **1–3** of increasing generation number (G0–G2) and a corresponding non-dendritic reference substance **4**

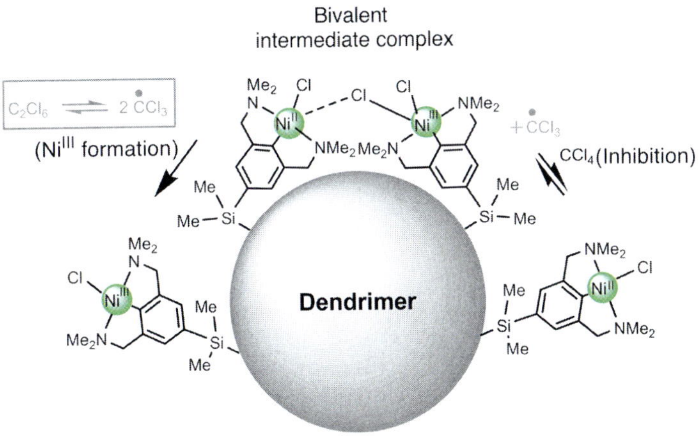

Fig. 6.28 Deactivation mechanism of the catalytic activity on the carbosilane dendrimer surface (schematic; according to van Koten *et al.*)

Contrary to expectations, the catalytic activity was found to decrease with increasing generation number. This was explained in terms of the increasing spatial demands of the nickel complex units bound in increasing numbers to the dendrimer periphery and increasingly limiting access to their active sites (see Fig. 6.28).

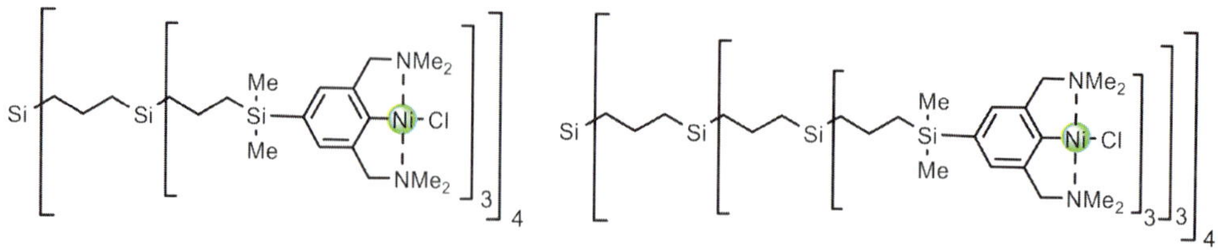

Fig. 6.29 Dendritic carbosilane-nickel complexes with decreasing catalytic activity owing to mixed complex formation

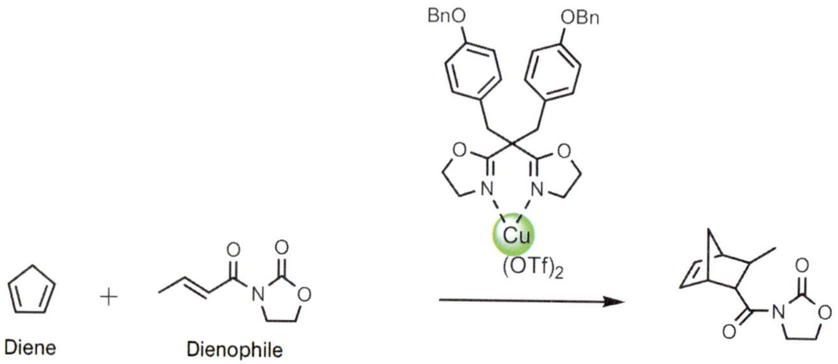

Fig. 6.30 Diels-Alder reaction, catalysed by a – non-dendritic – bis(oxazoline)-metal complex (according to Chow *et al.*)

The decrease in catalytic activity of the nickel-containing carbosilane dendrimer shown in Fig. 6.28 was attributed to the formation of mixed complexes with nickel in both oxidation states II and III on the dendrimer surface, which competes with the reaction with substrate radicals occurring in Kharash reactions (Fig. 6.29).

Bis(oxazoline)-metal complexes can catalyse Diels–Alder reactions between cyclopentadiene and crotonylimide (Fig. 6.30). A dendritic copper(II) catalyst was prepared by mixing corresponding dendritic ligands (each of different generation) and copper(II) triflate in a ratio of 1:1 in anhydrous dichloromethane. Expectedly, the dendritic ligands of higher generations needed a longer time (here 3–5 hours for G2, G3) for complete complexation than those of a lower generation (1.5 hours).

The dendritic bis(oxazoline) ligands used were of those of zeroth to third generation Fig. 6.31. They were made up of two components, the catalytic core unit consisting of bis-oxazoline and the dendron composed of polyether units [49].

It was found that the catalytic reactivity and the substrate binding profile depend upon the size and the generation of the dendrons. Since the reactions obey Michaelis-Menten kinetics, this family of catalysts was given the name "dendrizyme", alluding to enzymes. The dendrizyme/substrate binding constant

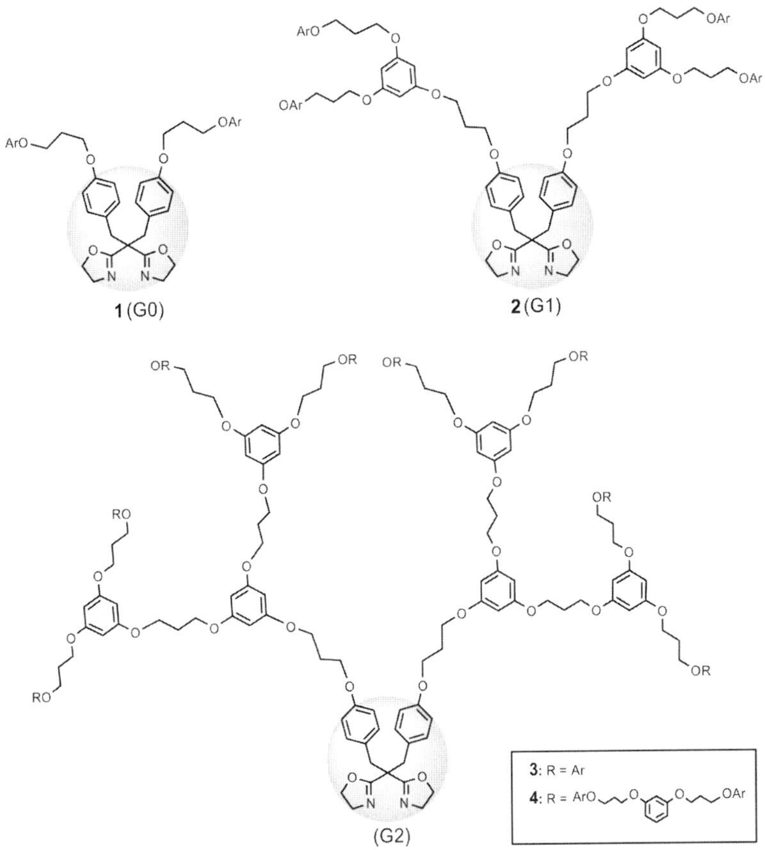

Fig. 6.31 Catalytically active dendritic bis(oxazoline) ligands **2–4** and reference compound **1** for Diels-Alder catalysis (according to Chow *et al.*)

decreases with increasing generation number of the dendrizyme. Such destabilisation is probably due to increasing deformation of the complex – as a result of increasing steric repulsion between the dendrons of higher generations (Fig. 6.32). Although the rate constant of the Diels-Alder reaction remains approximately the same from the zeroth to the second generation, it drops "suddenly" for the third generation. The rate-determining step is the Diels-Alder addition between the catalyst/dienophile complex and the cyclopentadiene; however, the reaction rate should be determined by the steric accessibility of catalyst niches.

The abrupt drop in reaction rate of the third-generation dendrizyme indicates that the catalytic sites are only accessible for G0 to G2, but are shielded by dendritic envelopment in the third generation.

Mutual steric repulsion of the large dendrons (Fig. 6.32) can be so strong that they are distended to the opposite side, that is to the catalytic site itself, to re-

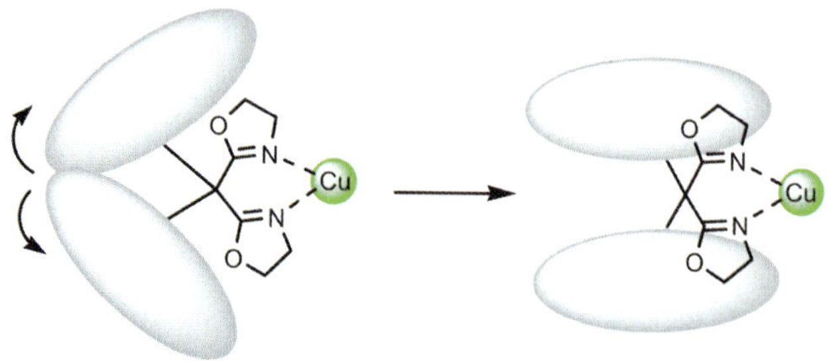

Fig. 6.32 Increasing steric shielding of the catalytic sites in the dendrizyme on transition to higher generations (schematic; according to van Koten *et al.*)

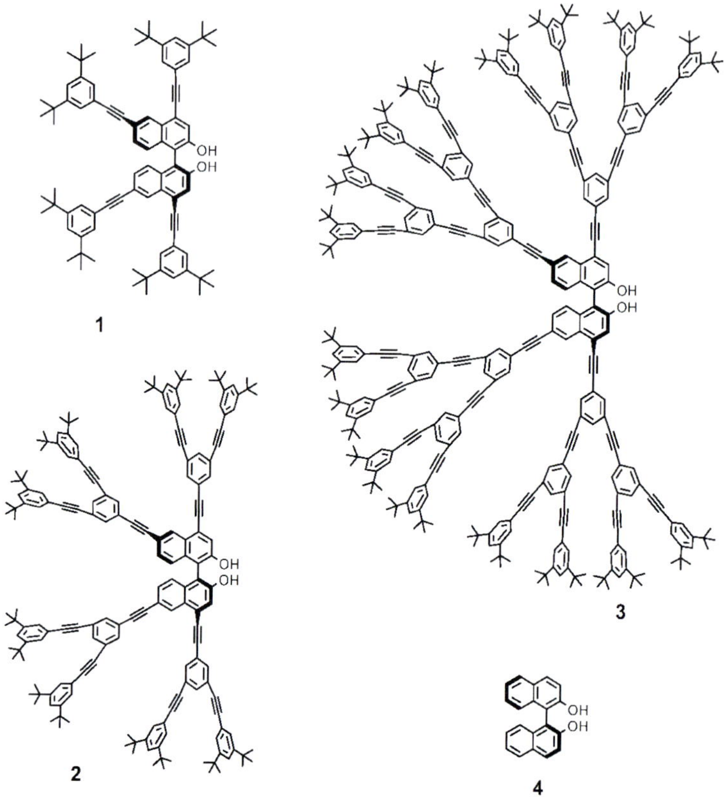

Fig. 6.33 Optically active dendrimers and BINOL 4 (according to Pu *et al.*)

duce their overlap. This backfolding process manifests itself in a decrease in steric accessibility of the catalytically active metal sites.

The catalytic properties of the dendritic BINOL ligand (*S*)-**1** (Fig. 6.33) already mentioned in Section 4.2.3.2 (Fig. 4.70) in the catalytic stereoselective reaction of benzaldehyde with diethylzinc were compared with those of unsubstituted

(*S*)-BINOL itself (**4**) as ligand [50]. It was seen that the complex formed by (*S*)-**1** with diethylzinc has a much greater catalytic activity than the reference substance. However, the enantiomeric excess (*ee*) is low in both cases: 5 mol% of (*S*)-**1** in toluene gives 98.6% conversion of benzaldehyde, while only 37% is attained with (*S*)-BINOL under the same conditions. The diethylzinc complex of dendronised BINOL is therefore a stronger Lewis acid than that of (*S*)-BINOL itself. Whereas the latter zinc complex probably exists in an aggregate state with bridging intermolecular Zn–O–Zn bonds, with a concomitant reduction in Lewis activity of the zinc site, the zinc complex derived from (*S*)-**1** does not form oligomers. The bulky and comparatively rigid dendritic arms prevent this from happening, but permit small molecules such as diethylzinc and benzaldehyde to penetrate into the interior of the ligand. On transition to $Ti(O\text{-}iPr)_4$ as Lewis acid, (*S*)-**1** proved to be a highly enantioselective ligand: 20 mol% of (*S*)-**1** and 1.4 eq. of $Ti(O\text{-}iPr)_4$ afford 100% conversion and an *ee* von 90:10 within 5 h on reduction of 1-formylnaphthalene with diethylzinc. Compared to (*S*)-BINOL there appear to be no significant differences in the catalytic properties, which is ascribed to structurally similar – probably monomeric – catalytically active species for the two ligands.

Related ligands for catalysis, namely BINAP ligands substituted with Fréchet dendrons (**5**, Fig. 6.34) were prepared by Chan *et al.* [51]. They form ruthenium(II) complexes *in situ*, whose activity in the stereoselective hydrogenation of 2-[*p*-(2-methylpropyl)phenyl]acrylic acid was investigated.

In contrast to some other dendritic catalysts, the reaction rate of the "dendron-embellished" BINAP catalysts increases with increasing generation. While the unsubstituted Ru(BINAP) catalyst gave an *ee* of 89.8:10.2 – at a turnover number (TON) of 6.3 h^{-1} and a conversion of 10.2% – the *ee* for the G0 dendron catalyst rises to 91.8:8.2 (at a similar TON and comparable conversion). On transition to higher generations the *ee* remained relatively constant, while the conversion rose to up to 69.3% and the TON to up to 21.4 h^{-1}. This significant dendritic effect is ascribable to the influence of the dendrons on the dihedral angle of the two naphthalene units of Ru(BINAP); it thus leads to a higher rate and a better *ee* of the reaction [52]. The G3 ligand could be recovered by precipitation and filtration and used in catalysis three more times with a similar activity and enantioselectivity.

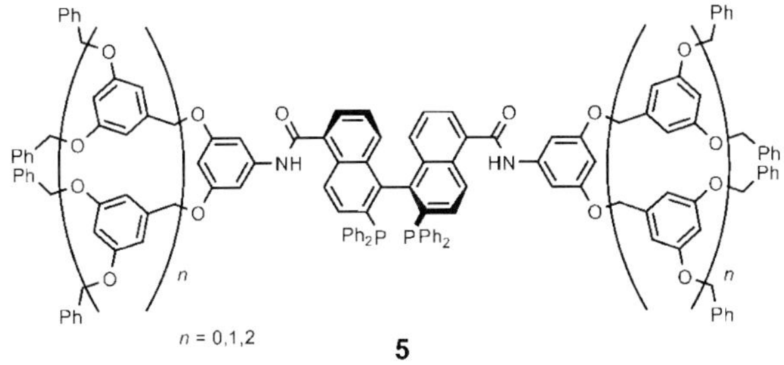

5

Fig. 6.34 Dendritic BINAP ligands (according to Chan *et al.*)

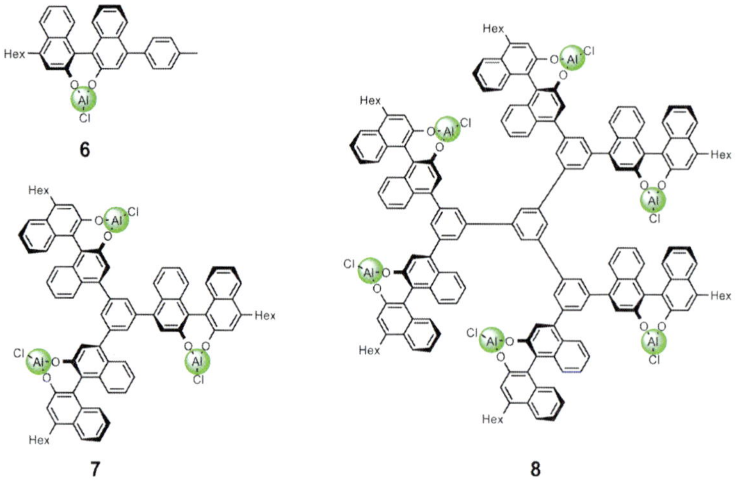

Fig. 6.35 BINOL catalysts (according to Chow and Wan)

Related branched catalysts with BINOL ligand units at the periphery were investigated by Chow and Wan (Fig. 6.35) [53] after model studies had revealed that the catalytic sites should be so far from one another that no intramolecular interaction was to be feared.

Measurements of molar rotation showed that this parameter is almost proportional to the number of chiral binaphthyl units and the molar rotation per binaphthyl unit varies only slightly. On catalysis of the Diels–Alder reaction of cyclopentadiene with 3-[(*E*)-but-2-enoyl]oxazolidin-2-one the branched catalysts **7** and **8** showed an approximately 25% higher reactivity than the monofunctional catalyst **6**; however, the former led to just a slight improvement of *ee* and *endo*-selectivity compared to **6**. It is thus inappropriate to speak of a dendritic effect on catalysis, although one does indeed exist in relation to the chiroptical properties.

A dendritic catalyst **9**, which like **7** and **8** has a chiral metal complex as core, was synthesised by Seebach *et al.* [54, 55]. The core building block was *α,α,α′,α′*-*t*etra*a*ryl-l,3-*d*ioxolan-4,5-*d*imethan*ol* (TADDOL), to which both chiral and achiral dendrons and those with peripheral octyl groups can be linked (Fig. 6.36, cf. Section 4.2.3, Fig. 4.62).

On use as homogeneous catalysts in the asymmetric reductive alkylation of benzaldehyde with diethylzinc to form secondary alcohols, the corresponding dendritic titanium-TADDOL complexes having either chiral or achiral dendrons gave enantiomeric excesses (*ee*) of up to 98.5:1.5 at a conversion of 98.7% (for the catalyst with G0 dendrons). With larger dendrons the reduction of the *ee* to 94.5:5.5 (G4) remained within reasonable limits, while the drop in conversion to 46.8% (G4) proved to be drastic. In comparison, the unsubstituted Ti-TADDOL complex gave an *ee* of 99:1 with complete conversion. This negative den-

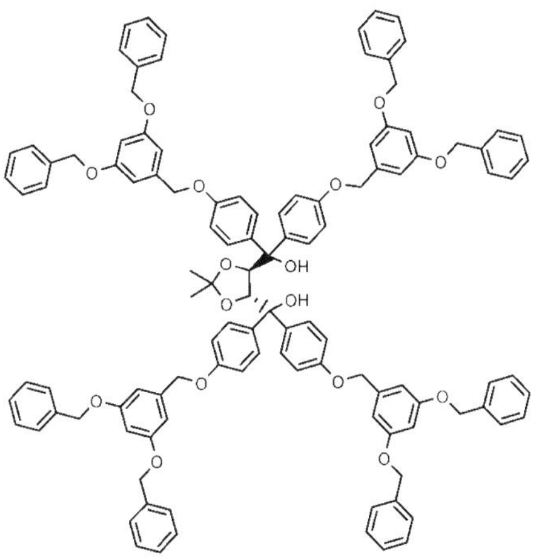

9

Fig. 6.36 Dendritic TADDOL ligand (according to Seebach *et al.*)

dritic effect with increasing dendron growth is attributable to increasing shielding of the catalytic site by the increasingly sterically demanding dendrons. Thus access to the catalytic site becomes more difficult for the substrates undergoing reaction and the conversion decreases. Independently of whether chiral information is present or absent in the dendrons, almost identical *ee* values are obtained. Additional chiral induction through the dendrons is precluded by the large distance between the metal site and the additional chiral building blocks.

Cobalt complexes of dendritic phthalocyanines (Fig. 6.37) showed a 20% lower catalytic activity (TON 339 min^{-1} for G2 dendrons) as catalysts for the oxidation of 2-mercaptoethanol than non-dendritic phthalocyanines [56]. By way of compensation, however, the dendritic catalysts proved to be more stable than non-dendritic ones, which is probably attributable to enclosure of the metallophthalocyanine core unit by the dendrons. This also prevents molecular aggregation of the phthalocyanines in polar solvents and thin films.

A poly(propylenamine) dendrimer (**11**, Fig. 6.37) functionalised with poly-(*N*-isopropylacrylamide) (PIPAAm) (see Section 4.1.2) was used as dendritic host for anionic cobalt(II)-phthalocyanine complexes (**a**, **b**) as guests, which are held together by supramolecular (electrostatic and hydrophobic) interactions [57]. These dendritic complexes were investigated as catalysts in the above-mentioned oxidation of thiols, where they show a remarkable temperature dependence: the reaction rate suddenly increases above 34°C. One attempted explanation assumes that the dendritic arms undergo phase separation and contraction above the Lower Critical Solubility Temperature (LCST). At this temperature the phthalocyanine complex site is more readily accessible for substrates and the reaction rate is therefore higher.

Moore and Suslick conducted "shape-selective catalysis" with metalloporphyrins **12** bearing sterically demanding polyester dendrons (**a**, **b**, Fig. 6.38) [58, 59].

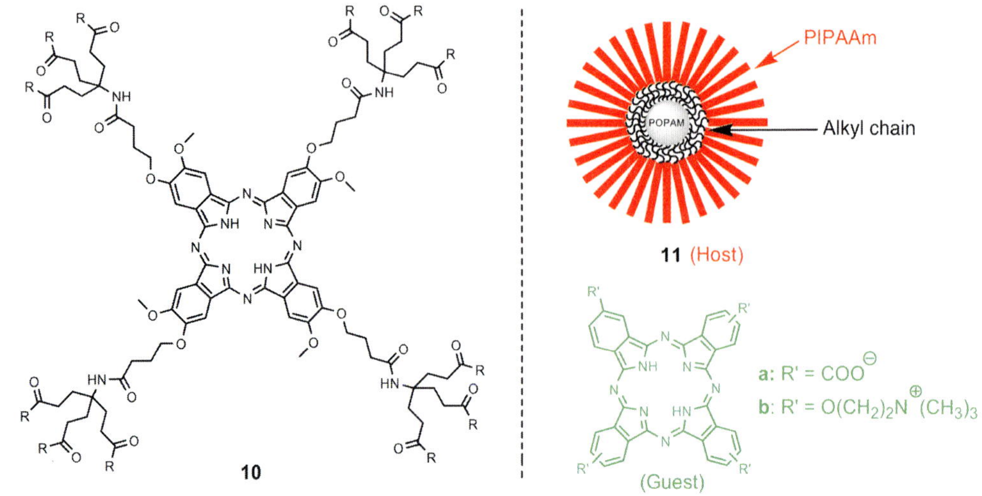

Fig. 6.37 Dendritic phthalocyanine **10** and supramolecular host-guest system **11 a, b** (according to Kimura *et al.*)

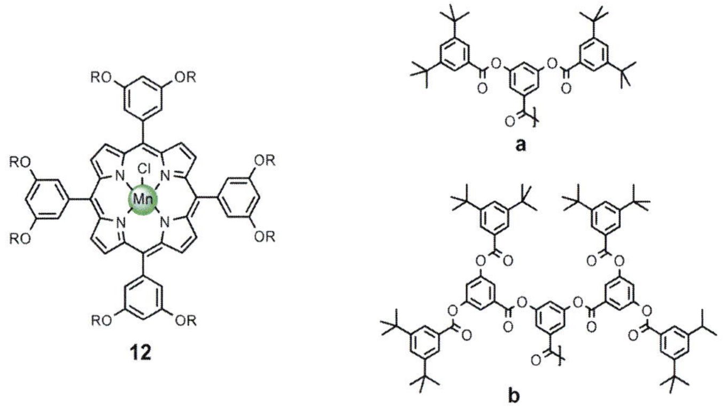

Fig. 6.38 Metallo-porphyrin dendrimers (according to Moore, Suslick *et al.*)

They investigated the influence of the size of the metallodendrimers on the substrate selectivity for two types of catalytic epoxidation reactions with iodosylbenzene as oxygen source and various alkenes as substrates. Dendritic complexes were compared with conventional porphyrin complexes (with magnesium as central ion). As the size of the metallodendrimer increased, a significant increase in (regio)selectivity was found for external and hence less hindered double bonds compared to internal (more strongly hindered) ones. Moreover, the dendritic catalysts showed a greater affinity for electron-rich alkenes. In addition, the metalloporphyrins were found to become increasingly stable to oxidation reactions with increasing generation number. This has a positive effect on catalysis in that the turnover numbers per metal ion increase. Even under the

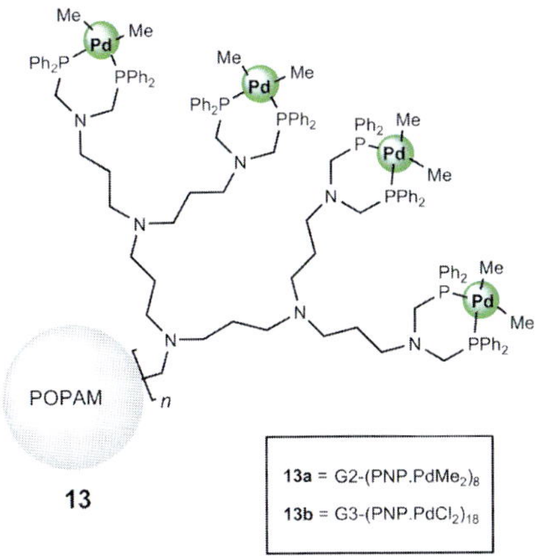

13

Fig. 6.39 Phosphine dendrimer catalyst (according to Reetz *et al.*)

conditions of epoxidation reactions, less than 10% of the catalyst were deactivated after 1000 reaction cycles.

The above examples clearly demonstrate that dendritic catalysts with catalytically active sites on the periphery represent a superior catalyst concept compared to those with a catalytically active site in the dendrimer core which will be subject to rapidly increasing shielding with increasing dendron size.

Reetz *et al.* described a DAB-based poly(propylenamine) dendrimer (DAB = 1,4-diaminobutane) armed with a kind of tweezers at the periphery, the ends of the branches being functionalised with diphenylphosphine groups (Fig. 6.39, see also Section 4.1.10 for more phosphodendrimers) [60]. A corresponding dendritic [$PdMe_2$] complex was tested as catalyst for Heck reaction [61] of bromobenzene with styrene to form stilbene. Remarkably, the activity of this dendritic catalyst was higher than that of the non-dendritic analogue. Turnover numbers (TON) of 54 h^{-1} compared to 16 h^{-1} for the classical catalyst were found for the third-generation dendrimer **13b**. The TON was not reported to be dependent upon the generation of the dendrimer. The dendritic scaffold exerts a positive influence on catalyst stability in homogeneous catalysis since, in contrast to former systems, the catalyst does not undergo significant decomposition with liberation of elemental palladium.

In the dendritic [Co(salen)] complexes prepared by Breinbauer and Jacobsen the dendrimer again serves as – covalent – support material for the catalytic entities attached to the periphery [62]. These dendritic Jacobsen catalysts were obtained by reaction of the corresponding PAMAM dendrimers with active ester derivates of chiral [Co(II)-(salen)] units according to standard peptide coupling methods. In "hydrolytic kinetic resolution" of vinylcyclohexane oxide the dendrimer **14** (Fig. 6.40) showed a dramatically increased reactivity compared to the commercially available "monomeric Jacobsen catalyst" [63–67]. Whereas the latter merely gave a conversion of less than 1% with an indeterminable *ee*, **14** afforded a conversion of 50% with an *ee* of 98:2.

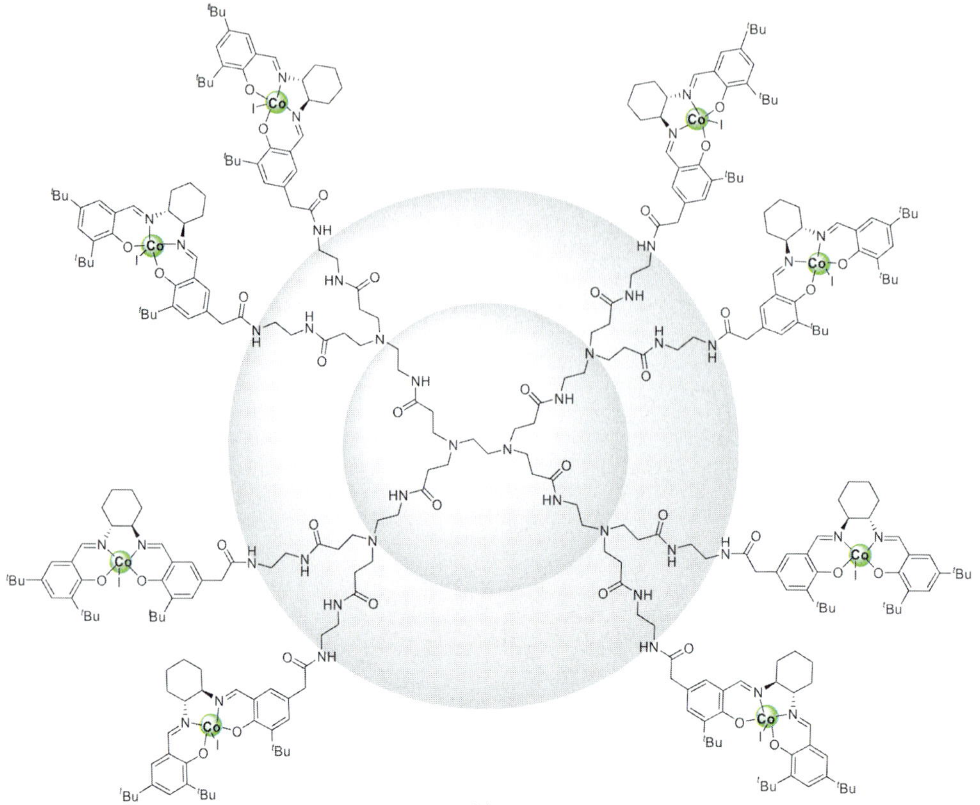

14

Fig. 6.40 Dendritic 8-Co-PAMAM-catalyst **14**

In order to ascertain the reason for the observed increase in reaction rate, the monomer **15**, with a representative group for the points of attachment of the catalytic units in the dendrimer, was synthesised as well as the "dimeric" model compound **16**, which "imitates" the "neighbourhood relationship" of two catalytic units within the branch of a PAMAM dendrimer (Fig. 6.41). In catalytic experiments it was found that the dendritic catalysts were considerably more reactive than the monomeric complex. Moreover, the dendritic catalysts have strikingly higher catalytic activities than the dimeric model compound.

6.3.2.2 **Metal-free dendritic catalysts**

The main emphasis has so far been placed on metal-containing catalysts, both in the field of dendritic catalysis and in organic catalysis in general. However, the current trend is increasingly towards catalysis with purely organic compounds. A number of example of such metal-free – and dendritic – catalysts,

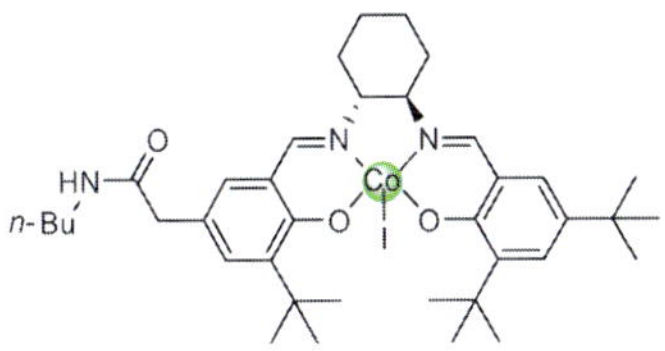

15

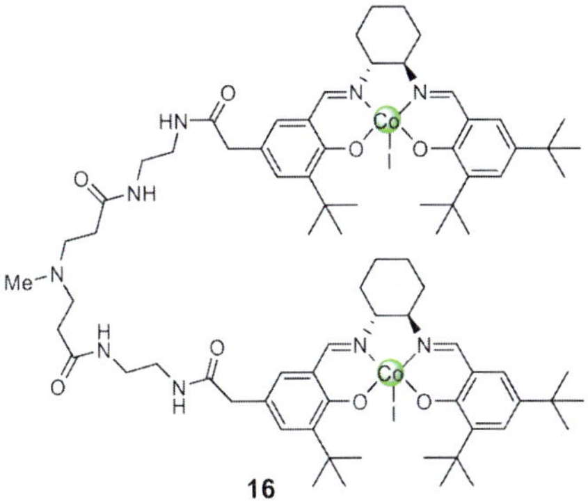

16

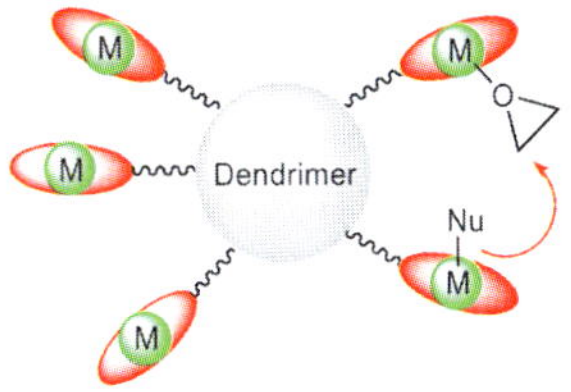

17

Fig. 6.41 Possible cooperative effect between the salen units of the dendritic catalyst **17** as well as monomeric (**15**) and dimeric model substances (**16**); M symbolises the metal site, and the (red) ellipses the salen ligands

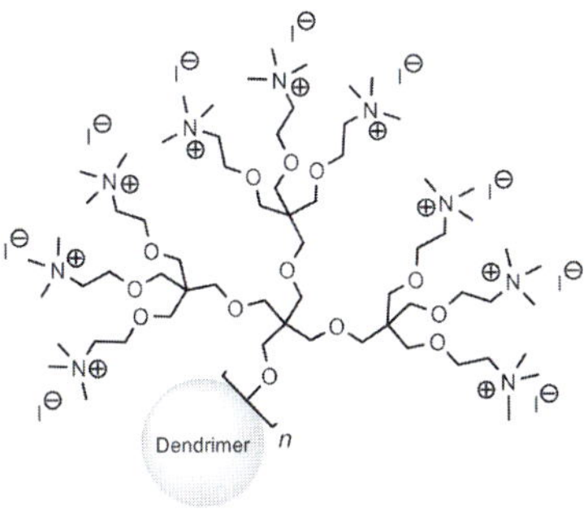

1

Fig. 6.42 Polyether dendrimer, functionalised with quaternary ammonium groups (according to Ford *et al.*)

which may be useful for shape-selective catalysis or as chiral auxiliaries, will be presented here.

Ford *et al.* functionalised polyether dendrimers with quaternary ammonium ions (**1** in Fig. 6.42) [68, 69].

The potential of these dendrimers for unimolecular decarboxylation of 6-nitrobenzisoxazole and *o*-iodobenzoate-catalysed bimolecular hydrolysis of *p*-nitrophenyldiphenylphosphate was investigated. In aqueous medium, polycationic dendrimers of this kind accelerate such reactions by stabilisation of the organic anions. At high concentrations these anions undergo ionic interactions with the polycationic periphery of the dendrimer. Thus the pseudo-micellar environment of the dendrimer promotes formation of nucleophilic organic anions. However, the corresponding POPAM complexes with Cu^{II}, Zn^{II}, and Co^{II} ions accelerated the same reactions by a factor of 1.3 to 6.3 compared to the metal-ion-free dendrimers [68].

Since internal quaternary ammonium ions of a dendrimer can serve as active sites for phase transfer or polyelectrolyte catalysis, Ford *et al.* synthesised corresponding amphiphilic polyamine dendrimers with octyl and triethylenoxy methyl ether chains [70]. The unimolecular concerted decarboxylation of sodium 6-nitrobenzisoxazole-3-carboxylate [71] served as a test of the catalytic ability of the dendrimers, *e.g.* **2** (Fig. 6.43). In the reaction considered, hydrogen bonding of protic solvent molecules with carboxylate anions lowers the reaction rate, whereas dipolar aprotic solvents increase the rate. A quaternary ammonium chloride dendrimer such as **2** accelerates the reaction by a factor of 200 to 500 compared to that in water. On assumption of a "single-site binding model", which resembles Michaelis–Menten enzyme kinetics and the Menger–Portnoy model [72, 73] for catalysis by surfactant micelles, it can be assumed that substrate binding is reversible, meaning that the substrate partitions between the aqueous phase and the dendrimer. The same research group had previously synthesised a dendrimer with 36 terminal alkyltrimethylammonium iodide groups and branches derived from pentaerythritol, yet without quaternary ammonium units [68]. The dendrimers with quaternary ammonium ions exhibit significantly higher values of the rate constant k_c for the decarboxylation proceeding in the interior of the dendrimer and also have a higher binding constant K than the dendrimers without quaternary ammonium groups. The higher binding constants for dendrimers such as **2** were unexpected because the anionic reactants in competition with the iodide ions of the dendrimers without quaternary ammonium ions are more strongly bound and thus more difficult to displace. The higher k_c values are, however, to be expected for dendrimers with quaternary ammonium ions since these are more hydrophobic owing to the octyl groups. For this reason, there are fewer water molecules in the interior which can stabilise the ground state of the reactant *via* hydrogen bonding. Dendrimers of higher generations than 2 show even higher reaction rates, meaning that the reactants are even less hydrated in such cases.

The amines **3** reported by Morao and Cossio also constitute neutral dendritic catalysts without metal sites; they consist of a simple amine core functionalised with Fréchet dendrons (Fig. 6.44) [74]. Such amines can catalyse the nitroaldol or Henry reaction [75] between aromatic aldehydes and nitroalkanes. Whereas neither the yield nor the stereoselectivity (*syn/anti* 1:1) of the reaction of *p*-nitrobenzaldehyde with nitroethane was found to change on use of different generations of dendritic catalysts, a distinctly negative dendritic effect was observed in the reaction of benzaldehyde with 3-nitro-l-propanol. Catalysts **3a** and **3b** gave

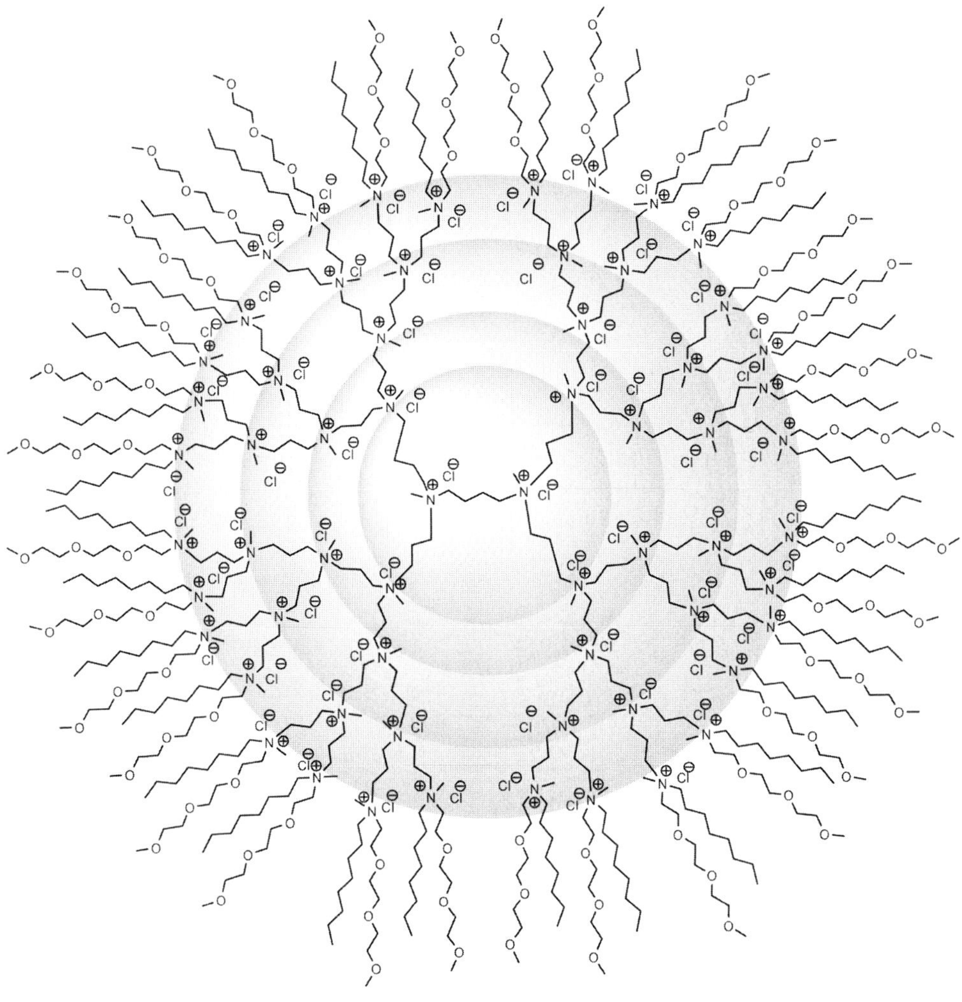

2

Fig. 6.43 DAB-based dendrimer with internal quaternary ammonium groups (according to Ford *et al.*)

conversions in excess of 95% within 4–6 hours, but the conversion dropped to 75% with the G2 catalyst **3c**, albeit within a period of 48 h. However, no significant changes occurred in the stereoselectivity; which remained unaltered at a *syn/anti* ration of 2:1. The catalytic activity decreases with increasing generation number because continued dendritic growth leads to ever-increasing steric/electronic shielding of the catalytic site by the dendrons. It is nevertheless remarkable that no negative dendritic effect is apparent in the first reaction investigated.

Soai *et al.* studied the influence of the flexibility of the dendritic support scaffold on catalytic properties for the chiral dendrimers **4** and **5** (Fig. 6.45). In the

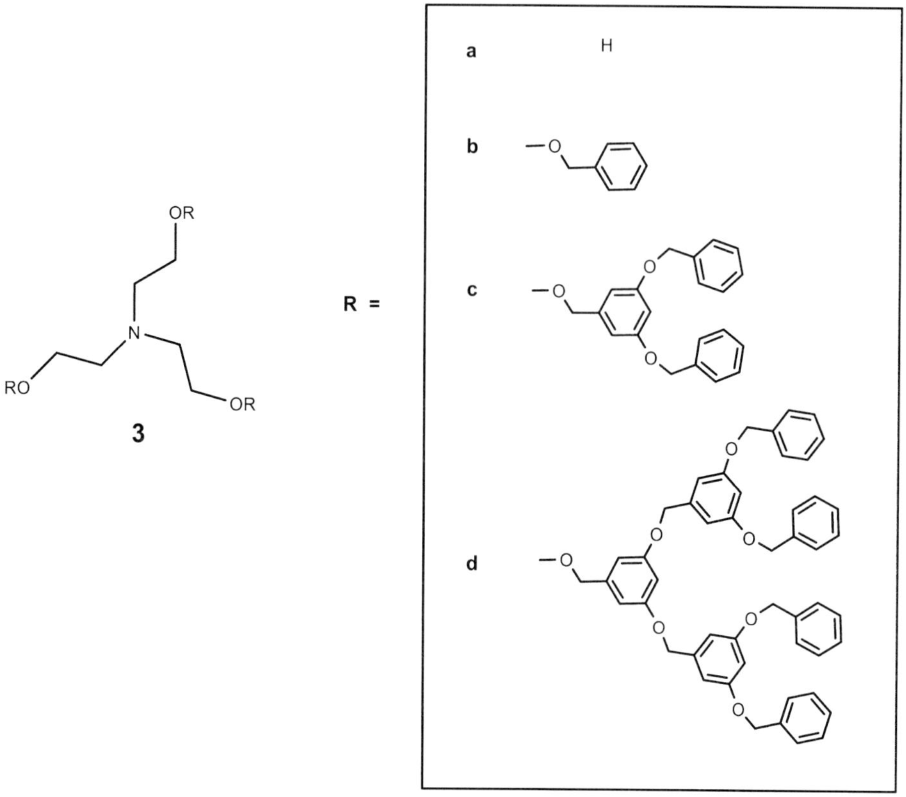

Fig. 6.44 Amine core enclosed by dendron substitution (according to Morao and Cossio)

case of 4 they used PAMAM as support for chiral ephedrine groups [76, 77]. However, the dendritic catalyst, for example 4, exerted only a modest effect on the enantiomeric excess of the product in the chiral addition of diethylzinc to *N*-diphenylphosphinylimine. A greater chiral induction was even achieved with the non-dendritic model compound , and that at significantly lower concentrations [78]. A negative dendritic effect was observed both for the yields and for the *ee* on going from the G1 to the G2 catalyst. On use of the comparatively rigid hydrocarbon scaffold 5 as support for the same ephedrine groups an *ee* of 86% (at a catalyst concentration of 3.3 mol%) was nonetheless obtained [79]. Overall the *ee* obtained with G1 and G2 catalysts varied only slightly, whereas a positive dendritic effect was noted for the yields on going from G1 to G2. The differing results for flexible and rigid systems (4 and 5, respectively) are attributed to the flexibility of the arms of the PAMAM dendrimer 4. This flexibility permits mutual interaction between the various ephedrine groups and thus prevents effective transfer of chiral information during C–C bond formation. In contrast, the rigidity of the hydrocarbon backbone in 5 reduces the interaction between chiral groups, thus favouring the stereoselective process.

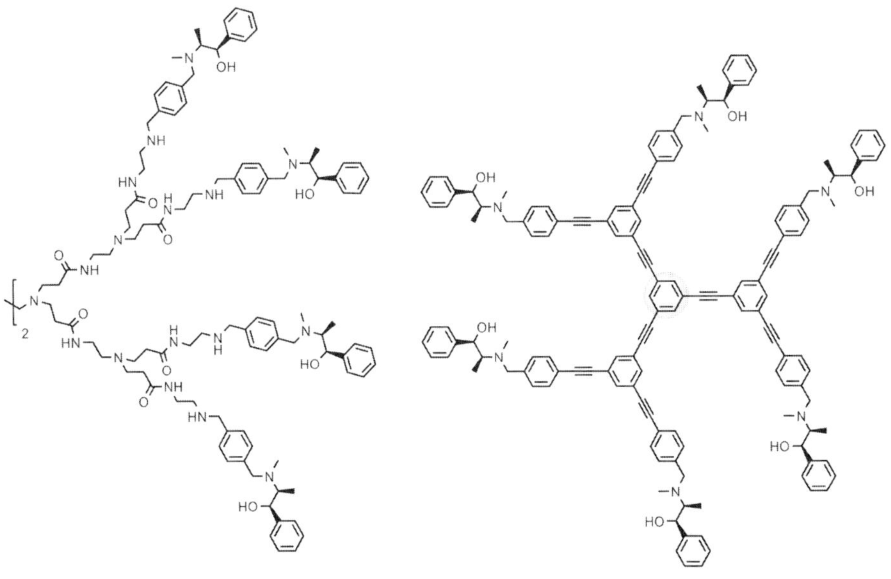

4 5

Fig. 6.45 Flexible (**4**) and more rigid (**5**) chiral dendrimer (according to Soai *et al.*)

Heme-containing monooxygenases such as cytochrome P450 [80] and peroxidases [81] catalyse numerous important oxidations in nature. The synthesis of compounds which imitate the structure and also the associated functionality of such enzymes therefore represents a challenge to the synthetic chemist. A number of porphyrin dendrimers [59, 82, 83] were prepared and successfully investigated with regard to their activities as mono-oxygenases [59, 82f, 83]. On the basis of their results in the area of dendritic cytochrome mimics [84], Diederich *et al.* synthesised three generations of dendritic Fe^{III}-porphyrins with covalently bound axial imidazole ligands as model compounds for heme-monooxygenases [85]. Formation of a pentacoordinated species with a free coordination site should be possible *via* the covalently bonded imidazole ligand. Such axially oriented nitrogen ligands are known to enhance the reactivity of metalloporphyrin-catalysed oxidations [86]. In nature, peroxidases also possess a mono-imidazole coordination site resembling those in the dendrimers **6a–c** (Fig. 6.46).

Diederich *et al.* had postulated that the highly reactive iron-oxo species, arising from oxygen transfer from the oxidant to the Fe^{III} site [87], should be greatly stabilised by enclosure within a dendritic superstructure. The catalytic potential of the dendrimers **6a–c** was determined in the epoxidation of alkenes [83a, 88] (1-octene and cyclooctene) and the oxidation of sulphides [83a] ((methylsulphanyl)benzene and diphenyl sulphide) to sulphoxides – in dichloromethane with iodosylbenzene as oxidising agent. Compared to the known metal-porphyrin catalysts, **6a–c** exhibit only low TON (7 and 28, respectively, for

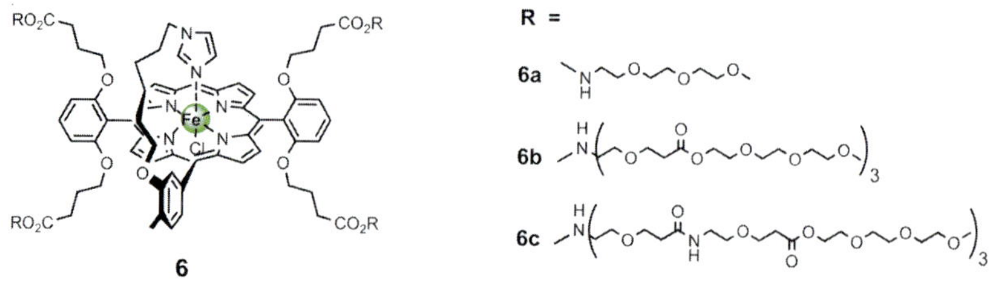

Fig. 6.46 Dendritic Fe^{III} porphyrins as model compounds for heme monooxygenases (according to Diederich *et al.*)

6a, and 25 and 57, respectively, for **6c**) in the oxidation reactions. This is attributed to a lack of stability of the metal-porphyrins towards self-oxidation. The only free *meso* position in the porphyrin macrocycle probably undergoes oxidative attack with ring opening [89]. Oxidation of the sulphides proceeds – as expected – more efficiently, with yields of up to 80% and selectivities exceeding 99%. A significant increase in the value of the TON was noted on going from G0 (38 and 30, respectively, for **6a**) to G2 (77 and 87, respectively, for **6c**). A notable positive dendritic effect was to be seen both for olefin epoxidation and for sulphide oxidations. The increasing extent of shielding by the dendritic coat imitates the function of the protein coat in natural heme proteins, slows down decomposition, and provides improved stability of the catalyst. Thus the second-generation dendritic catalyst **6c** proved to be the most efficient catalyst of this type.

The above brief overview of the current developmental status of dendritic catalysts reveals the enormous potential of structural and functional possibilities, and also underscores the need for further fundamental studies and new approaches to surpass routinely used conventional catalyst systems (see also Section 8.2).

6.3.3 Dendritic effects on electrochemical properties

6.3.3.1 Charge-transfer complex-based conductivity

PAMAM dendrimers of generations one to five were decorated on the periphery with naphthalenediimide units (Fig. 6.47) carrying cationic substituents assuring solubility in water. Subsequent reduction with sodium dithionite or formamide in aqueous medium [90] led to anion radical sites at the diimide units. These aggregate in solution to form π-stacks; in the solid state they form electrically conductive films [91].

It was discovered by ESR that every diimide group takes up an electron on reduction in water, and that thin dendritic films can be obtained whose conductivity depends upon atmospheric humidity. The conductivity increases with increasing humidity above the film and attains a value of up to 18 S/cm at 90% relative humidity.

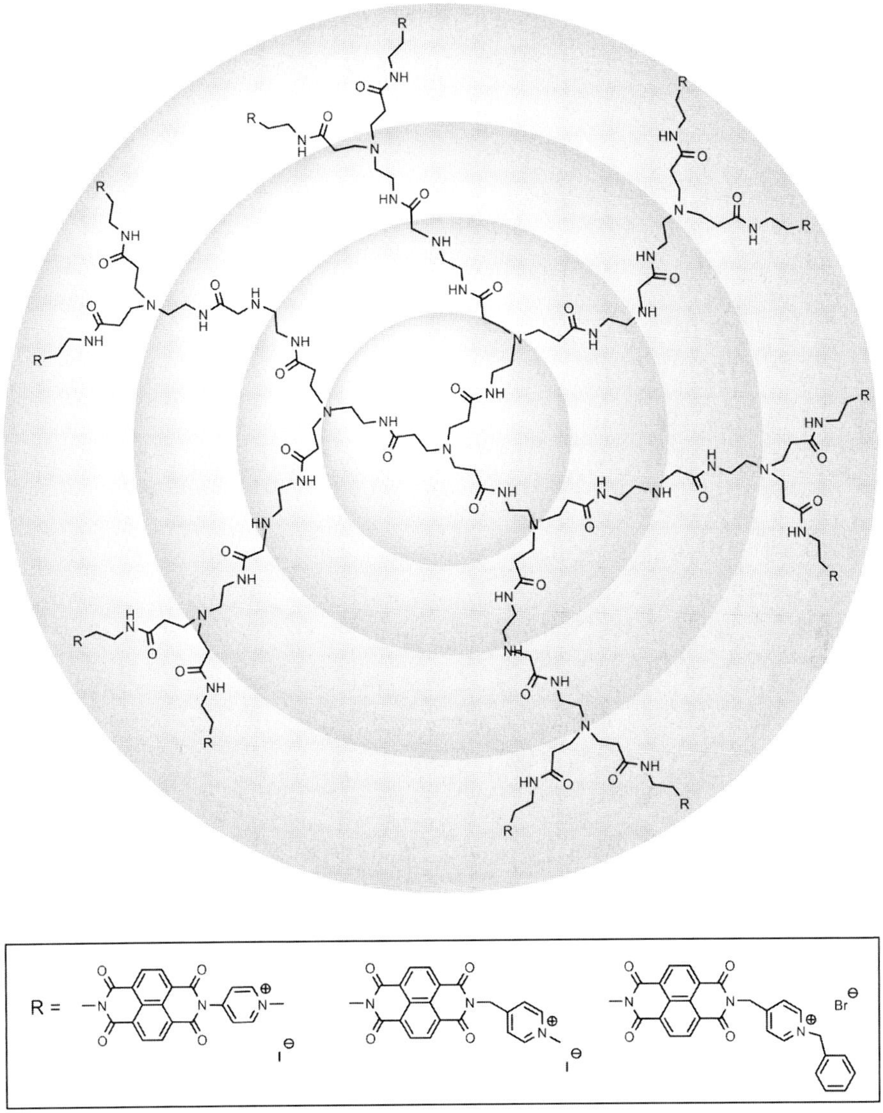

Fig. 6.47 G2 PAMAM dendrimer with three different cationic naphthalene diimide end groups (according to Miller, Tomalia *et al.*)

With regard to structure/property relations, Tomalia *et al.* assumed that the diimide groups in larger dendrimers come into close proximity. The first-generation dendrimer with only six diimide units displayed a lower conductivity, yet the fifth generation with 96 diimide units has no particular advantage over the second-generation dendrimer with just 12 diimide functions. It was therefore postulated that low π-stacks are formed in every dendrimer, with the result that the conductivity is similar for all the dendrimers investigated here. There is apparently sufficient flexibility within the PAMAM core to permit both intramolecular as well as intermolecular interactions between π-stacks [90].

6.3.3.2 **Redox gradients**

Blackstock *et al.* synthesised a redox-active polyarylamine dendrimer (Fig. 6.48) with electron transporting properties. It is made up of a 1,3,5-substituted benzene core unit with internal *p*-phenylenediamine spacers which are enclosed by arylamino groups [92]. The molecule exhibits C3 symmetry with nine pronounced redox functions in *meta* position. The hard-to-oxidise exterior triarylamino groups form a kind of coat around the more readily oxidised internal phenylenediamino groups. This leads to a radial redox gradient within the dendritic architecture. The gradient running from the periphery to the core probably supports the positive charge formed on the core in the course of oxidation processes and also inhibits discharge of the positive core during reduction reactions. Such polyarylamine dendrimers can therefore serve as three-dimensional "charge funnels" and as "charge depots" protecting any charge on the core.

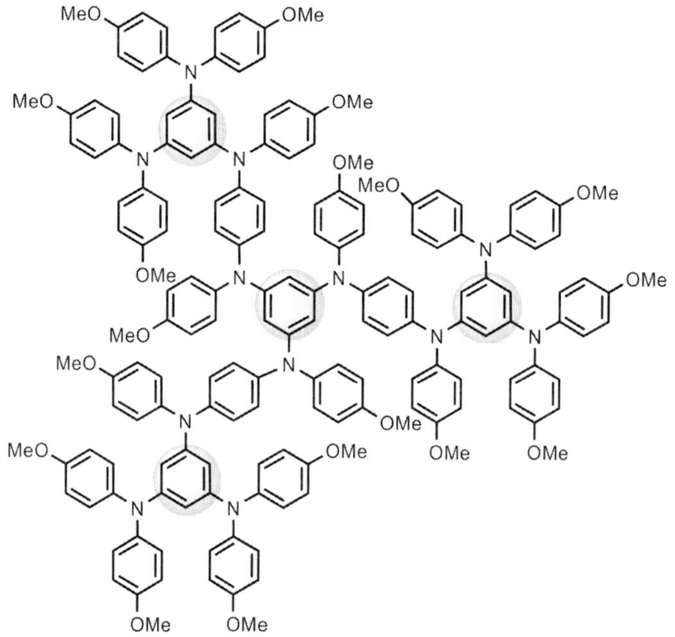

Fig. 6.48 Redox-active polyarylamine dendrimer (according to Selby and Blackstock)

6.3.3.3 Redox sensors

Astruc *et al.* prepared dendrimers functionalised with up to eighteen ferrocene units on the periphery (Fig. 6.49). These amido-ferrocene dendrimers are suitable as host molecules ("receptors") for small anions. The redox behaviour of the ferrocene end groups in the presence of diverse anions ($H_2PO_4^-$, HSO_4^-, Cl^-, Br^-, NO_3^-) could be followed by cyclic voltammetry (CV). The amido-ferrocene subunits interact with the anions *via* hydrogen bonds, the main effect being electrostatic attraction of the anionic guests owing to oxidation of the ferrocene groups – to ferrocenium cations. An anion-induced increase in redox potential was observed with increasing generation number. This dendritic effect is a consequence of the high surface density of the terminal redox sensor groups.

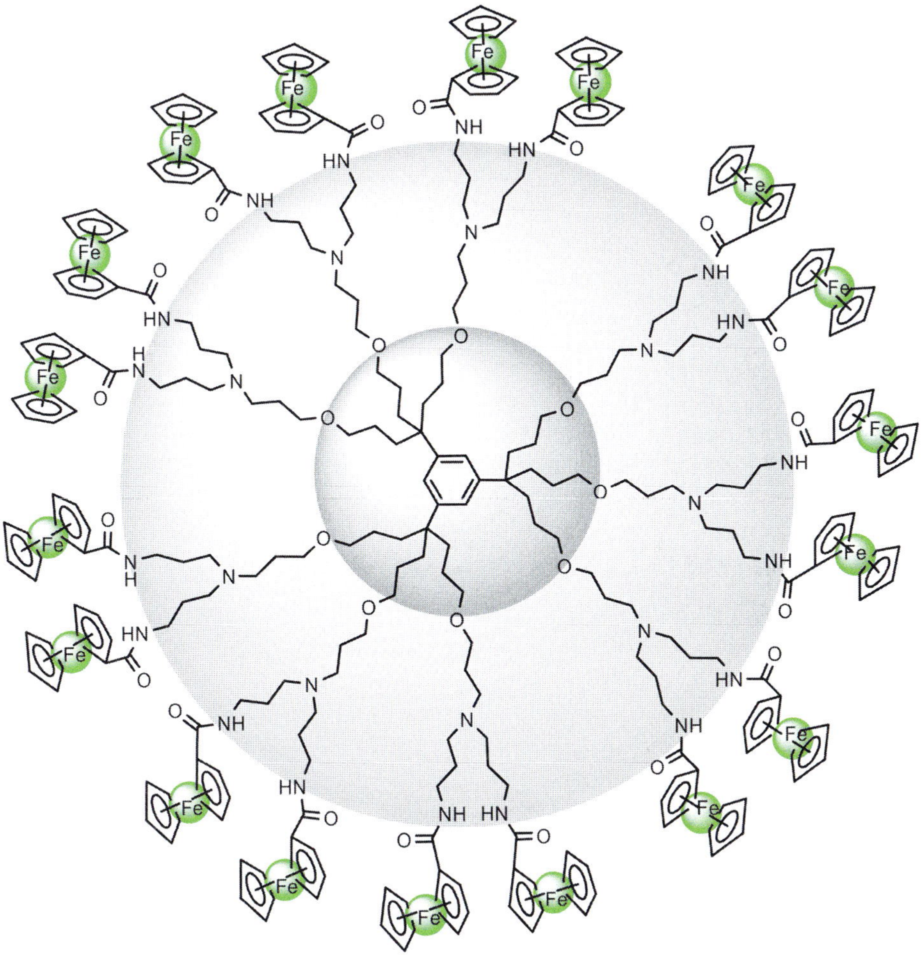

Fig. 6.49 Redox-active dendrimer with 18 terminal ferrocene units (according to Astruc *et al.*)

6.3.3.4 Redox potential and redox transfer kinetics

Modification of the electrochemical properties of a redox centre surrounded by dendritic fragments [93] can lead to two different dendritic effects. The first one is manifested in a shift of the redox potentials, the extent and direction depending upon the dendritic architecture and the solvent. Such behaviour was observed in dendritic iron-porphyrins [94]. The second effect is apparent in a delay of redox transfer kinetics and is characterised by a stepwise increase in the distance between the peaks in a cyclovoltammogram with increasing dendrimer generation number.

In the case of ruthenium(II)-bis(terpyridine) dendrimers (Fig. 6.50) of stepwise increasing dendrimer size, an influence of dendrimer size on the reversibility of redox processes could be deduced from size-exclusion chromatography and from the electrochemical properties [95].

6.3.3.5 Charge-separation processes

Dendritic zinc-porphyrins with terminal fullerene groups (Fig. 6.51) were examined for photoinduced electron-transfer properties. Such covalently bonded P_{Zn}–C_{60} dyads (P_{Zn}=porphyrin/zinc complex) **1** are regarded as good photoinduced electron-transfer units in which the charge-separation state, resulting from electron transfer from the porphyrin to the fullerene, offers an advantage owing to the low reorganisation energy of the fullerene [96].

Studies addressing the lifetime of the charge-separated state of these dendritic porphyrin/zinc-fullerene systems with up to seven porphyrin units ($7P_{Zn}$–C_{60}) revealed not only the – visible – light-harvesting properties of the large dendritic antennae for electron transfer to the focal fullerene, but also a delay in the return transfer of the electron. The lifetime of the charge-separated state proved to be much longer for the "heptamer" shown in Fig. 6.51 than for lower generations (monomer and trimer).

6.3.4 Summary of the dendritic effect

The feasibility of predicting whether a dendritic compound will indeed be subject to a dendritic effect will depend upon the molecular structure, especially the degree of branching, the generation number, the spacers, the target compound in question and its environment (temperature, solvent, guest substances). Flexible or rigid linear spacers often serve to weaken or even cancel dendritic effects. Dendritic effects on photo-physical and electrochemical properties are probably more easily planned for or even tailored to requirements, some time in the future. Clear prediction of such trends in the case of catalysis with dendrimers or of molecular recognition may be thwarted by the complexity of the chemical reaction mechanisms, which are often not completely understood even for the non-dendritic compounds. Depending upon the structure of the functionalised dendrimer, maximum effectiveness is attained for lower or subsequent generations whereas the effect may become negative for subsequent generations.

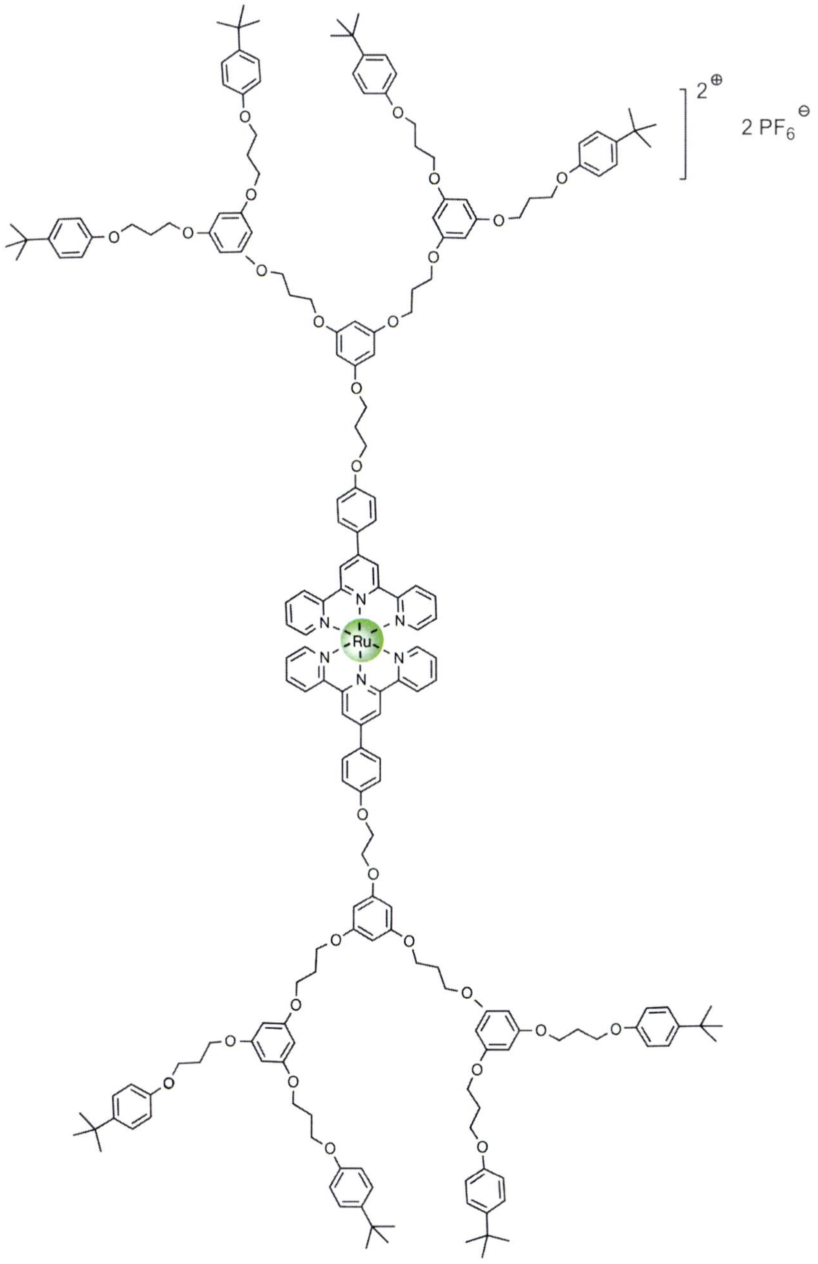

Fig. 6.50 Ruthenium(II)-bis(terpyridine) metallodendrimer (according to Chow *et al.*)

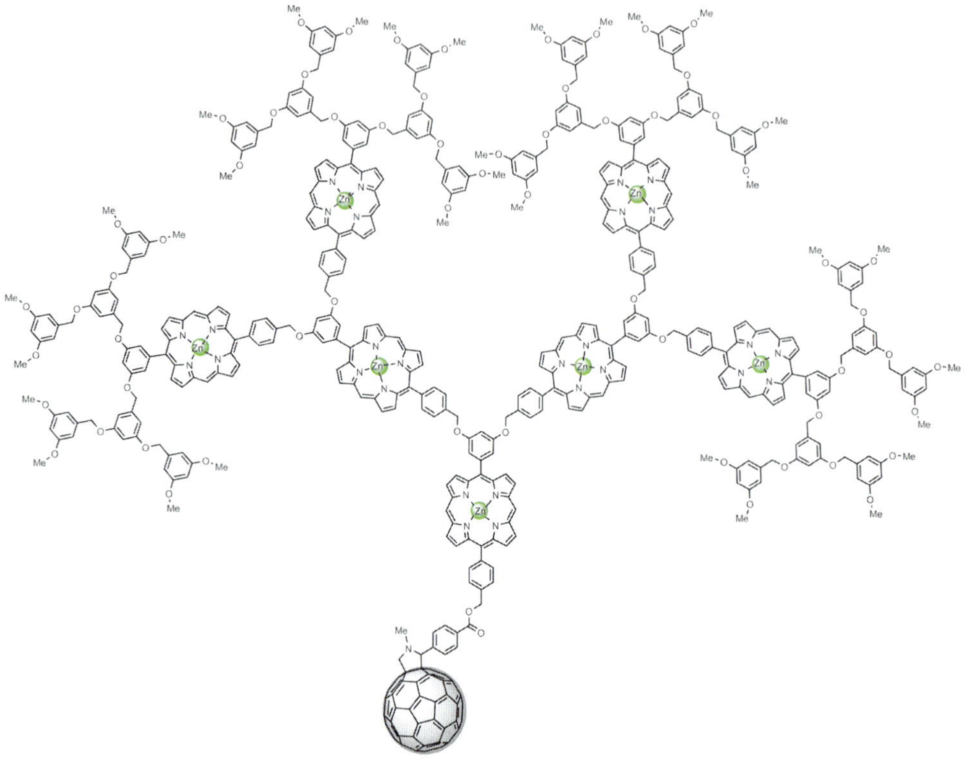

Fig. 6.51 Dendritically linked porphyrin/zinc complexes with fullerene at the focal point (according to Aida *et al.*)

Bibliography and Notes for Chapter 6 "(Special) chemical reactions of dendritic molecules"

Review articles are indicated by the words "Review(s)" or "Book(s)" in bold-faced type.

1 ***Book***: R. H. Grubbs, *Handbook of Metathesis*, Wiley-VCH **2003**, Vol. 1–3; ***Reviews***: A. Fürstner, *Angew Chem.* **2000**, *112*, 3140–3172; *Angew. Chem. Int. Ed.* **2000**, *39*, 3012–3043; M. Schuster, S. Blechert, *Angew. Chem.* **1997**, *109*, 2124–2144; *Angew. Chem. Int. Ed.* **1997**, *36*, 2036–2056; K. C. Nicolaou, P. G. Bulger, D. Sorlah, *Angew. Chem.* **2005**, *117*, 4564–4601; *Angew. Chem. Int Ed* **2005**, *44*, 4490–4527; R. R. Schrock, A. H. Hoveyda, *Angew. Chem.* **2003**, *115*, 4740–4782; *Angew. Chem. Int. Ed.* **2003**, *42*, 4592–4633; D. Astruc, *New. J. Chem.* **2005**, *29*, 42–56.

2 C. O. Liang, J. M. J. Fréchet, *Macromolecules* **2005**, *38*, 6276–6284.

3 S. Gatard, S. Nlate, E. Cloutet, G. Bravic, J.-C. Blais, D. Astruc, *Angew. Chem.* **2003**, *115*, 468–472; *Angew. Chem. Int. Ed.* **2003**, *42*, 452–456; ***Book***: D. Astruc, *Organometallic Chemistry and Catalysis*, Springer, Heidelberg **2007**.

4 S. B. Garber, J. S. Kingsbury, B. L. Gray, A. H. Hoveyda, *J. Am. Chem. Soc.* **2000**, *122*, 8168–8179.

5 P. Wijkens, J. T. B. H. Jastrzebski, P. A. Schaaf, R. Kolly, A. Hafner, G. van Koten, *Org. Lett.* **2000**, *11*, 1621–1624; S. B. Garber, J. S. Kingsbury, B. L. Gray, A. H.

Hoveyda, *J. Am. Chem. Soc.* **2000**, *122*, 8168–8179; H. Beerens, F. Verpoort, L. Verdonck, *J. Mol. Catal.* **2000**, *151*, 279–282; H. Beerens, F. Verpoort, L. Verdonck, *J. Mol. Catal.* **2000**, *159*, 197–201.

6 C. Ornelas, D. Mery, J.-C. Blais, E. Cloutet, J. R. Aranzaes, D. Astruc, *Angew. Chem.* **2005**, *117*, 7565–7570; *Angew. Chem. Int. Ed.* **2005**, *44*, 3752–3753.

7 A. K. Chatterjee, J. P. Morgan, M. Scholl, R. H. Grubbs, *J. Am. Chem. Soc.* **2000**, *122*, 3783–3784; A. K. Chatterjee, T.-L. Choi, D. P. Sanders, R. H. Grubbs, *J. Am. Chem. Soc.* **2003**, *125*, 11360–11370; ***Review***: R. H. Grubbs, *Acc. Chem. Res.* **2001**, *34*, 18–29.

8 ***Reviews***: a) G. Wulff; *Chem. Rev.* **2002**, *102*, 1–28; b) A. J. Hall, M. Emgenbroich, B. Sellergren, *Top Curr. Chem.* (Volume Eds. C. A. Schalley, F. Vögtle, K. H. Dötz) **2005**, *249*, 317–349; c) B. Sellergren, M. Lepistö, K. Mosbach, *J. Am. Chem. Soc.* **1988**, *110*, 5853–5860; d) F. L. Dickert, P. A. Lieberzeit, O. Hayden, *Nachrichten aus der Chemie*, November **2003**; e) K. J. Shea, *J. Am. Chem. Soc.* **2001**, *123*, 2072–2073; f) K. Haupt, *Anal. Chem.* **2003**, *75*, 376A–383A; g) ***Book***: M. Komiyama, T. Takeuchi, T. Mukawa, H. Asanuma, *Molecular Imprinting*, Wiley-VCH, Weinheim **2002**.

9 ***Reviews***: *Top. Curr. Chem* (Volume Eds. C. A. Schalley, F. Vögtle, K. H. Dötz) **2004**, 248; *Top. Curr. Chem.* (Volume Eds. C. A. Schalley, F. Vögtle, K. H. Dötz) **2005**, 249.

10 M. S. Wendland, S. C. Zimmerman, *J. Am. Chem. Soc.* **1999**, *121*, 1389–1390; S. L. Elmer, S. C. Zimmerman, *J. Org. Chem.* **2004**, *69*, 7363–7366; J. B. Beil, N. G. Lemcoff, S. C. Zimmerman, *J. Am. Chem. Soc.* **2004**, *12*, 13576–13577.

11 L. G. Schultz, Y. Zhao S. C. Zimmerman, *Angew. Chem.* **2001**, *113*, 2016–2020; *Angew. Chem. Int. Ed.* **2001**, *40*, 1962–1966.

12 S. C. Zimmerman, I. Zharov, M. S. Wendland, N. A. Rakow, K. S. Suslick, *J. Am. Chem. Soc.* **2003**, *125*, 13504–13518; S. C. Zimmerman, M. S. Wendland, N. A. Rakow, I. Zharov, K. S. Suslick, *Nature* **2002**, *418*, 399–403; see also Section 8.8.3.

13 Y. Kim, M. F. Mayer, S. C. Zimmerman, *Angew. Chem.* **2003**, *115*, 1153–1158; *Angew. Chem. Int. Ed.* **2003**, *42*, 1121–1126.

14 F. Vögtle, M. Plevoets, G. Nachtsheim, U. Wörsdörfer, *J. Prakt. Chem.* **1998**, *340*, 112–120; R. Herrmann, F. Vögtle, H.-P. Josel, B. Frommberger, G. Pappert, J. Issberner, Boehringer (Mannheim), Patent *DE 4439346A1*, **1994**.

15 R. Misra, R. Kumar, T. K. Chandrashekar, C. H. Suresh, A. Nag, D. Goswami, *J. Am. Chem. Soc.* **2006**, *128*, 16083–16091.

16 ***Review***: O. A. Mathews, A. N. Shipway, J. F. Stoddart, *Progr. Polymer Sci.* **1998**, *23*, 1–56; D. K. Smith, F. Diederich, *Chem. Eur. J.* **1998**, *4*, 1353–1361; S. Hecht, J. M. J. Fréchet, *Angew. Chem.* **2001**, *113*, 76–94; *Angew. Chem. Int. Ed.* **2001**, *40*, 74–91; *Review*: D. Seebach, P. B. Rheiner, G. Greiveldinger, T. Butz, H. Sellner, *Top. Curr. Chem.* (Volume Ed. F. Vögtle) **1998**, *197*, 125–164; M. A. Hearshaw, J. R. Moss, *J. Chem. Soc. Chem. Commun.* **1999**, 1–8.

17 P. Bhyrappa, J. K. Young, J. S. Moore, K. S. Suslick, *J. Am. Chem. Soc.* **1996**, *118*, 5708–5711; I. Marao, P. Cossio, *Tetrahedron Lett.* **1997**, *38*, 6461–6464; C. C. Mak, H.-F. Chow, *Macromolecules* **1997**, *30*, 1228–1230; H.-F. Chow, C. C. Mak, *J. Org. Chem.* **1997**, *62*, 5116–5127; P. B. Rheiner, H. Sellner, D. Seebach, *Helv. Chim. Acta* **1997**, *80*, 2027–2032; G. E. Oosterom, J. N. H. Reek, P. C. J. Kamer, P. W. N. M. van Leeuwen, *Angew. Chem.* **2001**, *113*, 1878–1901; *Angew. Chem. Int. Ed.* **2001**, *40*, 1828–1849.

18 J. W. Lee, Y. H. Ko, S.-H. Park, K. Yamaguchi, K. Kim, *Angew. Chem.* **2001**, *113*, 769–771; *Angew. Chem. Int. Ed.* **2001**, *40*, 746–749.

19 M. W. P. L. Baars, A. J. Karlsson, V. Sorokin, B. F. W. De Waal, E. W. Meijer, *Angew. Chem.* **2000**, *112*, 4432–4435; *Angew. Chem. Int. Ed.* **2000**, *39*, 4262–4265; U. Boas, S. H. M. Söntjens, K. J. Jensen, J. B. Christiansen, E. W. Meijer, *ChemBioChem* **2002**, *3*, 433–439.

20 V. Chechik, M. Zhao, R. M. Crooks, *J.* Am. Chem. Soc. **1999**, *121*, 4910–4911; B. González, C. M. Casado, B. Alonso, I. Cuadrado, M. Morán, Y. Wang, A. E. Kaifer, *Chem. Commun.*

1998, 2569–2570; G. R. Newkome, L. A. Godinez, C. N. Moorefield, *Chem. Commun.* **1998**, 1821–1822.

21 a) ***Overview***: N. Krasnogor, S. Gustafson, D. A. Pelta, J. L. Verdegay (eds.), *"Systems Self-Assembly"*: Multidisciplinary Snapshots. *Studies in Multidisciplinarity*, Elsevier, Amsterdam **2008**, vol. 5; b) S. Uppuluri, D. R. Swanson, L. T. Piehler, J. Li, G. L. Hagnauer, D. A. Tomalia, *Adv. Mater.* **2000**, *12*, 796–800.

22 A. Franz, W. Bauer, A. Hirsch, *Angew. Chem.* **2005**, *117*, 1588–1592; *Angew. Chem. Int. Ed.* **2005**, *44*, 2976–2979.

23 K. Hager, A. Franz, A. Hirsch, *Chem. Eur. J.* **2006**, *12*, 2663–2679.

24 W. Wang, A. E. Kaifer, *Angew. Chem.* **2006**, *118*, 7200–7204; *Angew. Chem. Int. Ed.* **2006**, *45*, 7042–7046; A. J. J. V. Heyen, C. C. Buron, Q. Tianshi, R. Bauer, A. M. Jonas, K. Müllen, F. C. De Schryver, S. De Feyter, *Small* **2008**, *4*, 1160–1167.

25 ***Reviews***: M. W. P. L. Baars, E. W. Meijer, *Top. Curr. Chem.* (Volume Ed. F. Vögtle) **2000**, *210*, 131–182; R. M. Crooks, B. I. Lemon III, L. Sun, L. K. Yueng, M. Zhao, *Top. Curr. Chem.* (Volume Ed. F. Vögtle) **2001**, *212*, 81–135; V. V. Narayanan, G. R. Newkome *Top. Curr. Chem.* (Volume Ed. F. Vögtle) **1998**, *197*, 19–77; F. Zeng, S. C. Zimmerman, *Chem. Rev.* **1997**, *97*, 1681–1712; D. K. Smith, F. Diederich, *Top. Curr. Chem.* (Volume Ed. F. Vögtle) **2000**, *210*, 181–227; R. M. Crooks, M. Zhao, L. Sun, V. Chechnik, L. K. Yueng, *Acc. Chem. Res.* **2001**, *34*, 181–190; A. Archut, G. C. Azzellini, V. Balzani, L. De Cola, F. Vögtle, *J. Am. Chem. Soc.* **1998**, *120*, 12187–12191; ***Book***: G. R. Newkome, C. N. Moorefield, F. Vögtle, *Dendritic Molecules: Concepts, Syntheses, Perspectives*, VCH, New York **1996**.

26 S.-K. Chang, A. D. Hamilton, *J. Am. Chem. Soc.* **1988**, *110*, 1318–1319.

27 A. Dirksen, U. Hahn. F. Schwanke, M. Nieger, J. N. H. Reek, F. Vögtle, L. De Cola, *Chem. Eur. J.* **2004**, *10*, 2036–2047; A. Dirksen, L. De Cola, *C. R. Chim.* **2003**, *6*, 873–882.

28 A. Archut, G. C. Azzellini, V. Balzani, L. De Cola, F. Vögtle, *J. Am. Chem. Soc.* **1998**, *120*, 12187–12191.

29 F. Vögtle, M. Gorka, R. Hesse, M. Maestri, V. Balzani, *Photochem. Photobiol. Sci.* **2002**, *1*, 45–51.

30 ***Book***: D. J. Cram, J. M. Cram, *Container Molecules and Their Guests* (Ed. F. Stoddart), Monographs in Supramolecular Chemistry, Royal Society of Chemistry, Cambridge **1994**; M. Yin, C. R. W. Kuhlmann, K. Sorokina, C. Li, G. Mihov, E. Pietrowski, K. Koynov, M. Klapper, H. J. Luhmann, K. Müllen, T. Weil, *Biomacromol.* **2008**, *9*, 1381–1389.

31 M. W. P. L. Baars, R. Kleppinger, M. H. J. Koch, S.-L. Yeu, E. W. Meijer, *Angew. Chem.* **2000**, *112*, 1341–1344; *Angew. Chem. Int. Ed.* **2000**, *39*, 1285–1288.

32 J. C. Garcia-Martinez, R. Lezutekong, R. M. Crooks, *J. Am. Chem. Soc.* **2005**, *127*, 5097–5103.

33 V. Balzani, P. Ceroni, S. Gestermann, M. Gorka, C. Kauffmann, F. Vögtle, *J. Chem. Soc. Dalton Trans.* **2000**, 3765–3771; F. Vögtle, S. Gestermann, C. Kauffmann, P. Ceroni, V. Vicinelli, V. Balzani, *J. Am. Chem. Soc.* **2000**, *122*, 10398–10404; V. Balzani, P. Ceroni, S. Gestermann, M. Gorka, C. Kauffmann, M. Maestri, F. Vögtle, *ChemPhysChem.* **2000**, 224–227.

34 G. R. Newkome, C. N. Moorefield, G. R. Baker, M. J. Saunders, S. H. Grossman, *Angew. Chem.* **1991**, *103*, 1207–1209; *Angew. Chem. Int. Ed.* **1991**, *30*, 1178–1180.

35 M. Liu, K. Kono, J. M. J. Fréchet, *J. Controlled Release* **2000**, *65*, 121–131.

36 V. Chechnik, M. Zao, R. M. Crooks, *J. Am. Chem. Soc.* **1999**, *121*, 4910–4911.

37 a) C. J. Hawker, K. L. Wooley, J. M. J. Fréchet, *J. Chem. Soc. Perkin. Trans. 1* **1993**, 1287–1297; S. Mattei, P. Seiler, F. Diederich, V. Gramlich, *Helv. Chim. Acta* **1995**, *78*, 1904–1912; S. Stevelsmans, J. C. M. van Hest, J. F. G. A. Jansen, D. A. F. J. van Boxtel, E. M. M. de Brabander-van den Berg, E. W. Meijer, *J. Am. Chem. Soc.* **1996**, *118*, 7398–7399; b) S. Xu, Y. Luo, R. Haag, *Macromol. Biosci.* **2007**, 968–974; c) C. Ornelas, E. Boisselier, V. Martinez, I. Pianet, J. R. Aranzaes, D. Astruc, *J. Chem. Soc., Chem. Comm.* **2007**, 5093–5095.

38 J. J. Michels, M. W. P. L. Baars, E. W. Meijer, J. Huskens, D. N. Reinhoudt, *J. Chem. Soc. Perkin. Trans. 2* **2000**, 1914–1918.

39 a) J. J. J. M. Donners, B. R. Heywood, E. W. Meijer, R. J. Nolte, C. Elissen-Roman, A. P. H. J. Schenning, N. A. J. M. Sommerdijk, *Chem. Commun.* **2000**, *19*, 1937–1938; J. N. H. Reek, A. P. H. J. Schenning, A. W. Bosman, E. W. Meijer, M. J. Crossley, *Chem. Commun.* **1998**, 11–12; ***Review***: A. W. Bosman, J. F. G. A. Jansen, E. W. Meijer, *Chem. Rev.* **1999**, *99*, 1665–1668; K. Kalyanasundaram, J. K. Thomas, *J. Am. Chem. Soc.* **1977**, *99*, 2039–2044; b) L. Tao, J. Geng, G. Chen, Y. Xu, V. Ladmiral, G. Mantovani, D. M. Haddleton, *J. Chem. Soc., Chem. Comm.* **2007**, 3441–3443.

40 P. R. Ashton, V. Balzani, M. Clemente-León, B. Colonna, A. Credi, N. Jayaraman, F. M. Raymo, J. F. Stoddart, M. Venturi, *Chem. Eur. J.* **2002**, *8*, 673–684.

41 V. Balzani, H. Bandmann, P. Ceroni, C. Giansante, U. Hahn, F.-G. Klärner, U. Müller, W. M. Müller, C. Verhaelen, V. Vicinelli, F. Vögtle, *J. Am. Chem. Soc.* **2006**, *128*, 637–648; V. Balzani, P. Ceroni, C. Giansante, V. Vicinelli, F.-G. Klärner, C. Verhaelen, F. Vögtle, U. Hahn, *Angew. Chem.* **2005**, *117*, 4650–4654; *Angew. Chem. Int. Ed.* **2005**, *44*, 4574–4578.

42 C. A. Schalley, C. Verhaelen, F.-G. Klärner, U. Hahn, F. Vögtle, *Angew. Chem.* **2005**, *117*, 481–485; *Angew. Chem. Int. Ed.* **2005**, *44*, 477–480.

43 G. M. Hübner, G. Nachtsheim, Q. Y. Li, C. Seel, F. Vögtle, *Angew. Chem.* **2000**, *112*, 1315–1318; *Angew. Chem. Int. Ed.* **2000**, *39*, 1269–1272.

44 The term *dendritic effect* has meanwhile become current usage for characterisation of generation-dependent physical and chemical phenomena in dendritic molecules. C. Valério, J.-L. Fillaut, J. Ruiz, J. Guittard, J.-C. Blais, D. Astruc, *J. Am. Chem. Soc.* **1997**, *119*, 2588–2589; P. K. Murer, J.-M. Lapierre, G. Greiveldinger, D. Seebach, *Helv. Chim. Acta* **1997**, *80*, 1648–1681.
Notably, the term *dendritic effect* is not be found in the index of the volume *Dendrimers and Other Dendritic Polymers* (Eds. J. M. J. Fréchet, D. A. Tomalia), Wiley, Chichester **2001** or in that of the 1st Edition of G. R. Newkome, C. Moorefield, F. Vögtle, *Dendritic Molecules: Concepts, Syntheses, Perspectives*, VCH, Weinheim **1996**, whereas it is given in the monograph G. R. Newkome, C. N. Moorefield, F. Vögtle, *Dendrimers and Dendrons: Concepts, Syntheses, Applications*, VCH, Weinheim **2001**.

45 G. P. Perez, R. M. Crooks, *The Electrochemical Society Interface* **2001**, 34–38.

46 ***Reviews*** in dendritic catalysts: a) R. van Heerbeek, P. C. J. Kamer, P. W. N. M. van Leeuwen, J. N. H. Reek, *Chem. Rev.* **2002**, *102*, 3717–3756; b) G. E. Oosterom, J. N. H. Reek, P. C. J. Kamer, P. W. N. M. van Leeuwen, *Angew. Chem.* **2001**, *113*, 1878–1901; c) B. Helms, J. M. J. Fréchet, *Adv. Synthesis & Catalysis* **2006**, 1125–1148.

47 D. A. Tomalia, P. R. Dvornic, *Nature* **1994**, *372*, 617–618.

48 A. W. Kleij, R. A. Gossage, J. T. B. H. Jastrzebski, J. Boersma, G. van Koten, *Angew. Chem.* **2000**, 112, 179–181; *Angew. Chem. Int. Ed.* **2000**, *39*, 176–178.

49 C. C. Mak, H.-F. Chow, *Macromolecules* **1997**, *30*, 1228–1230; H.-F. Chow, C. C. Mak, *J. Org. Chem.* **1997**, *62*, 5116–5127.

50 a) H. F. Chow, C. C. Mak, *Chem. Comm.* **1996**, 1185–1186; b) H. F. Chow, C. C. Mak, *J. Chem. Soc. Perkin Trans. 1* **1997**, 91–95; c) H. F. Chow, C. C. Mak, *Pure Appl. Chem.* **1997**, *69*, 483–488.

51 Q.-H. Fan, Y.-M. Chen, X.-M. Chen, D.-Z. Jiang, F. Xi, A. S. C. Chan, *Chem. Commun.* **2000**, 789–790.

52 For a similar effect see: a) Q.-H. Fan, C.-Y. Ren, C.-H. Yeung, W.-H. Hu, A. S. C. Chan, *J. Am. Chem. Soc.* **1999**, *121*, 7407–7408; b) T. Uemura, X. Zhang, K. Matsumara, N. Sayo, H. Kumobayashi, T. Ohta, K. Nozaki, H. Takaya, *J. Org. Chem.* **1996**, *61*, 5510–5516; c) V. Enev, C. L. J. Ewers, M. Harre, K. Nickisch, J. T. Mohr, *J. Org. Chem.* **1997**, *62*, 7092–7093.

53 H.-F. Chow, C.-W. Wan, *Helv. Chim. Acta* **2002**, *85*, 3444–3454; C. C. Mak, H.-F. Chow, *Macromolecules* **1997**, *30*, 1228–1230; C. C. Mak, H.-F. Chow, *J. Org. Chem.* **1997**, *62*, 5116–5127.
Regarding the influence of backfolding as an explanation for the unusual solubility behaviour of some metallo-dendri-

zymes, see: H. Brunner, *J. Organomet. Chem.* **1995**, *500*, 39–46.
54 D. Seebach, R. E. Marti, T. Hintermann, *Helv. Chim. Acta* **1996**, *79*, 1710–1740.
55 P. B. Rheiner, D. Seebach, *Chem. Eur. J.* **1999**, *5*, 3221–3236.
56 M. Kimura, Y. Sugihara, T. Muto, K. Hanabusa, H. Shirai, N. Koboyashi, *Chem. Eur. J.* **1999**, *5*, 3495–3500.
57 M. Kimura, M. Kato, T. Muto, K. Hanabusa, H. Shirai, *Macromolecules* **2000**, *33*, 1117–1119.
58 P. Bhyrappa, J. K. Young, J. S. Moore, K. S. Suslick, *J. Mol. Kat. A. Chemical* **1996**, *113*, 109–116.
59 P. Bhyrappa, J. K. Young, J. S. Moore, K. S. Suslick, *J. Am. Chem. Soc.* **1996**, *118*, 5708–5711.
60 a) M. T. Reetz, G. Lohmer, R. Schwickardi, *Angew. Chem.* **1997**, *109*, 1559–1562; *Angew. Chem. Int. Ed.* **1997**, *36*, 1526–1529; b) Use in a continuously operated membrane reactor: N. Brinkmann, D. Giebel, G. Lohmer, M. T. Reetz, U. Kragl, *J. Cat.* **1999**, *183*, 163–168.
61 a) R. F. Heck, "Vinyl Substitutions with Organopalladium Intermediates", in *Comprehensive Organic Synthesis* (B. M. Trost, I. Fleming, Eds.), Vol. 4, 833–863, Pergamon Press, Oxford **1991**; b) S. Chandrasekhar, C. Narsihmulu, S. S. Sultana, N. R. Reddy, *Org. Lett.* **2002**, *4*, 4399–4401; c) S. Li, Y. Lin, H. Xie, S. Zhang, J. Xu, *Org. Lett.* **2006**, 8, 391–394; d) G. Battistuzzi, S. Cacchi, G. Fabrizi, *Org. Lett.* **2003**, *5*, 777–780; e) J. Mo, L. Xu, J. Xiao, *J. Am. Chem. Soc.* **2005**, *127*, 751–760; f) A. L. Hansen, T. Skrydstrup, *Org. Lett.* **2005**, *7*, 5585–5587.
62 R. Breinbauer, E. N. Jacobsen, *Angew. Chem.* **2000**, *112*, 3750–3753; *Angew. Chem. Int. Ed.* **2000**, *39*, 3604–3607.
63 Literature concerning the Jacobsen catalyst: a) Preparation, see: J. F. Larrow, E. N. Jacobsen, Y. Gao, Y. Hong, X. Nie, C. M. Zepp, *J. Org. Chem.* **1994**, *59*, 1939–1942; J. F. Larrow, E. N. Jacobsen, *Org. Synth.* **1997**, *75*, 1–10; b) Epoxidation and aziridation of olefins, see: W. Zhang, J. L. Loebach, S. R. Wilson, E. N. Jacobsen, *J. Am. Chem. Soc.* **1990**, *112*, 2801–2803; W. Zhang, E. N. Jacobsen, *J. Org. Chem.* **1991**, *56*, 2296–2298; E. N. Jacobsen, W. Zhang, A. R. Muci, J. R. Ecker, L. Deng, *J. Am. Chem. Soc.* **1991**, *113*, 7063–7064; M. Bandini, P. G. Cozzi, A. Umani-Ronchi, *J. Chem. Soc. Chem. Commun.* **2002**, 919–927.
c) Variants of the Jacobsen catalyst: Refs. [65–67]
64 Literature devoted specifically to asymmetric epoxidation with Jacobsen catalysts: a) E. N. Jacobsen, *Acc. Chem. Res.* **2000**, *33*, 421–431; b) E. N. Jacobsen, W. Zhang, M. L. Güler, *J. Am. Chem. Soc.* **1991**, *113*, 6703–6704; c) M. Palucki, N. S. Finney, P. J. Pospisil, M. L. Güler, T. Ishida, E. N. Jacobsen, *J. Am. Chem. Soc.* **1998**, *120*, 948–954; d) W. Adam, K. J. Roschmann, C. R. Saha-Möller, D. Seebach, *J. Am. Chem. Soc.* **2002**, *124*, 5068–5073; e) T. G. Traylor, A. R. Miksztal, *J. Am. Chem. Soc.* **1989**, *111*, 7443–7448.
65 Concerning the Katsuki variant of the Jacobsen catalyst, see: a) R. lrie, K. Noda, Y. Ito, N. Matsumoto, T. Katsuki, *Tetrahedron Lett.* **1990**, *31*, 7345–7348; b) R. Irie, K. Noda, Y. Ito, T. Katsuki, *Tetrahedron Lett.* **1991**, *32*, 1055–1058; c) R. Irie, Y. Ito, T. Katsuki, *Synlett* **1991**, *2*, 265–266; d) R. Irie, K. Noda, Y. Ito, N. Matsumoto, T. Katsuki, *Tetrahedron: Asymmetry* **1991**, *2*, 481–494; e) N. Hosoya, R. Irie, Y. Ito, T. Katsuki, *Synlett* **1991**, *2*, 691–692.
66 Modifications of the Jacobsen catalyst according to Burrows: C. J. Burrows, K. J. O'Connor, S. J. Wey, *Tetrahedron Lett.* **1992**, *33*, 1001–1004.
67 Modifications of the Jacobsen catalyst according to Thornton: D. R. Reddy, E. R. Thornton, *J. Chem. Soc. Chem. Commun.* **1992**, 172–173.
68 J. J. Lee, W. T. Ford, J. A. Moore, Y. Li, *Macromolecules* **1994**, *27*, 4632–4634.
69 K. Vassilev, W. T. Ford, *J. Polym. Sci. A: Polym. Chem.* **1999**, *37*, 2727–2736.
70 Y. Pan, W. T. Ford, *Macromolecules* **2000**, *33*, 3731–3738.
71 J. J. Lee, W. T. Ford, *Macromolecules* **1994**, *27*, 4632–4634; D. S. Kemp, K. G. Paul, *J. Am. Chem. Soc.* **1975**, *97*, 7305–7312; J. J. Lee, W. T. Ford, *J. Org. Chem.* **1993**, *58*, 4070–4077; P. D. Miller, W. T. Ford, *Langmuir* **2000**, *16*, 592–596.

72 F. M. Menger, C. E. Portnoy, *J. Am. Chem. Soc.* **1967**, *89*, 4698–4703.

73 P. D. Miller, H. O. Spivey, S. L. Copeland, R. Sanders, A. Woodruff, D. Gearhart, W. T. Ford, *Langmuir* **2000**, *16*, 108–114.

74 I. Morao, F. P Cossio, *Tetrahedron Lett.* **1997**, *38*, 6461–6464.

75 R. Ballini, G. Bosica, *J. Org. Chem.* **1997**, *62*, 425–427.

76 T. Suzuki, Y. Hirokawa, K. Ohtake, T. Shibata, K. Soai, *Tetrahedron: Asymmetry* **1997**, *8*, 4033–4040.

77 I. Sato, T. Shibata, K. Ohtake, R. Kodaka, Y. Hirokawa, N. Shirai, K. Soai, *Tetrahedron Lett.* **2000**, *41*, 3123–3126.

78 For use in a continuously operated membrane reactor, see: A. W. Kleij, R. A. Gossage, R. J. M. Gebbink, N. Brinkmann, U. Kragl, E. J. Reyerse, M. Lutz, A. L. Spek, G. van Koten, *J. Am. Chem. Soc.* **2000**, *122*, 12112–12124.

79 The nickel-containing systems are also active in controlled radical polymerisation (ATRP). See: a) T. E. Patten, J. Xia, T. Abemathy, K. Matyjaszewski, *Science* **1996**, *272*, 866–868; b) C. Granei, Ph. Dubois, R. Jérôme, P. Teyssié, *Macromolecules* **1996**, *29*, 8576–8582; c) S. A. F. Bon, F. A. C. Bergman, J. J. G. S. van Es, B. Klumperman, A. L. German, "Controlled radical polymerization: towards control of molecular weight", in *Controlled Radical Polymerization* (K. Matyjaszewski, Ed.), *ACS Symposium Series* **1998**, *685*, 236–255.

80 R. E. White, M. J. Coon, *Ann. Rev. Biochem.* **1980**, *49*, 315–356; J. H. Dawson, M. Sono, *Chem. Rev.* **1987**, *87*, 1255–1276; M. Sono, M. P. Roach, E. D. Coulter, J. H. Dawson, *Chem. Rev.***1996**, *96*, 2841–2888.

81 H. B. Dunford, *Adv. Inorg. Biochem.* **1982**, *4*, 41–68; J. H. Dawson, *Science* **1988**, *240*, 433–439; A. M. English, G. Tsaprailis, *Adv. Inorg. Chem.* **1995**, *43*, 79–125.

82 a) P. J. Dandliker, F. Diederich, A. Zingg, J.-P. Gisselbrecht, M. Gross, A. Louati, E. Sanford, *Helv. Chim. Acta* **1997**, *80*, 1773–1801; b) K. W. Pollak, J. W. Leon, J. M. J. Fréchet, M. Maskus, H. D. Abruña, *Chem. Mater.* **1998**, *10*, 30–38; c) D.-L. Jiang, T. Aida, *J. Am. Chem. Soc.* **1998**, *120*, 10895–10901; d) S. A. Vinogradov, L.-W. Lo, D. F. Wilson, *Chem. Eur. J.* **1999**, *5*, 1338–1347; e) U. Puapaiboon, R. T. Taylor, *Rapid. Commun. Mass Spectrom.* **1999**, *13*, 508–515; f) M. Rimura, T. Shiba, M. Yamazaki, K. Hanabusa, H. Shirai, N. Kobayashi, *J. Am. Chem. Soc.* **2001**, *123*, 5636–5642.

83 ***Reviews***: a) B. Meunier, *Chem. Rev.* **1992**, *92*, 1411–1456; b) D. Mansuy, *Coord. Chem. Rev.* **1993**, *125*, 129–141; c) D. R. Benson, R. Valentekovich, S.-W. Tam, F. Diederich, *Helv. Chim. Acta* **1993**, *76*, 2034–2060; d) F. Bedioui, *Coord. Chem. Rev.* **1995**, *144*, 39–68; e) J. P. Collman, Z. Wang, A. Straumanis, M. Quelquejeu, E. Rose, *J. Am. Chem. Soc.* **1999**, *121*, 460–461; f) J. Yang, R. Breslow, *Angew. Chem.* **2000**, *112*, 2804–2806; *Angew. Chem. Int. Ed.* **2000**, *39*, 2692–2695; g) J. T. Groves, *J. Porphyrins Phthalocyanines* **2000**, *4*, 350–352; ***Review***: h) W.-D. Woggon, H.-A. Wagenknecht, C. Claude, *J. Inorg. Biochem.* **2001**, *83*, 289–300.

84 a) P. Weyermann, J.-P. Gisselbrecht, C. Boudon, F. Diederich, M. Gross, *Angew. Chem.* **1999**, *111*, 3400–3404; *Angew. Chem. Int. Ed.* **1999**, *38*, 3215–3219; b) P. Weyermann, J.-P. Gisselbrecht, C. Boudon, F. Diederich, M. Gross, *Helv. Chim. Acta* **2002**, *85*, 571–598.

85 P. Weyermann, F. Diederich, *Helv. Chim. Acta* **2002**, *85*, 599–617.

86 M. J. Gunter, P. Turner, *J. Mol. Catal.* **1991**, *66*, 121–141; Y. Naruta, K. Maruyama, *Tetrahedron Lett.* **1987**, *28*, 4553–4556.

87 J. T. Groves, T. E. Nemo, R. S. Myers, *J. Am. Chem. Soc.* **1979**, *101*, 1032–1033.

88 ***Reviews***: T. G. Traylor, *Pure Appl. Chem.* **1991**, *63*, 265–274; M. J. Gunter, P. Turner, *Coord. Chem. Rev.* **1991**, *108*, 115–161.

89 P. A. Adams, J. Louw, *J. Chem. Soc., Perkin Trans. 2* **1995**, 1683–1690; P. R. Ortiz de Mantellano, *Curr. Opin. Chem. Biol.* **2000**, *4*, 221–227.

90 L. L. Miller, R. G. Duan, D. C. Tully, D. A. Tomalia, *J. Am. Chem. Soc.* **1997**, *119*, 1005–1010.

91 J.-F. Penneau, L. L. Miller, *Angew. Chem.* **1991**, *103*, 1002–1003; *Angew. Chem. Int. Ed.* **1991**, *30*, 986–987; J.-F. Penneau, B. J. Stallman, P. H. Kasai, L. L. Miller,

Chem. Mater. **1991**, *3*, 791–796; C. J. Zhong, W. S. V. Kwan, L. L. Miller, *Chem. Mater.* **1992**, *4*, 1423; ***Review***: L. L. Miller K. R. Mann, *Acc. Chem. Res.* **1996**, *29*, 417–423.

92 T. D. Selby, S. C. Blackstock, *J. Am. Chem. Soc.* **1998**, *120*, 12155–12156.

93 H.-F. Chow, I. Y.-K. Chan, P.-S. Fung, T. K.-K. Mong, M. F. Nontrum, *Tetrahedron* **2001**, *57*, 1565–1572.

94 P. J. Dandliker, F. Diederich, M. Gross, C. B. Knobler, A. Louati, E. M. Sandford, *Angew. Chem.* **1994**, *106*, 1821–1824; *Angew. Chem. Int. Ed.* **1994**, *33*, 1739–1742; P. J. Dandliker, F. Diederich, J.-P. Gisselbrecht, A. Louati, M. Gross, *Angew. Chem.* **1996**, *107*, 2906–2909; *Angew. Chem. Int. Ed.* **1996**, *34*, 2725–2728.

95 H.-F. Chow, I. Y.-K. Chan, P.-S. Fung, T. K.-K. Mong, M. F. Nongrum, *Tetrahedron* **2001**, *57*, 1565–1572.

96 M.-S. Choui, T. Aida, H. Luo, Y. Araki, O. Ito, *Angew. Chem.* **2003**, *115*, 4194–4197; *Angew. Chem. Int. Ed.* **2003**, *42*, 2154–2157.

7
Characterisation and analysis

Unequivocal characterisation of dendrimers is somewhat complex, not least owing to their molecular size and symmetry, and therefore always relies on several analytical methods. In addition to various NMR techniques [1], which form the basis of dendrimer analysis, and elemental analysis, the principal methods used are various mass-spectrometric and chromatographic techniques, IR [2] and Raman spectroscopy, X-ray structure analyses, as well as electrophoretic techniques [3]. Small-angle X-ray, small-angle neutron, and Laser light scattering experiments are also attracting increasing attention, as are microscopic single-molecule observations. Provided appropriate components are present in the dendrimer, UV-Vis and fluorescence spectroscopy or chiroptical measuring methods also makes a useful contribution to dendrimer characterisation.

The following sections will present some of the most important analytical methods [4] employed in determining the chemical composition, the molecular mass, the morphology, the spatial geometry, and the homogeneity of dendrimers.

7.1
Chromatography

First of all, a number of chromatographic and electrophoretic separation methods will be considered because in most cases unequivocal identification will be impossible without prior work-up of the dendrimer sample. The term chromatography refers collectively to physical separation methods by which mixtures of substances can be resolved into their various components by repeated equilibration between a stationary and a mobile phase. We shall now present a detailed consideration of the chromatographic techniques which can be used to solve separation problems in dendrimer chemistry.

7.1.1
Liquid chromatography

In liquid chromatography the mobile phase (eluent) transports the dissolved sample through the stationary phase. Depending upon the polarity of the sta-

Dendrimer Chemistry. Fritz Vögtle, Gabriele Richardt and Nicole Werner

ISBN: 978-3-527-32066-0

tionary phase, a distinction is made between two methods: Normal phase (NP) chromatography and reversed phase (RP) chromatography. In NP chromatography a stationary polar phase such as silica gel (SiO_2) or aluminium oxide (Al_2O_3) and a non-polar mobile phase are used.

The separating effect results from the different strengths of interaction of the various components of a sample with the stationary phase.

In NP chromatography non-polar solvents (*e.g.* alkanes) are used as eluents; in cases of insufficient eluting strength the polarity of the eluent can be raised by addition of more polar solvent. More polar sample components have longer residence times in NP chromatography, *i.e.* they leave the column later because they are retained longer.

A silica-gel based stationary phase is also generally used in RP chromatography. However, alkylation of the hydroxyl groups renders the stationary phase apolar. The most commonly used eluents include polar solvents such as methanol, acetonitrile, or tetrahydrofuran or corresponding solvent mixtures. The separation mechanism of RP chromatography is complex and is best explained in terms of adsorption and partitioning effects.

In many cases, the constituents of a dendrimer sample are structurally very similar and differ only slightly in their physicochemical properties. A gradient-like alteration of the polarity of the solvent by increasing admixture of a more polar or less polar solvent frequently leads to better separation in such cases (gradient separation).

7.1.1.1 **Preparative liquid chromatography**

Preparative liquid chromatography is an important and comparatively inexpensive method in dendrimer chemistry for routine separation of the various components of relatively large samples (>1 g). The stationary phase is contained in columns whose diameter and length depend upon the size of the sample and the degree of separation required (column chromatography). The mobile phase (eluent) which transports the sample with all its components moves though the column under gravity or by application of additional pressure (flash chromatography) (Fig. 7.1).

Thin-layer chromatography (TLC) [5] is routinely used in dendrimer synthesis for purity checks and identification of individual constituents of a dendrimer sample and is suitable for monitoring column chromatographic separations. Since chemistry students are introduced to this analytical technique during their first semesters, we shall not go into further details here.

7.1.1.2 **High-performance liquid chromatography**

One of the most powerful separation techniques available is high-performance liquid chromatography (HPLC) [6]. It has a broad range of applicability which also encompasses non-volatile substances such as ionic compounds (*e.g.* amino acids, proteins, metal complexes) or high-molecular weight compounds, such as

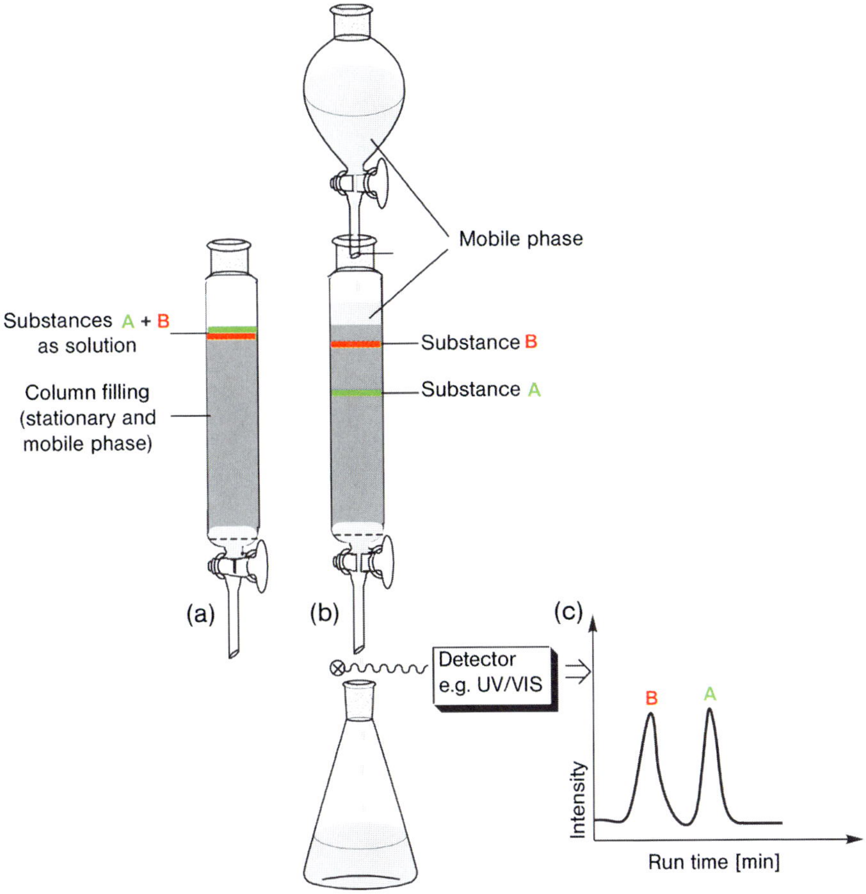

Fig. 7.1 Column chromatographic separation of a sample mixture containing two components A and B at normal pressure [based on C. Janiak, Lecture "Analytical Chemistry I", University of Freiburg]. a) After application of the analytical sample; b) components A and B are already spatially separated; c) record of separation (chromatogram)

polymers and dendrimers [7], which cannot be separated by gas chromatography.

In HPLC the sample is dissolved in a solvent that is miscible with the mobile phase and injected onto the column *via* a sample valve. Separation takes place on a chromatographic column which can be temperature controlled if required. The separated sample components are recorded by a detector interfaced to a computer (Fig. 7.2). HPLC exhibits a significantly higher separation performance than classical column chromatography (LC) introduced above. This is accomplished in HPLC by use of stationary phases having a small particle diameter ($d_p = 3$ to 10 μm for analytical separations). Since these particles strongly re-

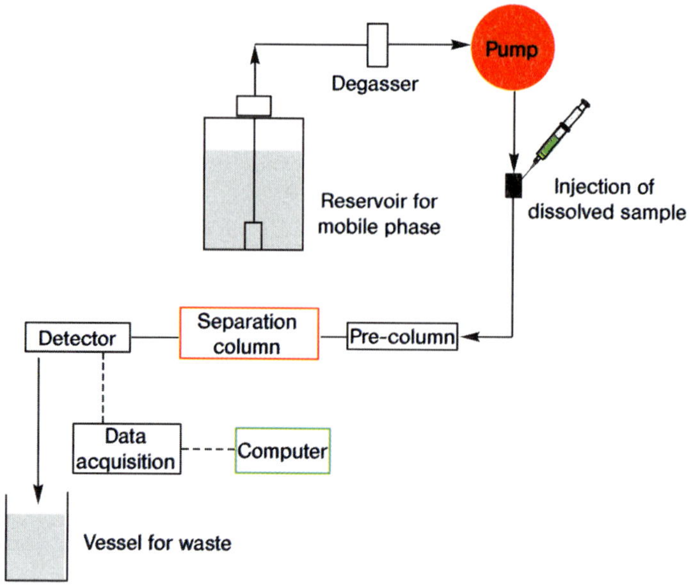

Fig. 7.2 HPLC set-up (schematic)

sist passage of the mobile phase, pressures of up to 400 bar are necessary to transport mobile phase through the separation column at acceptable flow rates.

HPLC has the advantage of being more than just an analytical technique; it also offers the possibility of preparative separation (*e.g.* of dendrimers) after appropriate adaptation. In recent years this method has been increasingly used for the identification and separation of dendrimers, with most of the separation problems relating to dendrimers having been solved with the aid of RP chromatography.

HPLC in dendrimer chemistry: Since separation is based primarily on adsorption between the peripheral groups of the dendrimer and the stationary phase, HPLC is more suitable for the separation of dendrimers with terminal defects or different end group functionalities [8]. The individual fractions separated by HPLC can be structurally characterised by mass-spectrometric methods (*e.g.* MALDI-MS, see below). Identification of all the defective compounds formed during a divergent dendrimer synthesis is one of greatest challenges facing dendrimer analysis. Molecules with comparatively small internal structural defects (*e.g.* missing branching units, intramolecular ring closures), which have only a slight effect on the retention time, are frequently seen only as shoulders in the chromatogram or remain hidden under the main peak.

Recent studies have shown that RP-HPLC in combination with MALDI-TOF mass spectrometry is suitable for preparative separation and characterisation of PAMAM dendrimers of different generations or with different peripheral groups [8]. Investigations performed so far indicate that the purity of a dendrimer can be at least semi-quantitatively evaluated on the basis of chromatograms.

In addition, chiral dendrimers (see Section 4.2) can be resolved with the aid of HPLC into their enantiomers, if the silica gel material used as stationary phase has optically active substances bound to its surface [9]. Since the chiral stationary phase (CSP) [10] undergoes different intensities of interaction with the enantiomeric dendrimers, they are retained to different degrees, and in the ideal case two completely separated (baseline separated) peaks are obtained. This separation technique was successfully applied *inter alia* to racemic mixtures of planar-chiral dendro[2.2]paracyclophanes, cycloenantiomeric dendro[2] rotaxanes, topologically chiral dendro[2]catenanes [11] as well as topologically chiral, dendritically substituted molecular knots (knotanes) [12] (Section 4.2.3).

7.1.2
Gel permeation chromatography

A chromatographic technique frequently used in polymer and dendrimer analysis is size exclusion chromatography (SEC) [13], which is often also called gel permeation chromatography (GPC). It is a straightforward method of determining the relative molar mass, the molar mass distribution, and the polydispersity index (PDI) with compatively modest consumption of material and time.

GPC is a further special form of liquid chromatography. The separation column is packed with porous, polymer gels (*e.g.* polystyrene gel) as stationary phase. The particle size of the packing material and the size distribution of the pores are well defined and uniform. In GPC molecules are separated according to their effective size in solution, *i.e.*, their hydrodynamic volume, and not according to their affinity for the support material.

Sample molecules that are too large to enter the pores of the support material, which is commercially available in various pore dimensions, are not retained and leave the column first. The required elution volume V_e is correspondingly small. Small molecules are retained most strongly because they can enter all the pores of the support material. Sample molecules of medium size can partly penetrate into the stationary phase and elute according to their depth of penetration into the pores (Fig. 7.3). No specific interactions should take place between the molecules of the dendrimer sample and the stationary phase in GPC since this can impair the efficiency of separation by the exclusion principal. After separation the eluate flows through a concentration-dependent detector (*e.g.* a UV/VIS detector) interfaced with a computer. One obtains a chromatogram which, to a first approximation, reflects the relative contents of molecules of molar mass M. If macromolecules of suitable molar mass and narrow molar mass distribution are available for calibration of the column, the relative "GPC molar mass" of the investigated dendrimer can be determined *via* the calibration function $\log(M) = f(V_c)$.

GPC in dendrimer chemistry: Since the principle of separation of GPC is based on the different sizes (hydrodynamic volumes) of the molecules, this is an ideal

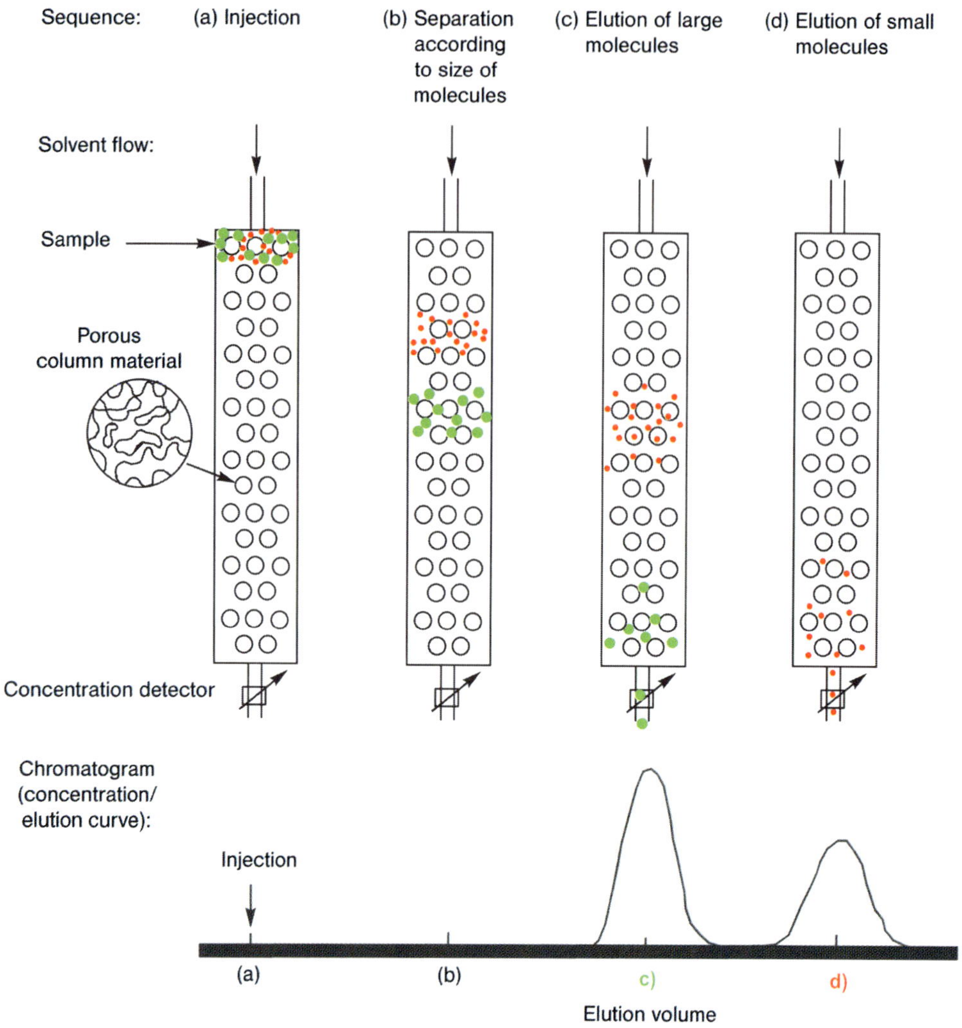

Fig. 7.3 Mode of operation of a GPC column (schematic)

technique for characterising different generations of dendrimers. In addition, it can also be used for detection of general impurities (*e.g.* residual monomers, plasticisers) or of defect structures by defective dendrimer growth (*e.g.* lower-generation dendrimers or dimers). The purity of the dendrimers studied can be quantitatively determined from the polydispersity index (M_w/M_n). In evaluating the polydispersity index it should be noted that dendrimers with small structural defects (*e.g.* intramolecular ring closures, missing end groups) do not differ significantly in size from structurally perfect dendrimers and therefore have little influence on the polydispersity values determined by GPC. The main responsibility for increasing polydispersity values with increasing dendrimer generation is borne by contamination with dendrimers having lower shell numbers and with dimers possibly formed by intermolecular cyclisation [14].

Conventional GPC with RI (refractive index) detection suffers from the serious drawback that it is a relative method of molar mass determination. The relative "GPC molar mass" determined corresponds to the true molar mass of the dendrimer investigated only if the size standard used also has a similar hydrodynamic volume to the sample molecule. However, the linear polymers generally used as size standards are frequently unsuitable for studies on highly branched macromolecules because the hydrodynamic volume depends not only on primary but also on secondary structural factors such as the conformation of the molecule. Thus various investigations have shown that in the case of spherical dendrimers the molar mass of higher generation dendrimers is underestimated as a result of their compactness (small hydrodynamic volume) [15]. These calibration problems can meanwhile be circumvented by use of a multi-angle light scattering (MALS) detector because this detector affords absolute values of the weight-average molar mass M_w of each fraction. However, successful use of this expensive detector requires a knowledge of the specific refractive index increment (d_n/d_c) of the dendrimer in the solvent.

7.2 Gel electrophoresis

In biochemistry gel electrophoresis is the method of choice for the separation of various kinds of macromolecules (*e.g.* nucleic acids, proteins). It is also used in dendrimer chemistry for separation and as a method of determining relative molar masses and for qualitative assessment of the purity of a dendrimer sample.

In gel electrophoresis the charged molecules present in the sample migrate under the influence of an electric field through a gel which contains an ionic buffer solution. They migrate towards the oppositely charged electrodes. The electrophoretic mobility of the individual molecules of a substance mixture through the porous gel varies with the charge, molar mass, and shape of the molecules. This leads to separation of a substance mixture into individual fractions seen as bands. The electrophoretic mobility is a characteristic parameter of a charged molecule. It is not only dependent upon the above-mentioned molecular parameters and the support material but is also influenced by the solvent used, by the electric field strength, and also by the ionic strength and the temperature of the buffer system used. The relative electrophoretic mobility and thus the relative molar mass of the substance studied can be estimated by comparison of its migration path with that of a molar mass standard applied in the same run.

Electrophoretic separation of dendrimers is usually performed with polyacrylamide gel applied to glass plates of carrier films (flat bed, slab gel electrophoresis). Polyacrylamide gel electrophoresis has been successfully used for the separation and quantification of PAMAM dendrimers with various core units, end groups, and generation numbers, with the possibility of working under acidic or basic conditions [16]. Conventional gel electrophoresis has the disadvantage of being comparatively time-consuming and labour-intensive.

A more recent biochemical technique, combining the strengths of gel electrophoresis and of HPLC, is capillary gel electrophoresis. Here the electrophoretic separation is performed in capillaries of small internal diameter. An advantage of capillary gel electrophoresis over flat bed and slab gel electrophoresis is that much stronger electric fields can be used, without interfering effects due to warming (*e.g.* distortion of the separation zones owing to thermal convection, drying out of the gel). This reduces analysis times. Moreover, apart from offering higher separation performance, capillary gel electrophoresis also permits direct on-capillary detection and automation of the separation process. For this reason, characterisation by capillary gel electrophoresis has acquired ever-increasing importance in recent years, also in connection with the use of dendrimers and their conjugates in biotechnology, clinical diagnostics, and pharmaceutical research.

7.3
NMR spectroscopy

To this day nuclear magnetic resonance (NMR) spectroscopy [17] remains the most important analytical method in chemistry because it permits determination of the structure and dynamics of molecules in solution. In addition, it also allows the study of intermolecular interactions as well as interactions with solvent molecules.

In contrast to other kinds of spectroscopy [18] (*e.g.* IR, UV), which generally provide information only about certain characteristics or functional groups of a molecule, NMR spectroscopy opens a window on individual atoms of a molecule. It is crucial for structure elucidation by means of NMR spectroscopy that the resonance frequency of an atomic nucleus in a real molecule measured during recording of a spectrum depends not only on the external magnetic field but also on the electronic (*i.e.* chemical) environment of the nucleus (chemical shift). The magnetic interaction between chemically non-equivalent atomic nuclei transmitted by covalent bonds leads to a fine structure in the spectra of the dissolved molecules. Covalent linkages of atoms can be recognised from this scalar (*J*) coupling. Measurement of dipolar coupling, the nuclear Overhauser effect (NOE), is useful for the determination of more complex molecular structures, since it leads to a change of intensity of the signals of spatially proximate but not chemically bonded atomic nuclei. Insofar as unequivocal assignment of the resonances has already been accomplished on the basis of (1D)- or (2D)-NMR measurement, such NOESY experiments permit quantitative determinations of internuclear distances for nuclei in different parts of the dendrimer molecule.

7.3.1
(1D)-NMR spectroscopic studies

Whereas solid-state NMR spectroscopy [19] (see Section 7.6.3.2) plays only a minor role in dendrimer characterisation, one-dimensional (1D-)NMR spectroscopy of dendrimers in solution still serves as a routine method in such characterisation.

The synthesis of dendrimers generally involves the repetition or iteration of two specific reaction steps for the construction of a generation. In dendrimers it is therefore generally necessary to distinguish between building blocks of the same kind or multiply present functional groups which vary only in their position within the dendrimer scaffold. In spite of their similar nature, they differ in their microenvironment owing to their different positions within the dendrimer scaffold. In the case of smaller dendrimer generations the sensitivity towards the chemical environment facilitates NMR-spectroscopic distinction between identical groups in the inner and outer shells. If heteroatoms are present in the dendrimer scaffold then not only ^{1}H-NMR and ^{13}C-NMR spectroscopy but also other NMR techniques (*e.g.* ^{15}N, ^{19}F, ^{29}Si, ^{31}P) can be used for characterising dendrimers. In a chiral dendrimer prepared by Seebach *et al.* the CF_3 groups distributed over the dendrimer scaffold proved to be excellent probes because the ^{19}F-NMR spectra permit distinction between interior and exterior CF_3 groups [20].

Characterisation of dissolved dendrimers by routine (1D)-NMR spectroscopy becomes increasingly difficult with increasing generation number. (1D)-NMR spectra of higher generations can give such broad signals, owing to the numerous identical groups in different microenvironments within the dendrimer scaffold, that precise assignment to the individual dendrimer shells is no longer possible. Defects in the dendrimer scaffold such as missing branching units or missing end groups can then hardly be detected. The structural perfection of a higher-generation dendrimer therefore cannot be unequivocally proven solely on the basis of routine (1D)-NMR spectroscopy. Yet the absence of specific groups resulting from the decreasing sensitivity of NMR spectroscopy at higher molar masses should not be assigned undue importance.

When assessing the purity of dendrimers it should also be noted that NMR spectroscopic methods approach their limit of detection at contamination levels of *ca.* 5%. Additional chromatographic method such as gel permeation chromatography (GPC, SEC; Section 7.1.2) or mass spectrometric methods (MALDI-MS, ESI-MS), as presented in Section 7.4, should also be employed in verification of structural perfection and purity of dendrimers.

7.3.2
Multidimensional NMR spectroscopy in dendrimer research

Multidimensional NMR spectroscopy ((2D)-NMR, (3D)-NMR) is also acquiring increasing importance in the characterisation of dendrimers. On inclusion of a further dimension in (2D)-NMR spectroscopy, it is possible, for example, to plot resonance frequencies and spin/spin coupling frequencies, which cannot be distinguished *a priori* in the (1D)-NMR spectrum, on different frequency axes and thus separate them. In many cases, unambiguous assignment of the chemical shifts only then became feasible. Thus (3D)-NMR experiments permitted realistic representation of the shells of a third-generation POPAM dendrimer and complete characterisation of the dendrimer structure in solution [21, 22].

High resolution multidimensional NMR experiments can provide the dendrimer chemist with a wealth of additional information extending far beyond the determination of the molecular structure. In the interpretation of (2D)-NOESY (NOESY = nuclear Overhauser enhancement spectroscopy) spectra, a knowledge of the spatial interrelationships between protons in different parts of the dendrimer scaffold can be acquired from proton-proton NOE interactions. At the same time, the prevailing conformation of the dendritic branches in the solvent used can be deduced from this information. Furthermore, studies of dendrimer/solvent interactions and the influence of solvent on the spatial structure of the dendrimer are also possible [22]. Thus the information content of such NMR experiments resembles that of small-angle scattering experiments on dissolved dendrimers (see Section 7.6).

Multidimensional NMR spectroscopy is also used in the elucidation of supramolecular host/guest interactions. Thus Meijer *et al.* applied (1D) and various (2D)-NMR-techniques [*e.g.* (2D)-NOESY, (2D)-TOCSY (TOCSY = Total Correlation Spectroscopy) NMR to the structural characterisation of a stable host/guest complex consisting of adamantylurea-functionalised POPAM dendrimer and cyanobiphenyl guest molecules [23]. Dendritic host/guest systems and hence also studies of the underlying intermolecular interactions in solution are playing an ever-increasing role, particularly in the context of increasing interest in dendrimers for medical applications (*e.g.* for drug delivery).

7.3.3
Diffusion NMR spectroscopy

Diffusions NMR spectroscopy (*e.g.* PGSE = Pulsed Gradient Spin Echo; STE = Stimulated Echo; DOSY = Diffusion Ordered Spectroscopy) is a straightforward and accurate method for determination of the self-diffusion coefficient of a molecule. Its principal use in dendrimer chemistry is for size determination of dissolved dendrimers since the self-diffusion coefficient is directly correlated with the hydrodynamic radius of the molecule *via* the Stokes–Einstein equation [24]. Although one-dimensional and multidimensional diffusion NMR experiments can thus make an important contribution to structural characterisation of dendrimers, they have been used comparatively rarely until recently [25, 26].

Diffusion NMR experiments performed by Newkome *et al.* show that the diffusion coefficients – and hence the hydrodynamic radii – can yield information about the influence of external factors (*e.g.* pH values) on the size and shape of dissolved dendrimers [26]. Since the spatial structure of a dendrimer in solution in turn directly affects its material properties, this NMR technique can also provide an important contribution to structure-activity relationships [27].

7.3.4
Dynamic NMR spectroscopy

Apart from the structure of biological macromolecules, their dynamic processes are known to be of great importance because the function of the molecule is often closely associated with conformational changes. In keeping with the increasing focus on medical applications, studies of molecular dynamics in solution are acquiring increasing significance in dendrimer research. The dynamics of dendritic branches can be investigated by measurement of ^{1}H- and ^{13}C-spin-lattice relaxation times (T_1). An essential condition is the complete assignment of all proton- and carbon resonances of the dendrimers to be studied on the basis of (1D)- and (2D)-NMR spectra. Since the mobility of a dendrimer segment is proportional to its T_1 value, the change of mobility of the various dendrimer segments and hence the relative density distribution of the dendrimer scaffold can be derived from the change in length of the relaxation times of the protons and carbon nuclei in the individual layers [28]. Thus, for example, the increase in ^{1}H- and ^{13}C-spin-lattice relaxation times on transition from the internal to the external dendrimer layers or dendrimer segments testifies to increasing mobility. At the same time, this behaviour is indicative of a decrease in the radial density distribution within the dendrimer scaffold from the centre to the periphery of the dendrimer [29].

7.4
Mass spectrometry

The only mass spectrometric methods available during the era of the first "cascade" synthesis in 1978 [30] were electron impact (EI) and field desorption (FD) mass spectrometry [31]. Fast atom bombardment (FAB) mass spectrometry is limited to fairly low mass ranges and not very suitable for compounds of low polarity. It was not until the development of new and gentle ionisation methods such as MALDI (matrix-assisted laser desorption ionization) [32] and ESI (electrospray ionization) [33] that the conditions were fulfilled for the start of intense research in the field of dendrimer chemistry. The following section will present the special features of these mass-spectrometric methods and their importance in dendrimer analysis.

7.4.1
Gentle ionisation methods: MALDI and ESI

In MALDI the ionisation energy is provided by a pulsed laser beam. The substance to be analysed is dissolved in a matrix (*e.g.* 2,5-dihydroxybenzoic acid; 2,4,6-trihydroxyacetophenone) which absorbs in the wavelength range of the laser used (Nd-YAG laser: $\lambda = 355$ or 266 nm; N_2 laser: $\lambda = 337$ nm). When the laser is fired at the sample, either the sample molecules absorb the energy directly (if chromophores are present), or energy transfer to the sample molecules occurs only after UV absorption by the matrix, effecting gentle ionisation.

In contrast to MALDI the principle of ESI is based on fine spraying of sample solution. The initially formed electrically charged droplets are rapidly desolvated, permitting the individual dendrimer species to be observed as isolated ions in the mass spectrum.

7.4.1.1 Study of dendrimers by MALDI and ESI-MS

MALDI and ESI usually generate (quasi)molecular ions in the form of protonated dendrimers and/or adducts with alkali metal ions. Such adduct formation can occur if the dendrimer structure contains heteroatoms and the alkali metal ions are present as impurities in the dendrimer sample. However, alkali metal or silver salts can also be added purposely to assist the ionisation of apolar dendrimers. In particular, MALDI-TOF-MS (matrix assisted laser desorption time-of-flight mass spectrometry) has proved to be a straightforward analytical method suitable for routine use in dendrimer research because it requires very small amounts of sample in the picomole to attomole range and is greatly superior to FAB-MS and also ESI-MS on ionisation in high mass ranges. In addition, readily interpretable spectra are obtained which afford unequivocal structural information since mainly singly charged quasimolecular ions are formed and hardly any fragmentations take place.

Although MALDI-MS plays an outstanding role in dendrimer analysis, additional use is also made of modern ESI mass spectrometers for monitoring syntheses, for determination of relative molecular masses, and for studying the purity and polydispersity of dendrimers, including those of higher generations [34].

MALDI-TOF-MS and ESI-MS number among the few analytical methods suitable for detailed studies of structural defects in dendrimers. If defects have been found in a dendrimer scaffold, tandem MS experiments (*e.g.* CID, Collision Induced Dissociation) coupled to MALDI or ESI ionisation permit a more detailed distinction between various types of defect structures on the basis of characteristic fragmentation patterns [35]. Owing to the high sensitivity and the low instrumental or experimental error of typically only 0.05% it is even possible to detect the absence of individual terminal groups. Meijer *et al.* used ESI mass spectrometry to scrutinise and quantify structural defects of POPAM [36] and PAMAM dendrimers [37]. In order to quantitatively determine the polydispersity and the purity of dendrimers, which is defined as the percentage of defect-free dendritic material, it is necessary not only to distinguish between the various molecular species in the sample but also to ionise them with similarly high efficiency and analyse them. The discrimination of higher masses precludes quantitative determination of the relative quantities of two species differing substantially in mass [38]. However, MALDI-TOF and ESI mass spectra generally permit semi-quantitative conclusions to be drawn about the purity of a dendrimer sample.

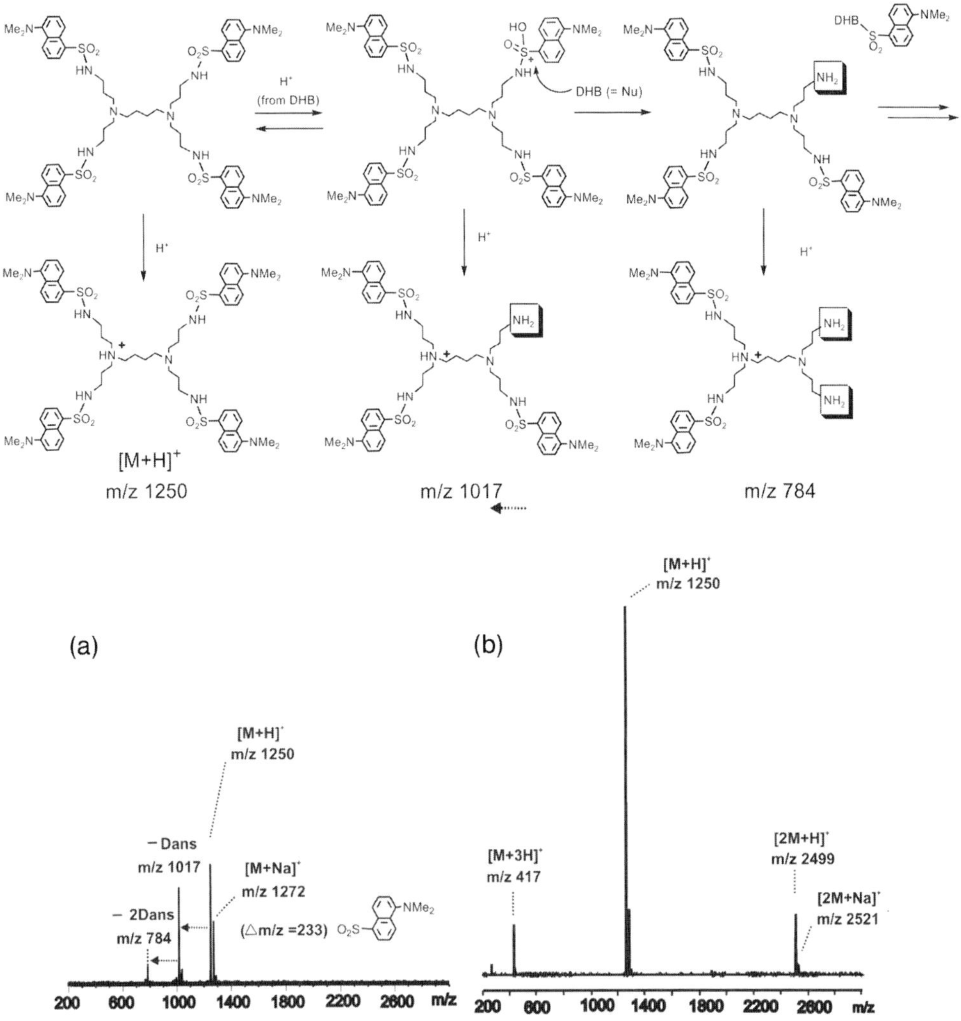

Fig. 7.4 a) The MALDI-TOF mass spectrum (matrix: 2,5-dihydroxybenzoic acid) gives an impression of incomplete functionalisation of the peripheral primary amino groups of the POPAM dendrimer with dansyl units; b) yet the ESI-FT-ICR mass spectrum confirms complete functionalisation of the dendrimer

Recent comparative studies on the various ionisation techniques in the case of POPAM dendrimers and POPAM dendrimers bearing peripheral sulphonamide groups nevertheless show that both ionisation methods can erroneously indicate a sample composition which does not correspond to reality [39].

Thus the interpretation of ESI-MS spectra of POPAM dendrimers recorded on an FT-ICR (Fourier-Transform Ion Cyclotron Resonance) mass spectrometer leads to drastic overestimation of defects in the sample molecules (Fig. 7.4). The

MALDI-TOF spectra of the sulphonamide dendrimers erroneously suggest formation of defect structures during synthesis; in fact, these are only generated during the ionisation process, primarily due to reaction of the peripheral sulphonamide groups with the acid matrix. An apparently unsatisfactory result should therefore always be checked by use of a second ionisation method or different MALDI matrices should be used to exclude misinterpretation of the spectra and to gain certainty about the true composition of the sample.

7.5
X-ray crystal structure analysis

The three-dimensional structure of dendrimers cannot be deduced directly from their two-dimensional structural formulae. However, single-crystal X-ray structural analyses do provide an insight into the three-dimensional molecular structure [40]. No other method permits the conformation of the structural components of the individual dendrimer branches to be determined so precisely – in the crystalline state. Frequently, however, the flexibility of the dendritic branches prevents the formation of a regular crystal lattice, thus precluding X-ray structure analysis. One way of significantly limiting the flexibility of the dendritic branches is to introduce terminal groups able to form hydrogen bonds. Thus, after introduction of urea-containing terminal groups, a first-generation POPAM dendrimer could be crystallised and subjected to X-ray crystallography [41].

Another problem is that dendrimers often crystallise as solvates or clathrates, that is with inclusion of solvent molecules. Such crystals are often rather unstable and decompose with release of the solvent. As a consequence of these properties of dendrimers, suitable single crystals have so far only been grown for just a few dendrimers up to the second generation at most.

Therefore, single-crystal X-ray structural analysis has hitherto been of limited use for structure elucidation of dendrimers. Most of the examples found in the literature [42a] are actually single-crystal X-ray structural analyses of rigid, dimensionally stable (shape resistent) polyphenylene dendrimers [42b] or of metallodendrimers [43] or polysilane dendrimers.

7.6
Small-angle scattering

Recent years have seen a growing focus of dendrimer research on the characterisation of material properties of dendrimers and their use in technical and medical applications. Many of the possible applications concern the use of dendrimers in solution. In the course of this development, Small-Angle Neutron Scattering [44] (SANS) and Small-Angle X-ray Scattering (SAXS) [45] have become increasingly important in dendrimer research. These scattering techniques permit non-destructive investigation of the structure, size, and shape of molecules in the size range of *ca.* 1 nm to 1 μm, not only in the solid state but also in solution. The following Sections offer a brief insight into the principle and performance of the SANS and SAXS techniques.

7.6.1
Principle of small-angle scattering

A conventional X-ray tube or ideally a synchrotron serves as radiation source in the SAXS technique. The neutrons required for SANS experiments are generated in a spallation source or a reactor. The principle of small-angle scattering is based on the interaction between an incident X-ray or neutron beam of wavelength λ and intensity I_0 with the scattering sites of the dissolved sample molecules. In contrast to X-rays, which are scattered by the electrons in the atomic shells, neutrons are scattered by the atomic nuclei. Since a dendrimer molecule is an ensemble of different scattering sites, the intensity of scattering is determined by the interference of the secondary radiation emanating from the various scattering sites of the molecule. If the intensity of the secondary radiation is measured with the aid of a position-dependent detector as a function of the angle of scattering it becomes possible to calculate structural information.

7.6.2
Capability of small-angle scattering

Neither SANS nor SAXS experiments afford atomic resolution of structure, as is obtained by single-crystal diffraction methods. However, the scattering data can provide information about the average three-dimensional structure and the average size of all the dissolved molecules present in the sample and their intermolecular interactions. Along with the radius of gyration R_g, which serves as a measure for the average size of the particles, other important parameters such as the average particle volume V_P, and the average molecular weight M_w can be calculated from the scattering intensity measured as a function of the scattering angle. Such statistical information about the bulk substance is often more important than any picture of a small detail provided by direct imaging methods such as electron microscopy (see Section 7.8).

Nevertheless, small-angle scattering techniques are not very suitable for the study of lower dendrimer generations because such dendrimers yield only weak scattering signals and a high uncertainty factor therefore exists. SANS and SAXS experiments are not only costly, but also make substantial demands on time and require relatively large amounts of substances (several 100 mg). These methods are therefore out of the question for routine use, such as for the determination of the molar mass of dendrimers.

One particular strength of SANS over SAXS is the differing scattering behaviour of isotopes towards neutrons, which is particularly pronounced for the hydrogen isotopes H and D. The true value of the SANS technique for structure analysis of dendrimers is due to this phenomenon. Thus, by deuterium label-

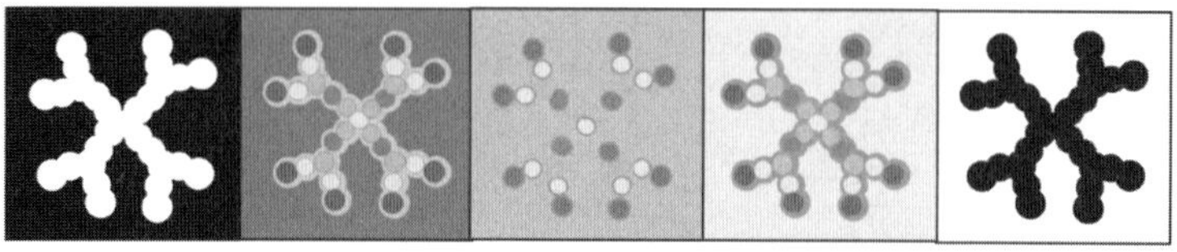

Fig. 7.5 Variation of contrast (schematic) by alteration of the isotopic composition of the solvent (*e.g.* H_2O/D_2O mixture; according to Ballauff *et al.*)

ling of defined components of the molecule (*e.g.* of the peripheral groups) or alteration of the isotopic composition of the solvent (*e.g.* H_2O/D_2O), the scattering contribution of a molecular component can be eliminated or strengthened at will, *i.e.* defined contrast settings are possible (Fig. 7.5).

Variation of contrast [46] is an important experimental technique in neutron small-angle scattering. Above and beyond size determination, it affords detailed insights into the internal structure of dissolved dendrimers and even permits the location of selected components of the molecule which have been previously labelled with deuterium.

7.6.3 Structural analysis of dissolved dendrimers with SANS and SAXS

The principal focus of interest has hitherto been on the group of flexible dendrimers, which embraces the major part of the dendrimers reported in the literature, also including the commercially available POPAM and PAMAM dendrimers having an aliphatic scaffold. In recent years valuable knowledge about the density profile, the three-dimensional structure, and the interactions of dendrimers in solution has been obtained with the aid of small-angle scattering techniques, as is apparent from the following overview.

7.6.3.1 Radial segment density distribution of flexible dendrimers

Controversial discussions have surrounded the three-dimensional structure of flexible dendrimers in solution during the past two decades. Now, on the basis of numerous SANS experiments using the contrast variation technique, the idea that isolated flexible dendrimers in good solvents do not take on the originally predicted "dense shell structure" but instead assume a "dense core structure" [47] appears to be gaining general acceptance. This means that in such dendrimers the segment density reaches a maximum at the centre of the molecule and decreases towards the periphery (cf. Fig. 7.6).

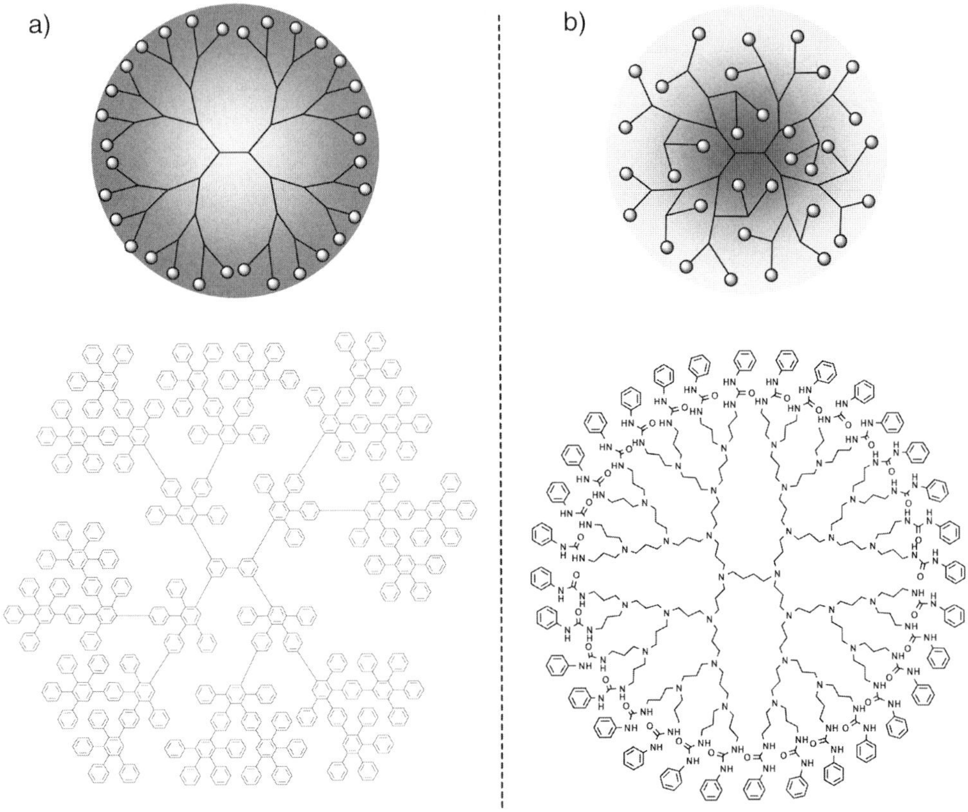

Fig. 7.6 Location of the end groups in a) rigid, dimensionally stable, and b) flexible dendrimers, investigated for the example of polyphenylene dendrimers and POPAM dendrimers with peripheral phenylurea groups (in *N*,*N*-dimethylacetamide); the grey scales also indicates the density profile (dark: high density; light: less dense arrangement of atoms in the molecule). Above: schematic; below: appropriate type of dendrimer

7.6.3.2 **Distribution of end groups**

The location of the end groups is also of importance with regard to possible applications because the rheological properties and surface activity of dendrimers in solution largely depend upon whether the end groups are distributed over the dendrimer scaffold or located in the periphery of the molecule. With the aid of the contrast variation technique it proved possible to locate the end groups within a fourth-generation flexible dendrimer scaffold. In order to distinguish the end groups from the rest of the dendrimer scaffold in SANS experiments, a high contrast was set between the end groups and the internal dendrimer scaffold by selective deuteration of the end groups. This not only permitted determination of the distribution of the end groups within the molecular scaffold but also afforded unequivocal experimental evidence of the theoretically predicted partial back-folding of terminal dendrimer segments in flexible dendrimers [48].

In contrast to the flexible dendrimers, those of the forth generation with a rigid molecular scaffold expectedly exhibit a dense shell structure [49], in which all the end groups are located in the periphery (Fig. 7.6). This could be demonstrated in contrast variation experiments together with solid state NMR spectroscopic studies on polyphenylene dendrimers [50].

The dendrimer type with a stilbene scaffold and long-chain end groups mentioned in Section 4.1.5.3 was shown to have a disc-like flattened molecular shape in solution by SAXS and SANS studies performed by Ballauff *et al.* [51].

NMR-spectroscopic investigations of dendrimers can provide comparable information. Thus Wooley *et al.* used REDOR (Rational Echo Double-Resonance) solid state NMR experiments for the study of the density distribution and backfolding of terminal groups in fifth-generation poly(benzyl ether) dendrimers [52].

The possibility of modifying the density profile of dendrimers in solution and of inducing conformational changes has particular implications for the use of dendrimers as "drug-carriers" in medicine. This could hold promise of the development of effective mechanisms for the controlled release of active ingredients. The discovery of possible control mechanisms is therefore one of the central problems which dendrimer researchers hope to solve with the aid of theoretical calculations, SANS and SAXS experiments, and multidimensional NMR techniques. Targeted modification of the chemical nature and size and/or shape of the end groups during synthesis and the behaviour of the different molecules in solution as a function of solvent quality and pH value are attracting special attention in this context.

7.6.3.3 Intermolecular interactions of flexible dendrimers in solution

Analysis of the effective interactions between flexible dendrimers in solution and their changes with increasing concentration in solution have been the subject of simulations, theoretical analyses, and individual SANS studies [53]. Ballauff, Vögtle *et al.* succeeded in determining the effective interaction between dissolved particles directly from experimental data [54]. Evaluation of the scattering data and subsequent comparison with theoretical models revealed that the fourth-generation flexible dendrimers investigated behave like soft particles in solution, in keeping with their fluctuating structure whose interaction can be described by a Gaussian potential up to the overlap concentration (concentration from the start of contact of two particles in solution).

The information gained by SANS and SAXS – frequently also in combination with simulations and theoretical analyses – can contribute to a more fundamental understanding of the three-dimensional structure, dynamics, and interactions of flexible dendrimers in solution and affords important information about the relation between size, shape, and internal composition of such dendrimers. The knowledge acquired is also valuable for the postulation of structure-reactivity relations. Such knowledge can help in better predicting the behaviour of flexible dendrimers in solution and their suitability for certain applications, and indicate starting points for the development of possible control mechanisms.

7.7
Scanning probe microscopy

On account of their controllable size, geometry, and functionality, dendrimers command interest for surface modification and for enlargement of (active) surfaces.

However, targeted control of surface structure requires an understanding of dendrimer/dendrimer interactions and dendrimer/surface interactions as well as a knowledge of the structure of the dendrimers in the solid state. Scanning probe microscopy methods (SPM) [55], such as scanning tunnelling microscopy (STM) [56] and atomic force microscopy (AFM) [57], can be used for visualisation of surface structures all the way to imaging of isolated individual molecules in diverse environments (air, gas, liquid) and for the study of their chemical and electronic properties. The SPM technique is based on scanning of molecules adsorbed onto the surface, with the resolutions attainable extending into the subnanometer range. Scanning probe microscopy has thus become indispensable for modern surface analysis and nanotechnology. In dendrimer research, STM and especially AFM have acquired increasing significance in the study of important structural parameters (*e.g.* size, conformation, rigidity) of dendrimers as well as the self-assembly of adsorbed dendrimers and dendrons [58]. The structures observed by AFM or STM in surface layers of dendritic molecules range from isolated single molecules *via* domain patterns and self-assembled monolayers to more complex architectures (*e.g.* nanofibres, three-dimensional clusters) [59].

The individual methods and their advantages and disadvantages are considered in greater detail in the following sections; a number of examples of STM and AFM studies on dendrimers are also presented.

7.7.1
STM and AFM

Dendrimers can be prepared for STM or AFM studies by a variety of methods. Controlled deposition on solid substrates (*e.g.* mica, glass, silicon, graphite, gold, among others) could be accomplished, *e.g.* by spin-coating [60] (uniform radial distribution of a dissolved substance on a rotating disc by centrifugal force) and by deposition from rapidly evaporating organic solvents [61].

In an STM experiment the dendrimer sample deposited on a conductive substrate (*e.g.* highly ordered pyrolytic graphite (HOPG)) is scanned line-by-line with a fine conductive microscopy tip. Depending upon the measuring mode, a piezoelectric scanner moves either the microscopy tip over the sample surface or the sample under the fixed tip. The microscopy tip approaches the sample

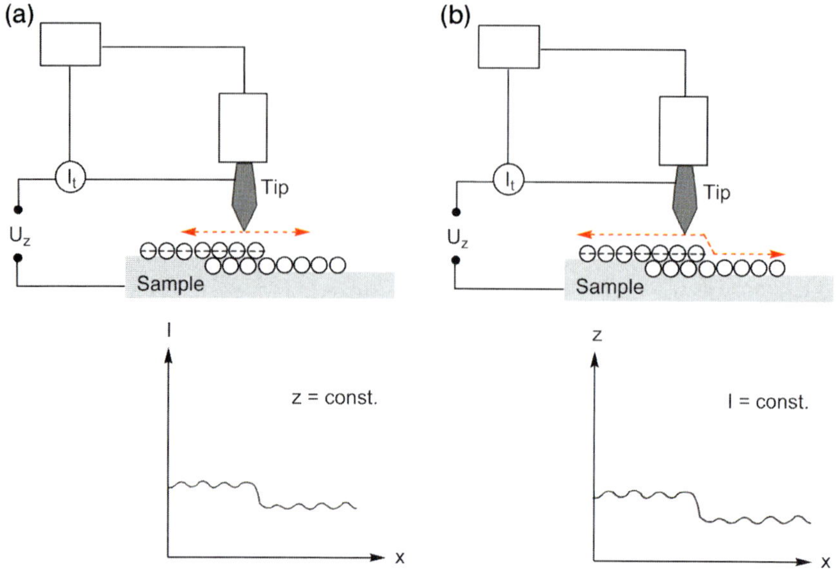

Fig. 7.7 Modes of operation of scanning tunnelling microscopy. a) At constant scanning distance (z); b) at constant tunnel current (I)

surface to be examined so closely that the electron clouds of their atoms "touch". If an electric potential is applied across the sample surface and the microscopy tip then a current will flow, according to the rules of quantum mechanics. The strength of this tunnel current depends upon the electronic structure of the sample surface and exponentially upon the distance between the tip and the sample. Measurement can be performed by the constant-height method (CHM) or the constant current method (CCM). In the CHM method the distance between the microscopy tip and the sample surface can be reconstructed on the basis of the strength of the tunnel current for each point (Fig. 7.7). The recorded values are shown in false colours in a computer-generated three-dimensional image of the surface. In the CCM procedure the three-dimensional image of the surface can be reconstructed from the position of the tip. The contrast in STM images reflects both the topography and the local electron density on a section of the sample surface.

Although the resolution of atomic force microscopy (AFM) is basically inferior to that of STM, the technique has the advantage that insulating materials can also be used as substrates. In AFM the forces acting between the tip and the sample surface are detected. The probe tip mounted on a flexible cantilever scans over the sample. AFM can be operated in contact mode, exploiting repulsive forces, as well as in non-contact mode, exploiting attractive forces. In the contact mode the probe tip is in direct contact with the sample surface (Fig. 7.8). Either the tip is passed over the sample surface at constant height (CHM,

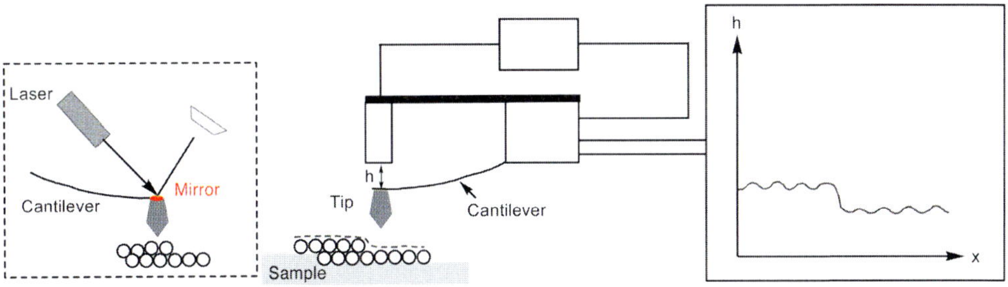

Fig. 7.8 Principle of atomic force microscopy (AFM)

see above) or the sample is scanned at constant force on the cantilever (CFM; constant force mode). In the CHM technique the three-dimensional image of the surface geometry is generated from the deflection of the cantilever detected by laser reflection.

In the CFM technique reconstruction is based on the movement of the piezoelectric element which controls the absolute distance between the tip and the sample. In the non-contact mode the cantilever is set in vibration slightly above its resonance frequency by another piezoelectric element and offered up to the sample without making direct contact. Attractive forces – above all van-der-Waals interactions – between tip and sample surfaces reduce the frequency of oscillation and the amplitude of the cantilever. This amplitude serves as a measure of the height of the object studied on generation of the topographic image and can likewise be detected by laser reflection. The tapping mode functions in a similar manner. Here an excitation frequency slightly below the resonance frequency of the cantilever generates a slightly greater amplitude with the result that the microscope tip sometimes touches the sample surface.

7.7.1.1 **AFM images of dendrimers**

AFM has so far been used to study, for example, self-assembling structures of liquid crystalline carbosilane dendrimers on silicon, glass, and mica surfaces [62], alkyl [63] and unsubstituted [64] polyphenylene dendrimers on graphite, and carboxylic acid-functionalised polyphenylene dendrimers on gold surfaces [65]. Systematic AFM investigations have also been carried out on adsorbed PAMAM dendrimers on mica, graphite, and gold surfaces to determine the molecular size. The diameters and heights of single molecules determined from AFM images have also been used to calculate the volumes of dendrimers of various generations [66]. The resulting volumes permitted estimation of the molar masses and the polydispersity of PAMAM dendrimers of various generations. In addition, the influence of the substrate and of the pH value of the solution on the size and shape of PAMAM dendrimers have also been discussed [60 b]. It became apparent during the determination of the structural parameters of various flexible dendrimers from AFM images that the nature of the substrate used may exert a decisive influence on the result. Thus it was observed that "soft" dendrimers (*e.g.* PAMAM dendrimers) undergo flattening on mica, gold, and platinum surfaces owing to their high flexibility and/or interactions with the substrate [66, 67]. This leads to underestimation of the height of the surface layers and to overestimation of the diameter of such dendrimers [68]. According to AFM studies by Müllen *et al.*, this problem does not occur with layers of rigid, dimensionally stable (shape resistent) polyphenylene dendrimers [64 a, 69].

A stable surface coating which is not destroyed by the microscope tip is a prerequisite for high resolution. Since "soft", flexible dendrimers are susceptible to deformation by the microscopy tip, the rigidity of the dendrimer also plays a role in the selection of the mode of operation of AFM. While theoretically offering less resolution than the contact mode, measurement in non-contact mode is far more gentle on the sample and thus suitable for dendrimers of various rigidities. The tapping mode generally offers an ideal compromise for gentle examination of flexible, soft dendrimers. It combines low applied forces on the surface of the sample with good resolution as well as average measuring speed and unites the advantages of the contact mode with those of the non-contact mode.

7.7.1.2 **STM images of dendrimers**

Because arene units give a high contrast in STM, dendrimers rich in aromatic groups such as Fréchet or polyphenylene dendrimeres are ideal for imaging by this technique. Phenanthroline- and hexaazatriphenylene-based heptanuclear Ru(II) dendrimers physisorbed on a graphite surface could be imaged at submolecular resolution by STM [70]. Other examples include STM studies performed on the self-assembly of thiophene dendrimers [71] and octyl-terminated Fréchet dendrons on graphite surfaces [72] as well as on metal ion-initiated self-assembly of PAMAM dendrimers with peripheral terpyridine units on a graphite surface [73].

7.8
Transmission electron microscopy

Transmission electron microscopy (TEM) is a universally applicable technique and is used in biology and medicine as well as in the materials sciences and dendrimer research. Modern TEM can generate images revealing distances on an atomic scale.

7.8.1
TEM

In transmission electron microscopy an electron beam is emitted from a cathode (*e.g.* tungsten wire) at high voltage. The wavelength of this electron beam depends upon the applied acceleration potential which accelerates the electrons in the direction of the anode. Working at high vacuum prevents absorption of the electron beam by gas particles. Powerful electromagnets act on the electron beam as lenses. On passage of electrons through the sample, which must be kept very thin, inelastic and elastic scattering occur at the sample atoms. The enlarged image of the sample, which is generated by interaction of the incident electron beam with sample atoms, is presented on a fluorescent screen and can be recorded on a photographic negative or a CCD camera.

7.8.1.1 TEM images of dendrimers

In principle, truly informative transmission electron micrographs of dendrimer molecules can only be obtained if the dendrimers remain stable on exposure to the high energy electron beam. However, the crucial problem encountered in electron-microscopic studies of dendrimers consists primarily in distinguishing isolated dendrimers from their surroundings. The lower the average atomic number of the atoms in the dendrimer scaffold, the weaker the contrast. That is why good TEM images are obtained mainly with electron-rich metallodendrimers [74]. TEM has also proved to be valuable in the study of the size and stability of dendritically stabilised palladium, gold, or silver nanoparticles [75] (see Fig. 7.9).

In the case of dendrimers containing no metal the scattering power of the sample, and hence the contrast, is normally insufficient for TEM images. As a rule, "negative staining" is therefore performed with such dendrimers, in that the density of the background is enhanced by addition of materials exhibiting pronounced electron-scattering properties (*e.g.* uranyl acetate, osmium tetrox-

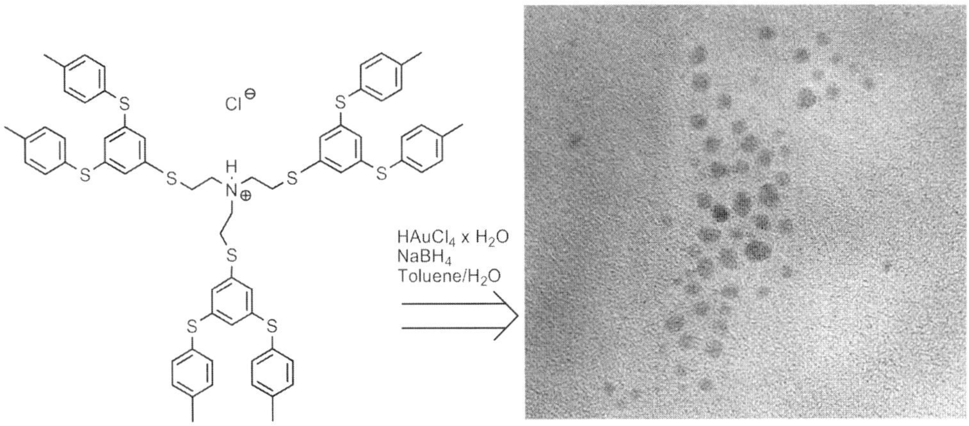

Fig. 7.9 Dendritic phase-transfer catalyst according to Vögtle *et al.* (left); transmission electron micrograph of the dendritically stabilised gold nanoparticle (black) after reduction of the gold salt solution in the organic phase

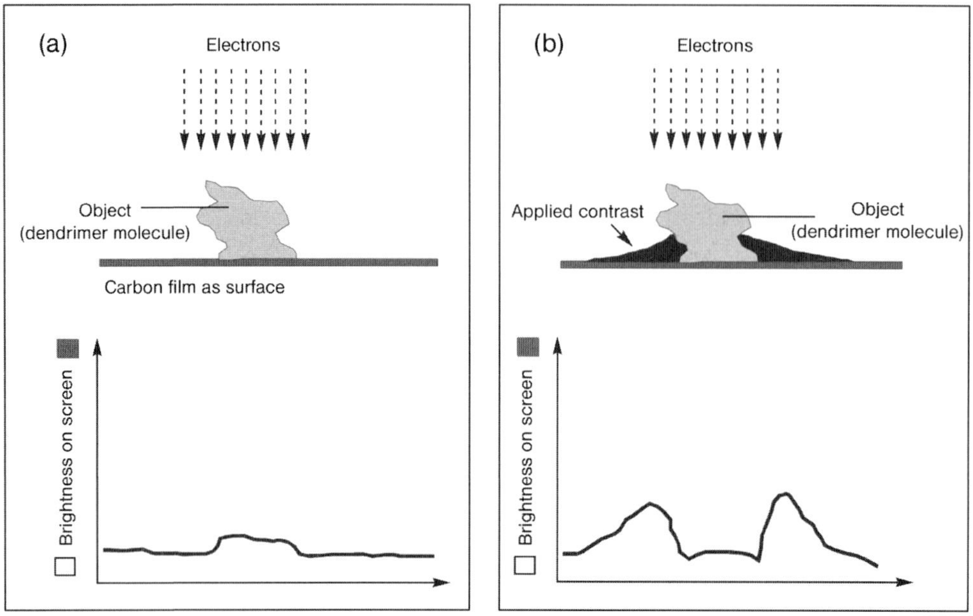

Fig. 7.10 Principle of negative staining of molecules on surfaces.
a) Without contrast; b) negative staining

ide), (Fig. 7.10). After appropriate "staining", single molecules of mainly higher-generation non-metallodendrimers could be imaged [76] and the shape and size of the molecules investigated. In general, TEM images correctly represent the two-dimensional form of the molecule but, in contrast to AFM images, afford little information about the height of the molecule. That is why the two microscopy techniques are frequently used in combination.

7.9
Chiroptical methods

Chiroptical methods used in dendrimer research exploit the optical activity as a characteristic property of chiral dendrimers for characterisation of their structures.

The term chiroptical basically refers to spectroscopic methods which afford values with opposite signs for the two enantiomers of a chiral compound [77]. Measurement of optical rotatory dispersion (ORD) and circular dichroism (CD) number among the most important chiroptical methods.

The following sections will first discuss the fundamentals of chiroptical methods and their application in order to subsequently summarise the chiroptical properties of dendritic structures as a set of general "rules" – insofar as this is at all possible.

7.9.1 Optical rotatory dispersion and circular dichroism

Optical rotatory dispersion (ORD) or circular birefringence is based on the differential refraction of left and right circularly polarised light by optically active media [78]. When linearly polarised light is incident upon an optically active medium, the two circularly polarised part-beams travel through the optically active medium at different velocities owing to their different refractive indices [79]. This leads to a phase shift of the two circularly polarised part-beams and thus to rotation of the plane of polarisation of the exiting light by an angle α. The extent of rotation is proportional to the difference of the refractive indices for the right and left circularly polarised light:

$$\alpha = (180 \times (\eta_l - \eta_r \times l)/\lambda_0$$

(α: angle of rotation in degrees; η_l, η_r refractive indices for left and right circularly polarised light; l: length of path through the medium; λ_0: wavelength *in vacuo*).

For the concentration-dependent specific rotations $[\alpha]$ measured in solution:

$$[\alpha]_\lambda^T = \alpha/(l \times c)$$

(α: angle of rotation in degrees; l: layer thickness in dm; c: concentration in g/mL)

Particularly in the case of comparative chiroptical studies on dendrimers belonging to a series it should be noted that large mass differences generally exist between the dendrimers and the specific rotations $[\alpha]$ change with the concentration of the solution examined. It is therefore often more expedient to consider molar rotations $[\Phi]$ instead of specific rotations $[\alpha]$.

$$[\Phi] = 0.01\,[\alpha] \times M$$

Measurement of the dependence of the angle of rotation α upon the wavelength outside the absorption region of the optically active substance under investigation yields simple ORD curves characterised by a monotonic increase or a monotonic decrease of the angle of rotation from long to short wavelengths of the measuring light. This behaviour is designated as "normal ORD".

However, if the dependence of the angle of rotation α upon the wavelength is measured in the region of the absorption band of the optically active substance under investigation, then superposition of an S-shaped component on the normal ORD curve is observed in this region. Circular dichroism effects are responsible for this anomalous ORD curve.

Unlike ORD, circular dichroism (CD) depends upon the differential absorption of the two circularly polarised part-beams of incident light by an optically active compound. The combined action of anomalous ORD and CD is summarised as the Cotton effect. Therefore Cotton effects can be determined both

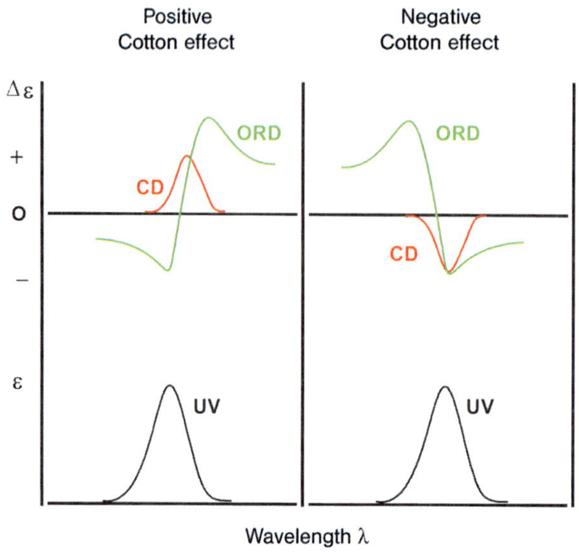

Fig. 7.11 Comparison of ORD (green) and CD bands (rot) of positive (upper left) and negative Cotton effects (upper right); the lower part shows the corresponding UV/Vis absorption spectrum (black). The wavelength range investigated normally covers 180–700 nm

by measurement of rotatory dispersion and by measurement of circular dichroism. An ORD curve is said to show a positive Cotton effect if the "peak" of the curve lies at higher wavelengths than the "trough" (Fig. 7.11).

Circular dichroism is measured by determining the difference $\Delta\varepsilon$ in absorption (ε) of left- and right-handed circularly polarized light and plotting it against the wavelength. At certain wavelengths the CD spectrum shows either a negative ($\Delta\varepsilon<0$) or a positive ($\Delta\varepsilon>0$) Cotton effect. However, circular dichroism is characterised not only by the sign, but also by the parameters $\Delta\varepsilon_{max}$ ($\Delta\varepsilon$ at the absorption maximum), λ_{max} (λ of the absorption maximum), and by the shape of the curve (*e.g.* fine structure) (Fig. 7.11). What proportion of the linearly polarised light is absorbed by the optically active substance, and to what extent, depends upon the molecular structure, that is upon the chirality of the stereocentres in the close proximity of the absorbing chromophore.

One major advantage of chiroptical methods is that they require less than mg amounts of substance which can be recovered after the measurement (non-destructive).

If two (or more adjacent) distinct chromophores exist in a chiral arrangement with respect to one another in a dendrimer molecule, the CD spectrum shows two intense Cotton effects of opposite signs which merge with each other [80]. This phenomenon, known as an exciton couplet, arises from the interaction of two electronic transition moments in a mutually chiral orientation, such as occurs, for example, in helical structures. In such cases non-destructive determination of the absolute configuration of the molecule (or supramolecule) even becomes possible with minimal amounts of sample (<1 mg) [11, 81].

7.9.2
Chiroptical studies on chiral dendritic structures

An understanding of the relation between the chirality of the dendritic building blocks and the chirality of the overall molecule is of crucial importance for the development of new dendritic materials whose properties and function depend on their chirality on the macroscopic level [82]. Most of the chiroptical studies on various types of chiral dendrimers were undertaken with the goal of elucidating this relation (see Section 4.2). Yet in spite of the numerous studies it remains difficult to predict chiroptical data because they depend upon a wide range of factors. As long as the analysis of chiroptical data fails to permit clear conclusions the anticipated utility of chiroptical studies cannot be fully exploited for structural characterisation of chiral dendrimers, in contrast to the situation with simple chiral molecules. Nevertheless, on the basis of studies undertaken so far the following general "rules" concerning the relation between the molecular chirality of dendritic building blocks and the macroscopic chirality of the dendrimers can be postulated. Information about the conformational order of dendrimers can be deduced from rule-compliant or anomalous behaviour:

- *The contribution of individual chiral building blocks to the overall rotation of the dendrimer remains almost constant for flexible dendritic structures regardless of the generation number.*

In the case of dendrimers and dendrons whose highly flexible structure can be described by a dynamic equilibrium of the conformers, the influence of dendritic branches does not generally give rise to any attenuation or enhancement of rotation.

Anomalous behaviour of the molar rotation depending upon the generation can basically be indicative of the existence of chiral substructures in the dendrimer branches. In many cases such anomalies merely occur as a result of constitutional differences between the chiral building blocks which are exposed to different local environments in the different parts of the dendrimer. CD-spectroscopic studies can often provide more detailed information about the underlying reasons.

- *The overall rotation of a dendrimer molecule is made up almost additively of the constant contributions of the individual chiral building blocks of the dendrimers.*

This indicates that the measured molar optical rotation is caused by the chiral building blocks and not by a stable conformational array of the dendrimer scaffold [83].

Deviations from this "additivity rule" occur if the contribution of individual chiral building blocks to the overall rotation changes due to generation-based constitutional differences between the chiral building blocks or the conformational equilibria of the free monomeric chiral building blocks undergo significant changes on incorporation into a dendrimer scaffold (see above rule) [84].

- *Steric packing effects can invalidate Rules 1 and 2.*

Steric hindrance of the dendritic branches can force the chiral building blocks in the dendrimer into a conformation which the corresponding free, monomeric building blocks would not normally adopt. In dendrimers with a chiral core, for example, this leads to a drop in molar rotation with increasing dendrimer generation [85] due to a weakening of the optical activity resulting from attachment of inherently achiral dendrimer branches. Any significant influence on the conformational equilibrium of the chiral monomeric building blocks should be apparent not only in the optical rotation values, but also from changes in the CD spectra.

- *Intramolecular interactions or interactions with other molecules can invalidate Rules 1 and 2.*

Specific attractive interactions between the dendrimer building blocks (*e.g.* hydrogen bonds between amino acid building blocks) or complexation of metal ions [86] by suitable functional groups in the dendrimer scaffold can favour the formation of chiral substructures and effect distortion or folding of the overall structure. Interactions with solvent molecules can also have a pronounced effect on the overall structure.

As shown by the above compilation, comparison of the molar rotations of different dendrimer generations (divided by the number of chiral building blocks in the dendrimer) with those of the corresponding, chiral monomers or appropriate model compounds affords information about the conformational order of the dendrimers.

Measurement of circular dichroism can even permit elucidation of relatively small structural changes. CD spectroscopy is also suitable for the solution of specific application-relevant questions. Studies of the sensor properties of chiral dendrimers make use of the fact that complexation of chiral guest molecules induces changes in the CD bands of the host dendrimers. Thus guest-selective chiroptical effects observed in titration experiments with enantiomeric guest molecules give an indication of the potential of the chiral dendrimer to act as an enantioselective sensor [87].

7.10 Summary

The following summary provides an overview of the properties and specific questions to be addressed in connection with the characterisation of dendrimers, and lists the appropriate analytical techniques.

1 Chemical composition
- Elemental analysis (C, H, N analysis)
- Mass spectrometry: ESI-MS, FAB-MS, elucidation on the basis of mass-spectrometric fragmentation patterns

2 Molar mass
- Mass spectrometry: MALDI-MS (also high generations), ESI-MS, and FAB-MS (low generations); relative methods
- Small-angle scattering: SANS, SAXS, SALLS (small-angle laser light scattering); absolute methods yielding the mean molar mass; weak scattering signals give rise to a high uncertainty factor for low generations
- Vapour-pressure osmometry
- SEC; relative method; non-availability of suitable size standards can frequently lead to inaccuracies

3 Dimensions
In the solid state:
- Scanning probe microscopy: AFM, STM (arene-rich dendrimers)
- Electron microscopy: TEM; metal-rich dendrimers, dendritically stabilised nanoparticles; contrasting otherwise necessary

In solution:
- Diffusion NMR spectroscopy
- SEC
- Small-angle scattering: SAXS, SANS (higher generations)
- Computer simulations

4 Structural defects
- Mass spectrometry: ESI-MS, MALDI-MS; dendrimers with low shell numbers, dimers, intramolecular ring closures, missing end groups, missing dendritic branches
- SEC (usually in combination with MALDI-MS); dendrimers with lower shell numbers, dimers; not suitable for detection of smaller defects such as missing end groups or intramolecular ring closures, since such defects have only a slight effect on the size of the molecule
- Small-angle scattering: SAXS, SANS; only suitable for higher dendrimer generations; SANS is particularly time-consuming and costly, and requires large amounts of sample compared to mass-spectrometric methods

5 Internal groups and end groups
- NMR spectroscopy: IR, ^{13}C; in the presence of heteroatoms, *e.g.* ^{15}N-, ^{19}F-, ^{29}Si-, ^{31}P-NMR
- IR spectroscopy
- Titration

6 Structure

- NMR spectroscopy: ^{1}H-, ^{13}C-NMR (ID-, 2D-, 3D-, 4D-NMR); in the presence of heteroatoms, *e.g.* ^{15}N, ^{19}F, ^{29}Si, ^{31}P
- Small-angle scattering: SAXS, SANS (higher generations), three-dimensional structure in solution

Bibliography and Notes for Chapter 7 "Characterization and analysis"

Review articles are indicated by the words "Review(s)" or "Book/Books" in bold-faced type.

1 L. J. Hobson, W. J. Feast, *Polymer* **1999**, *40*, 1279–1297.

2 A. D. Meltzer, D. A. Tirell, A. A. Jones, P. T. Inglefield, *Macromolecules* **1992**, *25*, 4541.

3 H. M. Brothers, L. T. Piehler, D. A. Tomalia, *J. Chromatogr. A* **1998**, *814*, 233–246; b) A. Ebber, M. Vaher, J. Peterson, M. Lopp, *J. Chromatogr. A* **2002**, *949*, 351–358.

4 ***Book***: C. Schalley (Eds.), *Analytical Methods in Supramolecular Chemistry*, Wiley-VCH, Weinheim **2006**.

5 ***Textbooks***: a) L. Kraus, A. Koch, S. Hoffstetter-Kuhn, *Dünnschichtchromatographie*, Springer-Verlag, Berlin, Heidelberg **1996**; b) B. Fried, J. Sherma (Eds.), *Practical Thin-Layer Chromatography – A Multidisciplinary Approach;* CRC Press, Boca Raton, New York, London, Tokyo **1996**.

6 ***Textbooks***: a) S. Lindsay, *Einführung in die HPLC*, Springer-Verlag, Berlin–Heidelberg **1996**; b) G. J. Eppert, *Flüssigchromatographie, HPLC-Theorie und Praxis*, Springer-Verlag, Berlin, Heidelberg **1997**; c) V. Meyer, *Praxis der Hochleistungs-Flüssigchromatographie*, Sauerländer, Frankfurt **1999**.

7 P. E. Froehling, H. A. Linssen, *Macromol. Chem. Phys.* **1998**, *199*, 1691–1695.

8 M. T. Islam, X. Shi, L. Balogh, J. R. Baker, Jr., *Anal. Chem.* **2005**, *77*, 2063–2070.

9 a) B. Ramagnoli, I. van Baal, D. W. Price, L. M. Harwood, W. Hayes, *Eur. J. Org. Chem.* **2004**, 4148–4157.

10 O. Trapp, G. Trapp, J. Kong, U. Hahn, F. Vögtle, V. Schurig, *Chem. Eur. J.* **2002**, *8*, 3629–3634.

11 C. Reuter, G. Pawlitzki, U. Wörsdörfer, M. Plevoets, A. Mohry, T. Kubota, Y. Okamoto, F. Vögtle, *Eur. J. Org. Chem.* **2000**, 3059–3067.

12 J. Recker, W. M. Müller, U. Müller, T. Kubota, Y. Okamoto, M. Nieger, F. Vögtle, *Chem. Eur. J.* **2002**, *8*, 4434–4442.

13 a) A. M. Striegel, R. D. Plattner, J. L. Willett, *Anal. Chem.* **1999**, *71*, 978–986; b) L. Bu, W. K. Ninidez, J. W. Mays, *Macromolecules* **2000**, *33*, 4445–4452.

14 M. T. Islam, X. Shi, L. Balogh, J. R. Baker, Jr., *Anal. Chem.* **2005**, *77*, 2063–2070.

15 a) D. A. Tomalia, A. M. Naylor, W. A. Goddard, III; *Angew. Chem* **1990**, *102*, 119–157; *Angew. Chem. Int. Ed.* **1990**, *29*, 138–175; b) C. Xia, X. Fan, J. Locklin, R. C. Advincula, A. Gies, W. Nonigez, *J. Am. Chem. Soc.* **2004**, *126*, 8735–8743.

16 A. Sharma, D. K. Mohanty, A. Desai, R. Ali, *Electrophoresis* **2003**, *24*, 2733–2739.

17 a) ***Textbooks***: H. Günther, *NMR-Spektroskopie*, Thieme, Stuttgart **1983**; E. Breitmaier, in *Untersuchungsmethoden in der Chemie*, H. Naumer, W. Heller (Eds.), Thieme, Stuttgart **1997**; H. Friebolin, *Ein- und zweidimensionale NMR-Spektroskopie*, Wiley-VCH, Weinheim **2006**; b) M. Pons (Ed.), *NMR in Supramolecular Chemistry*, Kluwer, Dordrecht **1999**.

18 ***Book***: D. H. Williams, I. Fleming, *Strukturaufklärung in der organischen Chemie*, Thieme, Stuttgart **1991**.

19 a) K. L. Wooley, C. A. Klug, K. Tasaki, J. Schaefer, *J. Am. Chem. Soc.* **1997**, *119*, 53–58; b) C. Klug, T. Kowalewski, J. Schaefer, T. Straw, K. Tasaki, K. Wooley, *Polym. Mat. Sci. Eng.* **1997**, *77*, 99–100; c) D. I. Malyararenko, R. L. Vold, G. L. Hoatsen, *Macromolecules* **2000**, *33*, 1268–1279; d) M. Wind, K. Saalwächter, U.-M. Wiesler, K. Müllen, H. W. Spiess,

Adv. Mater. **2001**, *13*, 752–756; e) M. Wind, K. Saalwächter, U.-M. Wiesler, K. Müllen, H.W. Spiess, *Macromolecules* **2002**, *35*, 10071–10086.

20 G. Greiveldinger, D. Seebach, *Helv. Chim. Acta* **1998**, *81*, 1003–1022.

21 a) S.S. Wijmenga, K. Hallenga, C.W. Hilberts, *J. Magn. Reson.* **1989**, *84*, 634–642; b) T.D. Spitzer, G.E. Martin, R.C. Crouch, J.P. Shockcor, B.T. Farmer, II, *J. Magn. Reson.* **1992**, *99*, 433–438; c) M. Chai, Y. Niu, W.J. Youngs, P.L. Rinaldi, *Macromolecules* **2000**, *33*, 5395–5398.

22 M. Chai, Y. Niu, W.J. Youngs, P.L. Rinaldi, *J. Am. Chem. Soc.* **2001**, *123*, 4670–4678.

23 D. Banerjee, M.A.C. Broeren, M.H.P. van Genderen, E.W. Meijer, P.L. Rinaldi, *Macromolecules* **2004**, *37*, 8313–8318.

24 ***Review***: Y. Cohen, L. Avram, L. Frish, *Angew. Chem.* **2005**, *112*, 524–560; *Angew. Chem. Int. Ed.* **2005**, *44*, 20–554.

25 a) H. Ihre, A. Hult, E. Söderlind, *J. Am. Chem. Soc.* **1996**, *118*, 6388–6395; b) C.B. Gorman, J.C. Smith, M.W. Hager, B.L. Parkhurst, H. Sierzputowska-Gracz, C.A. Haney, *J. Am. Chem. Soc.* **1999**, *121*, 9958–9966; c) J.M. Riley, S. Alkan, A. Chen, M. Shapiro, W.A. Khan, W.R. Murphy, Jr., J.E. Hanson, *Macromolecules* **2001**, *34*, 797–1809, d) S. Hecht, N. Vladimirov, J.M.J. Fréchet, *J. Am. Chem. Soc.* **2001**, *123*, 18–25; e) A.I. Sagidullin, A.M. Muzafarov, M.A. Krykin, A.N. Ozerin, V.D. Skirda, G.M. Ignateva, *Macromolecules* **2002**, *35*, 9472–9479; f) S.W. Jeong, D.F. O'Brien, G. Orädd, G. Lindblom, *Langmuir* **2002**, *18*, 1073–1076.

26 a) G.R. Newkome, J.K. Young, G.R. Baker, R.L. Potter, L. Audoly, D. Cooper, C.D. Weis, K. Morris, C.S.J. Johnson, *Macromolecules* **1993**, 26, 2394–2396; b) J.K. Young, G.R. Baker, G.R. Newkome, K.F. Morris, C.S. Johnson Jr., *Macromolecules* **1994**, *27*, 3464–3471.

27 a) C.B. Gorman, J.C. Smith, M.W. Hager, B.L. Parkhurst, H. Sierzputowska-Gracz, C.A. Haney, *J. Am. Chem. Soc.* **1999**, *121*, 9958–9966; b) J.M. Riley, S. Alkan, A. Chen, M. Shapiro, W.A. Khan, W.R. Murphy, Jr., J.E. Hanson, *Macromolecules* **2001**, *34*, 1797–1809.

28 a) Y. Tomoyose, D.-L. Jiang, R.-H. Jin, T. Aida, T. Yamashita, K. Horie, *Macromolecules* **1996**, *29*, 5236–5238; b) D.-L. Jiang, T. Aida, *Nature* **1997**, *388*, 454–456; c) D.-L. Jiang, T. Aida, *J. Am. Chem. Soc.* **1998**, *120*, 10895–10901; d) S. Hecht, J.M.J. Fréchet, *J. Am. Chem. Soc.* **1999**, *121*, 4084–4085.

29 Examples of studies of the molecular dynamics of dissolved dendrimers with the aid of NMR relaxation measurements: POPAM dendrimers: a) J.F.G.A. Jansen, E.M.M. Brabander-van den Berg, E.W. Meijer, *Science* **1994**, 265, 1226–1229; b) M. Chai, Y. Niu, W.J. Youngs, P.L. Rinaldi, *J. Am. Chem. Soc.* **2001**, *123*, 4670–5678; c) Y. Pan, W.T. Ford, *Macromolecules* **2000**, *33*, 3731–3738; PAMAM dendrimers: A.D. Meltzer, D.A. Tirrell, A.A. Jones, P.T. Inglefield, D.M. Hedstrand, D.A. Tomalia, *Macromolecules* **1992**, *25*, 4541–4548; polyether dendrimers: a) C.B. Gorman, M.W. Hager, B.L. Parkhurst, J.C. Smith, *Macromolecules* **1998**, *31*, 815–822; d) D.L. Jiang, T. Aida, *J. Am. Chem. Soc.* **1998**, *120*, 10895–10901; c) S. Hecht, J.M. Fréchet, *J. Am. Chem. Soc.* **1999**, *121*, 4084–4085; d) H.J. van Manen, R.H. Fokkens, N.M.M. Nibbering, F.C.J.M. van Manen, R.H. Fokkens, F.C.J.M. van Veggel, D.N. Reinhoudt, *J. Org. Chem.* **2001**, *66*, 4643–4650; carbosilane dendrimers: K.T. Welch, S. Arevalo, J.F.C. Turner, R. Gómez, *Chem. Eur. J.* **2005**, *11*, 1217–1227.

30 E. Buhleier, W. Wehner, F. Vögtle, *Synthesis* **1978**, 155–158.

31 a) C.G. Juo, L.L. Shiu, C.K.F. Shen, T.Y. Luh, G.R. Her, *Rapid Commun. Mass Spectrom.* **1995**, *9*, 604–609; b) G. Coullerez, H.J. Mathieu, S. Lundmark, M. Malkoch, H. Magnusson, A. Hult, *Surf. Interface, Anal.* **2003**, *35*, 682–692.

32 a) K.J. Wu, R.W. Odom, *Analytical Chemistry News & Features* **1998**, 456–461; b) L.J. Hobson, W.J. Feast, *Polymer* **1999**, *40*, 1279–1297; c) A.M. Striegel, R.D. Plattner, J.L. Willet, *Anal. Chem.* **1999**, *71*, 978–986; d) I.A. Mowat, R.J. Donovan, M. Bruce, W.J. Feast, N.M. Stainton, *Eur. Mass Spectrom.*

1998, *4*, 451–458; ***Textbook***: J. H. Gross, *Mass Spectrometry*, Springer, Berlin **2004**.

33 ***Review***: H. Frey, K. Lorenz, C. Lach, *Chem. in unserer Zeit* **1996**, *2*, 75–85; ***Textbook***: J. H. Gross, *Mass Spectrometry*, Springer-Verlag, Berlin **2004**.

34 Examples of MALDI-TOF-MS studies on dendrimers: a) H. S. Sahota, P. M. Lloyd, S. G. Yeates, J. Derrick, P. C. Taylor, D. M. Haddleton, *J. Chem. Soc., Chem. Commun.* **1994**, 2445–2446; b) T. Kawaguchi, K. Walker, C. L. Wilkins, J. S. Moore, *J. Am. Chem. Soc.* **1995**, *117*, 2159–2165; c) D. M. Haddleton, H. S. Sahota, P. Taylor, S. G. Yeates; *J. Chem. Soc., Perkin Trans. 1* **1996**, 649–656; d) B. L. Schwartz, A. L. Rockwood, R. D. Smith, D. A. Tomalia, R. Spindler, *Rap. Commun. Mass Spectrom.* **1995**, *9*, 1552–1555; e) J. W. Leon, M. Kawa, J. M. J. Fréchet, *J. Am. Chem. Soc.* **1996**, *118*, 8847–8859; f) G. Chessa, A. Scrivanti, R. Seraglia, P. Traldi, *Rapid Commun. Mass Spectrom.* **1998**, *12*, 1533–1537; g) P. B. Rheiner, D. Seebach, *Chem. Eur. J.* **1999**, *5*, 3221–3236; h) L. Bu, W. K. Nonidez, J. W. Mays, *Macromolecules* **2000**, *33*, 4445–4452; i) L. Zhou, D. H. Russell, M. Zhao, R. M. Crooks, *Macromolecules* **2001**, *34*, 3567–3573; j) H. Chen, M. He, X. Cao, X. Zhou, J. Pei, *Rap. Commun. Mass Spectrom.* **2004**, *18*, 367–370; examples of ESI-MS studies on dendrimers: a) G. J. Kallos, D. A. Tomalia, D. M. Hedstrand, S. Lewis, J. Zhou, *Rap. Commun. Mass Spectrom.* **1991**, *5*, 383–386; b) B. L. Schwartz, A. L. Rockwood, R. D. Smith, D. A. Tomalia, R. Spindler, *Rapid Commun. Mass Spectrom.* **1995**, *9*, 1552–1555; c) L. P. Tolic, G. A. Anderson, R. D. Smith, H. M. Brothers II, R. Spindler, D. A. Tomalia, *Int. J. Mass Spectrom.* **1997**, *165/166*, 405–418; d) U. Puapaiboon, R. T. Taylor, *Rap. Commun. Mass Spectrom.* **1999**, *13*, 508–515; e) J. W. Weener, J. L. J. van Dongen, E. W. Meijer, *J. Am. Chem. Soc.* **1999**, *121*, 10346–10355; f) A. M. Striegel, R. D. Plattner, J. L. Willett, *Anal. Chem.* **1999**, *71*, 978–986; g) S. Watanabe, M. Sato, S. Sakamoto, K. Yamaguchi, M. Iwamura, *J. Am. Chem. Soc.* **2000**, *122*, 12588–12589; h) Y. Rio, G. Accorsi, H. Nierengarten, J.-L. Rehspringer, B. Hönerlage, G. Kopitkovas, A. Chugreev, A. Van Dorsselaer, N. Armaroli, J.-F. Nierengarten, *New J. Chem.* **2002**, *26*, 1146–1154; i) D. Luostarinen, T. Partanen, C. A. Schalley, K. Rissanen, *Synthesis* **2004**, 255–262.

35 T. Felder, C. A. Schalley, H. Fakhrnabavi, O. Lukin, *Chem. Eur. J.* **2005**, *11*, 1–13.

36 J. C. Hummelen, J. L. J. van Dongen, E. W. Meijer, *Chem. Eur. J.* **1997**, *3*, 1489–1493.

37 *Review*: A. W. Bosman, H. M. Janssen, E. W. Meijer, *Chem. Rev.* **1999**, *99*, 1665–1688.

38 a) K. J. Wu, R. W. Odom, *Anal. Chem.* **1998**, *69*, 456–461; b) L. Bu, W. K. Nonidez, J. W. Mays, N. Beck, *Macromolecules* **2000**, *33*, 4445–4452.

39 B. Baytekin, N. Werner, F. Luppertz, M. Engeser, J. Brüggemann, S. Bitter, R. Henkel, T. Felder, C. A. Schalley, *Int. J. Mass Spectrom.* **2006**, *249*, 138–148.

40 P. Coppens, *Angew. Chem.* **1977**, *89*, 33–42; *Angew. Chem. Int. Ed.* **1977**, 16, 32–40; ***Textbooks***: a) J. Stähle in "*Untersuchungsmethoden in der Chemie; Einführung in die moderne Analytik*", Eds. H. Naumer, W. Heller, Thieme, Stuttgart **1997**; b) W. Massa, *Kristallstrukturbestimmung*, Teubner, Stuttgart **1994**.

41 W. Bosmann, M. J. Bruining, H. Kooijman, A. L. Speck, R. A. J. Janssen, E. W. Meijer, *J. Am. Chem. Soc.* **1998**, 8547–8548.

42 a) *X-ray crystal structures of 2nd-generation dendrimers:* A. Rajca, S. Janicki, *J. Org. Chem.* **1994**, *59*, 7099–7107; D. Seyferth, D. Y. San, *Organometallics* **1994**, *13*, 2682–2690; A. Sekiguchi, M. Nanjo, C. Kabuto, H. Sakurai, *J. Am. Chem. Soc.* **1995**, *117*, 4195–4196; B. Karakaya, W. Claussen, K. Gessler, W. Saenger, A.-D. Schlüter, *J. Am. Chem. Soc.* **1997**, *119*, 3296–3301; M. Brewis, G. J. Clarkson, V. Goddard, M. Helliwell, A. M. Holder, N. B. McKeown, *Angew. Chem.* **1998**, *110*, 1185–1187; *Angew. Chem. Int. Ed* **1998**, *38*, 1092–1094. *X-ray crystal structures of 1st-generation dendrimers:* H.-B. Mekelburger, K. Rissanen, F. Vögtle, *Chem. Ber.* **1993**, *126*, 1161–1169; J. B. Lambert, J. L.

Pflug, J.M. Denari, *Organometallics* **1996**, *15*, 615–625; J.W. Kriesel, S. König, M.A. Freitas, A.G. Marshall, J.A. Leary, T.D. Tilley, *J. Am. Chem. Soc.* **1998**, *120*, 12207–12215; C. Larré, D. Bressolles, C. Turrin, B. Donnadieu, A.-M. Caminade, J.-P. Majoral, *J. Am. Chem. Soc.* **1998**, *120*, 13070–13082; M. Nanjo, T. Sunaga, A. Sekiguchi, E. Horn, *Inorg. Chem. Commun.* **1999**, *2*, 203–206; R.A. Gossage, E. Muñoz-Martinez, H. Frey, A. Burgath, M. Lutz, A.L. Spek, G. van Koten, *Chem. Eur. J.* **1999**, *5*, 2191–2197; D. Ranganathan, S. Kurur, R. Gillardi, I.L. Karle, *Biopolymers* **2000**, *54*, 289–295; M. Brewis, G.J. Clarkson, M. Helliwell, A.M. Holder, N.B. McKeown, *Chem. Eur. J.* **2000**, *6*, 4630–4636; S. Harder, R. Meijboom, J.R. Moss, *J. Organomet. Chem.* **2002**, *689*, 1095–1101; K. Partner, M. Nieger, F. Vögtle, *Synlett* **2004**, 1167–1170; A.A. Williams, B.S. Day, B.L. Kite, M.K. McPherson, C. Slebodnick, J.R. Morris, R.D. Grandour, *Chem. Commun.* **2005**, 5053–5055; O. Lukin, V. Gramlich, R. Kandre, I. Zhun, T. Felder, C.A. Schalley, G. Dolgonos, *J. Am. Chem. Soc.* **2006**, *128*, 8964–8974; *X-ray crystal structures of zeroth-generation dendrimers:* J.L. Hoare, K. Lorenz, N.J. Hovestad, W.J.J. Smeets, A.L. Spek, A.J. Canty, H. Frey, G. van Koten, *Organometallics* **1997**, *16*, 4167–4173, A.W. Bosman, M.J. Bruining, H. Kooijman, A.L. Spek, R.A.J. Janssen, E.W. Meijer, *J. Am. Chem. Soc.* **1998**, *120*, 8547–8548; S. Leininger, P.J. Stang, *Organometallics* **1998**, *17*, 3981–3987; b) R.E. Bauer, V. Enkelmann, U.M. Wiesler, A.J. Berresheim, K. Müllen; *Chem. Eur. J.* **2002**, *8*, 3858–3864 and literature cited therein.

43 ***Review***: a) J.-P. Majoral, A.-M. Caminade, *Chem. Rev.* **1999**, *99*, 845–880, and literature cited therein; b) P.I. Coupar, P.A. Jaffrés, R.E. Morris, *J. Chem. Soc. Dalton Trans.* **1999**, 2183–2187, c) H. Schumann, B.C. Wassermann, M. Frackowiak, B. Omotowa, S. Schutte, J. Velder, S.H. Mühle, W. Krause, *J. Organomet. Chem.* **2000**, *609*, 189–195.

44 ***Book***: H.C. Benoit, J.S. Higgins, *Polymers and Neutron Scattering*, Clarendon Press, Oxford **1994**.

45 *Books*: a) G. Fournet, A. Guinier, *Small-Angle Scattering of X-Rays*, Wiley, New York **1955**; b) D.L. Svergun, L.A. Feigin, *Structure Analysis by Small-Angle X-Ray Scattering and Neutron Scattering*, Plenum Press, New York **1987**; ***Review***: c) B. Chu, B.S. Hsiao, *Chem. Rev.* **2001**, *101*, 1727–1762.

46 a) R.G. Kirste, H.B. Stuhrmann, Z. *Phys. Chem. NF* **1965**, *46*, 247–250; b) V. Luzatti, A. Tardieu, L. Mateu, H.B. Stuhrmann, *J. Mol. Biol.* **1976**, *101*, 115–127; c) L.A. Feigin, D.L. Svergun, *Structure Analysis by Small-Angle X-Ray Scattering and Neutron Scattering*, Plenum Press, New York **1987**; d) P. Hickl, M. Ballauff, A. Jada, *Macromolecules* **1996**, *29*, 4006–4014; e) P. Hickl, M. Ballauff, *Physica A* **1997**, 235, 238–247.

47 R.L. Lescanec, M. Muthukumar, *Macromolecules* **1990**, *23*, 2280–2288.

48 S. Rosenfeldt, N. Dingenouts, M. Ballauff, N. Werner, F. Vögtle, P. Lindner, *Macromolecules* **2002**, *35*, 8098–8105.

49 H.J. Hervet, P.G. de Gennes, *J. Phys. Lett. (Paris)* **1983**, *44*, L351–L360.

50 S. Rosenfeldt, N. Dingenouts, D. Pötschke, M. Ballauff, A.J. Berresheim, K. Müllen, P. Lindner, *Angew. Chem.* **2004**, *116*, 111–114; *Angew. Chem. Int. Ed.* **2004**, *43*, 109–112.

51 S. Rosenfeldt, E. Karpuk, M. Lehmann, H. Meier, P. Lindner, L. Harnau, M. Ballauff, *ChemPhysChem* **2006**, *7*, 2097–2104.

52 H.-M. Kao, A.D. Stefanescu, K.L. Wooley, J. Schäfer, *Macromolecules* **2000**, *33*, 6214–6216.

53 a) B. Ramzi, R. Scherrenberg, J. Brackmann, J. Joosten, K. Mortensen, *Macromolecules* **1998**, *31*, 1621–1626; b) A. Topp, B.J. Bauer, T.J. Prosa, R. Scherrenberg, E.J. Amis, *Macromolecules* **1999**, *32*, 8923–8931; c) C.N. Likos, S. Rosenfeldt, N. Dingenouts, M. Ballauff, P. Lindner, N. Werner, F. Vögtle, *J. Chem. Phys.* **2002**, *117*, 1869–1877.

54 C.N. Likos, S. Rosenfeldt, N. Dingenouts, M. Ballauff, P. Lindner, N. Werner, F. Vögtle, *J. Chem. Phys.* **2002**, *117*, 1869–1877.

55 P. Samori, *J. Mater. Chem.* **2004**, *14*, 1353–1366.

56 a) G. Binnig, H. Rohrer, C. Gerber, E. Weibel, *Phys. Rev. Lett.* **1983**, *50*, 120–123; ***Book***: b) *Scanning Tunneling Microscopy*, H.-J. Güntherodt, R. Wiesendanger (Eds.), Springer, Berlin **1994**.

57 a) E.T. Yu, *Chem. Rev.* **1997**, *4*, 1017–1044; b) S. Chiang, *Chem. Rev.* **1997**, *4*, 1015–1016; c) C.M. Lieber, Y. Kim, *Adv. Mater.* **1993**, *5*, 392–394; d) G. Binnig, C.F. Quate, C. Gerber, *Phys. Rev. Lett.* **1986**, *56*, 930–933.

58 ***Review***: J. Li, D.A. Tomalia in *"Dendrimers and Other Dendritic Polymers"* (Eds. J.M.J. Fréchet, D.A. Tomalia), Wiley-VCH, Weinheim **2001**.

59 a) D.J. Liu, S. De Feyter, P.C.M. Grim, T. Vosch, D. Grebel-Koehler, U.M. Wiesler, A.J. Berresheim, K. Müllen, F.C. De Schryver, *Langmuir* **2002**, *18*, 8223–8230; b) D. Liu, S. De Feyter, M. Cotlet, U.-M. Wiesler, T. Weil, A. Herrmann, K. Müllen, F.C. De Schryver, *Macromolecules* **2003**, *36*, 8489–8498.

60 a) W.T.S. Huck, F.C. J. M. van Veggel, S.S. Sheiko, M. Möller, D.N. Reinhoudt, *J. Phys. Org. Chem.* **1998**, *11*, 540–545; b) T.A. Betley, M.M. Banaszak Holl, B.G. Orr, D.R. Swanson, D.A. Tomalia, J.R. Baker, *Langmuir* **2001**, *17*, 2768–2773.

61 M. Sano, J. Okamura, A. Ikeda, S. Shinkai, *Langmuir* **2001**, *17*, 1807–1810.

62 S.A. Ponomarenko, N.I. Boiko, V.P. Shibaev, S.N. Magonov, *Langmuir* **2000**, *16*, 5487–5493. [63] S. Loi, U.-M. Wiesler, H.-J. Butt, K. Müllen, *Macromolecules* **2001**, *34*, 3661–3671.

64 D.J. Liu, H. Zhang, P.C.M. Grim, S.D. Feyter, U.M. Wiesler, A.J. Berresheim, K. Müllen, F.C. De Schryver, *Langmuir* **2002**, *18*, 2385–2391.

65 H. Zhang, P.C.M. Grim, D. Liu, T. Vosch, S. De Feyter, U.-M. Wiesler, A.J. Berresheim, K. Müllen, C. Van Haesendonck, N. Vandamme, F.C. De Schryver, *Langmuir* **2002**, *18*, 1801–1810.

66 J. Li, L.T. Piehler, D. Qin, J.R. Baker, D.A. Tomalia, *Langmuir* **2000**, *16*, 5613–5616.

67 a) A. Hierlemann, J.K. Campbell, L.A. Baker, R.M. Crooks, A.J. Ricco, *J. Am. Chem. Soc.* **1998**, *120*, 5323–5324; b) H. Tokuhisa, M. Zhao, L.A. Baker, V.T. Phan, D.L. Dermody, M.E. Garcia, R.F. Peez, R.M. Crooks, T.M. Mayer, *J. Am. Chem.. Soc.* **1998**, *120*, 4492–4501.

68 a) F. Morgenroth, C. Kübel, K. Müllen, *J. Mater. Chem.* **1997**, *7*, 1207–1211; b) A. Hierlemann, J.K. Campbell, L.A. Baker, R.M. Crooks, A.J. Ricco, *J. Am. Chem. Soc.* **1998**, *120*, 5323–5324; c) H. Zhang, P.C.M. Grim, P. Foubert, T. Vosch, P. Vanoppen, U.-M. Wiesler, A.J. Berresheim, K. Müllen, F.C. De Schryver, *Langmuir* **2000**, *16*, 9009–9014.

69 A.C. Grimsdale, K. Müllen, *Angew. Chem.* **2005**, *117*, 5732–5772; *Angew. Chem. Int. Ed.* **2005**, 35, 5592–5629, and literature cited therein.

70 L. Latterini, G. Pourtois, C. Moucheron, R. Lazzaroni, J.-L. Brédas, A. Kirsch-de Mesmaeker, F.C. De Schryver, *Chem. Eur. J.* **2000**, *6*, 1331–1336.

71 C. Xia, X. Fan, J. Locklin, R.C. Advincula, A. Gies, W. Nonidez, *J. Am. Chem. Soc.* **2004**, *126*, 8735–8743.

72 L. Merz, H.-J. Güntherodt, L.J. Scherer, E.C. Constable, C.E. Housecroft, M. Neuburger, B.A. Hermann, *Chem. Eur. J.* **2005**, *11*, 2307–2318.

73 D.J. Diaz, G.D. Storrier, S. Bernhard, K. Takada, H.D. Abruna, *Langmuir* **1999**, *15*, 7351–7354.

74 a) S.K. Hurst, M.P. Cifuentes, M.G. Humphrey, *Organometallics* **2002**, *21*, 2353–2355; b) A.M. McDonagh, C.E. Powell, J.P. Morrall, M.P. Cifuentes, M.G. Humphrey, *Organometallics* **2003**, *22*, 1402–1413.

75 a) R.W.J. Scott, A.K. Datye, R.M. Crooks, *J. Am. Chem. Soc.* **2003**, *125*, 3708–3709; b) M. Pittelkow, K. Moth-Poulsen, U. Boas, J.B. Christensen, *Langmuir* **2003**, *19*, 7682–7684.

76 a) G.R. Newkome, C.N. Moorefield, G.R. Baker, M.J. Saunders, S.H. Grossman, *Angew. Chem.* **1991**, *103*, 1178–1180; *Angew. Chem. Int. Ed.* **1991**, *30*, 1178–1180; b) C.L. Jackson, H.D. Chanzy, F.P. Booy, B.J. Drake, D.A. Tomalia, B.J. Bauer, E.J. Amis, *Macromolecules* **1998**, *31*, 6259–6265.

77 ***Textbooks***: G. Snatzke, N. Berova, K. Nakanishi, R.W. Woody (Eds.), Wiley-VCH, New York **2000**; K. Nakanishi, N. Berova, R.W. Woody (Eds.), *Circular Dichroism –*

Principles and Applications, VCH Publishers, New York–Weinheim **1994**; H. R. Christen, F. Vögtle, *Circular Dichroismus*, in *Organische Chemie, Von den Grundlagen zur Forschung*, Vol. II, Salle, Frankfurt am Main **1990**, pp. 301–334; N. Harada, K. Nakanishi, *Circular Dichroic Spectroscopy – Exciton Coupling in Organic Stereochemistry*, Oxford University Press **1983**; S. F. Mason, *Optical Activity and Chiral Discrimination*, Reidel, Dordrecht, Nato Advances Study Institutes Series, Series C, Vol. 48, **1978**; E. Charney, *The Molecular Basis of Optical Activity – Optical Rotatory Dispersion and Circular Dichroism*, Wiley, New York **1979**; S. F. Mason, *Molecular Optical Activity & The Chiral Discriminations*, Cambridge University Press, London **1982**; L. D. Barron, *Molecular Light Scattering and Optical Activity*, Cambridge University Press, London 1982; P. Crabbé, *ORD and CD in Chemistry and Biochemistry, An Introduction*, Academic Press, New York **1972**.

78 In right- or left circularly polarised light, the tip of the electrical field vectors describes a helix in a clockwise or anticlockwise sense, respectively.

79 Linearly polarised light can be understood as the superposition of two circularly polarised light rays of the same frequency, velocity, and amplitude (intensity), but of opposite polarisation (clockwise or anti-clockwise).

80 J. Recker, D. J. Tomcik, J. R. Parquette, *J. Am. Chem. Soc.* **2000**, 122, 10298–10307.

81 F. Vögtle, G. Pawlitzki, in H. Takemura (Ed.), *Cyclophane Chemistry for the 21st Century*, Research Signpost **2002**, *37*, 661(2), 55–90.

82 Terms such as macroscopic, nanoscopic, and mesoscopic chirality originate from studies by Mislow (A. B. Buda, T. Auf der Heyde, K. Mislow, *Angew. Chem.* **1992**, *104*, 1012–1031; *Angew. Chem. Int. Ed.* **1992**, *31*, 989–1007) and Avnir (O. Katzenelson, H. Z. Hel-Or, D. Avnir, *Chem. Eur. J.* **1996**, *2*, 174–181) and serve for definition of the chirality of large supramolecular and macromolecular systems, such as chiral clusters, aggregates, polymers, or dendrimers.

83 C.-O. Turrin, J. Chiffre, J.-C. Daran, D. de Montauzon, A.-M. Caminade, E. Manoury, G. Balavoine, J.-P. Majoral, *Tetrahedron* **2001**, *57*, 2521–2536.

84 J. R. McElhanon, D. V. McGrath, *Polymer Preprints* **1997**, *38*, 278–279.

85 M. J. Laufersweiler, J. M. Rohde, J.-L. Chaumette, D. Sarazin, J. R. Parquette, *J. Org. Chem.* **2001**, *66*, 6440–6452.

86 B. Buschhaus, F. Hampel, S. Grimme, *Chem. Eur. J.* **2005**, *11*, 3530–3540.

87 D. K. Smith, A. Zingg, F. Diederich, *Helv. Chim. Acta* **1999**, *82*, 1225–1241.

8
Special properties and potential applications

8.1
Introduction

As in other areas of science and technology, it is not easy to distinguish between established applications and highly promising future options in dendrimer chemistry; both aspects are therefore considered here in a single chapter. Dendrimer molecules are characterised by zones of different density. Depending upon the rigidity or the conformational mobility of their scaffold, they combine dense and less dense areas. They can form – more or less large and flexible – niches and cavities to accommodate solvent and to act as (selective) host compounds for guest substances.

In contrast to polymers, the structural precision and hence more uniform properties of dendrimers predestine them primarily for applications in the biomedical sector. A further reason for interest in these compounds is that effects such as luminescence can be enhanced by straightforward substitution and high local concentrations of certain structural elements and functionalities can be obtained.

Existing applications of dendrimers and dendritic polymers and those still in the planning stage are described together with the relevant properties. A number of characteristic general properties providing the basis for these applications are presented in a brief introductory overview. Other properties of specific dendrimer families are mentioned in Chapters 4 to 6.

Characteristics and properties (of structurally perfect) dendrimers:
- Nanoarchitecture with shell structure
- Structural precision
- Monodispersity
- Defined molecular size, defined number of end groups
- Low viscosity in solution
- Hydrophilicity/lipophilicity balance by design
- Rigidity/flexibility balance by design
- Accessible molecule surface
  (given not too high a generation and small end groups)
- Numerous functionalisation possibilities

Dendrimer Chemistry. Fritz Vögtle, Gabriele Richardt and Nicole Werner

ISBN: 978-3-527-32066-0

- Clear-cut accommodation of guest compounds
- Accessible by several-step to many-step (iterative) synthesis

Characteristics and properties of dendritic polymers:

- Straightforward, inexpensive synthesis
- Compact highly branched structure with variety of branching patterns
- Irregular globular shape, depending upon degree of branching
- Polydispersity
- Structural defects
- Highly reactive end groups
- Higher solubility than analogous linear macromolecules

Interest is presently focussed mainly on applications as additives, in catalysis, sensor technology, medical diagnostics and other areas, in which the current relatively high price of structurally perfect higher-generation dendrimers does not play a predominant role – because small amounts of substance often have far-reaching effects. In such areas, however, a perhaps highly demanding adaptation of the molecular structure to requirements may prove imperative in order to surpass existing methods, processes, and properties. On the other hand, the immense scope for variation of dendritic structures (type of scaffold, nature of branching, generation number, nanodimensions, solubility, lipophilicity/hydrophilicity balance, rigidity/flexibility balance, conformational dynamics, back-folding, niche/cavity formation, guest accommodation/selectivity, chirality, defects, *etc.*) and their functionalisation versatility permit such fine tuning. However, the time and experimental effort required have still not been invested for a number of applications.

8.2
Catalysis, membrane technology

Compared to polymers, dendrimer architectures offer favourable conditions for fixation of catalytically active moieties thanks to their monodispersity, variability, structural regularity of the molecular scaffold, and numerous functionalisation possibilities. Catalytic units can be fixed – multiply if required – on the periphery, in the core of a dendrimer, or at the focal point of a dendron. If the dendrimers are suitably functionalised at the periphery, appropriate metal complexes can be directly attached to the surface of the molecule. In contrast, dendrimers functionalised in the core or at the focal point shield the catalytically active site through their shell structure in a targeted manner, for example to attain substrate selectivity in the case of reactants of different sizes [1]. The corresponding concepts of "exodendral and endodendral fixation" of catalysts were introduced in the context of functionalistion of carbosilane, polyether, and polyester dendrimers [2]. Exodendral fixation refers to attachment of the catalytic units to the

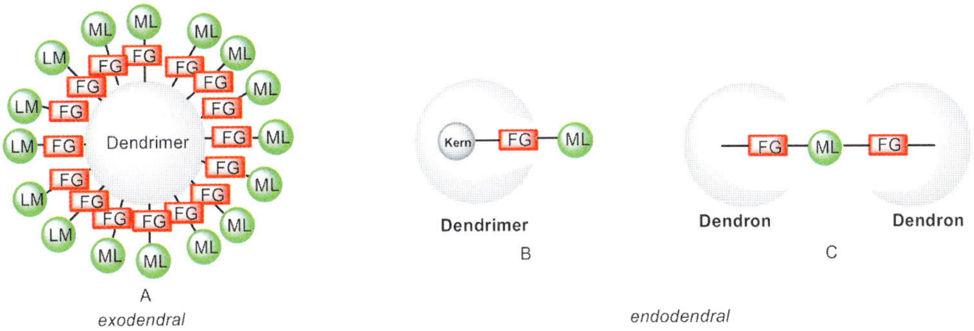

Fig. 8.1 Possible modes of fixation of catalytically active units (ML) to a dendrimer. A: exodendral fixation to the periphery of the dendrimer; B and C: endodendral fixation to the core of the dendrimer/dendron. The catalytic site can be, *e.g.*, a metal centre M (shown in green) bound to suitable functional groups (FG; shown in red); L symbolises additional complex ligand at the metal centre

ends of the branches whereas in the case of endodendral fixation the core of the dendrimer/dendrons is usually active (Fig. 8.1).

8.2.1
Dendrimers as catalyst supports

Formulae of dendrimers produced for catalytic purposes have already been shown in Chapter 4 and will not be repeated here, particularly since their possible suitability has merely been tested but no specific laboratory or industrial applications presently exist.

The spherical polyphenylene dendrimers developed by Müllen *et al.* [3] (see Section 4.1.5) do not aggregate on metal surfaces owing to their rigidity. This leaves more sites free for catalytic interactions. Compared to flexible polymers bearing the same metal, they offer easier access to the metal included as guest for catalytic processes. Such rigid polyphenylene dendrimers should therefore be suitable for the production of highly active catalysts. Moreover, the pronounced thermal stability of these hydrocarbon dendrimers permits organic reactions which would hardly be possible with polymeric catalysts.

Lüning successfully attached a concave arrangement of pyridine units to Fréchet-type dendrimers in homogeneous phase. A remarkable selectivity was thus achieved in base-catalysed addition of ketenes to alcohols and polyols (*e.g.* monosaccharides). The functionalised dendrimer catalysts exhibit a greater molar mass than conventional non-dendritic catalysts, thus permitting subsequent recycling of the catalyst by nanofiltration. These dendrimers are thus suitable as reagents for selective acylation of polyols [4].

Van Koten's group used a chemically inert, lipophilic carbosilane dendrimer scaffold as support material for fixation of up to 12 transition metal complex fragments. The covalently fixed fragments with nickel as catalytic site accelerated Kharasch addition of polyhaloalkanes to C-C double bonds. Recovery of the dendrimers was by ultrafiltration [5]. Studies on shape-selective alkene epoxidation with the aid of poly(phenyl ester) dendrimers (with manganese[III]-porphyrin core) as catalysts [1d, 6] showed second-generation dendrimers to have a much higher selectivity towards terminal double bonds than comparable manganese-complexed tetraphenylporphyrins.

Diaminobutyl dendrimers (DAB-POPAM) were functionalised with terminal diphenylphosphanyl groups and employed as catalysts in the Heck coupling of bromobenzene and styrene to give stilbene. Owing to their greater thermal stability, these dendritic palladium catalysts afforded higher yields than conventional palladium catalysts. In addition, the dendritic catalyst could be completely recovered by precipitation after addition of diethyl ether [7].

On account of their "effective shielding", dendritic catalysts generally proved to be significantly more stable than "normal" catalysts towards oxidative processes [1d].

Oxidation of thiols to disulphides with the aid of molecular oxygen can be catalysed by dendrimers having a cobalt-phthalocyanine core [1d, 8a]. Comparisons show that catalysts of the first dendrimer generation display a higher activity than those of the second generation. Such reduced catalytic activity can be explained in terms of restricted diffusion of the substrate into the dendrimer. On the other hand, catalysts of higher dendrimer generations exhibit greater stability because of better substrate uptake. Carbosilane dendrimers with up to twelve terminal metallocene groups (zirconocene, hafnocene, and titanocene) were utilised for methyl alumoxane-activated olefin (co)polymerisation and silane polymerisation. A high activity (5760 kg of polyethylene per mole of metal and hour) was obtained in the (co)polymerisation of ethene [1a].

For the enantioselective Aldol reaction with dendronised polymers see lit. [8b].

8.2.2
Catalytic dendrimers for membrane reactors

Functionalised dendrimers permit exploitation of the advantages of homogeneous and heterogeneous catalysis. Their spheroid architecture facilitates better recovery than comparable catalysts on polymer supports. They are distinctly easier to separate from the reaction mixture because the catalytic dendrimer molecules are larger than those of the resulting product. This attribute makes the dendrimers attractive as "nanoreactors".

For example, POPAM dendrimers of 1,3-diaminopropane type have been used in membrane reactors as supports for palladium-phosphine complexes serving as catalysts for allylic substitution in a continuously operated chemical membrane reactor. Good recovery of the dendritic catalyst support is of advantage in the case of expensive catalyst components [9]. It is accomplished here by ultra- or nanofiltration (Fig. 8.2).

Nanofiltration membranes are commercially available, *e.g.* with a retention capacity of 400 Da. The dalton unit serves as a measure of the separation ability

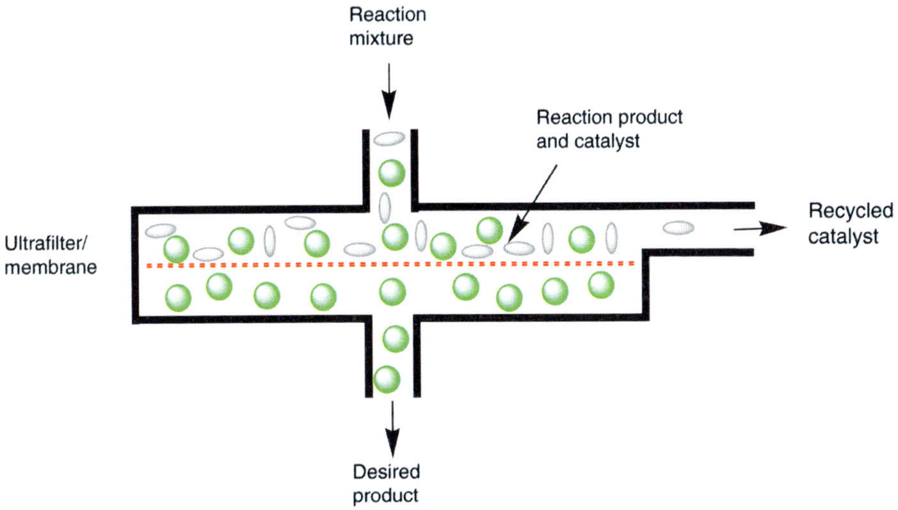

Fig. 8.2 Filtration apparatus for membrane reactors (schematic)

Table 8.1 Comparison of filtration processes

Filtration process	Molecular size [Da] or [kDa]	Pore size [µm]
Reverse osmosis	<100 Da	<0.001
Nanofiltration	100–1000 Da	0.001–0.01
Ultrafiltration	1000–500 Da	0.01–0.1
Microfiltration	>500 kDa	>0.1

of a membrane. Table 8.1 compares the separation ability and the pore size or the nature of the filtration process.

8.2.3
Dendrimers in enantioselective catalysis

Brunner *et al.* reported core-functionalised dendrimer metal catalysts ("dendrizyme") for use in enantioselective catalysis [1b, 10, 11]. Optically active dendrimer ligands were used, for example for the reaction of styrene with ethyl diazoacetate in a Cu(I)-catalysed enantioselective cyclopropanation. The starting compound for the preparation of these ligands was L-aspartic acid. However, (1*S*,2*S*)-2-amino-1-phenyl-1,3-propanol can also serve as reactant for the synthesis of optically active dendrimer ligands.

The aldimine chelate ligands formed from the primary amino functions present showed a slight increase of enantioselectivity. Further enlargement of the ligands, for example with (1*R*,2*S*)-ephedrine or (1*S*,2*S*)-2-(benzylamino)-1-phenyl-1,3-propanediol, did not lead to an increase of the enantiomeric excesses [12].

Addition of diethylzinc to benzaldehyde to give optically active secondary alcohols could be performed with the aid of POPAM dendrimers functionalised at their periphery with chiral amino alcohols. The number of alcohol end groups was raised to 64. Whereas low generations gave satisfactory enantioselectivities, they fell at higher generations: The fifth generation no longer shows any enantioselectivity [13]. Bolm *et al.* chose optically active dendritic catalysts functionalised with pyridyl alcohol units at the focal point of the Fréchet dendrons for addition of diethylzinc to benzaldehyde [14].

8.2.4
Dendrimers as phase transfer catalysts

Amphiphiles – molecules bearing both hydrophilic and hydrophobic groups – can aggregate to form micelles in water [15]. By analogy with micelles, appropriately designed dendrimer molecules exhibit different regions of the molecule (core/surface). Dendrimers with terminal carboxylate groups have been designated as "micellanols" [16] because they possess an apolar core and are soluble in aqueous media thanks to the carboxylate groups in their periphery. Functionalisation of a lipophilic dendrimer with hydrophilic end groups or chains leads to amphiphiles which permit accommodation of lipophilic guest compounds at the phase interface and can then transfer them into an aqueous medium [17, 1b].

The overall conclusion drawn from the above considerations is that it is not easy, using somewhat haphazard dendritic modification, to significantly improve existing catalysts which have been optimised over the course of decades. Given the wide scope for variation (dendrimer type, generation, cavity size, flexibility, philicity balance, see above), far more specific adaptations to the particular catalytic conditions are possible and necessary. The approaches to the topic adopted so far will hardly suffice for successful longer term planning.

8.3
Pigments, adhesives, additives in chemical materials

New applications are opening up through the introduction of pigments into dendrimers – *e.g.* for toner materials – and by admixture as additives to chemical materials.

8.3.1
Dendrimers as additives

Xerox Corp. patented a dry toner compound containing dendrimers as charge-enhancing species in the form of an additive [18]. In contrast to formerly used liquid toners, dry toners can be adapted in shape and size to the specific application by controlled construction (from the particle to the molecule). The uniform spheroid structure means that such toners can be applied efficiently and uniformly to the paper, which is important for colour printing. Toners with dendrimer additives require less material than their liquid counterparts.

In general, additives have to be highly effective, dosable in small amounts, and readily miscible with other components. Owing their molecular size, bleeding or loss by diffusion of the dendritic components is more difficult than for low-molecular-weight compounds [19].

8.3.2 Dendritic polymers for printing inks

On use as additives in printing inks, dendritic polymers ensure uniform adhesion of the ink to polar and non-polar foils. The hyperbranched compounds first attach themselves to the pigment particles. Owing to the large number of functional groups on the molecular surface of the dendrimers there are still enough anchor groups left over to effect good adhesion to the surface of the foils.

8.3.3 Dendritic polymers for paints

Applications of dendritic polymers are also found in the furniture and automotive industries in the form of polyurethane paints. These paints exhibit surface hardness, scratch resistance, chemical resistance, light fastness, and weathering resistance, as well as high gloss. One of the categories of polyurethane paints displays somewhat less flexibility alongside pronounced hardness and resistance to chemicals. A second group could be made more flexible, but is soft and less resistant to chemicals. Starting with the two paint types, use of hyperbranched polyisocyanates as crosslinking component with corresponding binders could provide a synergistic approach to a prototype paint with optimum properties [20].

8.3.4 Dendritic polymers as additives in foam formulation

Addition of highly branched polyesters affects viscosity and surface tension in foam formulations. This effect is attributable to modification of the branching density and the functional end groups. In some cases, improved rheological properties and faster curing of the foam formulations were also attained through optimisation of the structure. Highly branched (dendritic) polymers are also used as rheology modifiers influencing deformation and flow behaviour. This is also of interest in medicine for the study of changes in the flow properties of blood in the vascular system [21].

8.3.5
Network precursors for plastics

Their low viscosity and high molecular density favour the use of dendrimers in dental chemistry. In this area materials are required which form crosslinked polymer structures on photopolymerisation of tooth fillings at room temperature with minimum possible loss of volume to avoid crevice formation between filling and tooth. During each coupling step of photopolymerisation the van-der-Waals distance is reduced to the length of a covalent bond, normally causing volume shrinkage. Linear prepolymers are usually too viscous for this kind of application. Dendrimers could be used in the form of "network precursors" if the periphery is functionalised with appropriate groups [17].

8.3.6
Dendrimers as nanocapsules for dyes and for molecular imprinting

For "molecular encapsulation" of guest molecules Reinhoudt *et al.* synthesised a water-insoluble fifth generation POPAM dendrimer containing a total of 62 tertiary amino groups and whose periphery is "functionalised" with 64 apolar adamantyl groups (see Section 4.1). To ensure that this dendrimer (Fig. 8.3 a) is sol-

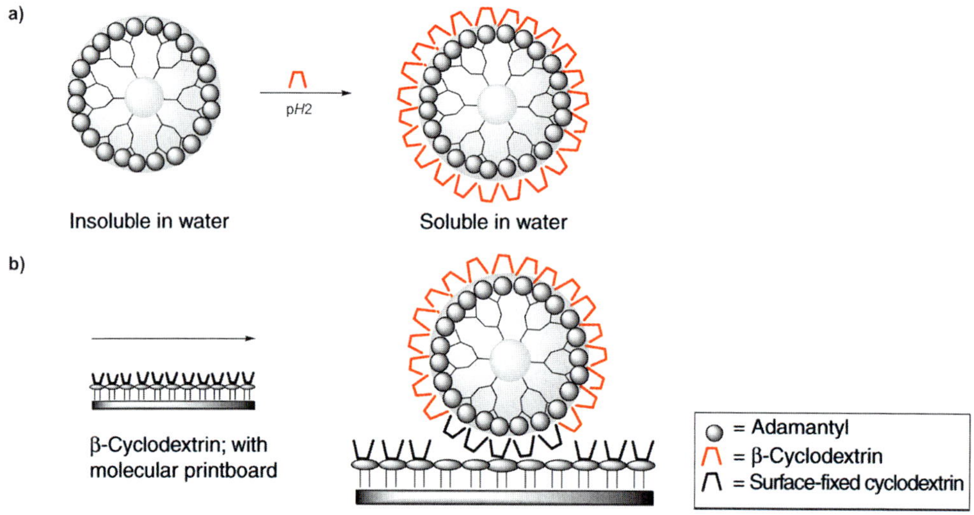

Fig. 8.3 a) Transformation of a fifth-generation adamantyl-terminated POPAM dendrimer into water-soluble dendrimer by covering the adamantyl end groups with *β*-cyclodextrin (in excess) (schematic: for the sake of simplicity neither all 64 adamantyl groups nor all branches of the dendrimer are shown; b) fixation of the dendrimer on a "molecular printboard" by "supramolecular exchange" of part of its peripheral *β*-cyclodextrin (red) by the *β*-cyclodextrin (black) located on the surface

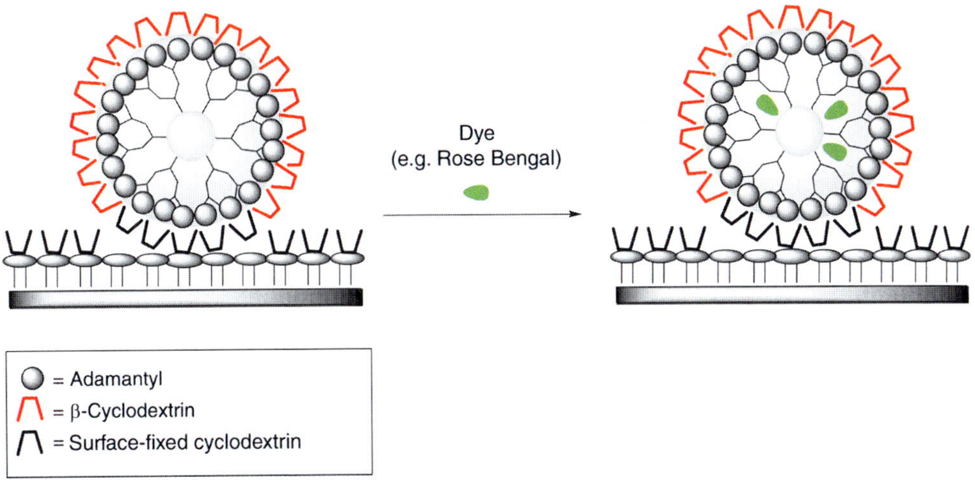

Fig. 8.4 Encapsulation of anionic dye molecules (green) by the dendrimer bound supramolecularly to the surface of the molecular printboard (schematic: only 23 of the 64 terminal adamantyl groups are shown and the branches are symbolised)

uble in water and is able to attract anionic guest molecules by electrostatic interactions and encapsulate them, the tertiary amine nitrogen atoms were protonated by addition of acid (pH 2). The hydrophobic adamantyl groups were subsequently supramolecularly complexed by *β*-cyclodextrin and a stable aqueous dendrimer solution thus obtained. Fixation of the POPAM dendrimer to a *β*-cyclodextrin (guest) layer (molecular printboard) on a glass surface (b) in Fig. 8.3 was accomplished by "supramolecular microcontact printing" [22].

Anionic dyes such as Bengal Rose and fluorescein are suitable as guest molecules for encapsulation since both are soluble on account of their acid groups and engage in electrostatic interactions with the – protonated – interior of the fixed dendrimer (Fig. 8.4). The fluorescent properties permit visualisation of inclusion of the surface-fixed dye molecule. The dendrimer thus acts as a "dendritic box" [23].

The above cyclodextrin-bound and surface-fixed dendrimers can be correspondingly "filled" with dye by "crossed microcontact printing". The dissolved dendrimer molecules remaining in solution can be filled with another dye from the solution and fixed analogously. In this manner, an alternatingly printed surface pattern was obtained. Encapsulation of the dye molecules in the surface-fixed dendrimers is reversible. The dendritic box can thus be filled, and also emptied again and refilled with another dye. Reinhoudt *et al.* succeeded in constructing a dendrimer array which permits filling of the dendrimer with dye at pH 2 and release of the encapsulated dye at pH 9: the tertiary amino groups of the dendrimers are deprotonated at pH 9 with the aid of a phosphate buffer. Renewed addition of an acid buffer solution reprotonates the amino groups, allowing another dye to be enclosed in the dendrimer. Alternating pH-controlled switching between a filled and a non-filled state is thus possible. In the future this system could also be used for the controlled release of active ingredients instead of dyes.

Irreversible inclusion of individual isolated molecules can permit their investigation in greater detail, for example in the case of radicals [24]. This principle of molecular encapsulation was used by Cram and Warmuth [25] for the study of molecularly encapsulated cyclobutadiene, arynes, and other guest molecules – which are somewhat unstable in the free form – in macrooligocyclic carcerands [26]. Similar applications based on the inclusion of molecules in dendrimers [27] are to be found in Section 8.7.1.

8.4 Dendrimers for displays and (opto)electronics

Cambridge Display Technology reported the development of organic dendrimer-based displays [28]. A transparent dendritic anode (indium tin oxide) is coated with several layers of light-emitting and electron-transporting materials (Fig. 8.5). After these layers comes a metal cathode. If current flows between the electrodes, the dendrimer material in the light-emitting layer is excited and emits light.

The advantage of dendrimer displays lies in their versatile design: Thus the dendritic structure permits independent modification of the dendrimer core (important for light emission), of the branching units (important for charge transfer), and of the end groups (responsible for conduction processes). With regard to efficiency, the dendrimers offer a way to transform small phosphorescent molecules into compounds which can be processed in solution. Since the end groups can be selected independently of the light-emitting dendrimer core, it is possible to adapt dendrimers to diverse process systems without impairing

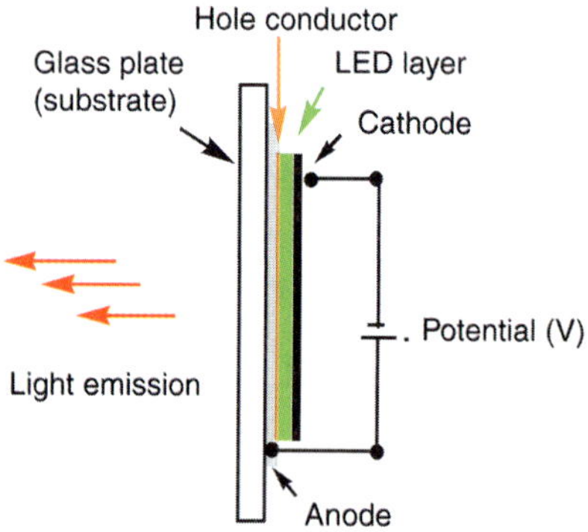

Fig. 8.5 Construction of PLED displays. The dendrimer is located in the green-coloured layer

the quality of light emission. Compared to polymer displays, dendritically modified polymeric light-emitting diodes (PLED) in displays were significantly brighter.

Molecules with stilbenoid chromophores are being used increasingly in light-emitting diodes (LED), in non-linear optics (NLO), and in optical data storage technology. Dendrimers with stilbenoid chromophores [29] could offer further advances because more chromophore groups can be brought together in a confined space in a dendrimer.

New materials for light emission in the blue region of the spectrum, based on low-generation polyphenylene dendrimers, were tested for use as organic light-emitting diodes (OLED) [30]. It was seen that the ratio of *p*-phenylene groups (with high fluorescence parameters) to 1,3,5-triphenylbenzene groups (with low fluorescence parameters) could be the reason for different relative quantum yields of photoluminescence [31]. Colour coordination of OLED was possible with the aid of dendrimers. While the first dendritic OLED only emitted green light, the influence of the chemical structure on the colour of the emitted light is now exploited for diodes which emit light of other wavelengths. Mixing of two dendrimers with the same surface structure but different cores emitting light of different wavelengths permitted a colour change from ultraviolet/violet to blue-green. The wavelength of the light can thus be changed without influencing the emission efficiency and operating voltage. Another advantage of dendrimers lies in the fine tuning of electronic and processing properties which circumvents the problem of phase separation in polymer mixtures [32].

8.4.1
Liquid-crystalline dendrimers

As the name implies, liquid-crystalline materials combine the properties of a crystal with those of a liquid in a very special way and are therefore of interest for display and data storage technology. Liquid-crystalline compounds usually consist of rod- or disc-shaped organic molecules which preferentially adopt a mutually parallel orientation [33]. A change of molecular orientation caused by application of an electrical potential transforms the optical properties and can be utilised for display applications.

A challenge for new developments is to produce materials from specific molecules which combine both the switching properties of lower-molecular-weight liquid crystals and the mechanical properties of plastics/chemical materials. Such elements of a liquid crystal can be joined to yield a larger molecular unit by means of a dendrimer scaffold (Fig. 8.6). Depending upon the nature of bonding, new liquid crystalline phases can arise. It is also conceivable that linkage to a dendrimer at low temperatures can prevent the formation of an ordered crystal, leading to the formation of a vitreous state on cooling. Vitreous liquid crystals may be of use as optical data storage materials [34].

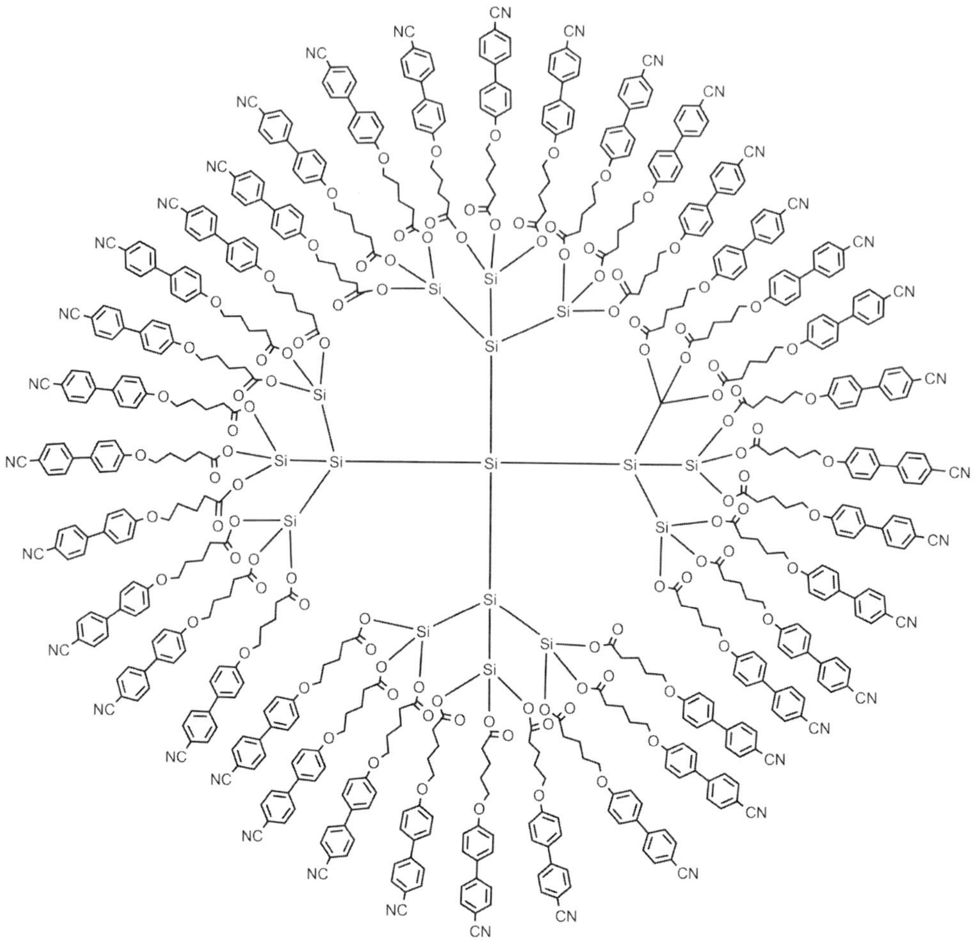

Fig. 8.6 Liquid-crystalline dendrimer (the central bonds have been lengthened for the sake of clarity)

8.5
Biomimetics, sensor technology, diagnostics (fluorescence)

8.5.1
Protein dendrimers

Dendritic architectures can be used to mimic protein folding, because the amino acids are crowded together in a confined space [35 a]. Synthesis permits incorporation of both natural and unnatural (abiotic) amino acids in such an "artificial protein" (peptide dendrimer), leading to materials that may be attractive for medicine. Given an appropriate nature and sequence of the amino acids,

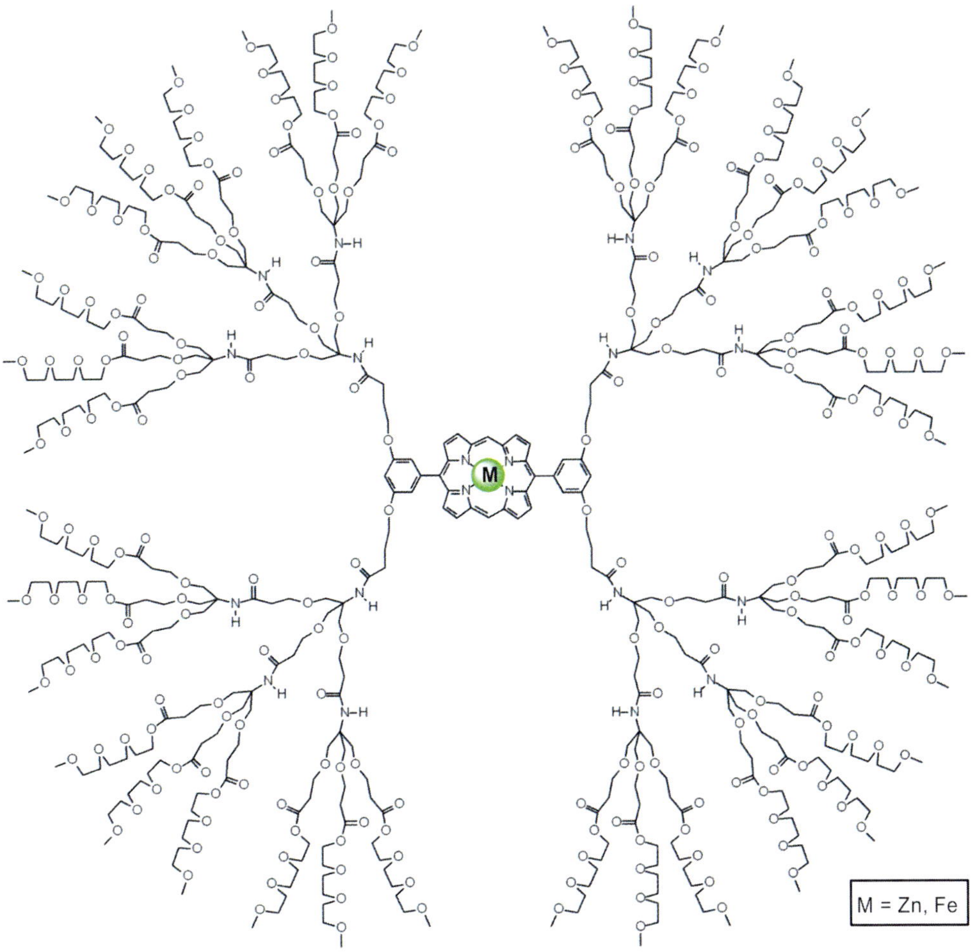

Fig. 8.7 Dendritic porphyrin-metal complex (according to Diederich *et al.*)

peptide dendrimers [35 b] can also exhibit enzyme-like activities (dendrizymes, see Section 8.2.3) [11 a]. Future research in this area could also lead to artificial (synthetic) enzymes (synzymes) [11 b]. Wedge-shaped dendritic peptide components were formed by solid-phase synthesis starting from lysine units. The multiple antigen proteins (MAP) obtained in this way contained between two and eight peptide chains and displayed significantly better immunogenicity than individual proteins [36].

Diederich *et al.* produced a mimic of the electron-transfer protein cytochrome C. Divergent synthesis afforded a water-soluble iron-porphyrin which possesses a covalently attached dendritic shell surrounding the electrically active heme core which is reminiscent of a protein shell (Fig. 8.7). In contrast to the dendritic porphyrins of Aida *et al.*, they consist of flexible dendritic poly(ether amide) units [37]. Studies of the first and second generations of this dendrimer revealed that the reduction potential is shifted towards positive values, given sufficient dendritic shielding. Increased shielding of the metal-complexed porphyrin reduces the contact with the aqueous solution and thus weakens or prevents interaction between porphyrin and solvent. The more all-embracing the dendritic shell, the lower the rate of electron transfer. Cytochrome C itself also shows similar properties [38].

8.5.2
Glycomimetics

Glycomimetics are synthetic carbohydrate analogues whose structure has been simplified and modified after the manner of biomolecules. Such active ingredients are used in the treatment of chronic inflammatory ailments such as rheumatism, dermatitis, and psoriasis.

Oligolysine dendrimers with terminal carbohydrate structures were functionalised by solid-phase synthesis for well-defined construction of "glycoclusters". Synthesis proceeded *via* coupling of lysine molecules. Eight amino groups are attached by peptide bonding to the periphery *via* a spacer-functionalised sialic acid derivative (Fig. 8.8).

In order to prevent infection, for example, with an influenza virus, it is necessary to block at least one out of two of the substances present on its surface: One of them is a neuraminidase, an enzyme that cleaves off sialic acid, and the other is a hemagglutinin, a sialic acid-specific lectin which aids the influenza viruses in docking onto oligosaccharides exposing sialic acid on the surface of host cells. Both compounds are prerequisite for the infection process.

Whereas glycodendrimers bind the influenza hemagglutinin in the micromolar range, a corresponding non-dendritic active substance requires millimolar concentrations to block the sialic acid-cleaving enzyme [39].

Glycodendrimers bearing both a self-assembling group and a group docking selectively onto the receptor (often called a ligand in biochemistry) have been used for the construction of non-covalent nanoparticles in water *via* self-assembly. The most favourable nanoparticle size can be attained by self-assembly of second-generation and third-generation dendrimers. Self-assembly no longer proceeds efficiently at higher generations. Corresponding particles can be deposited on a surface by solvent evaporation. In solution they act as non-covalent, polyvalent *in vitro* and *in vivo* receptor inhibitors.

Self-assembly of such dendrimers is a dynamic equilibrium process (Fig. 8.9). Conceivably, the non-covalent polyvalent ligands can be optimised with regard to their size and shape in the presence of natural multivalent receptors, as a kind of template for their own multivalent inhibitor, and thus open up a range of physiologically relevant polyvalent interactions [40 a].

For Oligosaccharide-modified PPI-dendrimers see lit. [40 b].

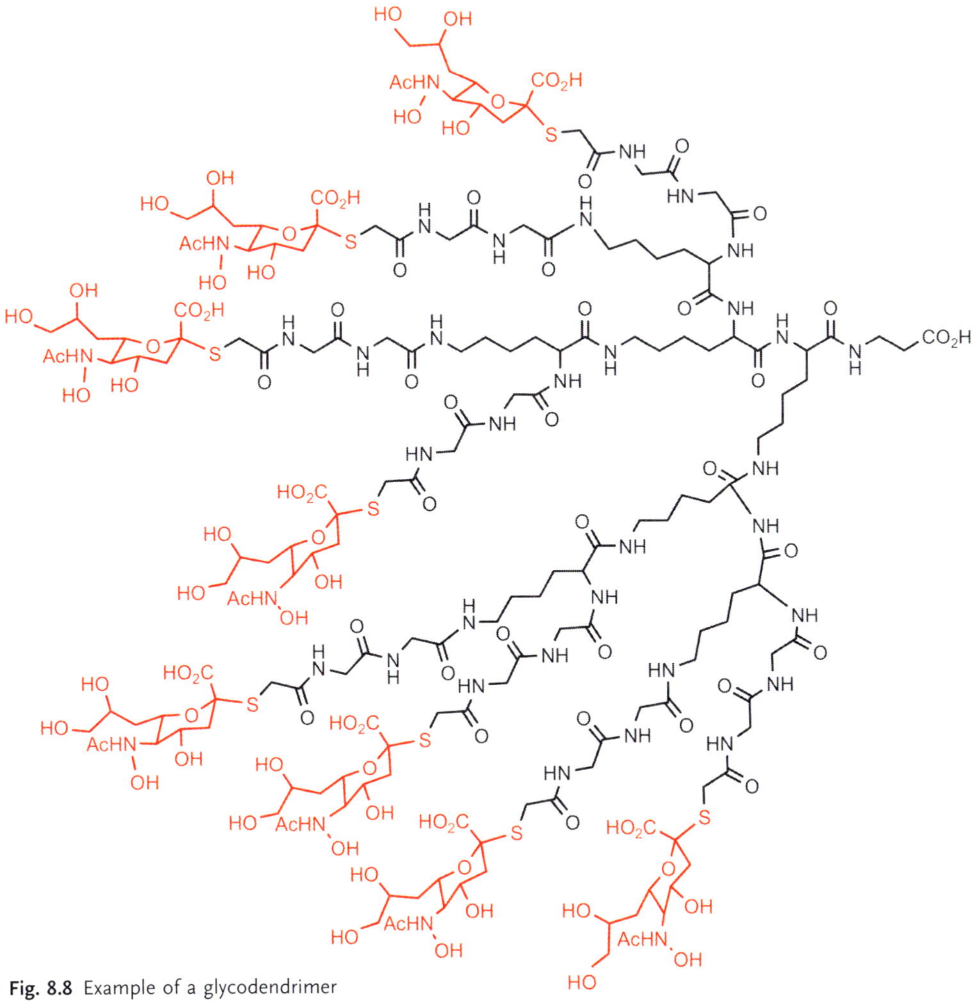

Fig. 8.8 Example of a glycodendrimer

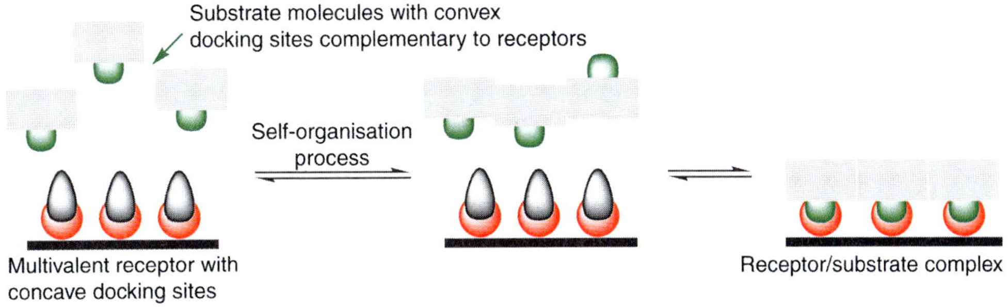

Fig. 8.9 Self-assembly of glycodendrimers (shown in green) to form non-covalent nanoparticles

8.5.3 Dendrimers in sensor technology

8.5.3.1 Quartz micro balance with dendritic sensor layers

The sensor principle of the Quartz Micro Balance (QMB) [41] is based on the micro gravimetric determination of the specific inclusion of a guest compound (ion, molecule) as analyte in a sensitive layer consisting of suitable host molecules (selectors; Fig. 8.10). Host substances such as dendrimers can be deposited by the electrospray method as a homogeneous layer on the micro balance quartz disc bearing a vapour-deposited gold coating. On application of an alternating potential of suitable frequency the coated microbalance quartz discs in the measuring chamber are made to vibrate at their resonance frequency while exposed to various gas-phase or vapour-phase (analyte) substances. The substance to be analysed is transferred by a temperature-controlled stream of nitrogen, carrying a well-defined concentration of substance, into the measuring chamber. Each impingement phase is followed by a pure nitrogen purging phase during which the analyte is desorbed. The frequency of the coated quartz

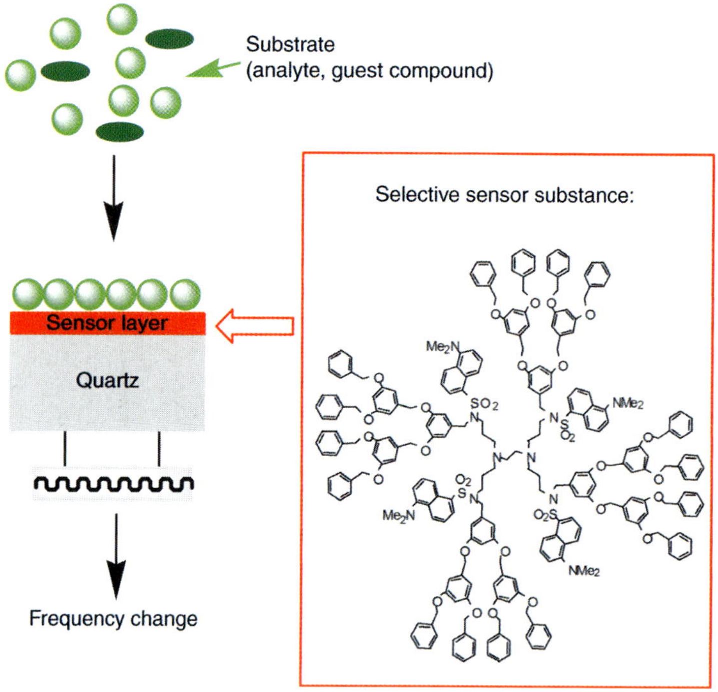

Fig. 8.10 Mode of operation of the quartz micro balance (left; schematic). Right: A dendrimer as guest-selective host compound (selector) for coating of the quartz crystal oscillator. Red: Selector layer; green: various analyte components

crystals is always measured as a function of time. Absorption of the substance analysed on one of the selector layers leads to a lowering of the resonance frequency of the quartz crystal oscillator, the magnitude of which is inversely proportional to the mass of analyte absorbed. The change of frequency is transformed into an electrical signal and analysed. If the sensor layer is sufficiently sensitive and selective for a certain chemical compound from a mixture of guest compounds then the concentrations of individual guests can be quantitatively determined in mixtures.

One area of application is the analysis of flavours in foods in order to ascertain, for example, the ripeness of fruit or the maturity of cheeses [42]. For this purpose dendrimers were used which can distinguish certain carbonyl compounds such as ketones, aldehydes, esters, and amides, also in mixtures. For example, measurement of the concentration of 2-heptanal is of importance specifically for determination of the degree of ripeness of apples. The concentration of this compound increases significantly with increasing ripeness [43]. This technique can also serve as an electronic nose for quality control of high-value products such as saffron, which in powder form may contain undesired contaminants (adulterants) such as curcuma, safflower, or marigold.

As described above, the comparatively rigid architecture of polyphenylene dendrimers makes them suitable as sensors for detection of volatile analytes. Because the selectivity of measurement with regard to a particular analyte is highly dependent upon the nature of the sensorially active (selector) surface of the quartz microbalance. Unsubstituted second-generation polyphenylene dendrimers were tested with various gaseous analytes such as benzaldehyde, nitrobenzene, acetophenone, and fluorobenzene, among others, and showed a high selectivity. However, chlorinated and unsubstituted aliphatic hydrocarbons, carbonyl compounds, and alcohols, amines are not detected. This selectivity of the polyphenylene dendrimers as selectors is ascribed to their aromatic scaffold, because it can undergo π-electron donor/acceptor interactions with guest molecules. Second to fourth generation polyphenylene dendrimers were also used as layer material and their substrate selectivity investigated. It became apparent that the number of guest molecules included depends upon the shape and size of the dendrimer and can be approximately predicted in certain cases. The sensitivity of detection with polyphenylene dendrimers is high, for example 5 ppm for acetophenone and aniline. These dendrimers characteristically also show long term stability, which is of advantage for the reproducibility of the measurements [44].

Poly(phenyl ester) dendrimers with six terminal platinum chelate complex units were used as highly selective sensors for sulphur dioxide, considered partly responsibly for the formation of smog and acid rain. In the presence of hydrogen peroxide or sunlight and ozone, SO_2 is oxidised to SO_3. The latter reacts spontaneously with water to form sulphuric acid. That is the reason why detection of minimal concentrations of sulphur dioxide with appropriate sensors is necessary. The arylplatinum(II) complexes attached to the molecular periphery of the dendrimers absorb sulphur dioxide with formation of a pentacoordi-

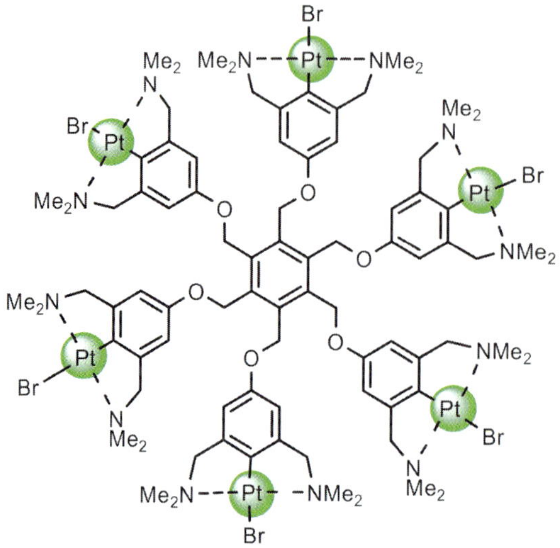

Fig. 8.11 Dendrimer with six terminal platinum-chelate complex units for SO_2 detection by the quartz micro balance (according to Albrecht, Schlupp, Bargon, van Koten *et al.*)

nated platinum adduct. In the case of the metallodendrimer shown in Fig. 8.11, a change of colour ensues immediately upon exposure of the dendrimer to an SO_2 atmosphere: The initially colourless complex turns orange on addition of SO_2. Spectroscopic analyses confirm the hexameric structure of the metallodendrimer: in the terminal platinum chelate complexes the sulphur dioxide molecules are bound to the platinum as fifth ligand [45].

Dendritic polymers are used as functional thin layers. Highly branched aromatic polyesters with polar end groups show good response behaviour in gas phase and liquid sensors [46]. Ease of recycling of the expensive platinum complexes (*e.g.* by nanofiltration) is a positive aspect of the use of such sensor dendrimers [47].

8.5.3.2 **Luminescent dendrimers as sensor materials**

Luminescence has already been considered in general terms in Chapter 5. Luminescent POPAM dendrimers of various generations with peripheral dansyl units were studied by Balzani and Vögtle *et al.* as sensor model systems with regard to the fundamental suitability of dendritic structures for multiplication of signalling groups (multi-labelling) [48].

Porphyrin dendrimers are suitable as sensors for small molecular and ionic analytes. Unsubstituted metal-free porphyrins often show poor solubility in water. However, if it proves possible to envelop them in hydrophilic dendrimers they can be used in water as fluorescent pH indicators because they exhibit distinct changes of their absorption and emission bands owing to protonation of

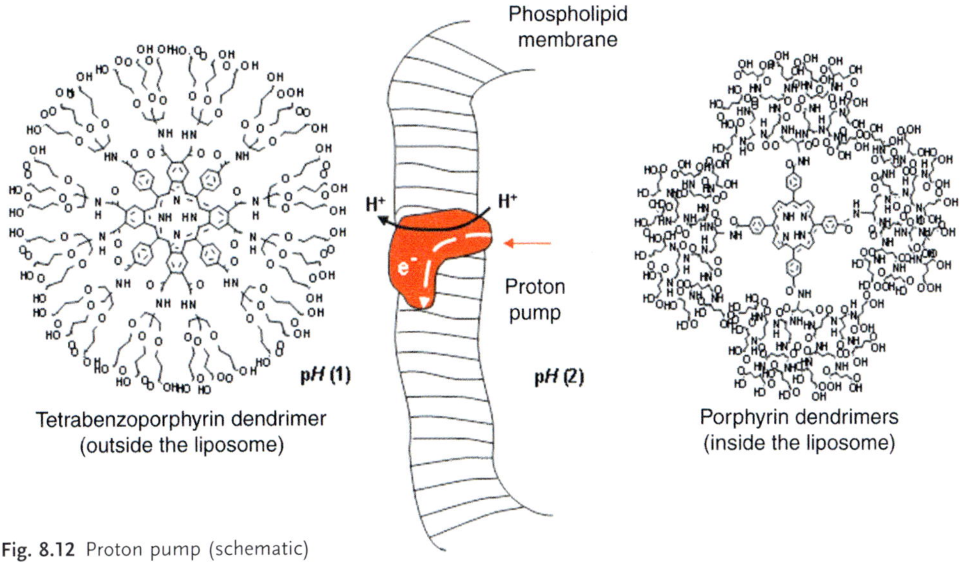

Fig. 8.12 Proton pump (schematic)

the nitrogen atoms. Photophysical studies on a Newkome-type first-generation poly(ester amide) dendrimer with a tetrabenzoporphyrin core and 36 carboxylate end groups as well as a polyglutamic acid-porphyrin dendrimer [49] showed that in both dendrimers the pK_a value of the porphyrin core was shifted into the physiological range. For further studies regarding a "biological proton pump" the polyglutamic acid dendrimer was artificially introduced into a phospholipid liposome suspended in a dendritic tetrabenzoporphyrin dendrimer solution. A change of pH value of the solution did not elicit a fluorescence response in the polyglutamic acid dendrimer, whereas the tetrabenzoporphyrin dendrimer located on the outside fluoresced immediately. However, if artificial proton channels are introduced into the liposome wall, both species show comparable responses to the pH value. Owing to the carboxylic groups and the size of the molecule, the porphyrin dendrimers are unable to penetrate the phospholipid membrane (cf. Fig. 8.12) [50].

The concept of bifunctionalisation led to a second-generation PAMAM dendrimer whose terminal amino groups bore eight anthracene entities and eight boronic acid groups (Fig. 8.13). This dendrimer could be used as a saccharide sensor. The boronic acid forms a complex with the saccharide, whereas the anthracene groups act as signalling units. As examples of saccharide analytes D-glucose, D-fructose, and D-galactose could be bound in the form of an intramolecular 2:1-boronic acid-saccharide complex. This gives rise to photoinduced electron transfer (PET). The resulting enhancement of the fluorescence intensity of the anthracene units depends upon the nature and number of bound saccharide guest molecules [51].

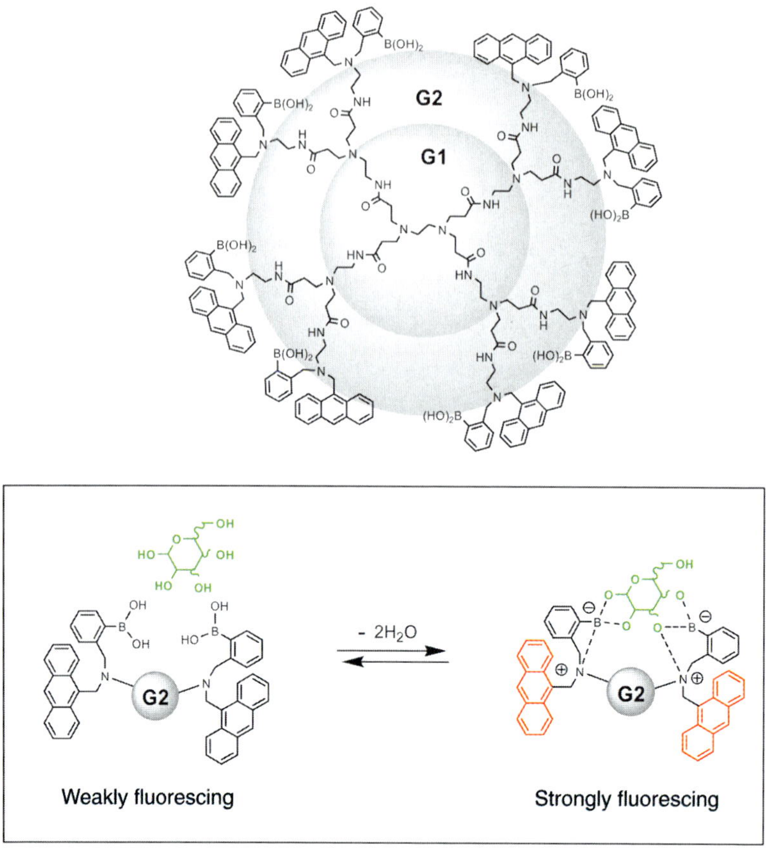

Fig. 8.13 Dendritic sugar sensor based on boronic acid/glycol interaction (G2 = 2nd generation; according to Shinkai *et al.*). For the sake of clarity only two of the eight end groups are shown in the boxed formula

8.5.3.3 Fluorescing PET sensors

The principle of fluorescing chemosensors is based on the interaction between two components: an ionophore (guest-selective receptor compound; selector) and a neighbouring fluorophore. These two components are covalently bonded to form a fluoroionophore unit. In the example depicted in Fig. 8.14 the cation receptor serves as electron donor, and the fluorophore as electron acceptor. If the fluorophore undergoes photochemical excitation, (photoinduced) electron transfer (PET) takes place from the HOMO to the LUMO. This process thus allows photoinduced electron transfer from the HOMO of the ionophore to the HOMO of the fluorophore with accompanying fluorescence quenching. Coordination of a metal cation M^{2+} by the receptor raises the redox potential of the fluorophore. The ensuing drop in energy level of the HOMO below the HOMO

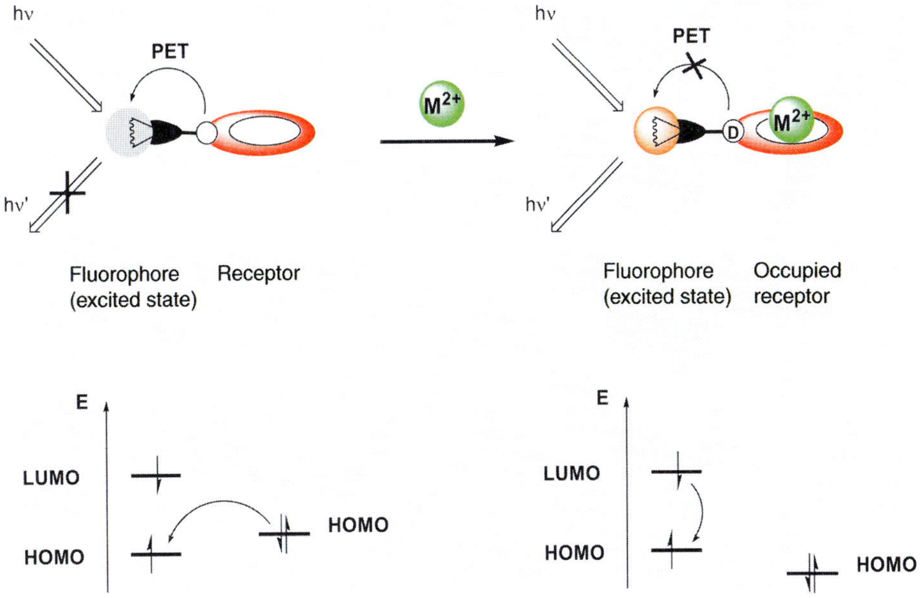

Fig. 8.14 PET sensor (schematic; according to de Silva *et al.*). Above: scheme of complex formation of the ligand portion of the fluoro-ionophore with the metal ion. Below: corresponding energy levels of the uncomplexed (left) and M^{2+}-complexed complex ligand (right)

level of the fluorophore prevents photoinduced electron transfer and leads to an increase in fluorescence intensity. The metal-occupied receptor (above right in Fig. 8.14) prevents photoinduced electron transfer, whereas the free receptor permits it. Switching of luminescence between the "on" and "off" states becomes possible, depending upon whether a metal ion is present or absent in the ionophore [52].

Balzani, Vögtle *et al.* demonstrated for dendritic cyclam compounds that signalling units are "dendritically amplified" by increasing the number of peripheral fluorophores. Opportunities for these PET sensor systems exist in the combination of selective coordination sites with the dendritic architecture. The mechanism resembles that described above, but can be influenced by the introduction of dendritic units of different generations into the sensor system. The distance between the terminal naphthalene groups of the dendron and the receptor (ion as guest in the cyclam core) is of crucial importance for electron transfer (ET) (Fig. 8.15) [53].

Aida *et al.* synthesised a dendritic zinc-porphyrin heptamer ($7P_{Zn}$-C_{60}) with a fullerene terminal entity as an electron-transfer system, which harvests visible light for electron transfer to the fullerene group (Section 6.3.3.5). Electron transfer from the porphyrin to the fullerene gives rise to a charge-separated state ($P_{Zn}^{\cdot+}$-$C_{60}^{\cdot-}$), whose lifetime was the subject of investigation. Comparative studies

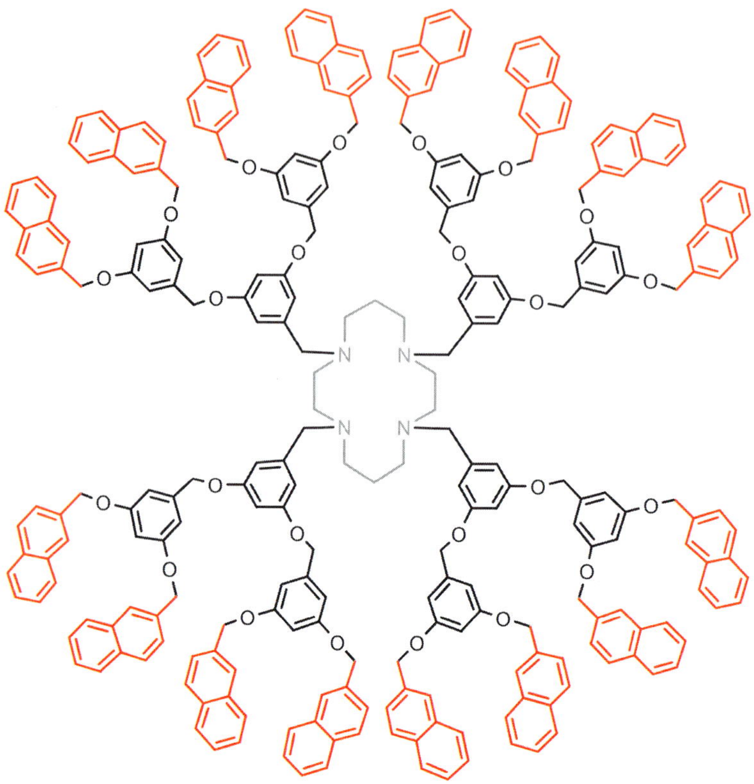

Fig. 8.15 Cyclam dendrimer (ionophore) with 16 terminal naphthalene fluorophores (according to Balzani, Vögtle *et al.*)

were performed on dendritic porphyrins in monomeric, trimeric, and hexameric form. The zinc-porphyrin heptamer ($7P_{Zn}$) was found not only to act as light-harvesting antenna, but also to delay transfer in the reverse direction, and thus to prolong the lifetime of the charge-separated state. These findings underscore the attractiveness of dendronised dye molecules for efficient conversion of solar energy into chemical potential [54].

A potassium-ion-selective, dendritic, fluorescing chemosensor, bearing three crown ether moieties in the periphery, shows a linear increase in fluorescence intensity with increasing potassium concentration (in acetonitrile). An important criterion for potassium chemosensors is their mode of action (selectivity) in the presence of large amounts of sodium. The tris-crown ether sensor shown in Fig. 8.16 is able to detect very small traces of potassium ions, even if large quantities of sodium ions are present in the same solution – such as in body fluids [55].

Such cation-selective sensors could be of use both in clinical analysis, for example during surgical operations, and in environmental analysis [56].

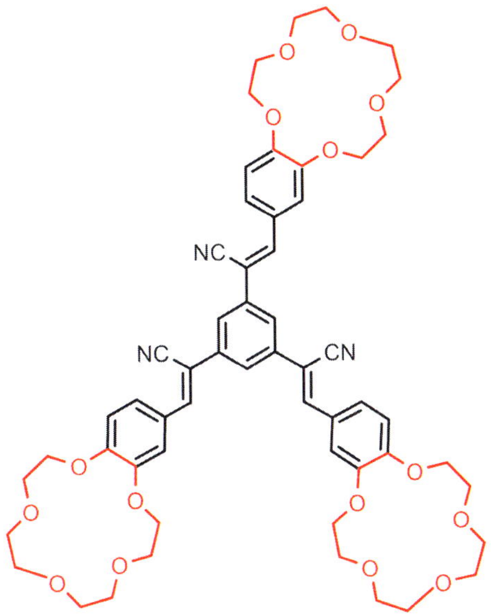

Fig. 8.16 Chemosensor with cation-selective benzo[15]crown-5 entities

8.6
Dendrimers in medical diagnostics

Dendrimers are seen to have applications in medicine, especially in diagnostics [57a]. Apart from good solubility in water, the possibility of introducing multiple special functionalities into the periphery of the nanoscale molecules permits high sensitivity to be attained and variation of the parameters over a wide range as required by the prevailing conditions.

8.6.1
Magnetic resonance imaging (MRI) processes

Magnetic resonance imaging (MRI) permits organs, blood vessels, or tissues to be visualised in the human body. To this end inhomogeneous, well-defined magnetic fields are generated which permit assignment of the nuclear magnetic resonance signal of the principal constituent of the body, namely water, to its point of origin, ultimately producing visible images. Introduction of contrast agents (usually by injection), which contain paramagnetic metal ions, significantly shortens the relaxation time of the water protons in the organ under investigation [57b]. Specifically for this imaging technique, Schering AG, Berlin, developed the dendritic contrast agent "Gadomer-24", a dendrimer with a trimesic acid core and branched second-generation lysine dendrons with 24 peripheral Gd(III)-cyclen complex entities located at the end of the branches (Fig. 8.17). The most significant advantage over polymer-bound Gd-chelate complexes –

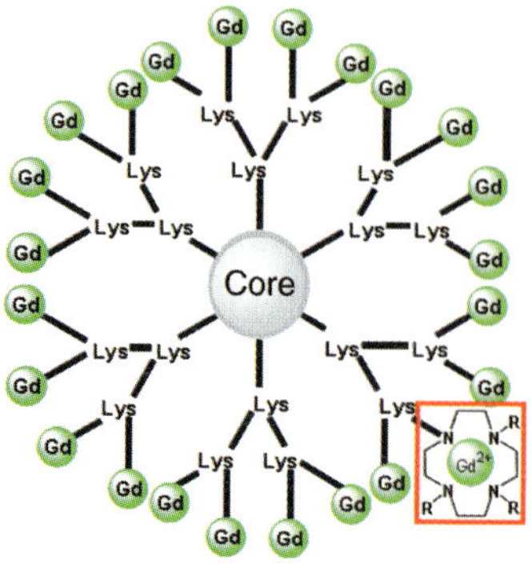

Fig. 8.17 Gadomer 24 (schematic): 24 gadolinium ions (green) are bound in 24 peripherally attached cyclen ligands; 18 lysine units form the dendritic scaffold. The molar mass is 17 kDa

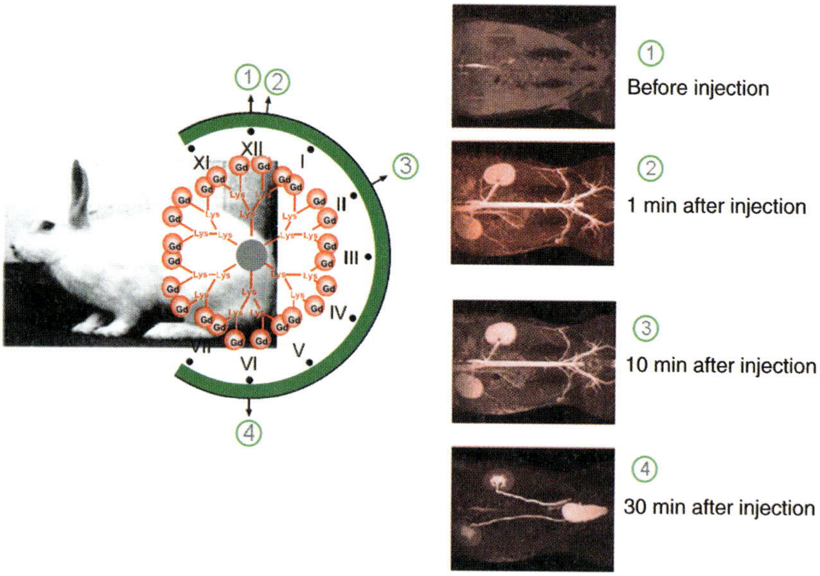

Fig. 8.18 Visualisation of the blood vessels of a rabbit with a dendritic Gadomer contrast agent (shown in red). The figure was kindly provided by Dr. H. Schmitt-Willich (Schering AG, Berlin)

apart from the precisely adjustable particle size – lies in the low toxicity because Gadomer-24 undergoes complete renal excretion.

Furthermore, this lysine dendrimer-bound gadolinium complex exhibits a higher *in-vivo* stability as a contrast agent and hence a longer residence time in tissue than comparatively simple commercially available gadolinium compounds and thus assures improved visualisation of organs, blood vessels, and tissue (Fig. 8.18) [58].

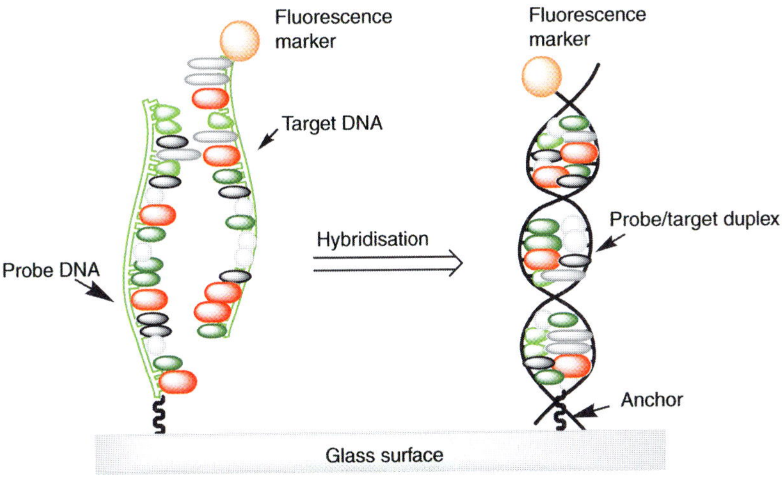

Fig. 8.19 DNA hybridisation with fluorescence markers (schematic)

8.6.2
DNA dendrimers as biosensors for DNA hybridisation

Access to nucleic acid dendrimers is initiated by a zip-fastener like dissociation of the DNA double strand by heating. The double strand separates into the two individual strands by thermal motion (denaturation). Subsequent association, hybridisation of complementary sequences, is followed by stepwise cross-linking to form DNA dendrimers, which can contain up to two million oligonucleotide-end group strands (Fig. 8.19). The latter can be labelled with fluorescence or radioactive markers.

Hybridisation can thus be used to produce probes, for example, for the detection of oligonucleotides in chick embryo tissue or of Epstein–Barr viruses (glandular fever) in transplant recipients – by detection of specific RNA strands. DNA dendrimers and corresponding detection devices have meanwhile become commercially available (Genisphere® company). Fluorescence-labelled polynucleotide dendrimers have also been used for signal intensification in DNA microchip technology [59].

8.7
Medical applications

Overviews on this topic were recently published with the book *"Dendrimer-Based Nanomedicine"* [60a].

8.7.1
Dendrimers as carriers for cytostatic agents

Dendritic drug-delivery systems are acquiring increasing significance in the development of new cancer drugs having fewer side effects [60b]. Thanks to a well-defined core-shell architecture [61] and the resulting clear distinction between the scaffold and the functionalised periphery, it is possible to introduce the active ingredient into the interior of the dendrimer (endoreceptor) or to attach it to the periphery (exoreceptor). Both possibilities have the benefit of offering a form of dendritic transport tailored to the active ingredient. For example, a longer residence time of the cytostatic agents in blood plasma is achieved in endocytosis because tumour blood vessel systems exhibit an enhanced permeability for macromolecules and a restricted discharge *via* the lymph system (EPR effect = Enhanced Permeability and Retention) [62]. On the one hand, the cytostatic agent can thus accumulate in tumour tissue (passive targeting); on the other hand, the toxicity for healthy tissues is reduced.

Tumour treatment with few side effects requires both encapsulation and targeted release of the cytostatic agent directly into the tumour cell. The latter can be accomplished by attaching the active ingredient *via* a covalent acid-labile bond because, in contrast to healthy tissue (pH 7.4), a relatively low pH value of *ca.* 5.5 is found in tumour cells (Fig. 8.20).

A prerequisite for the release mechanism is that the (dendritic) carrier should be soluble in water. The periphery of dendrimer molecules is usually hydrophobic and they are hence poorly soluble in aqueous media. Dendrimers with reverse polarity of the outer shell were therefore synthesised. For example, the ester functions of PAMAM dendrimers can be saponified so that carboxylate groups appear on the surface in place of the usual amino functions (ammonium at acidic pH values) [63]. The cytostatic agents adriamycin and methotrexate could be encapsulated in PAMAM dendrimers – attached to polyethylene glycol [63 d]. Dendrimers have also been used both as nanocarriers for cytostatic drugs and for simultaneous monitoring of drug uptake in the tumour cell. For this purpose actetylated PAMAM dendrimers having a diameter of <5 nm were conjugated with folic acid (as tumour recognition agent) and then coupled both to methotrexate as active ingredient as well as to fluorescein. These nanoparticles were injected into mice with implanted human KB tumours. In contrast to polymers without a recognition agent, the dendrimers containing folic acid remained concentrated in the tumour tissue for four days. Methotrexate could

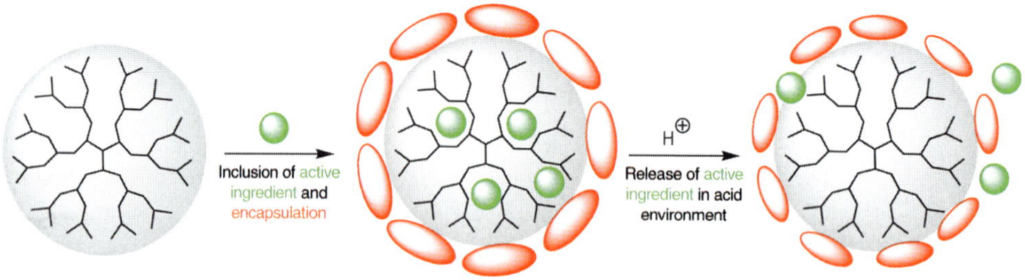

Fig. 8.20 Molecular encapsulation of a cytostatic agent in a dendrimer and release at sites with acid pH (according to Haag *et al.*)

thus be dosed in such a way that the anti-tumour properties were enhanced and the dark toxicity significantly reduced, a result that could not be achieved by simple administration of the active ingredient [64].

8.7.2
Gene therapy

In the fight against cancer and chronic illnesses gene therapy raises hopes extending beyond the realm of conventional medical methods. Two kinds of therapy are distinguished. In the *in-vivo* procedure the intact gene enclosed in a carrier (vehicle, "gene taxi") is introduced directly into the diseased area. In the *ex-vivo* variant, on the other hand, specific cells taken from the patient are genetically modified and cultured before being re-injected. Conventional gene therapy employs viral gene transfer. The advantages of this method are its high efficiency and specificity. However, it has the drawback of high immunogenicity; that is to say, reactions occur with antibodies against wild type viruses which are already present in the organism. Recent non-viral variants attempt to circumvent this disadvantage and to enhance efficiency while reducing toxicity.

PAMAM dendrimers were used for this purpose; their positive surface charge leads to a distinctly spheroid form at higher generations. In the pH range of 7–8, the terminal amino groups give rise to an overall positive charge. Therefore stable associates can be formed with negatively charged genes or DNA and transported into a cell or even into the cell nucleus. The dendrimers thus serve, as mentioned above, as a vehicle for transporting genes through membranes. Remarkably, it is precisely the non-perfect dendrimers that display significantly better transfection efficiencies, apparently because of their greater conformational flexibility and less pronounced steric effects (compared to perfect dendrimers) [65].

Several years ago, the Qiagen company launched the dendritic transfection reagent Poly-Fect® on the market. This is an "activated" PAMAM dendrimer. Activation consists in thermal treatment, in the course of which individual branches of the dendrimer are cleaved off (Fig. 8.21). Under physiological conditions its

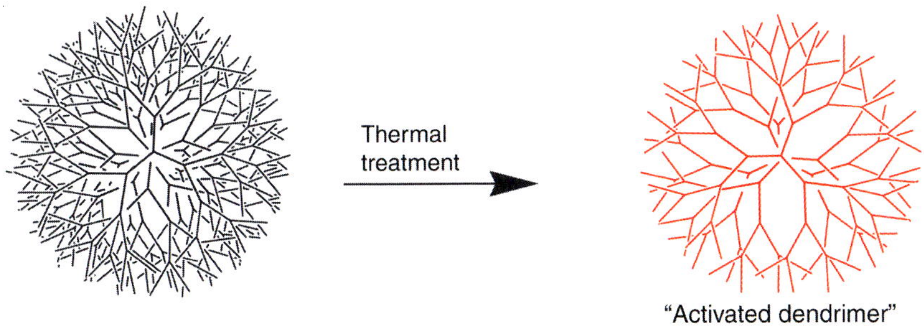

Fig. 8.21 PAMAM dendrimer thermolytically activated for gene transfer (schematic)

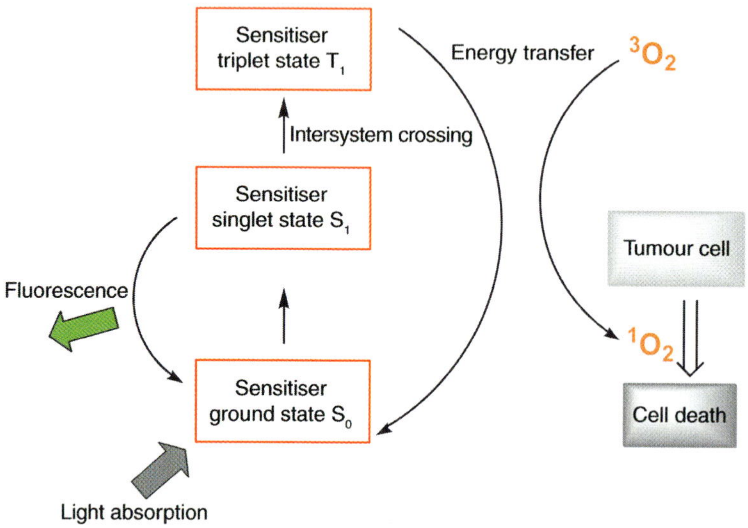

Fig. 8.22 Mode of action of photosensitisers in tumour therapy (schematic)

terminal amino groups are positively charged and interact with negatively charged phosphate groups of nucleic acid. The commercially available reagent is able to transform the DNA into a compact structure which greatly facilitates uptake in the eukaryotic cell. After having fused with the endosomes, the PolyFect® reagent effectively buffers pH inhibition of the lysosomal nuclease (Fig. 8.22). This raises the stability of the PolyFect-DNA complex and increases the transfer of intact DNA into the cell nucleus [66].

8.7.3
Photodynamic therapy

In photodynamic therapy, tumour tissue undergoes light-induced destruction in the presence of oxygen, after a photosensitiser has been injected to the diseased tissue. The light energy absorbed by the photosensitiser leads to formation of singlet oxygen which has a cytotoxic action (Fig. 8.22). The photosensitiser itself is subsequently excreted. It thus acts solely as catalyst. Low-molecular-weight photosensitisers showed toxic side effects which are detrimental for clinical applications.

On use as a photosensitiser, a third-generation polycationic porphyrin dendrimer synthesised by Aida *et al.* [67] with 32 quaternary ammonium groups and 32 carboxylic acid groups exhibits a significant singlet oxygen toxicity towards lung cancer cells and as well as a minimal dark toxicity (Fig. 8.23) [68]. Owing to their dendritic architecture, dendrimer porphyrins can transfer absorbed energy over a relatively large distance from the periphery to the porphyrin core, and thus represent potential photosensitisers for photodynamic therapy.

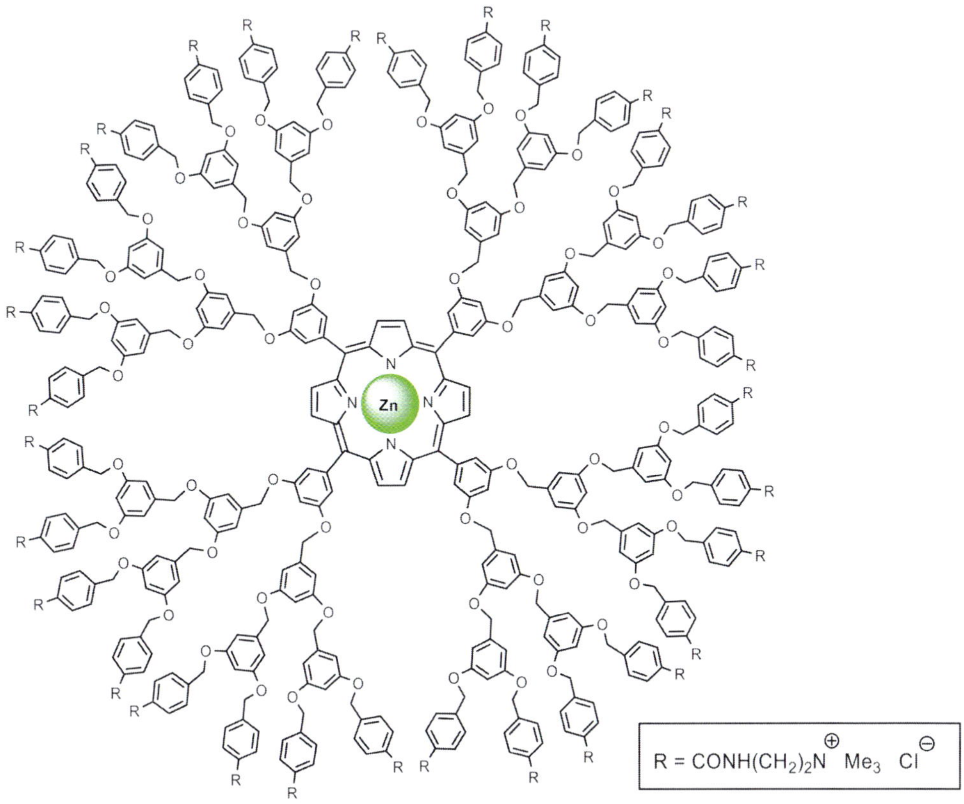

Fig. 8.23 Porphyrin dendrimer (according to Aida *et al.*)

Tetraphenylporphyrins [69] are hydrophobic compounds and are unable to penetrate unaided into the water-containing tissues of the human body. This only becomes possible on incorporation into a dendrimer or a liposome. Use in the human body has not yet been tested but preliminary experiments have been performed on ruptured red blood cells, erythrocyte ghosts. The blood cells were washed out, and the tetraphenylporphyrin was introduced into the membrane of the ghosts and subjected to photophysical investigation [70].

8.7.4
Dendrimers in prevention against HIV

The Australian company Starpharma has developed a vaginal microbiocide (VivaGel), as the first dendrimer-based pharmaceutical to receive official approval for clinical testing from the American Food & Drug Administration (FDA). Phase 1 clinical tests have been successfully completed, and the product, which causes neither irritation nor inflammation, is expected to come onto the market as a preventative against HIV infection in 2008 [71 a]. The mechanism of action of the substance utilises the polyvalent properties of the dendrimer. The functional end groups of the dendrimer can act as a "molecular adhesive tape". They enter into multiple interactions with the receptors located on biological structures such as cell membranes or viruses. The active ingredient of VivaGel is a fourth-generation polylysine dendrimer (see Section 4.1.4) with 32 naphthalene-disulphonate entities attached to its scaffold *via* amide bonds.

This polyionic structure can provide protection against HIV infection by binding to the gp120 glycoprotein receptor on the virus surface. The normal course of infection of healthy cells with HIV begins when the gp120 protein on the surface of the virus binds to the CD4 receptor on the surface of the healthy cell [71 b].

8.7.5 Culture of organs and tissue

The technique of tissue engineering involves culturing of living cells of an organism outside of the body, possibly combining them with extracellular components, and subsequently reimplanting them. The advantage of such implants lies in their acceptance by the immune system because the cultured cells are recognised as the body's own cells and are therefore not rejected. A drawback is the loss of functionalities, which is why only skin could so far be cultured. Increasing demand for tissue, for example for burns patients, and waiting lists for organ transplants call for alternatives and supplementary measures.

Grinstaff succeeded in growing bio-dendrimers *via* divergent synthesis with branched AB_2 monomers, which could be of future use in the clinical sector. Adducts of glycolic acid and lactic acid or derivatives of adipic acid are examples of the monomers used [72 a].

8.7.5.1 Wound healing

Various aspects of wound healing were investigated in an animal model relevant for clinical phase testing [72 b]. For this purpose, polyamidoamine dendrimers (PAMAM) were constructed, some functionalised with glucosamine groups, others with glucosamine-6-sulphate groups (Fig. 8.24). Glucosamine dendrimers act as immunomodulators, but glucosamine-6-sulphate dendrimers as anti-angiogenic substances.

After application of both dendrimers, rabbits which had undergone an eye operation for glaucoma showed better healing as well as less scar tissue formation and less inflammation [73].

S.I. Stupp reported studies on dendrimers for gel formation in bone fissures which have possible applications in the treatment of bone fractures [74].

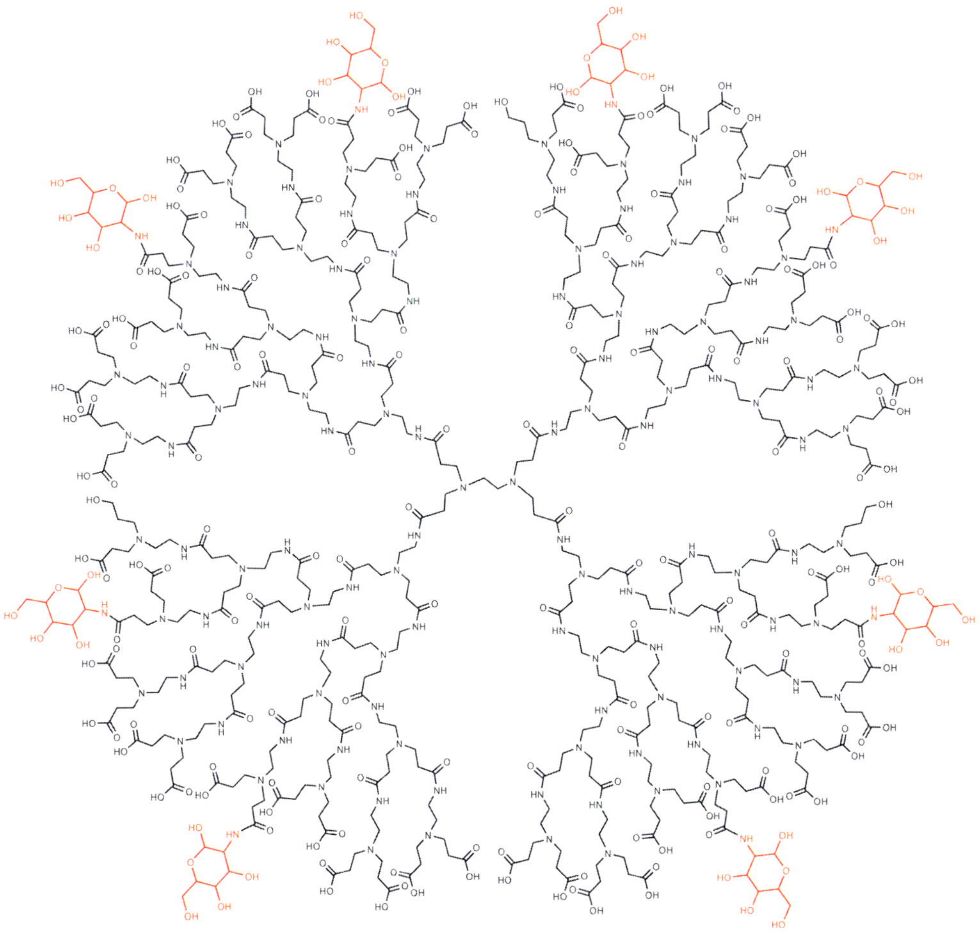

Fig. 8.24 PAMAM-based glucosamine dendrimer (according to Shaunak *et al.*)

8.7.6
Boron neutron capture therapy

In boron neutron capture therapy (BNCT) ^{10}B isotopes are introduced into the tumour and irradiated with neutrons. Slow neutrons are captured by the boron isotopes; the resulting ^{11}boron isotope decays with emission of high-energy α-radiation having a range of 9 µm, corresponding roughly to the diameter of a cell. This means that only that cell is destroyed in which capture takes place.

For the patient this means a prior injection of a non-radioactive compound containing the stable ^{10}B isotope which accumulates selectively in tumour tissue. In a second step, the person is exposed to low-energy neutrons which destroy the tumour as described above.

The incorporation of ^{10}B isotopes in water-soluble dendrimer scaffolds with coupling sites for tumour-targeting ligands is an innovative principle for the development of boron-rich stable compounds with high anti-tumour activity, not least because a higher concentration of ^{10}B in the tumour cell can be attained by attachment to the dendrimers and the dark toxicity is low. Corresponding boron-cluster-containing dendrimers for boron neutron capture therapy have already been synthesised [75]. In this therapy it is important to attain a maximum possible concentration of ^{10}B isotopes in the tumour tissue. The incorporation of boron compounds into PAMAM dendrimers was a highly promising concept, but alongside the good immunoreactivity an undesired high level of accumulation was also observed in the liver and the spleen [76]. A polylysine dendrimer with 80 terminal boron atoms coupled to antibody fragment gave encouraging results [77]. In order to raise the efficiency of tumour targeting with antibodies, a short polypeptide, the so-called epidermal growth factor (EGF), was coupled to fourth-generation PAMAM dendrimer. *In vitro* experiments showed a certain selectivity for the EFG receptor in the case of brain tumours [78 a].

8.8 Dendrimers in nanotechnology

Owing to the readily and precisely adjustable nanometre dimensions and the ease of multifunctionalisation, dendrimer molecules are predestined for a role in nanotechnology [60, 78 b, c].

8.8.1 Photoswitchable dendrimers

Starting from 1,3,5-tris(bromomethyl)benzene as core unit, Vögtle *et al.* constructed the first dendrimer with intramolecularly accumulated azobenzene entities in the periphery by convergent synthesis (Fig. 8.25); it expectedly exhibits complex photoswitchable properties owing to the possibility of intramolecular *E*/*Z* isomerisation [79, 80].

8.8.2 Dendrimers as impellers

Dendritic molecules may possibly be of use in nanoscale "impellers" or "valves". For this purpose azobenzene dendrons [80] with terminal hydroxyl groups (Fig. 8.27) were covalently fixed in the interior of well-defined silicon dioxide nanotubes. In the sol/gel process, the hydroxyl groups lead to incorporation of azobenzene groups into the SiO_2 nanotubes being formed (Fig. 8.26) [81].

Depending on the wavelength of the incident light, the azobenzenes undergo light-induced switching from the *E* to the *Z* and from the *Z* to the *E* conformer, which oppose diffusion of guest substances to different degrees. Given a suit-

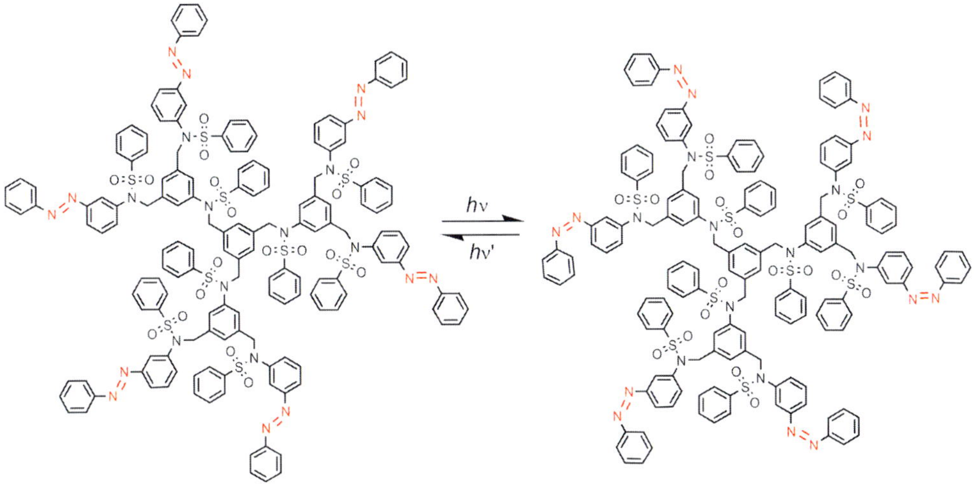

Fig. 8.25 The first photoswitchable dendrimer (according to Vögtle *et al.*); intermediates with partly isomerised azo groups are not shown

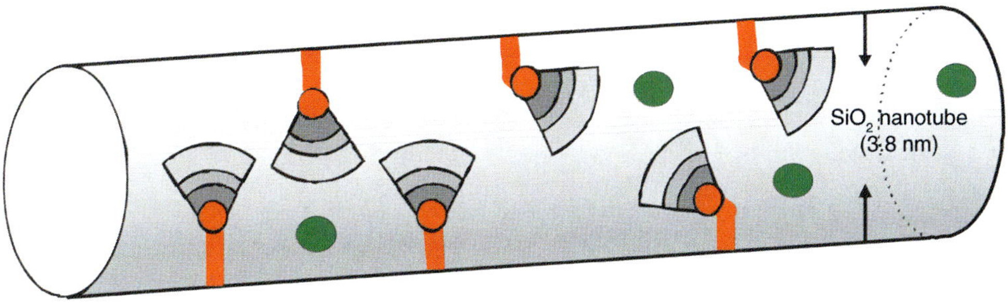

Fig. 8.26 Potentially switchable nanoimpeller (schematic); the cones stand for dendritic substituents; azo groups (red) are in the *E* (straight) or *Z* configuration (angled); guest substances are shown in green (according to Zink, De Cola, Vögtle *et al.*)

able size of the dendrons (Fig. 8.27) – shown as cones in Fig. 8.26 – light-driven passage of ions or molecules through the impeller might be possible. In the case of the *E* configuration, migration of molecules through the impeller is more likely to be subject to steric hindrance.

8.8.3
Dendrimers as nanotubes

Nanotubes are regarded as highly promising building blocks for a range of applications, for example in the areas of nanoelectronics. A prerequisite for the synthesis of dendrimer tubes [82] is the use of pore-filling guest compounds

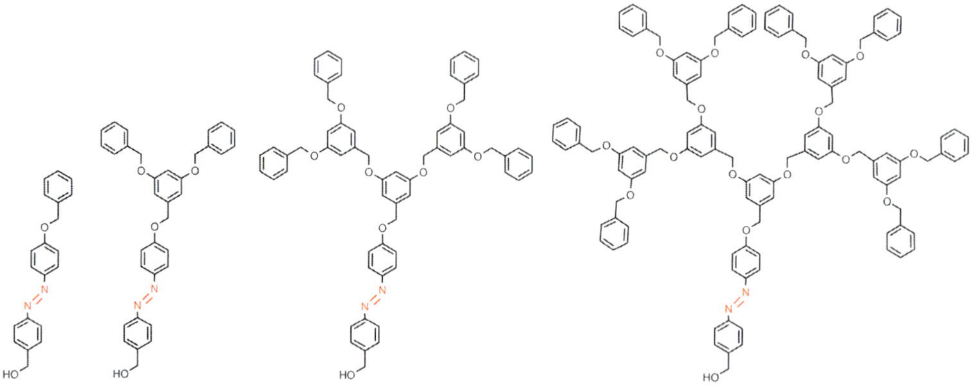

Fig. 8.27 Fréchet dendron-substituted azobenzenes of the zeroth to third generation with $-CH_2OH-$ functionalisation (according to Zink, De Cola, Vögtle *et al.*)

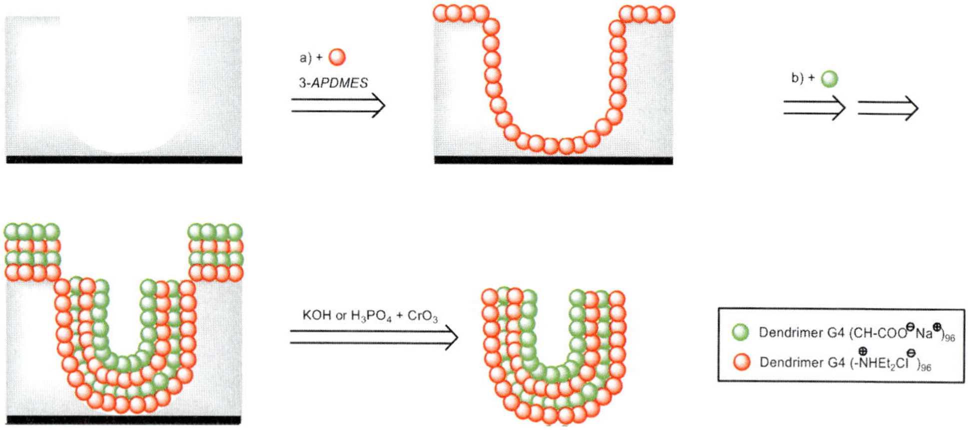

Fig. 8.28 Production of dendrimer nanotubes (3-APDMES = 3-aminopropyldimethoxysilane); according to Masuda *et al.* (schematic)

[83] as templates. For this purpose, the interior cavity walls of a well-defined pore-containing (concave) aluminium template [84] were lined with a layer of 3-aminopropyldimethoxysilane (3-APDMES), which acquires a positive charge at acid pH values (Fig. 8.28). The second layer applied was a fourth-generation *N,N*-substituted hydrazine dendrimer negatively charged through the presence of 96 acetate end groups, which was followed by dendrimer layers with alternating charges, consisting, for example, of the same dendrimer scaffold with 96 ammonium (salt) end groups. The two oppositely charged dendrimers were dissolved in water at pH 5.5 and, depending upon the layer to be applied, the tem-

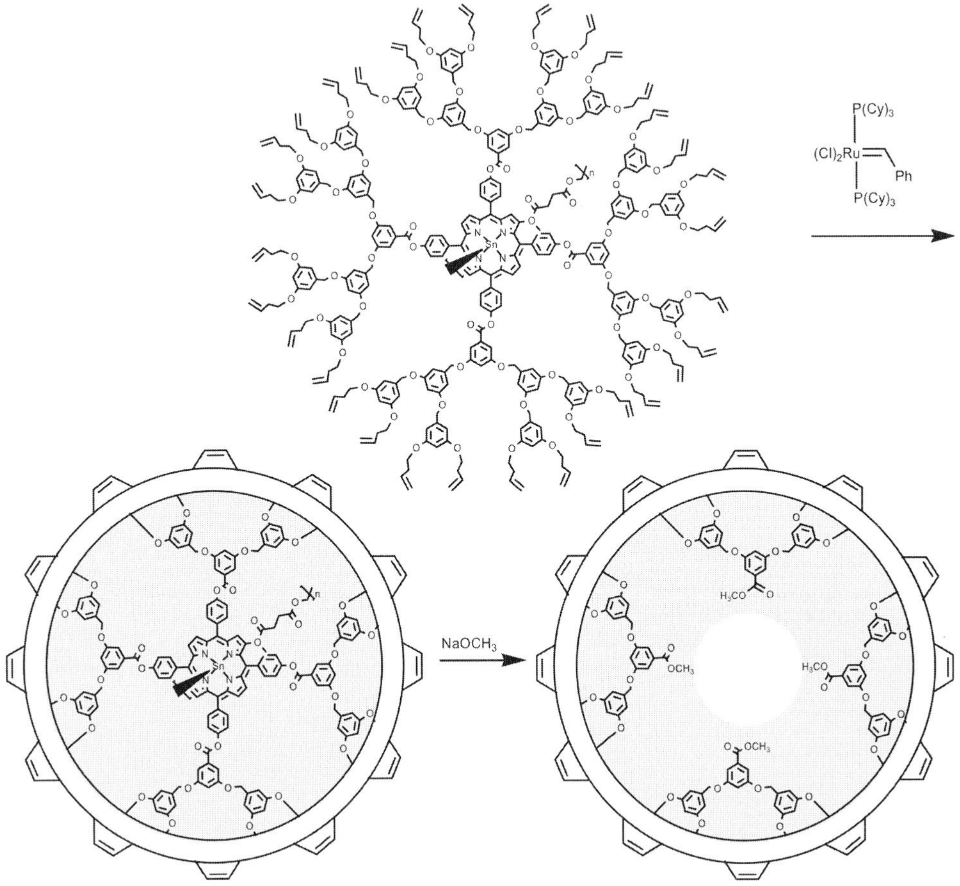

Fig. 8.29 Organic nanotubes by core removal from polylysine dendrimers (according to Zimmerman *et al.*)

plate alternatingly submersed in the appropriate dendrimer solution. The desired free dendrimer tubes were obtained by etching away the inorganic template with chromium(III) oxide (in phosphoric acid) or with potassium hydroxide solution.

The dendrimer tubes prepared by this multilayer process have only one open end and probably show a mechanical stability gradient, indicating that the effective number of layer-deposition steps in the proximity of the closed end is lower than elsewhere because the ionic dendrimer to be precipitated has limited access there. Production of uniform tubes would therefore require templates with two openings.

Dendrimer tubes, with an environment of numerous functional groups, could be considered for various applications. A conceivable approach would be to functionalise a porous membrane by lining the walls of the pores with dendrimers which can act as hosts for guest molecules. Such architectures could be useful for applications in the sensor sector or in the area of analytical separation science. Inclusion of other types of nanoparticles is also thinkable.

Construction of organic nanotubes starting from porphyrin dendrimers with core/shell architecture is also feasible. Figure 8.29 also shows how covalent nanotubes can be produced by removal of the dendritic component of the molecule. A coordination polymer is first synthesised from a dendritic metalloporphyrin with alkene end groups. This is subjected to intramolecular and intermolecular crosslinking by ring-closing metathesis at the periphery.

Applications of such nanotubes could make use of the possibility of varying the wall thickness of the tubes according to the generation of the dendrons used. The internal diameter of the tubes could also be designed and controlled by appropriate choice of core units. The functionality of the exterior and interior of the wall can be controlled by the choice of suitable starting materials [85].

8.8.4 Dendritic polymers as templates

Rapid progress in the area of microelectronics has created a need for highly functional new compounds. The symbiosis of high storage density and high response speeds for new microchips places demands on new materials, for example for insulating layers between conductor strips. Use of nanoporous organic layers has the advantage that dielectric constants less than 2 should be attainable. Dendrimers and highly branched polymers can act as potential porogens. Highly branched polyesters with thermolabile triazine groups in the main chain were incorporated as template in a temperature-stable matrix. The desired pores are formed on subsequent thermal crosslinking of the matrix with decomposition of the porogen. In an applications-oriented metal-insulator-metal arrangement, the dielectric constant of 2.7 could be lowered to a value of 2.2 [86 a].

The examples of applications of dendrimers are admittedly often only of a potential character, yet nevertheless reflect the broad range of possibilities covering practically all areas from materials of construction to medicine [86 b]. Precisely the latter area paces strict demands on applications with respect to purity and compatibility with the human organism.

Bibliography and Notes for Chapter 8 "Special properties and potential applications"

Review articles are indicated by the words "Review(s)" or "Book(s)" in bold-faced type.

1 ***Reviews***: a) G. E. Oosterom, J. N. H. Reek, P. C. J. Kamer, P. W. N. M. van Leeuwen, *Angew. Chem.* **2001**, *113*, 1878–1901; *Angew. Chem. Int. Ed.* **2001**, *40*, 1828–1849; b) D. Astruc, F. Chardac, *Chem. Rev.* **2001**, *101*, 2991–3023; c) D. Astruc, F. Lu, J. R. Aranzaes, *Angew. Chem.* **2005**, *117*, 8062–8083; *Angew. Chem. Int. Ed.* **2005**, *44*, 7852–7872; d) S. Hecht, J. M. J. Fréchet, *Angew. Chem.* **2001**, *113*, 76–94; *Angew. Chem. Int. Ed.* **2001**, *40*, 74–91.

2 M. B. Meder, I. Haller, L. H. Gade, *Dalton Trans.* **2005**, 1403–1415; A. Tuchbreiter H. Werner, L. H. Gade, *Dalton Trans.* **2005**, 1394–1402; SFB 623 of the University of Heidelberg: http://www.sfb623.uni-hd.de/

3 F. Morgenroth, E. Reuter, K. Müllen, *Angew. Chem.* **1997**, *109*, 647–649; *Angew. Chem. Int. Ed* **1997**, *36*, 631–634; F. Morgenroth, C. Kübel, K. Müllen, *J. Mater. Chem.* **1997**, *7*, 1207–1211; F. Morgenroth, A. J. Berresheirn, M. Wagner, K. Müllen, *Chem. Commun.* **1998**, 1139–1140; V. S. Iyer, K. Yoshimura, V. Enkelmann, R. Epsch, J. P. Rabe, K. Müllen, *Angew. Chem.* **1998**, *110*, 2843–2846; *Angew. Chem. Int. Ed.* **1998**, *37*, 2696–2699; F. Morgenroth, K. Müllen, *Tetrahedron* **1997**, *53*, 15349–15366; U. M. Wiesler, K. Müllen, *Chem. Commun.* **1999**, *22*, 2293–2294; ***Review***: U.-M. Wiesler, T. Weil, K. Müllen, *Top. Curr. Chem.* (Volume Ed. F. Vögtle) **2001**, *212*, 1–40.

4 U. Lüning, T. Marquardt *J. Prakt. Chem.* **1999**, *341*, 222–227; cf. J. P. W. Eggert, U. Lüning, *Eur. J. Org. Chem.* **2007**, 6046–6052.

5 J. W. J. Knapen, A. W. van der Made, J. C. de Wilde, P. W. N. M. van Leeuwen, P. Wijkens, D. M. Grove, G. van Koten, *Nature* **1994**, *372*, 659–663.

6 P. Bhyrappa, J. K. Young, J. S. Moore, K. S. Suslick, *J. Am. Chem. Soc.* **1996**, *118*, 5708–5711; P. Bhyrappa, J. K. Young, J. S. Moore, K. S. Suslick, *J. Mol. Catal. A* **1996**, *113*, 109–116; P. Bhyrappa, G. Vaijayanthimala, K. S. Suslick, *J. Am. Chem. Soc.* **1999**, *121*, 262–263.

7 a) M. T. Reetz, G. Lohmer, R. Schwickardi, *Angew. Chem.* **1997**, *109*, 1559–1562; *Angew. Chem. Int. Ed.* **1997**, *36*, 1526–1529; N. Brinkmann, D. Giebel, G. Lohmer, M. T. Reetz, U. Kragl, *J. Catal.* **1999**, *183*, 163–168; b) Dendronized solid phase: A. Dahan, M. Portnoy, *J. Amer. Chem. Soc.* **2007**, *129*, 5860–5869.

8 M. Kimura, Y. Sugihara, T. Muto, K. Hanabusa, H. Shirai, N. Kobayashi, *Chem. Eur. J.* **1999**, *5*, 3495–3500.

9 U. Kragl, C. Dreisbach, *Angew. Chem.* **1996**, *108*, 684–685; *Angew. Chem. Int. Ed.* **1996**, *35*, 642–644; N. Brinkmann, D. Giebel, G. Lohmer, M. T. Reetz, U. Kragl, *J. Catalysis* **1999**, *183*, 163–168.

10 ***Reviews***: D. Seebach, P. B. Rheiner, G. Greiveldinger, T. Butz, H. Sellner, *Top. Curr. Chem.* (Volume Ed. F. Vögtle) **1998**, *197*, 125–164; *general*: T. Ikariya, K. Murata, R. Noyori, *Org. Biomol. Chem.* **2006**, *4*, 393–406.

11 a) H. Brunner, *J. Organomet. Chem.* **1995**, *500*, 39–46; b) ***Review*** of "artificial enzymes" in general: R. Breslow (Ed.) *Artificial Enzymes*, Wiley-VCH, Chichester/Weinheirn **2005**; c) ***Overviews***: T. Darbre, J.-L. Reymond, *Acc. Chem. Res.* **2006**, *39*, 925–934; d) ***Book***: L. H. Gade (ed.), Dendrimer Catalysis, *Top. Organomet. Chem.* **2006**, 20.

12 H. Brunner, S. Altmann, *Chem. Ber.* **1994**, *127*, 2285–2296.

13 M. S. T. H. Sanders-Hoven, J. F. G. A. Jansen, J. A. J. M. Vekemans, E. W. Meijer, *Polym. Mater. Sci. Eng.* **1995**, *210*, 180–182.

14 C. Bolm, N. Derrien, A. Seger, *Synlett* **1996**, 387–388.

15 ***Books***: J.-M. Lehn, *Supramolecular Chemistry. Concepts and Perspectives*, VCH, Weinheim **1995**; J. W. Steed, J. L. Atwood, *Supramolecular Chemistry*, Wiley, Chichester **2000**; F. Vögtle, *Supramolekulare Chemie*, Teubner, Stuttgart 1989, **1992**; *Supramolecular Chemistry*, Wiley-VCH, Chichester **1993**; J.-H. Fuhrhop, J. Köning, *Membranes and Molecular Assemblies: The Synkinetic Approach*, The Royal Society of Chemistry, Cambridge **1994**; K. Ariga, T. Kunitake, *Supramolecular Chemistry – Fundamentals and Applications*, Springer, Berlin/Heidelberg **2006**.

16 G. R. Newkome, C. N. Moorefield, G. R. Baker, A. L. Johnson, R. K. Behera, *Angew. Chem.* **1991**, *103*, 1205–1207; *Angew. Chem. Int. Ed.* **1991**, *30*, 1176–1178; G. R. Newkome, G. R. Baker, C. N. Moorefield, M. J. Saunders, *Polym. Prepr.* **1991**, *32*, 625–626; G. R. Newkome, C. N. Moorefield **1992**, *U. S. Pat.* 5154853.

17 ***Review***: H. Frey, K. Lorenz, C. Lach, *Chemie in unserer Zeit* **1996**, *30*, 75–85.

18 F. M. Winnick, J. M. Duff, G. G. Sacripante, A. R. Davidson (Xerox Corp.), USA 5256516 A 931026, **1993**; *Chem. Abstr.* **1994**, *120*, 90707i.

19 www.xeroxtechnology.com

20 www.colour-europe.de/pf_812_forschung_nano-4.htm; B. Bruchmann, *Baummoleküle im Nanomaßstab: Dendrimere für neue Drucksysteme und Autolacke.*

21 B. Voigt, *Chemie in Dresden*, **2004**, 94–99.

22 S. Onclin, J. Huskens, B.J. Ravoo, D.N. Reinhoudt, *Small* **2005**, *8–9*, 852–852; ***Review***: *Molecular Machines, Top. Curr. Chem.* (Volume Ed.T.R. Kelly) **2005**, *262*.

23 J.F.G.A. Jansen, E.M.M. de Brabander-van den Berg, E. Meijer, *Science*, **1994**, *266*, 1226–1229; J.F.G.A. Jansen, E.W. Meijer, *J. Am. Chem. Soc.*, **1995**, *117*, 4417–4418; see also ***Review***: D.J. Cram, J.M. Cram, *Container Molecules and Their Guests*, RSC, Cambridge **1994**.

24 J.F.G.A. Jansen, R.A.J. Jansen, E.M.M. de Brabander-van-den Berg, E.W. Meijer, *Adv. Mater.* **1995**, *7*, 561–564.

25 R. Warmuth, *Angew. Chem.* **1997**, *109*, 1406–1409; *Angew. Chem. Int. Ed.* **1997**, *36*, 1347–1350; R. Warmuth, *J. Inclusion Phenom.* **2000**, *37*, 1–38; ***Overview***: R. Warmuth, J. Yoon, *Acc. Chem. Res.* **2001**, *34*, 95–105; D.J. Cram, M.E. Tanner, R. Thomas, *Angew. Chem.* **1991**, *103*, 1048–1051; *Angew. Chem. Int. Ed.* **1991**, *30*, 1024–1027.

26 D.J. Cram, S. Karbach, Y.H. Kim, L. Baczynskyj, G.W. Kalleymen, *J. Am. Chem. Soc.* 1985, *107*, 2575–2576; D.J. Cram, M.E. Tanner, R. Thomas, *Angew. Chem.* **1991**, *103*, 1048–1051; *Angew. Chem. Int. Ed.* **1991**, *30*, 1024–1027.

27 M.W.P.L. Baars, P.E. Froehling, E.W. Meijer, *Chem. Commun.* **1997**, 1959–1960.

28 http://www.cdtltd.co.uk

29 H. Meier, M. Lehmann, *Angew. Chem.* **1998**, *110*, 666–669; *Angew. Chem. Int. Ed.* **1998**, *37*, 643–645; H. Ma, A. K.-Y. Jen, *Adv. Mater.* **2001**, *13*, 1201–1205.

30 D. Hertel, C.D. Müller, K. Meerholz, *Chemie in unserer Zeit* **2005**, *39*, 336–347; ***Book***: K. Müllen, U. Scherf (Eds.), *Organic Light-Emitting Devices, Synthesis, Properties, and Applications*, Wiley-VCH, Chichester, Weinheim **2006**.

31 L.A. Khotina, L.S. Lepnev, N.S. Burenkova, P.M. Valetsky, A.G. Vitukhnovsky, *J. Luminescence* **2004**, *110*, 232–238.

32 J.P.J. Markham, E.B. Namdas, T.D. Anthopoulos, I.D.W. and G.J. Richards, P.L. Burn, *Appl. Phys. Lett.* **2004**, *85*, 1463–1465.

33 ***Reviews***: F. Vögtle, *Supramolekulare Chemie*, Teubner, Stuttgart **1989**, **1991**; *Supramolecular Chemistry*, Wiley-VCH, Chichester **1993**. U. Finkenzeller, *Spektrum der Wissenschaft* **1990**, 54–62.

34 H.-S. Kitzerow, B. Westermann, *ForschungsForum* of the University of Paderborn 6–**2003**, 86–88; A. Sunder, M-F. Quincy, R. Mülhaupt, H. Frey, *Angew. Chem.* **1999**, *111*, 3107–3110; *Angew. Chem. Int. Ed.* **1999**, *38*, 2928–2930.

35 a) Cf. H. Shao, J.W. Lockmann, J.R. Parkette, *J. Amer. Chem. Soc.* **2007**, *129*, 1184–1185; b) A. Esposito, E. Delort, D. Lagnoux, F. Djojo, J.-L. Reymond, *Angew. Chem.* **2003**, *115*, 1419–1421; *Angew. Chem. Int. Ed.* **2003**, *42*, 1381–1383; E. Delort, T. Darbre, J.-L. Reymond, *J. Am. Chem. Soc.* **2004**, *126*, 15642–15643; D. Lagnoux, T. Darbre, M.L. Schmitz, J.-L. Reymond, *Chem. Eur. J.* **2005**, *11*, 3943–3950; J. Kofoed, T. Darbre, J.-L. Reymond, *Org. Biomol. Chem.* **2006**, 3268–3281.

36 C. Rao, J.P. Tam, *J. Am. Chem. Soc* **1994**, *116*, 6975; K. Sadler, J.P. Tam, *Rev. Mol. Biotechnol.* **2002**, *90*, 195–229; cf. also J.P. Mitchell, K.D. Roberts, J. Langley, F. Koenten, J.N. Lambert, *Bioorg. Med. Chem. Lett.* **1999**, *9*, 2785–2788; I. van Baal, H. Malda, S.A. Synowsky, J.L.J. van Dongen, T.M. Hackeng, M. Merkx, E.W. Meijer, *Angew. Chem.* **2005**, *117*, 5180–5185; *Angew. Chem. Int. Ed.* **2005**, *44*, 5052–5057.

37 J.P. Collman, L. Fu, A. Zingg, F. Diederich, *Chem. Commun.* **1997**, 193–194.

38 P.J. Dandliker, F. Diederich, M. Gross, C.B. Knobel, A. Louati, E.M. Sanford, *Angew. Chem.* **1994**, *106*, 1821–1824; *Angew. Chem. Int. Ed.* **1994**, *33*, 1739–1742; P.J. Dandliker, F. Diederich, J.-P. Gisselbrecht, A. Louati, M. Gross, *Angew. Chem.* **1995**, *107*, 2906–2909; *Angew. Chem. Int. Ed.* **1995**, *34*, 2725–2728; P. Weyermann, J.-P. Gisselbrecht, C. Boudon, F. Diederich, M. Gross, *Angew. Chem.* **1999**, *111*, 3400–3404; *Angew. Chem. Int. Ed.* **1999**, *38*, 3215–3219.

39 T. K. Lindhorst, *Chemie in unserer Zeit* **2000**, *34*, 38–52.

40 a) G. Thoma, A. G. Katopodis, N. Voelcker, R. O. Duthaler, M. B. Streiff, *Angew. Chem.* **2002**, *114*, 3327–3330; *Angew. Chem. Int. Ed.* **2002**, *41*, 3195–3198; b) D. Appelhaus, Y. Zhong, H. Komber, P. Friedel, U. Oertel, U. Scheler, N. Morgner, D. Kuckling, S. Richter, B. Brutchy, B. Voit, *Macromol. Biosci.* **2007**, 373–383.

41 R. Schumacher, *Chemie in unserer Zeit* **1999**, *33*, 268–278.

42 C. Heil, G. R. Windscheif, S. Braschohs, F. Flörke, J. Gläser, M. Lopez, J. Müller-Albrecht, U. Schramm, J. Bargon, F. Vögtle, *Sensors and Actuators B 61*, **1999**, 51–58; see also C. Ziegler, W. Göpel, H. Hämmerle, H. Hatt, G. Jung, L. Laxhuber, H.-L. Schmidt, S. Schütz, F. Vögtle, A. Zell, *Biosensors & Bioelectronics* **1998**, *13*, 539–371; BASF AG: Method and Apparatus for the Detection of Fumigants in Air Samples, EP 0202206, 28.02.02.

43 U. Herrmann, T. Jonischkeit, J. Bargon, U. Hahn, Q.-Y. Li, C. A. Schalley, E. Vogel, F. Vögtle, *Anal. Bioanal. Chem.* **2002**, *372*, 611–614.

44 M. Schlupp, T. Weil, A. J. Berresheim, U. M. Wiesler, J. Bargon, K. Müllen, *Angew. Chem.* **2001**, *113*, 4124–4129; *Angew. Chem. Int. Ed.* **2001**, *40*, 4011–4015.

45 M. Albrecht, M. Schlupp, J. Bargon, G. van Koten, *Chem. Commun.* **2001**, *18*, 1874–1875.

46 B. Voigt, *Chemie in Dresden*, **2004**, 94–99; G. Belge, D. Beyerlein, C. Betsch, K.-J. Eichhorn, G. Gaulitz, K. Grundke, B. Voigt, *Anal. Bioanal. Chem.* **2002**, *374*, 403–411.

47 M. Albrecht, G. van Koten, *Adv. Mater.* **1999**, *11*, 171–174; M. Albrecht, R. A. Gossage, M. Lutz, A. L. Spek, G. van Koten, *Chem. Eur. J.* **2000**, *6*, 1431–1445.

48 V. Balzani, P. Ceroni, S. Gestermann, C. Kauffmann, M. Gorka, F. Vögtle, *J. Chem. Soc., Chem. Commun.* **2000**, 853–854; F. Vögtle, S. Gestermann, C. Kaufmann, P. Ceroni, V. Vicinelli, V. Balzani, *J. Am. Chem. Soc.* **2000**, *122*, 10398–10404.

49 O. S. Finikova, V. V. Rozhkov, M. Cordero, S. A. Vinogradov, 3rd International Dendrimer Symposium, Berlin, September **2003**, Abstracts pp. 54–55.

50 O. Finikova, A. Galkin, V. Rozhkov, M. Cordero, C. Hägerhäll, S. Vinogradov, *J. Am. Chem. Soc.* **2003**, *125*, 4882–4893.

51 T. D. Janes, H. Shinmori, M. Takeuchi, S. Shinkai, *J. Chem. Soc., Chem. Commun.* **1996**, 705–706.

52 A. P. Silva, S. A. de Silva, *J. Chem. Soc., Chem. Commun.* 1986, 1709–1710; A. P. de Silva, H. Q. N. Gunaratne, T. Gunnlaugson, A. J. M. Huxley, C. P. McCoy, J. T. Rademacher, T. E. Rice, *Adv. Supramol. Chem.* **1997**, *4*, 1–53; A. P. de Silva, H. Q. N. Gunaratne, T. Gunnlaugson, A. J. M. Huxley, C. P. McCoy, J. T. Rademacher, T. E. Rice, *Chem. Rev.* **1997**, *97*, 1515–1566; C. S. Foote, *Acc. Chem. Res.* **1998**, *31*, 199–324; M. Sauer *Angew. Chem.* **2003**, *115*, 1834–1835; *Angew. Chem. Int. Ed.* **2003**, *42*, 1790–1793.

53 C. Saudan, V. Balzani, M. Gorka, S.-K. Lee, M. Maestri, V. Vicinelli, F. Vögtle, *J. Am. Chem. Soc.* **2003**, *125*, 4424–4425; C. Saudan, V. Balzani, P. Ceroni, M. Gorka, M. Maestri, V. Vicinelli, F. Vögtle, *Tetrahedron* **2003**, *59*, 3845–3852; C. Saudan, V. Balzani, M. Gorka, S.-K. Lee, J. van Heyst, M. Maestri, P. Ceroni, V. Vicinelli, F. Vögtle, *Chem. Eur. J.* **2004**, *10*, 899–905.

54 M.-S. Choi, T. Aida, H. Luo, Y. Araki, O. Ito, *Angew. Chem.* **2003**, *115*, 4194–4197; *Angew. Chem. Int. Ed.* **2003**, *42*, 4060–4063; M.-S. Choi, T. Yamazaki, I. Yamazaki, T. Aida, *Angew. Chem.* **2004**, *116*, 152–160; *Angew. Chem. Int. Ed.* **2004**, *43*, 150–158.

55 W.-S. Xia, R. H. Schmehl, C.-J. Li, *Eur. J. Chem.* **2000**, 387–389.

56 A. Koller, *Appl. Fluoresc. Technol.* **1989**, *1*, 1–8; A. Mayer, S. Neuenhofer, *Angew. Chem.* **1994**, *106*, 1097–1126; *Angew. Chem. Int. Ed.* **1994**, *33*, 1044–1072.

57 a) S.-E. Stiriba, H. Frey, R. Haag, *Angew. Chem.* **2002**, *114*, 1385–1390; *Angew. Chem. Int. Ed.* **2002**, *41*, 1329–1334; b) ***Overviews***: A. Kundu, B. Keppler, *Nachr. Chemie* **2007**, *55*, 843–846; N. Tirelli (ed.), Special Issue "Branched Macromolecular Structures and their Bio-applications", *Macromolecular*

Bioscience 2007 (issue 8; Wiley-VCH); c) Dendritic dipeptide pores: M. Peterca, V. Percec, A. E. Dulcey, S. Nummelin, S. Korey, M. Ilies, P. A. Heiney, *J. Amer. Chem. Soc.* **2006**, *128*, 6713–6720.

58 ***Reviews***: M. Fischer, F. Vögtle, *Angew. Chem.* **1999**, *111*, 934–955; *Angew. Chem. Int. Ed.* **1999**, *38*, 884–905; G. R. Newkome (Ed.), *Advances In Dendritic Macromolecules* **2002**, Volume 5; C. S. Winalski, S. Shortkroff, R. V. Mulkern, E. Schneider, G. M. Rosen, *Magnetic Resonance in Medicine* **2002**, *48*, 965–972; *Contrast Agents I, Top. Curr. Chem.* (Volume Ed. W. Krause), **2002**, *221*; *Contrast Agents II, Top. Curr. Chem.* (Volume Ed. W. Krause), **2002**, *222*; *Contrast Agents III, Top. Curr. Chem.* (Volume Ed. W. Krause), **2005**, *252*; H.-J. Weinheim, W. Ebert, B. Misselwitz, B. Radüchel, H. Schmitt-Willich, J. Platzek, *Eur. Radiol.* **1997**, *7*, 196; contrast agents from Schering AG, Berlin, information at: http://www.schering-diagnostics.de/phys/products/mri/info.htm

59 S. E. Stiriba, H. Frey, R. Haag, *Angew. Chem.* **2002**, *114*, 1385–1390; *Angew. Chem. Int. Ed.* **2002**, *41*, 1329–1334.

60 a) ***Book***: I. Majoros, J. Baker, Jr. (eds), *Dendrimer-Based Nanomedicine*, World Scientific Publishing Co., London **2006**; ***Overview***: V. Marx, *Nature Biotechnology* **2008**, *26*, 727–732; b) ***Book***: U. Boas, J. B. Christensen, P. M. H. Heegaard, *Dendrimers in Medicine and Biotechnology*, RSC Publishing, Cambridge **2006**.

61 a) R. Haag, *Angew. Chem.* **2004**, *116*, 280–284; *Angew. Chem. Int. Ed.* **2004**, *43*, 278–282; M. Krämer, J.-F. Stumbé, H. Türk, S. Krause, A. Komp, L. Delineau, S. Prokhorova, H. Kautz, R. Haag, *Angew. Chem.* **2002**, *114*, 4426–4431; *Angew. Chem. Int. Ed.* **2002**, *41*, 4252–4256; L. Fernandez, M. Gonzalez, H. Cerecetto, M. Santo, J. J. Silber, *Supramol. Chemistry* **2006**, *18*, 633–643; b) Polyglycerol dendrimers for drug and dye transport: M. Radowski, A. Schukla, H. von Berlepsch, Ch. Böttcher, G. Pickaert, H. Rahage, R. Haag, *Angew. Chem.* **2007**, *119*, 8312–8315; *Angew. Chem. Int. Ed.* **2007**, *46*, 8164–8167.

62 Y. Matsumura, H. Maeda, *Cancer Res.* **1986**, *46*, 6387–6392; H. Maeda, Y. Matsumura, *CRC Crit. Rev. Ther. Drug Carrier Syst.* **1989**, *6*, 193–210.

63 a) C. Kojima, K. Kono, K. Maruyama, T. Takagishi, *Bioconjugate Chem.* **2000**, *11*, 910–917; b) M. Liu, K. Kono, J. M. J. Fréchet, *J. Controlled Release* **2000**, *65*, 121–131; c) R. Haag, J.-F. Stumbe, A. Sunder, H. Frey, A. Hebel, *Macromolecules* **2000**, *33*, 8158–8166; d) M. W. P. L. Baars, R. Kleppinger, M. H. J. Koch, S.-L. Yeu, E. W. Meijer, *Angew. Chem.* **2000**, *112*, 1341–1344; *Angew. Chem. Int. Ed.* **2000**, *39*, 1285–1288; e) L. J. T. Twyman, A. E. Beezer, R. Esfand, M. J. Hardy, J. C. Mitchell, *Tetrahedron Lett.* **1999**, *40*, 1743–1746.

64 J. F. Kukowska-Latallo, K. A. Candido, Z. Cao, S. S. Nigavekar, I. J. Majoros, T. P. Thomas, L. P. Balogh, M. K. Khan, J. R. Baker, Jr., *Cancer Res.* **2005**, *65*, 5317–5324.

65 ***Reviews***: H. Stephan, H. Spies, B. Johannsen, K. Gloe, U. Hahn, F. Vögtle, *Mensch und Umwelt* **2001/2002**, 29–33; M. Weber, *Nachr. Chem.* **2000**, *48*, 18–23; R. Haag, *Angew. Chem.* **2004**, *116*, 280–284; *Angew. Chem. Int. Ed.* **2004**, *43*, 278–282.

66 a) PolyFect Transfection Reagent Handbook 09/**2000**; J. P. Behr, *Chimia* **1997**, *51*, 34–36; b) Details of binding and transport of RNA and DNA to (of) dendrimers: J. H. Zhou, Y. Y. Wu, N. Hafdi, J.-P. Behr, P. Erbacher, L. Peng, *J. Chem. Soc., Chem. Comm.* **2006**, 2362–2364; M. A. Kostiainen, G. R. Szilvay, D. K. Smith, M. B. Lindner, O. Ikkala, *Angew. Chem.* **2006**, *118*, 3618–3622; *Angew. Chem. Int. Ed.* **2006**, *45*, 3538–3542; c) M. Guillot, S. Eisler, K. Weller, H. P. Merkle, J.-L. Gallini, F. Diedrich, *Org. Biomol. Chem.* **2006**, 766–769; ***Overview***: V. Marx, *Nature Biotechnology* **2008**, *26*, 729–732.

67 N. Nishiyama, H. R. Stapert, G.-D. Zhang, D. Takasu, D.-L. Jiang, T. Nagano, T. Aida, K. Katuoka, *Bioconjugate Chem.* **2003**, *14*, 58–66.

68 D.-L. Jiang, T. Aida, *J. Am. Chem. Soc.* **1998**, *120*, 10895–10901.

69 ***Book***: L. Kaestner, *Tetraphenylporphyrine – Farbstoffe für die photodynamische Therapie*, Logos-Verlag, Berlin **1997**.

70 http://www.medizin-netz.de/science/pdt.htm.

71 a) B. Halford, *Chemical & Engineering News* **2005**, *13*, 30–36; http://nanotech-wire.com; b) J.F. Bernejo, P. Ortega, L. Chonco, R. Eritja, R. Samaniego, M. Müllner, E. de Jesús, F.J. de la Mata, J.C. Flores, R. Gómes, A. Muñoz-Fernándes, *Chemistry Eur. J.* **2007**, *13*, 483–495

72 M.W. Grinstaff, *Chem. Eur. J.* **2002**, *8*, 2839–2846; M. Walthier, S.M. Johnson, T. Kim, M.W. Grinstaff, *Bioconjugate Chem.* **2006**, *17*, 873–876; see also J. Kessler, S. Stupp: stupp.northwestern.edu/news/index.ttml.

73 S. Shaunak, S. Thoas, E. Gianasi, A. Godwin, E. Jones, I. Teo, K. Mireskandari, P. Luthert, R. Duncan, S. Patterson, P. Khaw, S. Brocchini, *Nature Biotechnology* **2004**, *22*, 977–984; M. Walthier, S.M. Johnson, T. Kim, M.W. Grinstaff, *Bioconjugate Chem.* **2006**, *17*, 873–876.

74 S.I. Stupp, lecture at EURESCO Symposium "Supramolecular Chemistry" in Obernai, October **2005**.

75 G.R. Newkome, J.N. Keith, G.R. Baker, G.H. Escamilla, C.N. Moorefield, *Angew. Chem.* **1994**, *106*, 701–703; *Angew. Chem. Int. Ed.* **1994**, *33*, 666–668; J. Nemoto, J. Cai, Y.J. Yamamoto, *J. Chem. Soc., Chem. Commun.* **1994**, 577–580; D. Armspach, M. Cattalini, E.C. Constable, C.E. Housecroft, D. Philips, *Chem. Commun.* **1996**, 1823–1824; W. Yang, R.F. Barth, D.M. Adams, A.H. Soloway, *Cancer Res.* **1997**, *57*, 4333–4339.

76 B. Qualmann, M.M. Kessels, H.-J. Musiol, W.D. Sierralta, P.W. Jungblut, L. Moroder, *Angew. Chem.* **1996**, *108*, 970–973; *Angew. Chem. Int. Ed.* **1996**, *35*, 909–911.

77 R.F. Barth, D.M. Adams, A.H. Soloway, F. Alam, M.V. Darby, *Bioconjugate Chem.* **1994**, *5*, 58–66.

78 a) J. Capala, R.F. Barth, M. Bendayana, M. Lauzon, D.M. Adams, A.H. Soloway, R.A. Fenstermarker, J. Carlsson, *Bioconjugate Chem.* **1996**, *7*, 7–15; b) ***Books***: J.W. Steed, D.R. Turner, K.J. Wallace, Core Concepts in Supramolecular and Nanochemistry, Wiley, Chichester **2007**; J. Atwood, J.W. Steed (ed.), *Organic Nanostructures*, Wiley-VCH, Weinheim **2008**; ***Overviews***: V. Marx, *Nature Biotechnology* **2008**, *26*, 729–732; C.P. Martin, M.O. Blunt, E. Vaujour, A. Fahmi, A. D'Aléo, L. De Cola, F. Vögtle, P. Moriarty, in N. Krasnogor, S. Gustafson, D.A. Pelta, J.L. Verdegay (eds.), *"System Self-Assembly"*: Multidisciplinary Snapshots. *Studies in Multidisciplinarity*, Elsevier, Amsterdam **2008**, vol. 5, 1–20; c) K.K. Perkin, K.M. Bromley, S.A. Davis, A. Hirsch, C. Böttcher, S. Mann, *Small* **2007**, *3*, 2057–2060; d) S.A. Backer, I. Suez, Z.M. Fresco, M. Rolandi, J.M.J. Fréchet, *Langmuir* **2007**, *23*, 2297–2299.

79 H.-B. Mekelburger, K. Rissanen, F. Vögtle, *Chem. Ber.* **1993**, *126*, 1161–1169; P. Sierocki, H. Maas, P. Dragut; G. Richardt, F. Vögtle, L. De Cola, F.A.M. Brower, J.I. Zink, *J. Phys. Chem. B* **2006**, *110*, 24390–24398; For other azobenzene dendrimers see chapter 5.2.2; ***Overview***: T. Nagasaki, S. Shinkai, *J. Incl. Phenom. Macrocycl. Chem.* **2007**, 58, 205–219.

80 *Advances in Dendritic Macromolecules* (Ed. G.R. Newkome), Volume 5, Elsevier, Amsterdam **2002**; O. Villavicencio, D.V. McGrath 1–44; S. Wang, X. Wang, L. Li, R.C. Advincula, *J. Org. Chem.* **2004**, *69*, 9073–9084.

81 P. Sierocki, H. Maas, P. Dragut, G. Richardt, F. Vögtle, L. De Cola, F. (A. M.) Brouwer, J.I. Zink, *J. Phys. Chem. B* **2006**, *110*, 24390–24398; EU project LIMM: Light Induced Molecular Movements, IST-2001-35503 (**2002–2005**); S. Angelos, E. Choi, F. Vögtle, L. De Cola, J.I. Zink, *J. Phys. Chem. C* **2007**, *111*, 6589–6592; cf. ***Books*** on general aspects: Y. Zhao, *Smart Light-Responsive Materials*, Wiley-Blackwell, Indianapolis 2009; Ch. Dugave (ed.), *cis-trans Isomerisation in Biochemistry*, Wiley-VCH, Weinheim **2006**; For photomobile polymer materials and light driven plastic motors, see: M. Yamada, Y. Yu, M. Kinoshita, C.J. Barret, T. Ikeda, *Angew. Chem.* **2008**, *120*, 5064–5066; *Angew. Chem. Int. Ed.* **2008**, *47*, 4986–4988; cf. M. Groß, *Chem. Unserer Zeit* **2008**, *42*, 254.

82 D.H. Kim, P. Karan, P. Göring, J. Leclaire, A.-M. Caminade, J.-P. Majoral, U. Gösele, M. Steinhart, W. Knoll, *Small* **2005**, *1*, 99–102.

83 C. R. Martin, *Science* **1994**, *266*, 1961–1966; C. R. Martin, *Acc. Chem. Res.* **1995**, *28*, 61–68.

84 H. Masuda, K. Fukuda, *Science* **1995**, *268*, 1466–1468; H. Masuda, F. Hasegaa, S. Ono, *J. Electrochem. Soc.* **1997**, *144*, L1340–L1342; K. Nielsch, J. Choi, K. Schwirn, R. B. Wehrspohn, U. Gösele, *Nano Lett.* **2002**, *2*, 677–680.

85 Y. Kim, M. F. Mayer, S. C. Zimmerman, *Angew. Chem.* **2003**, *115*, 1153–1158; *Angew. Chem. Int. Ed.* **2003**, *42*, 1121–1126.

86 a) M. Eigner, B. Voigt, K. Estel, J. W. Bartha, *e-polymers* **2002**, 28; B. Voigt, *Chemie in Dresden*, **2004**, 94–99; b) Sol-gel phase transitions through aggregation of nano fibres with hydrophilic PEG-dendrimers: K.-S. Moon, H.-J. Kim, E. Lee, M. Lee, *Angew. Chem.* **2007**, *119*, 6931–6934; *Angew. Chem. Int. Ed.* **2007**, *45*, 6807–6810; cf. D.-J. Hong, E. Lee, M. Lee, *J. Chem. Soc., Chem. Comm.* **2007**, 1801–1803; ***Overview***: V. Marx, *Nature Biotechnology* **2008**, *26*, 729–732.

Outlook

As demonstrated on the pages of this book, dendritic molecules of low and high molecular weight and multiply branched substituents have meanwhile established a firm presence throughout the entire realm of chemistry. Numerous teams all over the world are actively pursuing research in this area. Several types of dendrimers of various generations have become commercially available. Individual applications have already been launched on the market, others are in the development stage.

What can we expect of dendrimer chemistry in the future? While most of the dendrimers hitherto synthesised are limited to five generations at most, and the highest generation number so far attained is a little over ten, in the longer term we can look forward to accessing pure dendrimers of higher generations in the two-figure or perhaps even three-figure range, with a correspondingly large number of peripheral groups, and to their detailed characterisation thanks to improved analytical techniques.

Further elements of the periodic system will be incorporated into the core, branches, and periphery of dendrimers. High-purity dendritic silicones still appear to be absent. Whereas purely aromatic dendritic hydrocarbons have been described, aliphatic C_xH_y dendrimer are hardly known. This will require efficient non-symmetrical C-C coupling reactions; posing a challenge to develop fundamentally new synthetic approaches yielding minimal amounts of by-products.

There will be advances in the chemical functionalisation of dendrimers with regard to dendron building blocks, especially non-trivial, multiple, selective, and orthogonal functionalisation of the periphery. Chemically functional – including chiral – and physically functional entities are increasingly being introduced into dendritic molecules in a targeted fashion. The manifold possibilities of structural variation and adaptation have their benefits and drawbacks: The former lie in the opportunity to optimise certain properties; the latter are seen in the considerable effort often required in the absence of relevant experience and theoretical models permitting accurate prediction of structure/property relationships (including chirality).

There is also a lack of straightforward standard and low-cost syntheses for higher generations of certain dendritic scaffolds, such as those in which nitrogen atoms are responsible for branching. As our knowledge of dynamic non-

Dendrimer Chemistry. Fritz Vögtle, Gabriele Richardt and Nicole Werner

ISBN: 978-3-527-32066-0

covalent and adaptive systems [*] increases, greater attention will be focussed on supramolecular dendrimers.

Dendritic branching, starting from shape-persistent macrocycles, with branches extending into the interior of the rings instead of protruding outwards (*i.e.* intraannular branches), could become accessible.

The targeted production of dendroisomers and their exact analysis raise hopes that it will become possible to design such large molecules rich in "inner workings" with a view to obtaining specific properties, such a solubility, melting point, thermal stability, cavities.

Attempts will be made to gain access to accurate purity or defect information for individual dendrimers, which will permit uniform classification of a substance as a pure cascadane or hyperbranched compound with a higher or lower degree of dispersion.

Major applications could lie in nanotechnology, paints, surface coatings, catalysts, film formation, optoelectronics, gelators, diagnostics.

There are clearly ample opportunities and challenges for research, development, and applications in this fascinating field. Only novel molecular and supramolecular architectures will lead to fundamentally new high-chem materials.

Owing to the wide range of topics addressed in dendrimer chemistry and the many points of contact with other disciplines, future developments will also be driven by door-opening advances achieved in neighbouring fields.

[*] J.-M. Lehn, Lecture presented at the University of Bologna on December 4, 2006: "From supramolecular chemistry to constitutional dynamic chemistry"; J.-M. Lehn, "From supramolecular chemistry towards constitutional dynamic chemistry and adaptive chemistry", *Chem. Soc. Rev.* **2007**, *36*, 151–160.

Subject Index

a

adamantanetetracarboxlic acid 135
adamantyl group 296 f.
additive 294
additivity rule 279
adhesive 294
aggregation 97 ff.
alkaloid derivative 152
– dendritically substituted 152
allosteric transmission 163
3-aminopropyldimethoxysilane (3-APDMES) 322
amphifullerene 112 f.
– hexakis-malonic ester adduct 112
antenna effect 176 ff.
antennae 68
arborol 4
– peripherally tryptophan-substituted 158
arene
– dendrimer 93
arene component
– condensed 93
arylplatinum(II) complex 305
asymmetric catalysis 161
– chiral dendrimer 161
atomic force microscopy (AFM) 271 ff.
– image of dendrimer 273
atropine unit 151
O-(7-azabenztriazol-1-yl)-*N,N,N'*,N'-tetramethyluronium hexafluorophosphate, *see* HATU
azobenzene
– Fréchet dendron-substituted 322
azobenzene unit
– methyl orange 189
– pentafluorophenyl ester-activated 64
– photoisomerisation 188

b

barbiturate guest molecule 209
bathochromic shift 212
bifunctionalisation 61 ff., 307
– non-covalent local 65
(*S*)-1,1'-bi-2-naphthol 148
BINAP ligand
– dendritic 229
(*S*)-BINOL 229
– catalyst 230
bioluminescence 114
biomimetics 300 ff.
biosensor 313
biphenyl dendrimer
– amphiphilic 106
bis(methyldiphenylphosphin)amine 53
boron neutron capture therapy (BNCT) 319
branching unit
– functionality 10
– metal complex 34

c

carbohydrate dendrimer 108
carbosilane dendrimer 120 ff.
– crown-ether decorated 124
– cyclic core unit 126
– deactivation mechanism 225
– dendritic dumbbell 122
– glycosidic 122
– nickel-loaded 225
– thiophene 123
carbosiloxane 124
carbosiloxane dendrimer 123
carboxylate 114
cascadane 14 ff.
– nomenclature 17
cascadaplexes 15
16-Cascade:1,4-diaminobutane[4-N,N,N',N']:(1-azabutylidene)2:aminopropane 16
cascade molecule 1, 15
cascadyl 17

Dendrimer Chemistry. Fritz Vögtle, Gabriele Richardt and Nicole Werner

ISBN: 978-3-527-32066-0

catalysis 224 ff., 290
– dendritic effect 224
– enantioselective 293
– phase transfer 294
– shape-selective 231
catalyst
– exodendral and endodendral fixation 290
catalyst support 291
cauliflower dendrimer 125
cell transfection 315
charge depot 242
charge funnel 242
charge separation 244
charge transfer interaction 223
chemical composition 281
chemical reaction
– covalent 195
– dendritic molecule 195 ff.
– metathesis 195 ff.
chemosensor 311
– fluorescing 308
chiral amplification 161 ff.
chiral attenuation 161
chiral dendrimer 143 ff., 239
– application 146 ff., 158 ff.
– asymmetric catalysis 161
– chiral core and achiral branching scaffold 146 ff.
– chiral end group 158
– chiral spacer or chiral branching unit 156 ff.
– classification 143
– scaffold 156
chiral dendritic structure
– chiroptical study 279
chirality
– information 164
– periphery 158
– transmission 150
chiroptical method 276 ff.
chiroptical study 145
chiroselective host/guest recognition 162
chromatography 253 ff.
– high performance liquid 254
– liquid 253 f.
chromoionophore 66
chromophore 66 f.
circular dichroism (CD) 276 ff.
click chemistry 31, 54
cobalt(II)
– phthalocyanine complex 231
– salen complex 233
cobaltocenium unit
– terminal 85
complexation 33
– coordinative 133
– supramolecular 133
compression 212
conductivity
– charge-transfer complex-based 240
container molecule 212
contrast agent 311
convergent synthesis 27
– double-stage convergent method 29
coordination-chemical synthesis 33
coordinative complexation 133
copper
– dendritic catalyst 226
– metallodendrimer with Cu(I)-phenanthroline core 134
core
– functional 50
– functionality 10
core unit
– metal complex 33
Cotton effect 151 ff., 278
cross-linker 201
cryptochirality 149 ff.
cyclam core unit 184
cyclam dendrimer 310
β-cyclodextrin 217 ff., 296 f.
cytochrome mimic
– dendritic 239
cytochrome C 112
cytostatic agent 313

d

DAB-dendr-(NHCONH-Ad)$_n$ 206
DAB dendrimer 136, 237
DAB-POPAM 292
dansyl fluorescence 214 f.
DCC (dicyclohexylcarbodiimide) 108
defect 13
degree of branching 13
dendrimer 1 ff.
– achiral 81 ff., 144
– additive 294
– adhesive 294
– AFM image 273
– analysis 253 ff.
– biocompatible 146
– carbohydrate-coated 108
– carbon/oxygen-based 102
– carbosilane 225
– catalyst support 291

– catalytic 292
– characteristics 289 f.
– characterisation 253 ff.
– chiral, *see* chiral dendrimer
– DAB-based 233 ff.
– DNA 313
– flexible 268 ff.
– Fréchet, *see* Fréchet dendrimer
– functional, *see* functional dendrimer
– guest molecule 217
– hybrid 89
– impeller 320
– interior coordination site 136
– intermolecular interaction 270
– intramolecular interaction 280
– ionic 111 ff.
– liquid-crystalline 299
– locally bifunctionalised 65
– luminescent 306
– medical application 311 ff.
– membrane reactor 292
– metal core 133
– metal at the branching point 137
– metal in the dendritic branches 137
– metallo-porphyrin 232
– monofunctional 49 ff.
– multichromophoric 67 ff.
– multifunctional 49, 61, 72 ff.
– multiple receptor unit 207
– naphthyl-decorated 185
– nickel 224
– olefin metathesis 199
– optically active 228
– peripherally complexed metal 135
– peripherally homogeneous 52
– phase transfer catalysis 294
– phosphorus-based 114
– photoisomerisation 176
– photoswitchable 191, 320
– pigment 294
– polyanionic 111 ff.
– polycarboxylic acid 200
– polycationic 114 ff.
– porphyrin-based 109
– property 289 ff.
– protein 301
– pseudorotaxane-terminated fourth-generation 205
– radial segment density distribution 268
– redox active 243
– ruthenium-carbene complexed 197
– segmented-block 28
– self-assembly 207 f.
– silane 117 ff.
– silicon-based 117 ff.
– soft 274
– STM image 274
– supramolecular 208
– surface-block 28, 62
– synthesis 81 ff.
– TEM image 275
– tube 323
– type 71 ff., 81 ff.
dendrimer effect 12
dendrimer growth
– internal modification on conclusion 58
– modification 55
– peripheral group 54
dendrimer metamorphosis 58
dendrimer periphery
– non-covalent modification 204
dendrimer scaffold
– functional unit 55
dendrimer structure
– small-angle neutron scattering (SANS) 268
– small-angle X-ray scattering (SAXS) 268
dendriplexes 15
dendritic anode 298
dendritic antenna 92
– light harvesting 178
dendritic architecture 7
dendritic blade 153
dendritic box 148, 211, 297
dendritic catalysis 224 ff.
– metal-containing 224 ff.
– metal-free 234 ff.
dendritic chirality effect 161
dendritic effect 151, 221, 244
– catalysis 224
– electrochemical property 240
– inclusion of guest 223
dendritic energy funnel 180
dendritic hydrocarbon 93
dendritic micelle 214 ff.
dendritic molecule
– chirality 161
– covalent introduction of functionality in the interior 203
– definition and classification 15
– guest inclusion in dendritic host molecule 207
– photophysical property 169 ff.
– starfish-like 117
– synthetic method 25
– nomenclature 16

dendritic multiporphyrin system 182
dendritic network 137
dendritic peptide 36
dendritic polymer 295, 306, 324
dendritic protective shield 175
dendritic ruthenium complex 181
dendritic sensor layer 304
dendritic sugar sensor 308
dendritic stopper group 220
dendritic unit 19
dendrizyme 226 ff.
dendrocleft 153
dendro isomer 42 ff.
dendroknot 149 f.
dendron 9 ff., 103
– first generation (G1) 220
– inclusion 219
dendronised linear polymer (denpol) 39 ff.
– cylindrical shape 42 f.
dendrophane 149
denpol 42
dense-core model 12
dense-core structure 268
dense-shell model 12
dense-shell structure 268
Dexter mechanism 172
diagnostics 300
diamine 1
1,4-Diaminobutane[*N,N,N′,N′*]:{4-azabutyl(4,4)}$^{G_1,G_2}{}_{4n,8n}$:3-aminopropyl$_{16}$-cascadane 17
1,4-Diaminobutane[*N,N,N′,N′*]:{4,7-diaza-3-oxoheptyl(7,7)}$^{G_1,G_2}{}_{4n,8n}$:3-aminoethyl$_{16}$-cascadane 21
diaminobutyl unit 136
diaza-Cope rearrangement 153
dicobalt-octacarbonyl 137
dicyclohexylcarbodiimide, *see* DCC
diene 95 ff.
dienophile 95 ff.
diffusion ordered spectroscopy (DOSY) 262
dimension 281
N-(3-dimethylaminopropyl)-2-naphthalene-1-sulfonamide 185
dinitrile 1
DIPEA (diisopropylethylamine) 108
diphenylphosphine group 132
dispersity 13
display 298
divergent synthesis 25 f.
DNA
– dendrimer 313
– hybridisation 313
donor/acceptor interaction 223
double-exponential method 29
double-stage convergent method 29
drug carrier 270, 313 f.
dye 296 f.

e

E/Z isomerisation 320
E/Z photoreaction 188
electrochemical property
– dendritic effect 240
electronics 298
enantioselective catalysis
– dendrimer 293
enantioselective receptor 153
end group 281
– distribution 269
endoreceptor 314
energy transfer (ET) 172 ff.
– dipole-dipole interaction 172
– host → guest 210
– radiative emission 172
– through the dendritic shell to the azobenzene unit 191
eosin Y 210
ESI-MS 263 ff.
excimer 171
excited dimer 171
exoreceptor 314

f

ferrocene 132, 243
– chiral 157
– core 218
– optically active 135
ferrocenyldiphosphine ligand 135
fluorescence 170, 214 ff., 308
– dansyl 214 f.
– sensor 155
– yield 171
fluorescent dendrimer 175
fluorescent dimer 173
fluorescent guest 174 f.
fluoroionophore 308
fluorophore 173, 308 f.
foam formulation 295
Förster mechanism 172
fractal 17
fractal structure 8
Fréchet amphiphile
– benzene-based 106
Fréchet dendrimer 5, 102 ff.

Fréchet dendron 103, 147, 218 ff.
Fréchet-type stopper 220
fullerene
– porphyrin/zinc complex 246, 309
fullerene dendrimer 179
fullerene dendron (fullerodendron) 113
functional dendrimer 49 ff., 71
– functional core 71
– functional core and internal functional unit 73
– functional core, internal functional units, and functionalised molecular periphery 74
– functional molecular periphery 71
– globally bifunctionalised periphery 72
– internal functional unit 71
– locally bifunctionalised molecular periphery 73
– multifunctional 72 ff.
– synthesis 71 ff.
functional unit
– dendrimer scaffold 55
– different functionality 67
– in different parts of the molecule 67
functionalisation
– convergently grown dendrimer/dendron 54
– divergently synthesised dendrimer 52
– follow-on 53
– terminal group 52
functionality
– branching unit 10
– core 10

g
Gadomer-24 311 f.
galabiose 122
gel electrophoresis 259
gel permeation chromatography (GPC) 257
gene therapy 315
generation 10
– limiting 11 ff.
β-D-glucopyranosyl group 218
glucosamine dendrimer
– PAMAM-based 319
glycinylurea guest molecule
– POPAM dendrimer 206
glycocluster 302
glycodendrimer 108, 302
– self-assembly 303
glycomimetics 302
glycopeptide dendrimer 110
– first-generation 110
'graft-to' method 39
'graft-from' method 39
Grubbs catalyst 202
guest molecule
– π-electron donor/acceptor interaction 305
– molecular tweezer 220
– viologen-type 220
guest inclusion
– dendritic host molecule 207
– diffusion 214
– steric compression 212

h
Hamilton receptor 207
– dendrimer 210
– macrocyclic 209
– peripheral 209
HATU (*O*-(7-azabenztriazol-1-yl)-*N*,*N*,*N*′,N′-tetramethyluronium hexafluorophosphate) 108
heme monooxygenase 240
high performance liquid chromatography (HPLC) 254
HIV protection 317 f.
host/guest energy transfer 210
host/guest interaction
– stereoselective 153 f.
host/guest recognition
– chiroselective 162
host/guest system
– supramolecular 232
host molecule 243
Hoveyda catalyst 198
hydrocarbon cascade scaffold
– quarter-directional 216
hydrocarbon dendrimer 5
hyperbranched molecule 15
hyperbranched polymer 38
hypercore 29, 103 ff., 214
– second generation 103
hypermonomer method 30

i
impeller 320
imprint 201
ink
– printing 295
internal group 281
intradendrimer quenching effect 92
ionic dendrimer 111 ff.
– porphyrin 184

ionisation method 263 f.
ionophore 310
– ligand 66
iptycene 93
E/Z isomerisation 320

j
Josiphos ligand 135

k
kick-line analogy 163

l
lanthanide ion 36, 134
layer-block dendrimer 29
light-emitting diode (PLED) 299
limiting generation 11 ff.
liquid-crystalline dendrimer 299
liquid-crystalline phase
– stilbenoid dendrimer 97
luciferase 114
luminescence 114, 169
– type 169
luminophore 173
D-lysine 36
L-lysine 36, 63

m
macrocyclisation 207
macromonomer method 41
magnetic resonance imaging (MRI) 311
majority rules model 164
MALDI-MS 263 ff.
Mandala dendrimer 112
Mandelbrot set 8
Mark–Houwink relation 11
mass spectrometry 263 ff.
medical application 313
medical diagnostics 311
membrane reactor 292
– catalytic dendrimer 292
membrane technology 290
metal complex
– branching unit 34
– core unit 33
metallo-carbosiloxane dendrimer 126
metallodendrimer 33 ff., 133, 231 f.
– catalyst 198
– Cu(I)-phenanthroline core 134
– dicobalt-hexacarbonyl unit 137
– optically active ferrocene entity 135
– ruthenium(II)-bis(terpyridine) 245
metalloporphyrin 231 f.
– shape-selective catalysis 231
metathesis 195 ff.
methyl orange 189
– push/pull azobenzene derivative 189
N-methylnaphthalenesulfonamide 185
micellanoate 216
micellanol 294
micellar molecule 111
microdomain 153
mismatch destabilisation 165
modifier 131
molar mass 259, 281
molar rotation value 148
molecular encapsulation 314
molecular imprinted polymer (MIP) 201
molecular imprinting 201 ff., 296 f.
– dendrimer 202, 296 f.
molecular periphery
– bifunctionalised 62
molecular recognition 102 ff.
molecular tweezer 219 ff.
monoamine 1
monodispersity 14
multifunctionalisation 59, 70 ff.
multiple antigen peptide (MAP) 33, 301
multiporphyrin system
– dendritic 182

n
nanocapsule 296
nanofiltration 292
nanoimpeller 320
nanoreactor 292
nanotechnology 320
nanotube 321 f.
– organic 323
naphthalene fluorophore 310
naphthyl-decorated dendrimer 185
neodymium(III) ion
– polylysine dendrimer 90
Newkome dendrimer 133
Newkome nomenclature 16
nickel
– carbosilane dendrimer 225
nickel complex
– dendritic carbosilane 226
– *para*-phenylene-spaced porphyrin heneicosameric 109
nickel dendrimer 224
NIPAM-PAMAM dendrimer 88
NMR spectroscopy 260 f.
– 1D 261
– diffusion 262

– dynamic 263
– multidimensional 261 f.
nodal nomenclature 19
non-linear optics (NLO) 299
norbornene 198
nuclear Overhauser effect (NOE) 260

o
octopus molecule 2
olefin metathesis 196 ff.
oligo(dibenzyl sulphone) dendrimer 58
oligo-lysine dendrimer
– dansyl-decorated 92
optical rotatory dispersion (ORD) 276 f.
optoelectronics 298
organ culturing 318
organic light-emitting diode (OLED) 299
orthogonal synthesis 28

p
paint 295
palladium nanoparticle 213
PAMAM (poly(amidoamine)) 3, 52, 85 ff., 238
– activated 315
– 8-Co-PAMAM catalyst 234
– generation 87
– glucosamine dendrimer 319
– D-glucose 111
– limiting generation 88
– nitrilotripropionamide core unit 87
– palladium nanoparticle 213
– second generation 20, 241
– tetrafunctional ethylenediamine core 88
– trifunctional core unit 86
– wound healing 318
[2.2]paracyclophane 149
parallel monomer-combination method 56
pentafluorophenyl ester-activated azobenzene unit 64
peptide dendrimer 300
perfection 13
peripheral group
– prior to dendrimer growth 54
periphery
– functional 51
– globally bifunctionalised 72
– locally bifunctionalised molecular 73
PET, *see* photoinduced electron transfer
phase transfer catalysis 294
phenylacetylene dendrimer 95
L-phenylalanine 158
phloroglucin branching unit 157
phosphine dendrimer 233
phosphodendrimer 130 ff.
phosphorescence 171
phosphorus-based dendrimer 114, 130 ff.
– polycationic 115
photodynamic therapy 316
photoexcitation 211
photoinduced electron transfer (PET) 113, 307 ff.
– fluorescing PET sensor 308
photoisomerisation 176, 211
– azobenzene unit 185, 211
– $E \rightarrow Z$ and $Z \rightarrow E$ 211
– reversible 186
E/*Z* photoreaction 188
photosensitiser 316
phthalocyanine
– dendritic 232
pigment 294
PIPAAm, *see* poly(*N*-isopropylacrylamide)
plaster cast 201
plastics 296
platinum-siloxane complex 121
PNIPAAm, *see* poly(*N*-isopropylacrylamide)
poly(amidoamine), *see* PAMAM
polyanionic dendrimer 111 ff.
polyarene dendrimer 93
– hyperbranched 102
polyarylamine dendrimer 242
poly(aryl ether) dendrimer 146
– Fréchet-type 146
polybenzene dendrimer 93 ff.
– convergent synthesis 5 ff.
– hyperbranched 102
poly(benzyl ether) dendrimer 52 ff., 148 f.
– axially chiral binaphthyl core unit 148
poly(benzyl ether) dendron 52
polycarboxylic acid dendrimer 200
polycationic dendrimer 114 ff.
– convergent synthesis 116
polydispersity 14
polydispersity index (PDI) 14
polyester dendrimer 104 ff.
poly(ester amide) dendrimer 307
polyetheramide shell 182
polyether dendrimer 102, 235
polyether dendron 104
poly(ethyleneimine) (PEI) 39
poly(*N*-isopropylacrylamide) (PIPAAm, PNIPAAm) 89, 231
polylysine dendrimer 2, 90 ff.
– butyloxycarbonyl group 90
– neodymium(III) ion 90

– organic nanotube 323
polymer
– dendritic 295
– dendronised linear 39
– hyperbranched 113
polymer-analogous method 39
polymeric light-emitting diode (PLED) 299
polyolefin 200
poly(phenyl ester) dendrimer 305
polyphenylene
– hyperbranched 114
– linear 102
polyphenylene dendrimer 60, 93 ff.
– divergent route 95 ff.
polypropyleneimine (PPI), *see also* POPAM (poly(propylenamine)) 81 ff.
polypyridine ligand 34
polypyridine-osmium complex 137
polypyridine-ruthenium complex 137
polysilane dendrimer 119
– first generation 119
polysiloxane dendrimer 129
POMAM 89
POPAM (poly(propyleneamine)), *see also* poly(propyleneimine) (PPI) 16, 52, 81 ff.
– amphiphilic switchable 64
– barbiturate docking 209
– chiral Boc-protected amino acids in the molecular periphery 160
– DAB-POPAM 292
– DAB based 233
– decorated 83
– divergent synthesis 82
– generation 82 ff.
– glycinylurea guest molecule 206
– methyl orange-functionalised third-generation 190
– per-monosubstituted dansyl-POPAM dendrimer 84
– peripheral azobenzene unit 185
– peripheral Hamilton receptor niche 209
– peripheral oligo-ethylene glycol ether unit 213
– second-generation stilbene-decorated 101
– sulphonamide-substituted G2-G4 187
– terminal aza-crown ether and ferrocene unit 86
– terminal cobaltocenium unit 85
porphyrin
– dendrimer 183, 306 ff.
– dendritic multiporphyrin system 182
– Fe(III) 239 f.
– metal complex 301
– metallo-porphyrin dendrimer 232
– polycationic dendrimer 316
– 5,10,15,20-tetrakis(3,5-dihydroxyphenyl)porphyrin 203
– 5,10,15,20-tetrakis(4-hydroxyphenyl)porphyrin 203
– zinc complex 183, 244 ff.
porphyrin-based dendrimer 109
porphyrin heneicosameric nickel complex 109
– *para*-phenylene-spaced 109
postmodification 54 ff., 71 f.
premodification 56, 71
printing ink 295
propeller conformation 162
protective shell/coating 119, 132 ff.
protein
– artificial 300
– dendrimer 300
proton pump 307
pseudo prefix 204
pseudorotaxane moiety 204
pulsed gradient spin echo (PGSE) 262

q

quartz micro balance 304
– dendritic sensor layer 304
quasi-rotaxane 219

r

radial segment density distribution 268
receptor 243
receptor unit 207
– dendrimer 207
REDOR (rational echo double resonance) solid state NMR 270
redox gradient 242
redox potential 244
redox sensor 243
redox transfer kinetics 244
rhenium complex 210
ring-closing metathesis 202
ring-opening metathesis polymerisation (ROMP) 198
rotaxane 205, 220 ff.
– second generation (G2) and third generation (G3) Fréchet-type stopper 220
ruthenium complex 229
– carbene-complexed dendrimer 197
– heteroleptic 176
– homoleptic 176
– luminophoric 182

– photoactive 33
– Ru(BINAP) catalyst 229
– $[Ru(bipy)_3]^{2+}$ 34, 174 ff.
– ruthenium(II)-bis(terpyridine) metallo-dendrimer 244 f.

s
salen
– cobalt complex 233 ff.
– dendronised 153
– ligand 154
scanning probe microscopy (SPM) 271
scanning tunnelling microscopy (STM) 271
– constant current method (CCM) 272
– constant-height method (CHM) 272
segmented-block dendrimer 28
self-assembly
– dendrimer 207
– glycodendrimer 303
– metal-directed 35
– supramolecular 208
sensor
– material 306
– technology 304 ff.
sergeant-and-soldier principle 164
Shiga toxin 121
Sierpinski triangle 8 f.
silane dendrimer 117 ff.
– second generation 119
silicon-based dendrimer 117 ff.
silicon-based polymer 128
– hyperbranched 128
siloxane
– hyperbranched 129
siloxane dendrimer 127
– fourth generation 128
single-site binding model 236
small-angle scattering 266 f.
– capability 267
– principle 267
– small-angle neutron scattering (SANS) 266
– small-angle X-ray scattering (SAXS) 266
solid phase synthesis 31
9,9′-spirobi[9*H*-fluorene]unit 153
starbust
– supramolecular 218
starburst dense packing 14
starburst limit effect 14
starburst-limited generation 3
Staudinger index 11
stereoselective host/guest interaction 153 f.
stilbene
– two-photon tracer 132
stilbenoid dendrimer 97 ff.
stimulated echo (STE) 262
stopper
– Fréchet type 220
structural defect 281
structure 282
sulphonyl chloride 65
supertriptycene 93
supramolecular complexation 133
supramolecular dendrimer 332
supramolecular (host/guest) interaction 204
supramolecular microcontact printing 297
supramolecular regioselectivity 36
supramolecular starbust 218
supramolecular synthesis 35
surface-block dendrimer 28, 62, 72
synthesis
– convergent 27, 50
– divergent 25 f., 50
– double-exponential method 29
– double-stage convergent 29
– orthogonal 28
– parallel 56
– solid phase 31
– supramolecular 35, 50
– two-stage 56

t
TADDOL (*a,a,a′,a′*-tetraaryl-1,3-dioxolan-4,5-dimethanol) 230
– dendritic 231
template 201 ff.
– dendritic polymer 324
terminal functional group 9
5,10,15,20-tetrakis(3,5-dihydroxyphenyl)porphyrin 203
5,10,15,20-tetrakis(4-hydroxyphenyl)porphyrin 203
thin-layer chromatography (TLC) 254
(*R,R*)-threitol 157
tissue culture 318
titanium
– TADDOL complex 230
transition metal ion 34
transmission electron microscopy (TEM) 274 f.
– dendrimer image 275
tris(hydroxymethyl)methane 147
tumour therapy 314 ff.
tweezer, *see* molecular tweezer

two-component gelator 36
two-stage synthesis 56

u
ultrafiltration 292
unimolecular micelle 214 ff.
– dendritic 214

v
valve 320
Vero toxin 121
viologen guest compound 220 ff.
viscosity
– intrinsic 11
– limiting 11

w
wound healing 318

x
X-ray crystal structure analysis 266

z
zinc complex
– porphyrin 183, 244 ff.
– porphyrin heptamer 310

KLEINE TEXTE FÜR VORLESUNGEN UND ÜBUNGEN
HERAUSGEGEBEN VON HANS LIETZMANN

152

EPISTULA APOSTOLORUM

NACH DEM ÄTHIOPISCHEN UND KOPTISCHEN TEXTE

HERAUSGEGEBEN

VON

HUGO DUENSING

BONN
A. MARCUS UND E. WEBER'S VERLAG
1925

EINLEITUNG

Der koptische Text dieses dem 2. Jahrhundert n. Chr. angehörenden, bis dahin gänzlich unbekannten Apokryphons wurde von Carl Schmidt im Jahre 1895 in Kairo auf dem Institut de la Mission Archéologique entdeckt. Winzige lateinische Fragmente erkannte Bick in einem Wiener Palimpseste, der aus dem Kloster Bobbio stammt. Die allein vollständige äthiopische Version wurde 1913 von Guerrier in der Patrologia orientalis vol. 9 fasc. 3 herausgegeben. Die koptische Version des aus dem 4. oder 5. Jahrhundert stammenden Kairiner Papyrus ist der älteste und als direkter Abkömmling des griechischen Originales wichtigste Textzeuge, dessen Wert nur dadurch stark beeinträchtigt wird, dass 15 Blätter der ursprünglichen Hs fehlen. Hinsichtlich der Vollständigkeit übertrifft den koptischen der äthiopische Text, der aber nur in jungen, dem 18. Jahrhundert angehörenden Hss vorliegt. Sein Wert ist nicht nur durch die Jugendlichkeit der Überlieferung, sondern auch dadurch beschränkt, dass er nur eine Afterversion darstellt. Zwischen ihm und dem griechischen Original liegt als Zwischenglied sicher wohl ein arabischer Text, wenn nicht gar zwischen diesem und dem grieschischen als weiteres Zwischenglied ein koptischer anzunehmen ist. Trotz der hieraus sich ergebenden Einschränkungen ist der äthiopische für eine Rekonstruktion der ursprünglichen Schrift höchst wertvoll, weil er nicht nur die Lücken, die der Kopte lässt, ausfüllt, sondern auch diesem gegenüber manchmal das Ursprüngliche bewahrt hat.

Der Titel der Schrift, welche so lange verschollen gewesen ist, hat sich nicht erhalten, doch kann er erschlossen werden. In den lateinischen Fragmenten findet sich als Randüberschrift epistula; die notwendige Genetivergänzung, die auf der fehlenden, bzw. stark beschnittenen Gegenseite gestanden haben muss, fehlt. Aber sie lässt sich mit Hilfe des Einganges der äthiopischen Version wiedergewinnen. In diesem enthalten die Worte „Brief des Kollegiums der Apostel, der Jünger Jesu Christi, den für alle (bestimmten)" den ursprünglichen Titel, und danach kann man den verkürzten lateinischen Titel wohl als epistula apostolorum ansetzen. Eine katholische Epistel will das Werk sein, und das ist es auch insofern, als die hier vertretenen Anschauungen sich im wesentlichen auf dem Boden des Gemeinchristentums des 2. Jahrhunderts halten, wenn auch hin und wieder ein fremder Einschlag, wohl gnostischer Herkunft, zu spüren ist. Als Epistel lässt sich das Werk aber eigentlich nicht sehr gut bezeichnen. Sein eigentlicher Inhalt sind Zukunftsoffenbarungen des Auferstandenen, die dieser im Wechselgespräch mit seinen Jüngern mitteilt. Aber um diese dem derzeitigen orbis christianus mitteilen zu können, war wohl nur die Form möglich, dass die Jünger in einem an die ganze Christenheit gerichteten Schreiben darüber berichteten. In der äthiopischen Überlieferung tritt die Schrift nicht selbständig, sondern — zusammen mit einer apokalyptischen Rede Jesu an seine Jünger in Galiläa, die ihr vorhergeht — als Bestandteil des Testamentum Domini auf, s. das Explizit S. 42.

Der Übersetzung des äthiopischen Textes liegt die Ausgabe von Guerrier zugrunde, welche eine Wiedergabe des Textes der Hs Or. 793 des Britischen Museums ist (Bezeichnung: L), doch ist vielfach von diesem Texte abgewichen und ihm der Text der anderen Hss, deren Lesarten Guerrier verzeichnet, nämlich der drei Hss der d'Abbadieschen Kollektion No. 51, 90 und 199 (Bezeichnung: ABC), sowie der Text der Stuttgarter Hs Cod. Orient. fol. No. 49, dessen Varianten Wajnberg notiert hat, vorgezogen (S). Für den koptischen Text ist die Übersetzung Schmidts benutzt, doch nicht ohne mannigfache Verbesserungen, die Sethe und Rahlfs beigesteuert haben, und denen ich selbst einige wenige hinzugefügt habe. Soweit die Korrekturen in den GGA 1922 S. 242 ff. verzeichnet sind, werden sie hier als begründet vorausgesetzt.

Erklärung der Schriftarten und Zeichen.

Mediaeval Nur im äthiopischen Text.

Kursiv *Nur im koptischen Text.*

Romanisch Gemeinsames Gut.

⟨ ⟩ Ergänzungen von Lücken des koptischen Textes.

( ) Erläuterungen, wenn die Worte in den Klammern mit denselben Typen gedruckt sind wie die Umgebung.

[ ] Zu Tilgendes.

Wo die Verschiedenheit nicht zu gross war, sind beide Überlieferungsformen zusammengearbeitet. Wenn in diesen Partien sich (vielfach nebeneinander stehende) Ausdrücke in verschiedenem Druck stossen, so sind die Worte in gleichem Druck zusammenzunehmen und mit den Worten des beiden Texten gemeinsamen Gutes zu lesen, also einmal die Worte der äthiopischen Überlieferung mit dem gemeinsamen Satz, dann die der koptischen mit diesem zusammen. In diesen zusammengearbeiteten Partien sind die Lücken des Kopten nicht durch Winkelklammern angedeutet, sondern einfach nach der äthiopischen Überlieferung, die sich dann meist mit der koptischen deckt, ausgefüllt. Wer sich über den Umfang der Ergänzungen des koptischen Textes in diesen Partien unterrichten will, muß zu Schmidts Ausgabe greifen.

Literatur.

Carl Schmidt, Gespräche Jesu mit seinen Jüngern nach der Auferstehung (Texte und Untersuchungen III 13) Leipzig 1919. — ZNW Bd. 20, 1921, S. 174f. GGA 1922 S. 241—252 (hier ein Teil der im folgenden gebrachten Verbesserungen). Hennecke, Ntliche Apokryphen, 2. Aufl. S. 146—150, woselbst auch eine Inhaltsangabe der Schrift.

1. (**12.**) Was Jesus Christus seinen Jüngern als einen Brief offenbart hat und wie Jesus Christus offenbart hat den Brief des Kollegiums der Apostel, der Jünger Jesu Christi, den für alle (bestimmten)[1], der wegen der Pseudapostel Simon und Kerinth geschrieben worden ist, damit niemand sich ihnen anschliesse, weil in ihnen eine List ist, mit welcher sie die Menschen töten, auf dass ihr fest seid und nicht wankt, nicht erschüttert werdet und nicht abweicht vom Worte des Evangeliums, das ihr gehört habt. Wie wir (es) gehört, (im Gedächtnis) behalten und für alle Welt aufgeschrieben haben, so vertrauen wir (es) euch, ihr unsere Söhne und Töchter, in Freude an im Namen Gottes des Vaters, des Herrschers der Welt, und in Jesus Christus. Die Gnade mehre sich über euch!

2. (**13.**) (Wir) Johannes, Thomas, Petrus, Andreas Jakobus, Philippus, Bartholomäus, Matthäus, Nathanael, Judas Zelotes und Kephas[2], wir haben geschrieben[3] an die Kirchen des Ostens und des Westens, gen Norden und Süden, indem wir euch erzählen und verkünden das von unserm Herrn Jesus Christus, wie wir †geschrieben† und ihn gehört und betastet haben, nachdem er auferstanden war von den Toten und wie er uns offenbart hat Grosses, Staunenerregendes, Wahrhaftes.

3. (**14.**) Das wissen wir: unser Herr und Heiland Jesus Christus (ist) Gott[4], Sohn Gottes, der gesandt worden ist von Gott, dem Herrscher der ganzen Welt, dem Schaffer und Schöpfer dessen, was mit jedem Namen benannt wird, der über allen Herrschaften ist, Herr der Herren und König der Könige, der Gewaltige[5] der Gewaltigen, der Himmlische, der über Cherubim und Seraphim[6] ist und zur Rechten des Thrones des Vaters sitzt; der durch sein Wort den Himmeln gebot und die Erde, und was auf ihr ist, erbaute und das Meer begrenzte, dass es nicht seine Grenze überschreite, und

[1] = katholischen. [2] Die Eigennamen sind im Urtext durch „und" verbunden. [3] Briefstil = wir schreiben. [4] C S om Gott. [5] Mit Or. 795 ist ḫajjāla statt ḫajla zu lesen. [6] A om und Seraphim.

(macht, daß)[1] Tiefen und Quellen sprudeln und auf der Erde fließen Tag und Nacht; der die Sonne, den Mond und die Sterne am Himmel gründete und der Licht und Finsternis schied, der der Hölle gebot und im Augenblick entbietet den Regen zur Winterszeit und Nebel, Reif und Hagel und die Tage (?) zu ihrer Zeit; der erschüttert und festigt; der den Menschen nach seiner Gestalt und seinem Bilde geschaffen hat; der durch die Patriarchen und Propheten in Bildern geredet hat und in Wahrheit durch den[2], den die Apostel verkündigt und die Jünger betastet haben. Und Gott, der Herr, der Sohn Gottes — wir glauben: das Wort, welches aus der heiligen Jungfrau Maria Fleisch wurde, wurde in ihrem Schosse vom heiligen Geiste (verursacht) getragen, und nicht durch Lust des Fleisches, sondern durch den Willen Gottes wurde es geboren und wurde in Bethlehem in Windeln gewickelt und offenbart und dass es grossgezogen wurde und heranwuchs, indem wir (es) sahen.

4. (15.) Dies tat unser Herr Jesus Christus, der von Joseph und von Maria seiner Mutter dahin gegeben wurde, wo er die Schrift lernen sollte. Uud der, welcher ihn lehrte, sagte zu ihm, während er ihn lehrte, „Sag: Alpha!“ Er antwortete und sagte zu ihm: „Sage du mir zuerst, was Beta ist.“ Und . . . wahr . . . . . was getan wurde[3].

5. (16.) Darauf war eine Hochzeit in Kana von Galiläa. Und man lud ihn ein mit seiner Mutter und seinen Brüdern. Und Wasser machte er zu Wein, und Tote weckte er auf und machte, dass die Lahmen gingen; dem die Hand verdorrt war, streckte er sie wieder aus, und die Frau, welche zwölf Jahre an Blutfluss litt, berührte den Saum seines Gewandes und ward sofort gesund; und indem wir nachdachten und uns wunderten über die preiswürdige Tat, die er vollbrachte, sagte er zu uns: „Wer hat mich berührt?“ Und wir sagten zu ihm: „O Herr, das Menschengedränge hat dich angerührt.“ Und er antwortete und sagte zu uns: „Ich habe gemerkt, dass eine Kraft von mir ausgegangen ist.“ Zu der Zeit trat vor ihn jenes Weib, antwortete ihm und sprach zu ihm: „Herr, ich habe dich berührt.“ Und er antwortete und sagte zu ihr: „Geh, dein Glaube hat dich gesund gemacht.“ Dann machte er die Tauben hören und die

[1] Etwas derartiges ist zu ergänzen. [2] Vor zaḥawārjāt ist ba einzuschieben. [3] Der letzte Satz ist unübersetzbar.

Blinden sehen, die Besessenen befreite er[1] und die Aussätzigen reinigte er. Und der Dämon Legion, den ein Mann hatte, begegnete Jesus, schrie und sagte: „Bevor der Tag unseres Verderbens herangekommen ist, bist du gekommen, uns zu vertreiben." Der Herr Jesus aber schalt ihn und sprach zu ihm: „Geh von diesem Mann aus, ohne ihm irgend etwas zuzufügen." Und er ging in die Säue hinein und versenkte sie im Meer, und sie erstickten. Darauf wandelte er auf dem Meer, und es bliesen die Winde; und er schalt sie, und die Wogen des Meeres beruhigten sich. Und als wir, seine Jünger, keinen Denar hatten, sagten wir zu ihm: „Meister[2], was sollen wir machen hinsichtlich des Steuereinnehmers?" Und er antwortete und sagte zu uns: „Einer von euch werfe die Angel, die Reuse[3], in die Tiefe und ziehe einen Fisch heraus und er wird in ihm einen Denar finden. Den[4] gebt dem Steuereinnehmer für mich und euch." Als wir darauf kein Brot ausser fünf Broten und zwei Fischen hatten, gebot er den Leuten sich zu lagern, und es stellte sich ihre Zahl auf fünf Tausend ausser den Kindern und Frauen, denen[5] wir Brotstückchen vorlegten; und sie wurden satt, und es blieb (davon) übrig, und wir trugen zwölf volle Körbe von Brocken weg, indem wir fragten und sagten: „Welche Bewandtnis hat es mit diesen[6] fünf Broten?" Sie sind ein Bild unseres Glaubens betreffs des grossen Christentums und d. h. an den Vater, den Herrscher der ganzen Welt, und an Jesum Christum, unsern Heiland, und an den heiligen Geist, den Parakleten[7], und an die heilige Kirche und an die Vergebung der Sünden.

6. (**17.**) Und dies offenbarte und zeigte uns unser Herr und Heiland und wir euch[8] gleicherweise, damit ihr Genossen an[9] der Gnade des Herrn und unseres Dienstes und an unserer Herrlichkeit seid, indem ihr auf das ewige Leben sinnt. Seid ohne zu wanken fest in der Erkenntnis und Erforschung[10] unseres Herrn Jesu Christi, und er wird sich gnädig erweisen und retten immerdar und in alle nimmer endende Ewigkeit.

---

[1] Wörtlich: trieb er aus. [2] Nach ABCS eingefügt. [3] Falsche Glosse zu „Angel". [4] Das Pluralsuffix am Verbum ist in das Singularsuffix zu verbessern. Allenfalls kann „Denar" kollektivisch als „Geld" verstanden sein. [5] la'ella „denen" ist wohl in la'ellū „diesen" zu verbessern. [6] Statt 'ella ist 'ellū zu lesen. [7] ABC om den Parakleten. [8] „euch" nach ABCS eingefügt. [9] Statt „Genossen an" nach anderer Lesart „gegründet auf". [10] Lies wabaṭajjeqō.

7. (18.) **Kerinth und Simon sind gekommen**[1], **die Welt zu durchwandeln. Diese sind aber Feinde unseres Herrn**[2] **Jesu Christi,**

<table>
<tr>
<td>die in Wirklichkeit diejenigen abwendig machen[3], die glauben an das wahre Wort und an die Tat d. h. Jesum Christum. Seht euch also vor und hütet euch vor ihnen; denn in ihnen[4] ist Bedrängnis und Befleckung und Tod. Das Ende davon[5] wird zu Verderben und Gericht werden[6].</td>
<td>denn sie verkehren die ⟨Worte⟩ und die Tat d. h. Jesum Christum. Von diesen nun haltet ⟨ihr⟩ euch fern; denn Tod ist nämlich in ihnen und grosse Verderbensbefleckung, diesen, denen das Gericht zuteil werden wird ⟨und das⟩ Ende und das Verlorensein ewiglich.</td>
</tr>
</table>

8. (19.) Sieh also *deswegen* **haben wir nicht gezögert**

<table>
<tr>
<td>mit dem wahren Zeugnis von unserm Herrn und Heiland[7] Jesus Christus, wie er gehandelt hat, während wir ihn[8] sahen, und wie er beständig unsere Gedanken bei uns sowohl erklärte als wirkte.</td>
<td>euch zu schreiben betreffs des Zeugnisses von unserm Heiland Christus dasjenige, was er getan hat, während wir auf ihn blickten ⟨und?⟩ wiederum ⟨noch?⟩ in den Gedanken und Werken,</td>
</tr>
</table>

<table>
<tr>
<td>9. (20.) Diesen, für den wir[9] Zeugen sind, kennen wir als den, der gekreuzigt ist in den Tagen des Pontius Pilatus und des Fürsten Archelaus, der zwischen zwei Räubern gekreuzigt ist, und mit ihnen zusammen hat man ihn vom Holz[10] des Kreuzes abgenommen, und er ist begraben an einem Orte, der qarānejō (κρανίου) heisst, wo-</td>
<td>dieser, über den ⟨wir⟩ Zeugnis ablegen, dass der Herr ist dieser, der gekreuzigt ist durch Pontius Pilattius und Archelaus inmitten des Räuberpaares

⟨und⟩ der begraben ist an einem Orte, der ⟨Schädelstätte⟩ heisst. Es sind gegangen</td>
</tr>
</table>

---

[1] L. nach Kopt. maṣ'ū statt waṣ'ū. Beides geht in den Lesarten zu Henoch 42_{2-3} ebenfalls durcheinander. [2] So S u. Kopt. Die anderen Hss „Gottes". [3] Oder: verkehren. [4] „in ihnen" nach AC eingesetzt. [5] Verbessert man das Singularsuffix hū in das Pluralsuffix hōmū, so heisst es: „ihr Ende". [6] Es ist oben angenommen, dass kōna la hier wie Gen. 2_7, Jerem. 6_{10}, 15_{16} in der Bedeutung „werden zu" gebraucht ist. Aber der Text ist wohl nicht in Ordnung. [7] „Heiland" nach BCS (in verschiedener Stellung) eingefügt. [8] „ihn" nach A. [9] „wir" nach C. [10] „Holz" fehlt in AC.

hin[1] drei Frauen: Sara und Martha und Maria Magdalena gingen. Sie trugen Salbe, um (sie) auszugiessen

zu jenem Orte (drei) Frauen: *Maria, die zu* Martha *Gehörige*[2], und Maria (Magd)alena. Sie *nahmen* Salbe, *gossen sie*[3]

auf seinen Leib, indem sie weinten und trauerten über das, was geschehen war.

Und sie näherten sich dem Grabe und fanden den Stein (da), wohin man ihn abgewälzt hatte vom Grabe, und sie öffneten seine Tür

Als sie aber sich dem Grabe *genähert hatten, blickten sie hinein*

und fanden seinen *den* Leib nicht.

10. (21.) Und wie sie *aber* trauerten und weinten, erschien ihnen der Herr und sprach zu ihnen: „*Wen beweinet ihr?* Weinet *nun* nicht *mehr!* Ich bin's, den ihr suchet. Es gehe aber eine von euch zu euren Brüdern und sage ihnen: Kommet, unser *der* Meister ist auferstanden von den Toten." Und Maria *Martha* kam zu uns und sagte *es* uns. Und wir sprachen zu ihr:

„Was ist uns und dir[4], o Weib? Wer gestorben und begraben ist, kann denn der (wieder) leben?"

„*Was willst du mit uns*[4], o Weib? *Der* gestorben *ist, [ist]* begraben, *und wäre es möglich, dass er* lebe?"

Und wir glaubten ihr nicht, dass unser *der* Heiland von den Toten auferstanden wäre. Darauf ging sie zu unserm *dem* Herrn zurück und sprach zu ihm: „Niemand von ihnen hat mir geglaubt inbetreff deiner Auferstehung, *dass du lebst.*" Und er sprach zu ihr: „Es möge eine andere von euch hingehen *zu ihnen,* indem[5] sie dies wiederum zu ihnen sagt." Und Sara *Maria* kam und

tat uns dasselbe kund, und wir beschuldigten sie der Lüge.

sagte es uns wiederum, und nicht haben wir ihr geglaubt.

Und sie kehrte zu unserm *dem* Herrn zurück und *auch* sie sagte *es* ihm wie Maria.

---

[1] Statt bōhaba ist bahaba zu lesen. [2] Ändert man mit Schm. das ta (= die der) vor Martha in au „und", so sind auch im Kopt. die drei Frauen vollzählig genannt. [3] Sachlich unrichtig. Das richtige Verständnis hat Äth. [4] Der Äth. hat die ursprüngliche Form des Textes hier bewahrt, der Kopte aber sinngemäß übersetzt. [5] Streiche kama vor 'enza mit A C.

11. (22.) Und darauf sprach der Herr zu Maria und *auch* zu ihren Schwestern: „Lasst uns zu ihnen gehen!“ Und er kam und fand uns drinnen *verhüllt*[1].

Und wir zweifelten und glaubten nicht. Wie ein Gespenst kam er uns vor, und wir glaubten nicht, dass er es wäre. Aber er war es.

Er rief uns heraus; wir aber dachten, es wäre ein Gespenst, *(und)* nicht glaubten wir, dass es *der Herr* wäre.

Und also *darauf* sprach er zu uns: „Kommet und fürchtet euch nicht! Ich bin euer Lehrer *Meister*, den du, Petrus, ehe der Hahn krähte, dreimal verleugnet hast, und jetzt verleugnest du wiederum?“ Und wir *aber* kamen zu ihm, denkend und zweifelnd *in* ⟨*unserm*⟩ *Herzen*, ob er es *vielleicht* auch wäre. Und *da* sprach er zu uns: „Weshalb zweifelt ihr *noch* und warum seid ihr ungläubig?[2] Ich bin es, der euch gesagt hat das inbetreff meines Fleisches, meines Todes und meiner Auferstehung. Und damit ihr wisset[3], dass ich es bin, so lege, Petrus, deine Hand und deine Finger[4] in die Nägelmale[5] meiner Hände, und du *selbst*, Thomas, *lege deine Finger* in *die Lanzenstiche* meine*r* Seiten, und du *aber* auch, Andreas,

sieh, ob mein Fuss auf die Erde tritt und (dabei) eine Spur hat.

betrachte meine Füsse und sieh, ob sie nicht die Erde berühren.

Denn es steht geschrieben im Propheten:

„Ein Gespenst, ein Dämon[6] aber hat keine Spur auf der Erde.“

„Eines Dämonsgespenstes Fuss möge[7] nicht haften auf der Erde.“

12. (23.) Wir *aber* also betasteten ihn, *damit wir erkennten, ob* dass er wahrhaftig[8] im Fleisch auferstanden wäre. *Und* darauf fielen wir vor ihm auf unser Angesicht,

---

[1] Oder nach S (vgl. AC) „fischend“. [2] Der Äth. ist nach AC herzustellen, indem man mit Wajnberg za'īta'amenū statt za'īta'amna liest. Das heisst dann wörtlich: „Weshalb zweifelt ihr und was ist's, dass ihr nicht glaubt?“ Die kleine Differenz zwischen Äth. und Kopt., dass im ersteren ἔτι fehlt, im letzteren (καὶ) τί, erklärt sich durch einen verschiedenen griechischen Text. In dem hinter der äth. Überlieferung stehenden Text stand das ἔτι hinter καὶ und wurde hinter diesem zu τὶ verkürzt. [3] Für den Äth. ist die Übersetzung „erkennet“ näherliegend. [4] (und) deine Finger nur in BC und Kopt. [5] Äth. Singular. [6] Stellt man beim ersten Substantiv den stat. constr. her, so heisst es es wie im Kopt. „das Gespenst eines Dämons“. [7] Bedeutet vielleicht: pflegt. [8] „wahrhaftig“ ist mit dem Lat. und Äth. in den Nebensatz zu ziehen.

baten ihn um Verzeihung und flehten ihn an dafür, daß wir ihm nicht geglaubt hatten.

und ⟨*be*⟩*kannten unsere Sünden, dass wir ungläubig gewesen seien.*

Darauf sprach *der* unser Herr und unser Heiland zu uns: „Steht auf, und ich will euch offenbaren, was auf Erden [1] und was oberhalb der Himmel ist *und das, was in den Himmeln ist* und eure *Ruhe* Auferstehung, die im Himmelreich, um derentwillen *denn* mein Vater *hat mir die Macht gegeben* mich gesandt hat, euch und diejenigen, die an mich glauben, hinaufzuführen."

13. (24.) Und was er uns offenbart hat, ist dies, wie er zu uns sprach: „Während ich vom [3] Vater des Alls [4] her kam [3], indem ich an den Himmeln vorüberging, wobei ich die Weisheit des Vaters anzog und in seine Kraft von seiner Kraft her mich kleidete, war ich in [6] den Himmeln. Und indem ich an den Engeln und Erzengeln vorüberging in ihrer Gestalt und wie einer [7] von ihnen, ging ich vorüber an den Klassen, Herrschaften und Fürsten, indem ich das Maß der Weisheit des Vaters [8], der mich gesandt hat, besaß. Und die [9] Erzengel Michael,

Was er *aber* uns [2] offenbart hat, *ist, was er sagt: „Es geschah aber, als ich im Begriff war, herzukommen* von dem Vater des Alls, *da* [5] *ging ich vorüber* an den Himmeln, *da zog ich an* die Weisheit des Vaters *und zog an die Kraft seiner Macht (?).* Ich befand mich in den Himmeln, *und die Erzengel und die Engel — ich ging vorüber* in ihrer Gestalt, *als wäre ich* einer aus ihnen, *unter den Herrschaften und Gewalten. Ich durchschritt sie,* indem ich die Weisheit dessen, der mich gesandt hat, besaß. *Der Oberanführer der Engel aber ist* [10] Michael *und*

---

[1] AC dagegen: das der Himmel. Stellt man dies hinter „was oberhalb usw.", so deckt sich dieser Text ungefähr mit dem Kopt. [2] neï ist in nen „uns" verbessert. [3] So AC. [4] Nach dem Kopt. ist ab in aba und kuellō in kuellū zu verbessern. Sonst wäre es zum Folgenden zu ziehen: „indem ich durch alle Himmel hinging". [5] Diese notwendige Übersetzung steht im Widerspruch zu der lat. ⟨transie⟩ns caelos und äth. Überlieferung, die hier beide eine Subjunktion haben. [6] Es ist mit Wajnberg lasamajat in bas′ korrigiert, um die Übereinstimmung mit dem Kopt. herzustellen. Aber die Richtigkeit des Textes ist auch so recht fraglich. Der unveränderte äth. Text lautet: „ward ich den Himmeln, den Engeln und Erzengeln, indem ich an ihnen vorüberschritt, in ihrer Gestalt und wie einer von ihnen", d. h. ich ward, verwandelte mich in die Gestalt der Himmel, Engel und Erzengel. Dann geht der Text weiter: „An den Klassen, Herrschaften, Fürsten" usw. [7] Nach ABCS einzufügen. [8] AC+ „und in seine Kraft mich kleidete". [9] Lat. Singular. [10] Kopt. fehlerhaft. Die Reihenfolge der Namen ist dieselbe wie im Lat.

Gabriel, Raphael und Uriel folgten mir[1] bis zum fünften Himmelsfirmament[2], indem ich ihnen vorkam wie[3] einer von ihnen. Derartige Kraft ist mir vom Vater gegeben. Dann machte ich, dass die Erzengel mit der Stimme außer sich gerieten[4] und hingingen zum Altar des Vaters und [dem Vater] dienten [in] ihr[em][5] Werk, bis dass ich zu ihm zurückkehre. Dieses[6] tat ich so in der Ähnlichkeit (Gestalt) seiner Weisheit.

Gabriel *und* Uriel *und* Raphael, *sie aber* sind mir nachgefolgt bis zum fünften Firmament, *indem sie nämlich dachten in ihrem Herzen, daß ich* einer von ihnen *wäre. Der* Vater *aber hat* mir gegeben die Kraft von dieser Beschaffenheit. *Und an jenem Tage schmückte ich* die Erzengel *mit einer wunderbaren* Stimme, *damit sie* hineingingen zum Altar des Vaters *und* dienten *und vollendeten den Dienst,* bis dass ich zu ihm *gehe. So habe ich es gemacht durch die Weisheit der Ähnlichkeit*[7].

Denn ich ward alles[8] in jedem bei ihnen, damit ich

den Willen der Barmherzigkeit	*die Veranstaltung*

des Vaters .... und[9] die Herrlichkeit dessen, der mich gesandt hat, vollendend[10] zu ihm zurückgehe[11].

14. (**25**) Wisset ihr *nämlich,* dass der Engel Gabriel gekommen ist und der Maria die *(gute)* Botschaft gebracht hat?" Und wir *selbst antworteten,* sprachen zu ihm: „Ja Herr." Und *da* antwortete er und sprach zu uns: „Erinnert ihr euch denn nicht, dass ich vordem *vor einer kleinen Weile* zu euch gesagt habe, dass ich den Engeln wie ein Engel[12] geworden bin *(und) ich machte alles in jedem?*" Und wir sprachen zu ihm: Ja, Herr." *Da antwortete er* und sprach zu uns: Damals *an jenem Tage nämlich, wo ich angenommen habe die*[13] Gestalt des Erzengels Gabriel, erschien ich der[14] Maria und redete mit ihr und ihr Herz nahm *mich* auf, sie glaubte[15]

[1] Lat.+ palam. [2] „Firmament" fehlt in C wie im Lat. [3] „wie" nach C eingesetzt. [4] Oder vielleicht: „durch eine Stimme erschraken". [5] Das Eingeklammerte fehlt in ABC wie auch zum Teil im Lat. [6] Nach A. [7] Der Kopt. stimmt mit dem Lat. per sapientia(m) [s]imilitudinis überein. [8] So mit Schm. nach VII, 4 zu emendieren. Sonst würde es heissen: im All in jedem. [9] Nach der Emendation Schmidts. [10] Der Kopt. hat allerdings wörtlich: welcher mich gesandt hat, dass ich ihn vollende. [11] Im Äthiop. ist der Subjunktiv 'egbā' herzustellen. [12] Kopt. Engel unter den Engeln. [13] Äthiop. (erschien ich) in der Gest. [14] L.+ Jungfrau, S+ heiligen Jungfrau. [15] L.S „den Glauben" statt „sie glaubte".

und lachte, und **ich,** das Wort, **ging** in sie **ein** und **ward Fleisch,** und ich selbst **war** für mich selbst **Diener** und[1] in der Gestalt eines Engelsbildes[2],

ich formte mich (?), **ich ging** *hin***ein** *in ihren Leib, ich* **ward Fleisch,** *da ich mir allein war* **Diener** *in bezug auf Maria in Wahrnehmbarkeit von Engelsgestalt;*

so werde ich *es* **tun, nachdem**[3] ich zu meinem *dem* Vater **gegangen bin.**

15. (26.) Und **ihr** begeht also das Gedächtnis meines Todes, d. h. das Pascha;

Und **ihr** *gedenket meines Todes. Wenn nun das Pascha stattfinden wird,*

dann wird man einen von euch, die ihr bei mir stehet, **um meines Namens willen ins Gefängnis**[4] **werfen, und er wird**

sehr betrübt und traurig sein, weil, während ihr das Pascha feiert, er, der in Gewahrsam ist[5], es nicht mit euch gefeiert hat.

⟨sein⟩ in Trauer und Sorge, dass ihr ⟨das⟩ Pascha feiert, während er sich im Gefängnis befindet und ⟨fern⟩ von euch ist; denn er wird trauern, dass er nicht das Pascha ⟨mit⟩ euch feierte.

Und **ich werde** *nämlich* **meine Kraft in der Gestalt eines**[6] *des* **Engels** *Gabriel* **senden, und die Tore**[7] **des Gefängnisses werden sich öffnen,** und **er wird heraus kommen**[8] und kommen **zu euch,**

um **mit euch** zu wachen und zu ruhen. Und **wenn ihr** beim Hahnenschrei meine **Agape**

er wird eine Nacht des Wachens **mit ⟨euch⟩** *zubringen und bei euch bleiben, bis der Hahn kräht.*

---

[1] Um Übereinstimmung mit dem Kopt. herzustellen, müsste man dies „und" streichen. Aber der Äth. hat hier vielleicht das Ursprüngliche. Das Verbindende zwischen Kap. 14 und 15 ist dies, dass Christus nach seinem Heimgang zum Vater auch in Engelsgestalt auf Erden wieder tätig sein wird. [2] —bildes nach BC eingefügt ('ar'ajā). [3] Verb. 'emdeḫrēhū „danach" in 'emdeḫra „nachdem". [4] Im äthiop. Text ist mit ABCS bēta vor mōqeh z. l. [5] Hier wiederholt der äthiop. Text das Wort „weil" und die Hss AC davor „er wird betrübt sein" [C+ trauern]. Nähert sich diese Lesart auch dem Kopten, so ist sie doch nur zu verwerten, wenn das za vor westa mōqeh gestrichen wird. Dann würde der Passus lauten: „weil, während ihr das Pascha feiert, er sich im Gewahrsam befindet, und er wird betrübt sein [C+ und trauern], weil er das Pascha (so AC) nicht mit euch gefeiert hat." C hat hinter „mit euch" noch „gewesen ist", möglicherweise ein versprengter Rest des im Kopt. vorher stehenden Satzes: „und ⟨fern⟩ von euch ist". [6] Oder nach L: „meines", aber auch ohne Gabriel. [7] Äthiop. Singular. [8] „herauskommen" nur in ABC erhalten. Wahrscheinlich ist jemaṣe' „wird kommen" als aus jewaẓe' entstandene falsche Dublette zu streichen.

und mein Gedächtnis [1] vollendet, so wird man ihn wiederum nehmen und ihn zum Zeugnis ins Gefängnis werfen, bis dass er herausgeht, um zu predigen, wie ich euch befohlen habe. Und wir sprachen zu ihm: „Herr, hast du denn nicht das Trinken des Pascha vollendet? Liegt es uns denn ob, es wiederum zu tun?"

Und er sprach zu uns: *nämlich* bis *zu dem Tage, (wo)* werde mit meinen Wunden."

Wenn ihr *aber* vollendet *das* Gedächtnis, *das da ist für mich, und meine* [2] Agape, *so wird er wiederum* ins Gefängnis *geworfen werden* zum Zeugnis, bis dass er *von dort* heraus *kommt (und) predigt das, was ich euch gegeben habe.*" Wir sprachen *aber* zu ihm: „Herr, *ist es etwa wiederum eine Notwendigkeit, dass wir den Kelch nehmen und trinken?*"

„Ja, *eine Notwendigkeit ist es* dass ich vom Vater kommen *denen, die um meinetwillen getötet sind.*"

16. (27.) Und wir sprachen *aber* zu ihm: „Herr, groß ist dies, was du zu uns sagst und uns *nämlich zuvor* offenbar*t ha*st. In was für einer Kraft und *oder* in was für einer Gestalt wirst [3] du kommen?" *Er antwortete aber, indem er zu uns sagte* [4]: „Wahrlich *nämlich* ich sage euch: Ich werde *nämlich* kommen wie die Sonne, die erglänzt, so *und* werde ich, indem ich siebenmal mehr als sie in *meinem* Glanze (Herrlichkeit) leuchte, während ich auf dem Flügel [5] der Wolke [5] getragen werde in Glanz, und indem *das Zeichen des Kreuzes* mein Kreuz vor mir hergeht, auf die Erde *herab*kommen, dass ich richte die Lebendigen und die Toten."

17. (28.) Und wir sprachen *aber* zu ihm: „Herr,

wieviel Jahre noch?" Und er sprach zu uns: „Wenn das hundertundfünfzigste Jahr vollendet ist, zwischen [6] Pfingsten und Pascha

nach noch wieviel Jahren *wird dies geschehen?* Er sprach zu uns: „Wenn das *Hundertstel und das Zwanzigstel* [7] vollendet *sein wird,* zwischen Pfingsten und *dem Fest der Ungesäuerten*

wird stattfinden die Ankunft meines [8] *des* Vaters." Und wir *aber* sprachen zu ihm: „Herr [8], jetzt *nun, was* hast du zu uns

---

[1] tazkāreja wie so oft falsch der Nom. für den Akk. [2] L. taagape. [3] Äth. schärfer: steht es dir bevor, zu kommen? [4] Äth. nur: Und er sprach zu uns. [5] Kopt. Plural. Das Verb. des Kopt. nicht sicher zu ergänzen. [6] So nach AC. Die übrigen Hss „in den Tagen". [7] = $^{6}/_{100}$. [8] Auch im Lat.

gesagt: Ich werde kommen — und wiederum sagst[1] du zu uns: Es wird kommen, der mich gesandt hat?"[2] Und *da* sprach er zu uns: „Ich bin ganz im *meinem* Vater und der *mein* Vater in mir"[3].

Darauf sprachen wir zu ihm: „Wirst du uns wirklich bis zu deiner Ankunft verlassen? Wo werden wir einen Lehrer finden?" Und er antwortete und sprach zu uns: „Wisset ihr nicht, dass ich bis jetzt sowohl hier als dort war bei dem, der mich gesandt hat?" Und wir sprachen zu ihm: „Herr, ist es denn möglich, dass du sowohl hier als auch dort seiest?" Und er sprach zu uns: „Ganz bin ich im Vater und der Vater in mir

nach seinem[4] Bilde und nach seiner Gestalt und nach seiner Kraft und nach seiner Vollkommenheit und nach seinem Lichte, und ich bin sein vollkommenes Wort[5].

hinsichtlich der Ähnlichkeit der Gestalt und der Macht (?) und der Vollkommenheit und des Lichtes und des vollendeten Masses und hinsichtlich der Stimme. Ich bin *das* Wort.

18. (29.) Dies ist, als er gekreuzigt worden, gestorben und auferstanden war, indem er dieses sagte, und das Werk, das im Fleisch vollbracht wurde dadurch, dass er gekreuzigt wurde, und seine Himmelfahrt — dies ist die[6] Vollendung der Zahl, und die Wunder und sein[7] Bild und alles Vollkommene werdet ihr an mir sehen in betreff der Erlösung, die durch mich geschieht,

Ich bin ihm geworden eine Sache, d. h. . . . . . . vollendet nach dem Typus; ich bin geworden am ach[ten] (Tage), welches ist der Herrntag. Die ganze Vollendung *aber der Vollendung* werdet ihr sehen in betreff der Erlösung, die *geschehen ist in bezug auf mich,*

und *ihr werdet mich sehen,* wie ich gehen werde *zum Himmel,* zu dem *meinem* Vater, *der sich befindet* in den Himmeln[8]. Aber siehe *nun,* ein neues Gebot gebe ich euch, daß ihr

---

[1] Äth. hast du gesagt. [2] Kopt. wörtlich: der, welcher mich gesandt hat, ist der da kommen wird. [3] Von hier an hat der Kopte einige Zeilen übersprungen (Homoicoteleuton). [4] Es ist oben 'esma in 'em korrigiert. Wer das ablehnt, muss übersetzen: denn (ich bin) sein Ebenbild sowohl nach seiner Gestalt als nach usw. [5] Statt des Satzes: „und — Wort" haben AC die Worte „und nach der vollkommenen Stimme (Wort)". [6] Streiche ba mit ABC. [7] Nach CS dagegen: typum earum. [8] Äth. kann auch übersetzt werden: in den Himmel.

einander liebet und[1] gehorchet einander[2] und (dass) beständig Friede zwischen euch herrsche. Liebet eure Feinde, und was ihr nicht wollt, dass man euch tue, das tut auch ihr keinem andern.

19. (**30.**) Und dies predigt[3] sowohl als lehret (es) diejenigen, welche an mich glauben, und predigt über das Himmel[4]-reich meines Vaters und wie mein Vater[5] mir die Vollmacht gegeben hat, (C+ so gebe ich [sie] euch), dass ihr die Kinder des himmlischen Vaters herzubringt. Predigt und sie sollen glauben! Ihr (seid es), denen es obliegt, seine Kinder in den Himmel zu führen." Und wir sprachen zu ihm: „Herr, dir ist es möglich, das, was du uns gesagt hast, zu tun. Wie werden aber wir (es) können?" Und er sprach zu uns: „Wahrlich, ich sage euch: predigt und redet, wie ich (denn) mit euch sein werde, denn es gefällt mir wohl, mit euch zu sein, damit ihr mir[6] Miterben des Himmelreiches dessen, der mich gesandt hat, werdet. Wahrlich, ich sage euch, ihr werdet mir Brüder und Genossen sein, denn mein Vater hat an euch und an[7] denen, die durch euch an mich glauben, Wohlgefallen. Wahrlich, ich sage euch, eine derartige und eine solch große Freude hat (euch) mein Vater[8] bereitet, dass sie Engel und Herrschaften zu schauen und zu sehen begehrten und begehren werden, aber man wird ihnen nicht gestatten, die Größe meines Vaters zu sehen." Und wir sprachen zu ihm: „Herr, wie (?) ist das, was du uns sagst?" Und er sprach

zu uns: **Ihr werdet ein Licht sehen,** das mehr leuchtet als Licht und vollkommener ist als das Vollkommene[9]. Und der Sohn wird durch den Vater, das Licht, vollendet werden — denn der Vater ist vollkommen —[11], welchen	**Ihr werdet** *nämlich* **sehen ein Licht,** *(das) erhabener ist als das, welches leuchtet* . . . . . .[10]

[1] Kopt., der hier die direkte Rede hat, bricht mit diesem Wort ab. Es fehlt ein Blatt. [2] om AS. [3] Ausser S haben die Hss vor „predigt" „und". [4] „Himmel" om. ABC. [5] „mein Vater" om. ABS. [6] Nach BC+ meine. [7] So nach AS. Statt „an" haben die andern Hss „es gibt solche, (welche)". [8] B statt dessen „Gott". [9] B: „und das Vollkommenste des Vollkommenen". [10] Hier hat Kopt., wie es scheint, einen verkürzten Text. [11] Will man nicht so durch Parenthesierung helfen, so dass das folgende „welchen" auf den Sohn bezogen werden kann, dann muß man wohl mit Littmann bei Schm. S. 262 im

Tod und Auferstehung vollenden, und die eine Vollendung übertrifft die andere. Und die Rechte des Vaters bin ich ganz, ich bin in ihm, der vollendet.	*die Vollendung, die vollendet ist in (?) ..... ich bin ganz die ⟨Rechte⟩ des Va⟨ters ....⟩ als ich, welcher ist die Vollkommenheit.*

Und **wir** Zwölf *aber* **sprachen zu ihm: „Herr, in Allem wardst du uns Heil und Leben. Von einer solchen Hoffnung** redest *verkündigst* **du zu uns?"** [1] Und **er sprach zu uns: „Habt Vertrauen und seid** guten Mutes *ruhigen Herzens!* **Wahrlich, ich sage euch,** eine solche *eure* **Ruhe wird euch zuteil werden, wo es nicht Essen noch Trinken,**

weder Trauer noch Singen [2]	*weder Frohlocken noch Trauern,*

weder irdisches Gewand **noch Vergänglichkeit gibt** *denen, die in ⟨ihm⟩ sind.* Und nicht an der unteren Schöpfung *nämlich* **habt ihr** so **Anteil, sondern werdet** zur unvergänglichen meines Vaters gehören [3], ihr, die ihr nicht vergehen werdet. Wie ich beständig im Vater [4] **bin, so auch ihr in mir."** Und **wir sprachen wiederum zu ihm: „In welcher Gestalt?** *in der Art* **von den Engeln oder** *auch* **von Fleisch?"** Und **er antwortete** uns dieserhalb und **sprach zu uns: „***S⟨iehe⟩,* **ich habe euer Fleisch angezogen, in dem ich geboren und** getötet *gekreuzigt* und begraben **und auferstanden bin durch meinen himmlischen Vater, damit erfüllet würde**

was durch den Propheten David über meinen Tod und über meine Auferstehung gesagt ist: Herr, wie	*die Prophezeiung ⟨des⟩ Propheten David in betreff dessen, was er ver⟨kündigt⟩ hat über mich und ⟨meinen⟩ Tod und meine Auferstehung, indem ⟨er sagte⟩: Herr,*

zahlreich sind geworden, die mich bedrängen *⟨streiten⟩ mit mir; und* **viele haben sich erhoben wider mich.**

Viele sagen zu meiner Seele: Nicht gibt es für ihn Rettung seines Gottes *bei Gott.* **Du aber, Herr, bist meine Zuflucht** [5], **meine Ehre und der mein Haupt erhebt.**

Mit meiner Stimme schrie ich zu Gott, *zum Herrn,* **und er erhörte mich** vom Berg seines Heiligtums.

Relativsatz „Tod und Auferstehung" streichen und übers. „welcher vollendet". [1] Möglich ist auch die Unterordnung dieses Satzes unter den vorhergehenden mit „indem". [2] Oder „Sorge" statt „Singen". [3] Kopt. „werdet nehmen von". Kopt. hat hier starke Lücken. [4] Kopt. „in ihm". [5] Kopt. wörtlich: der mich aufnimmt zu dir.

Ich legte mich nieder [1] **und entschlief** und **ich stand (wieder) auf, denn**

Gott richtete mich auf. **Ich fürchte mich nicht vor** Tausenden **von Völkern,** die mich umgeben und sich wider mich erhoben haben. **Erhebe dich, Herr,** mein Gott, und rette mich!	*du, Herr, nimmst dich meiner an.* **Ich** *werde* **mich nicht fürchten** *vor einer Myriade* **von Völkern,** *die wider mich kämpfen rings um mich herum.* **Erhebe dich, Herr!** *Rette mich, mein Gott!*

Denn du hast geschlagen alle, die mir feindlich sind ohne Grund; und **du hast die Zähne der Sünder zerbrochen. Des** Gottes *Herrn* **ist die Rettung und**

dein Segen (sei) über deinem Volke. Alles, was durch die Propheten geredet ist, ist also ausgeführt, geschehen und vollendet durch mich, denn ich habe in ihnen (oder: durch sie) geredet,	*sein Wohlgefallen über sein Volk. Wenn aber alle Worte, die geredet sind durch die Propheten, erfüllt sind durch mich, denn* **ich** *bin in ihnen gewesen,*

um wieviel mehr wird das, was ich selbst **euch** kundgetan habe *sage,* **wahrhaftig** *[das, was ich euch sage]* [2] wirklich **geschehen, damit der, welcher mich gesandt hat, verherrlicht werde von euch** [3] und **von denen, die an mich glauben.**

20. (31.) Nachdem **er** *aber* **dies zu uns gesprochen hatte, sprachen wir zu ihm: „Herr, in allem hast du dich uns gnädig erwiesen und uns gerettet,** dies alles **hast du uns offenbart.** *Noch* eins **wiederum möchten wir dich fragen, wenn du es uns gestattest."** Und **er antwortete** und **sprach zu uns: „Ich weiss,**

dass ihr zuhört und zu hören euch sehnt; betreffs dessen, was ihr wünscht, fraget mich! [4] Siehe, fragt [4] mich und behaltet das Gehörte im Gedächtnis, und es wird mir angenehm sein, mit euch zu reden.	*nämlich, dass ihr werdet fragen* [5] *und euer Herz zufrieden ist, wenn ihr mich hört. Betreffs dessen aber, was ihr wünscht, fragt mich danach und ich will schön mit euch reden.*

---

[1] Kopt. wörtlich: ich schlief ein. Kopt. und Äth. haben aber beide als Text ἐκοιμήθην vor sich gehabt, das sie verschieden verstanden. [2] Diese nur im Kopt. vorhandenen Worte sind Dittographie. [3] Hs L „von mir". [4] Die Hss haben: ihr fragt mich. Aber es ist der Imperativ herzustellen. [5] Textfehler.

21. (32.) *Denn* wahrlich, ich sage euch: wie der *mein* Vater **m i c h** von den Toten auferweckt hat, gleicherweise werdet auch **i h r** im Fleisch auferstehen, und er wird euch hinaufführen[1] über die Himmel an den Ort, von dem ich von Anfang an (= früher schon) zu euch geredet habe, *an den Ort,* den *euch* bereitet hat, der mich gesandt hat. Und

deshalb habe ich alle Barmherzigkeit vollendet: ohne gezeugt zu werden, bin ich von Menschen geboren[2] und, ohne Fleisch zu haben, habe ich Fleisch angezogen und bin aufgewachsen, damit (ich) euch, die ihr im Fleisch gezeugt werdet . . . .[3].	*also werde* ich alle *Heilsveranstaltungen vollenden: der ich bin ungezeugt und (doch) gezeugt* von Menschen, *der ich bin* ohne Fleisch, *(und doch) habe ich getragen das* Fleisch, *denn deswegen bin ich gekommen,* damit *ihr*[4]

und ihr in der Wiedergeburt die Auferstehung in eurem Fleisch erhaltet[5], einem Gewande[6], das nicht vergehen wird, mit allen, die hoffen und glauben an den, der mich gesandt hat; denn so hat mein Vater an euch Wohlgefallen gefunden, und denen, welchen ich will, gebe ich die Hoffnung des Reiches." Darauf sprachen wir zu ihm: „Gross ist, wie du hoffen lässt und redest." Er antwortete und sprach zu uns: „Glaubet[7], dass alles, was ich euch sage, geschehen wird!" Und wir antworteten ihm und sprachen zu ihm[8]: „Ja, Herr." Und er sprach zu uns: „Wahrlich, ich sage euch, daß ich alle Gewalt von meinem Vater empfangen habe, damit ich die in der Finsternis Befindlichen ins Licht zurückführe und die in der Vergänglichkeit Befindlichen in die Unvergänglichkeit und die im Irrtum Befindlichen in die Gerechtigkeit und die im Tode Befindlichen ins Leben und damit die in Gefangenschaft Befindlichen entfesselt werden, weil (?) das, was von seiten der Menschen unmöglich ist, von seiten des Vaters[9] möglich ist. Ich bin die Hoffnung der Hoffnungslosen, der Helfer derer, die keinen Helfer haben, der Schatz

[1] Kopt. Passiv: ihr werdet hinaufgenommen werden. [2] Kann auch „gezeugt" übersetzt werden. [3] Es fehlt ein Verbum wie „wiedergebäre". [4] Hier bricht der Kopte ab. Es fehlen etwa 10 Zeilen. [5] So nach A C. Die Lesart tetnaš'ū von LBS bedeutet „ihr werdet auferstehen". [6] Apposition zu „in eurem Fleisch". [7] Im Urtext muß statt des Imperativs gestanden haben: „Glaubet ihr?" [8] Von den hier beginnenden 4 Zeilen des Blattes XV des Kopten haben sich nur die oben durch den Druck gekennzeichneten einzelnen Wörter erhalten. Darauf fehlen 8—9 Zeilen. [9] A: bei Gott.

der Bedürftigen [1], der Arzt [2] der Kranken, die Auferstehung der Toten."

22. (33.) Nachdem er dies zu uns gesagt hatte, sprachen wir zu [3] ihm: „Herr, steht es wirklich dem Fleisch bevor, mit der Seele und dem Geist (zusammen) gerichtet zu werden, und wird (die eine Hälfte davon) [4] *zwar* im Himmelreich ruhen und die andere *aber* ewiglich, indem sie (noch) leben, gestraft werden?" Und er sprach *aber* zu uns: „Wie lange noch [5] fragt und forscht ihr?" [6]

23. (34.) Und wir sprachen wiederum zu ihm: „Herr, es ist aber *nämlich* nötig *uns, dass wir erfragen aus dir,* weil [7] du uns [7] befohlen hast *befiehlst,* dass wir predigen, verkünden und lehren, damit wir

indem wir genau von dir hören, gute Verkündiger werden und sie lehren, dass sie an dich glauben.	*selbst genau von dir erfahren und nützliche Prediger* werden, *und die, welche durch uns lehren werden,* an dich glauben.

Deshalb befragen [8] wir dich *oftmals.*

24. (35.) Er antwortete *uns* und [9] sagte zu uns: „Wahrlich, ich sage euch,

das Fleisch jedes Menschen wird auferstehen mit seiner Seele [lebendig] [10] und seinem [11] Geiste."	*die Auferstehung des Fleisches wird geschehen, indem die Seele mit dem Geist in ihm ist.*

Und wir sprachen zu ihm; „Herr, ist es *denn* möglich, *dass* das, was

vergangen und zerstreut ist, lebendig werde --	*aufgelöst und vernichtet ist, heil werde —*

nicht als ob wir leugnen *ungläubig sind,* fragen wir *dich — oder ist es unmöglich dir —,* sondern wir glauben wirklich, dass das, was du sagst, (auch) geschehen ist und geschehen wird."

---

[1] AC: Armen. [2] S: die Medizin. [3] Von hier an haben sich einige Wörter von Blatt XVI des Kopten erhalten. [4] Die Worte in Klammern ergänzt, um dem unverständlichen Text einen Sinn abzugewinnen. Text wörtlich: mit der Seele gerichtet zu werden, und der Geist wird im Himmelreich ruhen und eine Hälfte davon usw. [5] So C. Kopt. wörtlich: bis zu welchem Tage. A: weshalb. L: was, bitte. [6] Äth. kann auch futurisch übersetzt werden. [7] Nach A B C. [8] So ist der Kopte mit Schmidt zu verb. [9] Kopt. untergeordnet: indem .... [10] Nach A getilgt. [11] Nach S.

Und er sprach zu uns, indem er zürnte:	*Und er zürnte uns, indem er zu uns sagte:*

„Ihr Kleingläubigen, wie lange noch[1] fragt ihr mich?

Und forschet (nur) ohne Schmerz nach dem, was ihr zu hören wünscht!	*Aber was ihr wünseht, sagt es mir, und ich selbst werde es euch sagen ohne Neid. Nur*

haltet meine Gebote und tut, was ich euch sage, *und wendet nicht euer Antlitz von Jemandem ab, damit auch ich nicht mein Antlitz von euch abwende, sondern* ohne Zögern[2] und ohne Scheu[3] und ohne Ansehen der Person dienet auf dem engen, geraden[4] und schmalen Wege. *Also*[5] und dadurch wird *mein* der Vater *selbst* in jeder Hinsicht[6] frohlocken über euch."

25. (**36.**) Und wir sprachen wiederum zu ihm: „Herr,

siehe, wir haben dich zum Spott[7] mit dem vielen Fragen."	*bereits schämen wir uns nämlich, dass wir ⟨dich⟩ oftmals befragen und dich belästigen."*

Da antwortete ⟨er⟩ und sprach zu uns: „Ich weiss *nämlich,* dass ihr in Glauben und aus mit eurem ganzen Herzen mich fragt. Und *deswegen* freue ich mich über euch. *Denn* wahrlich, ich sage euch, ich ergötze mich[8], und mein Vater, *der* in mir *ist,* freut sich, dass ihr so forscht und *mich* fragt. Eure Scheulosigkeit *aber* lässt[9] mich frohlocken, und euch selbst gereicht es zum Leben." Und als er *aber* dies zu uns gesagt hatte, wurden wir froh,

weil er in Sanftmut zu uns gesprochen hatte	*⟨dass⟩ wir ihn befragten,*

und wir sprachen wiederum[10] zu ihm: „Unser Herr, in allem

hast du dich uns gnädig erwiesen und gewährst[11] uns Leben, denn alles, was wir dich gefragt haben, sagst du uns."	*machst du uns lebendig und erbarmst dich ⟨unser. Jetzt⟩ nun wirst du uns kundtun das, was wir ⟨dich⟩ fragen werden?"*

Darauf sprach er zu uns: „Fällt *ist* denn das Fleisch oder der Geist hin *vergänglich?"*[12] Und wir sprachen zu ihm:

[1] Nach AB ist hier ausgelassen: „glaubt ihr nicht". [2] Im Äth. nach ABCS. [3] Im Äth. nach A. Die anderen Hss: ohne zu beschämen. [4] Kopt. „geraden" vor „engen". [5] Genau: So ist es auch mit meinem Vater. Er wird .... [6] Fehlt in S. [7] Nach Konjektur (nezanaguegue statt nezanage'). Letzteres würde bedeuten: Wir sind für dich (= in deinen Augen) wahnsinnig. [8] Kopt. lückenhaft. Es ist mit Wajnberg die I. Person Imperf. IV, I anzunehmen. [9] Im Äth. ist eine leichte Verbesserung nötig. [10] Fehlt in C. [11] „Wenn man mit S tāh" liest. [12] Kopt. wörtlich: ⟨Was ist's⟩ denn, ⟨das ver⟩geht? Ist es ⟨das

„Das Fleisch." Und *da (?)* sprach er zu uns: „Was *also* hingefallen ist, wird wieder aufstehen, *und was verloren, wird gefunden werden,* und was krank ist, wird gesund, damit daran *an diesem Sobeschaffenen* gepriesen ⟨*offenbar*⟩ werde *die Herrlichkeit* mein*es* Vat*ers*. Wie er *es* an mir getan hat, so *werde* ich *selbst es tun* an euch und allen [1], die da an mich glauben."

26. (**37.**) Wahrlich *aber* ich sage euch, das Fleisch wird auferstehen mit der Seele lebendig, damit

sie bekennen und gerichtet werden gerecht mit [2] dem Werk,	*statthabe ihre Verantwortung* ⟨*an*⟩ *jenem* ⟨*Tage*⟩ *in betreff dessen*

was sie getan haben, es sei Gutes oder Böses, auf dass es werde zu einer Auswahl und Darstellung für die, welche geglaubt [3] und *welche* getan haben das *die* Gebot*e* meines Vaters, der mich gesandt hat. *Und also* darauf wird das gerechte Gericht stattfinden *in Strenge;* denn so will es mein Vater und er sprach zu mir: Mein Sohn, am Tage des Gerichts *nämlich* sollst du dich *einerseits* vor dem *den* Reichen nicht scheuen und *andrerseits* den Armen nicht schonen [4], sondern übergib gemäss der Sünde eines jeden ihn ewiger Strafe.

Denjenigen aber, die mich geliebt haben und mich lieben [5] und	*Meinen Geliebten* aber,

die mein*es Vaters* Gebot*e*, *der mich gesandt hat,* getan haben, werde ich Ruhe im *des* Leben*s* verleihen im Reiche meines Vaters, welcher im Himmel ist [6],

siehe, schaut, was für eine Macht er mir verliehen hat, und er hat mir gegeben, dass, ... was ich will und wie ich gewollt habe, ... und denen ich Hoffnung erweckt habe.	*und sie werden schauen, was er mir verliehen hat, und er hat mir die Macht gegeben,* ⟨*damit ich*⟩ *tue, was ich will, und damit ich gebe den ..... und denen, welchen zu geben und* ⟨*zu verleihen*⟩ *ich beschlossen habe.*

Fleisch⟩ oder der Geist? Ein Rest dieser Konstruktion scheint auch in der Lesart der äth. Hss ABC, die dem Fragesatz ein „Was" vorausschicken, zu stecken. [1] ABS euch allen. [2] Fehlt in ABC. [3] Perf. nach AC. Kopt. einfach „Gläubigen". [4] Kopt. wohl: mit den Armen nicht Erbarmen haben. So auch die äth. Hss BC, die „dich erbarmen" haben. [5] „und mich lieben" fehlt in AC. [6] Kopt. und ABS.

27. (**38.**) Und deswegen *nämlich* bin ich hinabgestiegen[1] zu ⟨*dem Orte des*⟩

Abraham, Isaak und Jakob, zu euren Vätern, den Propheten, und habe ihnen Botschaft gebracht, damit sie aus der Ruhe, die unterhalb, in die Himmel kämen[2], und habe ihnen gegeben die rechte Hand[3] der Taufe[3]

Lazarus, und habe gepredigt ⟨den Gerechten und⟩ den Propheten, damit sie aus der Ruhe, die unterhalb, herauskommen, und hinaufgehen zu der, die sich befindet ⟨oberhalb⟩, ... rechte ⟨Han⟩d auf sie ....

des Lebens und Vergebung und Erlass *Befreiung* aller Bosheit, wie *ich es* euch auch *getan habe,* und von jetzt an auch denen, die an mich glauben. Wer aber an mich glaubt und meine Gebote nicht tut, obwohl er an meinen Namen glaubt *bekannt hat,* hat keinen Nutzen *davon*[4].

Vergeblich ist er einen Lauf gelaufen[5]. Sein Ende ist für das Verderben und für die Strafe großen Schmerzes (bestimmt), denn er hat gesündigt gegen mein Gebot.

Er ist ⟨gelaufen einen⟩ eitlen Lauf. Denn Solche werden im Verlorensein sein und ⟨Verderben, weil sie haben⟩ missachtet meine Gebote.

28. (**39.**) Euch aber habe ich gegeben, dass ihr Kinder des Lichtes in Gott seiet und rein seiet von aller Bosheit und aller Macht des Gerichtes[6], und denjenigen, die durch euch an mich glauben, will ich ebenso tun, und[7] wie ich euch gesagt[8] und verheissen habe, dass er aus dem Gefängnis herausgehen und aus den Fesseln und den

⟨*Um so mehr aber ihr*⟩, *die Kinder des Lebens, ich habe* ⟨*euch*⟩ *erlöst von allem Bösen und von* ⟨*der Gewalt der*⟩ *Archonten und alle,* die durch euch ⟨an⟩ mich glauben. *Denn was* ich ⟨euch⟩ verheissen habe, *werde* ⟨*ich*⟩ *auch ihnen geben, damit sie* ⟨aus dem⟩ Gefängnis kommen und der Fessel der Archonten und dem *gewaltigen* Feuer."

---

[1] Nach A C ist für den Äth. hier ausgelassen „und habe geredet". [2] Ich verbessere 'emtāḥtū 'erafta in 'emzatāḥtū 'eraft und jamṣe'ū in jemṣe'ū. [3] Vielleicht auch: mit der rechten Hand die Taufe. [4] Äth. kann auch übersetzt werden: „dem wird das Glauben an meinen Namen nichts nützen". [5] Nach S: badīṛa badara. Dasselbe Verbum ist 1 Kor 9_{24} in der äthiopischen Bibel gebraucht. [6] Schon hier ist mit Wajnberg zu vermuten, dass kuenonē „Gericht" aus makuānent „Fürsten, Archonten" entstellt ist. [7] Zu tilgen. [8] Nach S 'ebēlakemū übersetzt.

Lanzen[1] und dem schrecklichen Feuer befreit werden soll."

Und **wir** ⟨*antworteten und*⟩ **sprachen zu ihm**: „**Herr,** in jeder Hinsicht **hast du uns** *nämlich* erfreut und uns **Ruhe gegeben,**

denn in Treue und in Zuverlässigkeit hast du unseren **Vätern und den Propheten** verkündigt und ebenso uns und jedem."

und hast gegeben ... in Wundern zu ..... des Glaubens, wirst du nun dieses selbst predigen ....., nachdem du gepredigt hast den ⟨**Vätern**⟩ **und den Propheten?**"

Und *da* **sprach er zu uns**: „**Wahrlich** *nämlich,* **ich sage euch**:

euch und alle, die glauben, und auch die, welche noch glauben werden an den, der mich gesandt hat, **werde ich** aufsteigen lassen **in die Himmel, an den Ort, welchen zubereitet hat den Auserwählten** und Auserwähltesten **der Vater,** der geben wird die **Ruhe,** die er verheißen hat, **und das ewige Leben.**

⟨*alle*⟩, *die geglaubt haben an mich* ⟨*und die*⟩ *glauben* ⟨*werden*⟩ *an den. der mich gesandt hat, werde ich hinauf*⟨*führen*⟩ *in die Himmel, an* **den Ort,** ⟨**welchen**⟩ *mein* **Vater** ⟨**zubereitet**⟩ **hat den Auserwählten,** *und ich werde euch geben das Reich, das auserwählte, in der* **Ruhe und das ewige Leben.**

29. (40.) Diejenigen aber, die gegen mein Gebot gesündigt haben[2], anderes lehren, abziehen **und hinzufügen,** und für ihre **eigene** Ehre wirken, indem sie abwendig machen **diejenigen, welche recht an mich glauben,** [S+ werde ich dem Verderben übergeben.]"

Diejenigen aber, die ⟨*meine*⟩ *Gebote übertreten haben und gelehrt haben in anderen Lehren, indem sie .... das Geschriebene* **und hinzufügen .... eigene ....,** *indem sie lehren in andern Worten* ⟨**diejenigen, welche**⟩ **recht an mich glauben,** *wenn* ⟨*sie*⟩ *durch Derartige zu Fall gebracht werden, ... ewige Strafe.*"

Und wir **sprachen zu ihm**: „**Herr,** wird **denn** eine andere **Lehre** und Bedrängnis[3] existieren?"

Wir sprachen ⟨*aber* **zu ihm**⟩: „**Herr,** ⟨*werden*⟩ **denn** *Lehren von* ⟨*anderen*⟩ *existieren ausser dem, was* ⟨*du*⟩ *uns gesagt hast?*"

---

[1] Es ist mit Wajnberg 'emkuenāt in 'emmakuānent = „von den Fürsten, Archonten" zu verbessern. Aber auch damit ist der ursprüngliche Text noch nicht ganz wiedergewonnen. [2] Perf. nach A B C wie im Kopt. Die übrigen Hss haben konform dem Folgenden das Imperf. [3] Oder sollte ḥemmamē aus hemz „Gift" entstellt sein?

Er sprach zu uns: „Wie die Guten und Schönes Vollbringenden (so) werden (auch) die Bösen offenbar werden [1]. Und dann wird eintreten ein gerechtes Gericht nach ihrem Werk gemäss dem, wie sie gehandelt haben, und man wird sie dem Verderben übergeben."

. . . . . uns: „Es ist nämlich notwendig, dass sie existieren, damit ⟨die⟩ Bösen und die Guten offenbar werden. Und also wird offenbar werden das Gericht *an denen, welche diese Werke tun, und ⟨gemäss ihren⟩ Werken werden sie gerichtet und dem Tode übergeben werden."*

Und *wiederum* sprachen wir zu ihm: „Herr, selig sind wir, die wir dich sehen und *dich* hören, wie du zu uns redest *derartiges,* und deren *denn unsere* Augen haben gesehen solch grosse Krafttaten *Wunder,* die du getan hast." Und er antwortete und sprach zu uns: „Aber vielmehr selig *sind* werden sein die, welche mich nicht *ge*sehen und (doch) glauben *geglaubt haben,* denn sie *Solche* werden genannt werden Kinder des Reiches und ⟨sie⟩ werden sein Vollkommne in dem Vollkommenen; diesen werde ich sein ewiges Leben im Reiche meines Vaters." Und wir sprachen wiederum zu ihm: „Herr, wie wird man glauben können, dass du *gehen und* uns verlassen wirst, wie du *zu uns* sagst: es wird kommen eine Zeit *ein Tag* und eine Stunde,

wo es dir bevorsteht, zu deinem Vater zu gehen?

dass ich hinaufgehe zu *meinem* Vater."

30. (41.) Er *aber* antwortete und sprach zu uns: „Gehet und prediget den zwölf Stämmen Israëls und *prediget auch* den Heiden und Israel und dem *ganzen* Lande Israels, von dem [2] Osten und *bis zu* dem [2] Westen,

dem Norden und dem Süden,

vom ⟨Süden *bis zum*⟩ Norden,

und (so) viele werden an mich, an den Sohn Gottes, glauben." Und wir *aber* sprachen zu ihm: „Herr, wer wird uns glauben und *oder* wer wird auf uns hören und wie können ⟨*indem*⟩ wir tun und lehren und erzählen die Wunder und Zeichen und Krafttaten [3], *die* wie du (sie) getan hast?" Und er antwortete *darauf* und sprach zu uns: „Geht und predigt [und lehret] [4] betreffs [der Ankunft und] [2] der Barmherzigkeit

---

[1] Mit Hilfe der Hss-Varianten ließe sich der entstellte Text so herstellen, daß er lautet: „damit die Gutes und Böses Tuenden offenbart werden". Dies nähert sich dem Kopt., es fehlt aber immer noch der Hauptsatz. [2] Äthiopisch la kann auch „gen" übersetzt werden. [3] Im Kopt. ist die Reihenfolge: Krafttaten, Zeichen, Wunder. [4] Die eingeklammerten Worte stehen in B ausserhalb des Textes, und in AC fehlen

meines Vaters *und was* wie mein Vater **getan hat durch**[1] **mich, werde auch ich tun durch euch**[1]**, indem ich** mit **in** *euch* **bin, und ich werde euch geben meinen Frieden und aus meinem Geiste**[2] **werde ich euch geben** [m][3]**eine Kraft,**

[damit es euch widerfahre][4], dass sie glauben. Auch ihnen wird diese Kraft gegeben und vererbt werden, damit sie sie den Heiden geben.

auf dass ihr prophezeit ihnen zum ewigen Leben. Den andern aber werde ich selbst wiederum geben meine Kraft, auf dass sie lehren die übrigen Völker[5].

31. (**42.**) Und siehe, ihr werdet einen Mann treffen, dessen Name Saul ist, d. h. verdolmetscht Paulus. Er ist Jude, beschnitten nach der Vorschrift des Gesetzes, und er wird meine Stimme vom Himmel hören mit Schrecken, Furcht und Zittern; und seine Augen[6] werden verfinstert und durch eure Hand mit Speichel[7] bekreuzt werden. Und alles tut ihm, wie ich euch getan habe. Andern übertragt es. Und dieser Mann — sogleich werden seine Augen aufgetan werden[8], und er wird Gott, meinen himmlischen Vater, preisen. Und er wird stark werden im Volk und wird predigen und lehren, und viele werden ergötzt werden, wenn sie (es) hören, und werden gerettet werden. Darauf wird man ihn hassen und in die Hand seines Feindes ausliefern, und er wird vor sterblichen [und vergänglichen][9] Königen bekennen, und das Ende des Bekenntnisses zu mir wird über ihn kommen; dafür, dass er mich verfolgt und gehasst[10] hatte, wird er zu mir bekehren[10] und predigen und lehren, und er wird unter meinen[11] Auserwählten sein ein auserwähltes Rüstzeug und eine Mauer, die nicht fällt. Der Letzte der Letzten wird Prediger für die Heiden werden, vollkommen im (oder: durch) den Willen meines Vaters. Wie ihr aus den Schriften erfahren habt, daß eure Väter, die Propheten, über mich geredet haben und es an mir erfüllt ist — dieses Wahrhaftige sprach er[12] —, so werdet auch ihr ihnen[13] zum Führer und jedes Wort, welches

---

die Worte „der Ankunft und" ganz. Der ursprüngliche äthiopische Text stimmt also mit dem Kopten. [1] Äthiopisch vielleicht „an". [2] Äthiop. „meinen Geist und". [3] In B fehlt das Pronomen „mein". [4] Fehlt in B. [5] Im Kopt. fehlen von hier an 6 Blätter. [6] Das erste 'a'jentīhū ist mit BC gestrichen. [7] „mit Speichel" nach ABC. [8] S: er wird auftun. [9] Nur in A. [10] gehasst nach ABCS eingesetzt. Ebenso: „wird er zu mir bekehren". [11] „meinen" nach ABCS. [12] Nach S das wa vor jebēlana gestrichen wie auch das Suffix „zu uns". [13] Statt „ihnen" wird wahrscheinlich das Singularsuffex „ihm" einzusetzen sein. Wie das Folgende zeigt, gehen die Worte auf Paulus.

ich zu euch geredet habe und welches ihr über mich geschrieben habt, daß ich das Wort des Vaters bin und der Vater in mir ist, so werdet auch ihr an jenem Mann, wie es euch geziemt: lehret und erinnert (ihn), was in den Schriften über mich gesagt ist und erfüllt worden ist, und dann wird er den Heiden zum Heil sein.

32. (**43.**) Und wir sprachen zu ihm: „Meister, haben wir mit ihnen zusammen e i n e Hoffnung auf das Erbe?" Er antwortete und sprach zu uns: „Sind die Finger der Hand gleich oder die Ähren auf dem Felde oder geben die fruchttragenden Bäume gleiche Frucht? Bringen sie nicht je nach ihrer Natur Frucht?" Und wir sprachen zu ihm: „Herr, redest du wieder im Gleichnis zu uns?" Er sprach zu uns: „Seid nicht betrübt! Wahrlich, ich sage euch, meine Brüder seid ihr, Genossen im Himmelreich bei meinem Vater, denn so hat es ihm gefallen. Wahrlich ich sage euch, auch denjenigen, die ihr lehrt und die an mich glauben, werde ich diese Hoffnung geben."

33. (**44.**) Und wir sprachen wiederum zu ihm: „Wann Herr[1], werden wir jenen Mann treffen, und wann wirst du zu deinem Vater und auch unserm Gott und Herrn gehen?" Und er antwortete und sprach zu uns: „Jener Mann wird aus dem Lande Cilicien ausziehen nach Damaskus in Syrien, um die Kirche, die euch zu schaffen obliegt[2], zu zerreißen. Ich bin es[3], der durch euch (zu ihm) redet; und er wird eilends kommen[3]. In diesem Glauben wird er [C: + stark] sein, damit erfüllt werde das Wort des Propheten, der da spricht: Sieh', aus dem Lande Syrien will ich zu rufen anfangen ein neues Jerusalem, und Zion werde ich mir unterwerfen und es wird gefangen werden, und die Unfruchtbare[4], die keine Kinder hat, wird kinderreich[5] sein und wird Tochter meines Vaters genannt werden, mir aber meine Braut, denn also hat es gefallen dem, der mich gesandt hat. Jenen Mann aber werde ich abwenden, dass er nicht hingeht und den bösen Plan vollbringt, und durch ihn wird Ehre meines Vaters eintreten. Denn[6] nachdem ich fortgegangen bin und bei meinem Vater weile, werde ich

[1] „Herr" nach ABCS eingesetzt. [2] Lies mit S hallawakemū. [3] Das za vor we'etū ist unverständlich. Der Nachsatz kann auch bedeuten: „und es wird eilends eintreffen (= sich erfüllen)." [4] Steht inkorrekt im Akkusativ, bzw. Stat. constr. [5] So nach C wallāda. [6] Nach ABC eingefügt.

vom Himmel her mit ihm reden und es wird alles geschehen, wie ich es euch über ihn vorhergesagt habe.“

34. (45.) Und wir sprachen wiederum zu ihm: „Herr, so Bedeutsames hast du zu uns geredet und uns verkündigt [1] und hast uns nie ausgesprochenes Großes offenbart und in jeder Hinsicht hast du uns beruhigt und hast dich uns gnädig erwiesen. Denn nach deiner Auferstehung offenbartest du uns dies alles, damit wir wirklich gerettet würden. Du sagtest uns aber nur, daß Zeichen und Wunder am Himmel und auf Erden geschehen würden, ehe das Ende der Welt einträte. Daß wir es also erkennen, lehre uns!“ Und er sprach zu uns: „Euch [2] werde ich lehren und (zwar) nicht nur, was an euch geschehen wird, sondern (auch) an denjenigen, die ihr lehren werdet und die glauben werden, und es gibt solche, die diesen Mann (?) hören und an mich glauben werden [3]. In jenen Jahren und in jenen Tagen wird dies geschehen.“ Und wir sprachen wiederum zu ihm: „Herr, was ist denn das, was geschehen wird?“ Und er sprach zu uns: „Dann [4] werden die Gläubigen und auch die, welche nicht glauben, ein Horn am [5] Himmel sehen und das Gesicht großer Sterne, die, während es Tag ist [6], sichtbar sind, und einen Drachen [7] vom Himmel, indem er bis zur Erde reicht, und indem Sterne, die wie Feuer sind [8], herabfallen, und große Hagelschlossen von heftigem (?) Feuer [9], und wie Sonne und Mond miteinander streiten, und beständig der Schrecken von Donner und Blitzen, Donnerkrachen und Erdbeben, wie Städte einstürzen und bei ihrer Zerstörung Menschen sterben, beständig Dürre infolge Ausbleibens des Regens, eine große Pest und ein ausgebreitetes und häufiges [10] schnelles Sterben, so dass denen, die sterben, das Begräbnis fehlen wird; und es wird das Hinausgehen (= Hinausgetragenwerden?) [11] von Kindern und Verwandten auf e i n e m Bett (= Bahre) geschehen. Und der Verwandte wird sich seinem Kinde nicht zuwenden noch das Kind seinem Verwandten, und ein Mensch wird

---

[1] Nach ABCS. [2] kama vor kījākemū ist mit ABC zu streichen. [3] Die in AS fehlenden Worte: wabō ’ellaze be’esi jesam‘ū waja’amnū sind unverständlich. Wajnbergs Übersetzung ist selbst bei Änderung von bō in ba unmöglich, weil „dieser Mann“ im Nominativ steht, man also noch mehr ändern müsste. [4] ’amēha zu lesen nach der Parallelstelle 39 [3]. [5] ABCS = vom. [6] Nach AC: ’enza ma‘ālet we’etū. [7] So nach AS (C) mit Änderung des Nominativs in den Akkusativ. LB: Wunder. [8] Den Relativsatz haben AC nicht. [9] Der Text ist nicht in Ordnung. [10] Nach ABC. [11] In der Parallelstelle bei Guerrier S. 40 Z. 1 steht: Es wird Streiten der Kinder mit ihren Eltern in e i n e m Bette eintreten.

sich seinem Nächsten nicht zuwenden [1]. Die Verlassenen aber, die verlassen wurden, werden auferstehen und die sehen, welche sie verlassen haben, indem sie sie hinausbrachten [2], weil Pest (war). Alles ist Hass und Bedrängnis und Eifersucht, und dem einen [3] wird man nehmen und dem andern schenken, und schlimmer wie dies wird sein, was nach diesem kommt.

35. (46.) Dann wird mein Vater wegen der Bosheit der Menschen in Zorn geraten; denn zahlreich sind ihre Versündigungen, und der Schauder vor ihrer Unreinheit ist sehr wider sie, deren Leben in Verderbnis ist [4]." Und wir sprachen zu ihm: „Herr [5], was (wird) denn denen (zuteil), die auf dich hoffen?" Und er antwortete und sprach zu uns: „Wie lange seid ihr weiter trägen Herzens? Wahrlich, ich sage euch, wie der Prophet David über mich [6] und die Meinigen gesprochen hat, so will er es auch über die [6], welche an mich glauben werden. Es werden aber in der Welt Betrüger [7] und Lästerer [8] der Gerechtigkeit sein, und sie wird treffen die Prophezeiung Davids, der da spricht: „Ihre Füße sind eilend zum Blutvergiessen [9], und ihre Zunge flicht Betrug [10], und Schlangengift ist unter ihren Lippen [11]. Und ich sehe dich, wie du mit einem Dieb umherläufst, und mit einem Hurer ist dein Anteil [12]. Indem du ferner dasitzst, schmähst du deinen Bruder und [indem] [13] du legst einen Fallstrick dem Sohne deiner Mutter. Was meinst du, soll ich sein wie du?" [14] Und nun sehet, wie der Prophet Gottes über alles geredet hat, damit erfüllet werde alles [15], was zuvor gesagt worden ist."

36. (47.) Und wiederum sprachen wir zu ihm: „Herr, werden dann nicht die Heiden sagen: Wo ist ihr Gott?" Er antwortete und sprach zu uns: „Daran werden die Auserwählten offenbar werden, daß sie, nachdem sie durch eine solche Not bedrängt worden sind, hinausgehen." Und wir sprachen zu ihm: „Wird ihr Ausgang aus der Welt (eintreten) durch eine Pest, die sie gequält hat?" Und er sprach zu

---

[1] Nach AC(S) übersetzt. [2] Nach der Parallelstelle S. 40 sollte man „nicht hinausbrachten" erwarten. [3] Mit ABC 'emze statt 'emza, Verb. mit C im Plural! [4] Die Übersetzung des sicher verderbten Textes ist ein Notbehelf. [5] Nach ABC eingefügt. [6] Nach ABC. [7] Es ist oben der Nominativ ḥabbālejān hergestellt. [8] So nach AL. BC: Feinde. [9] Sprüche 1_{16}. [10] Psalm 50 (49) 19b. [11] Psalm 140 (139) $_4$. [12] Der Nominativ ist zu lesen. [13] Zu streichen. [14] Psalm 50 (49) $_{10\ 20\ 21}$. [15] kuellū in S ist besser als kuellō der anderen Hss.

uns: „Nein, sondern, wenn sie Qual leiden, wird ihnen solches Leiden zu einer Prüfung werden, ob sie Glauben haben und ob sie dieses meines Wortes gedenken und mein Gebot tun. Diese werden auferstehen, und ihr Verweilen wird wenige Tage dauern, damit gepriesen werde, der mich gesandt hat, und ich mit ihm. Denn er hat mich zu euch gesandt. Dieses sage ich zu euch. Ihr aber sagt (es) Israel und den Heiden, damit sie hören; auch sie sollen gerettet werden und an mich glauben und der Not der Pest entgehen. Und wer der Not des Todes entgangen ist[1], einen solchen wird man nehmen und ihn im Gefängnis verwahren unter Peinigung wie der eines Diebes." — Und wir sprachen zu ihm: „Herr, werden sie gleich werden den Ungläubigen, und wirst du ebenso[2] die bestrafen, die der Pest entgangen sind?" Und er sprach zu uns: „An meinen Namen glaubend, haben sie das Werk der Sünder getan[3], wie Ungläubige haben sie gehandelt." Und wir sprachen wiederum zu ihm: „Herr, haben die, die entgangen sind, in diesem Teil kein Leben?"[4] Er antwortete und sprach zu uns: „Wer die Verherrlichung meines Vaters getan hat, der ist die Wohnung meines[5] Vaters"[6].

37. (**48.**) Und wir sprachen zu ihm: „Herr, lehre uns, was nach diesem geschehen wird!" Und er sprach zu uns: „In jenen Jahren und Tagen[7] wird Krieg über Krieg (sein), und die vier Ecken der Welt werden erschüttert werden und werden sich gegenseitig bekriegen. Darauf (wird eintreten) eine Wolkenbewegung, Finsternis und Dürre und Verfolgung derer, die an mich glauben, und der Auserwählten; darauf Uneinigkeit, Streit und Bosheit des Handelns[8] gegeneinander. Darunter gibt es solche, die an meinen Namen glauben und (trotzdem) der Bosheit folgen und eitle Lehre lehren. Und diesen wird man folgen und wird sich ihrem Reichtum, ihrer Verworfenheit, ihrer Trunksucht und ihrem Bestechungsgeschenk unterwerfen, und Ansehen der Person wird unter ihnen herrschen.

[1] Perf. nach ABC. [2] Das im Text stehende Wort ist entstellt; der Sinn muß aber wohl der oben angegebene sein. [3] Es ist jedenfalls hāṭe'ān zu lesen. Ob man davor za oder mit A gebra liest, ist für den Sinn belanglos. [4] Die Übersetzung dieses Satzes unterliegt Bedenken. [5] Nach AC: bei meinem Vater. [6] Statt auf Io 14_2 ist vielmehr auf 14_{23} zu verweisen, falls der Text überhaupt in Ordnung ist. Bei der Lesart von AC müsste man „haben", „finden" oder Ähnliches ergänzen. [7] Hier setzt ein koptisches Fragwort mit wenigen Worten ein, s. Schmidt S. 117 ff. [8] ABC: der Lebensweise.

38. (49.) Diejenigen aber, welche das Antlitz Gottes schauen wollen und welche die Person der sündigen Reichen[1] nicht berücksichtigen und sich nicht scheuen vor den Menschen, die sie verführen[2], sondern sie zu tadeln trachten — diese werden beim Vater verwundet sein[3], wie auch die, welche ihren Nächsten zu tadeln trachten, gerettet werden. Dieser ist ein Sohn der Weisheit und des Glaubens. Wenn er aber nicht ein Sohn der Weisheit geworden ist[4], so wird er hassen und verfolgen und nicht sich zuwenden

seinem Bruder und wird verachten und ihn abweisen.	*seinen Nächsten, wird sich auf ihn stürzen und ihn* .....

Die aber, welche wandeln in **Wahrheit**[5] **und in der Erkenntnis des Glaubens** an mich

und die Erkenntnis der Weisheit haben und Geduld um der Gerechtigkeit willen, indem man sie, die nach der **Armut** streben, verachtet und sie (trotzdem) aushalten, gross ist ihr Lohn. Die man schmäht, **quält** und verfolgt, **indem sie bloss** sind, und **indem** man **hochmütig gegen** sie **ist** und indem sie hungern und dürsten und **weil sie ausgehalten haben** — selig werden sie sein in den Himmeln und **werden** immerdar **mit mir sein. Wehe denen, die** sie hassen und verachten! Und **ihr Ende** ist für **das Verderben.**"	*indem sie besitzen die Liebe zu mir — sie haben nämlich erduldet Schmähungen*[6] — *die werden stolz sein, indem sie wandeln in* **Armut** *und indem sie die ertragen, welche sie hassen* ⟨*und*⟩ *die, welche sie schmähen. Man hat sie* ⟨*ge*-**quält**⟩, **indem sie bloss** *waren*, **indem** *die Menschen* **hochmütig gegen sie** *waren, indem sie wandeln in Hunger und Durst; aber* **weil sie ausgehalten haben**[7], ... *die Seligkeit der Himmel, sie* **werden** *selbst* **mit mir sein** *ewiglich.* **Wehe** *aber dagegen denen, die wandeln in Hochmut und Prahlerei, denn* **ihr Ende** *ist* **das Verderben** *(Verlorensein).*"

39. (50.) Und **wir sprachen** *aber* **zu ihm: „Herr,**

wird dies allen widerfahren?"	*das deinige ist dies, damit du uns nicht über sie kommen lässt.*"

[1] Oder: der reichen Sünder. [2] ABC: irren. [3] Fehler. Vielleicht mit Guerrier in queṣūlāna „gekrönt" zu verbessern. [4] Mit S ist kōna statt kōnū zu lesen. [5] Das äthiop. Wort kann auch „Gerechtigkeit" bedeuten. [6] ὕβρις nach Hermas mand. 8_{10} verstanden des Äthiopen wegen. Schmidt: Misshandlungen. [7] Hinter cha ist offenbar etwas ausgefallen.

Und er *antwortete aber und* sprach zu uns: „Wie wird das Gericht der Gerechtigkeit geschehen

den Sündern und den Gerechten?"

entweder der Gerechte oder der Ungerechte?"

Und wir sprachen zu ihm: „Herr,

wird man an jenem Tage nicht zu dir sagen; Du hast führen lassen zu Gerechtigkeit und Sünde, hast geschieden[1] Finsternis und Licht[2], Böses und Gutes?"

an jenem Tage *nämlich* wird man zu dir sagen: *Du hast nicht verfolgt die* ⟨*Gerechtigkeit*⟩ *und* ⟨*die*⟩ *Ungerechtigkeit, das Licht und die Finsternis, das Böse und das Gute."*

Und *darauf* sprach er zu uns: *„Ich werde ihnen antworten, indem ich sage:* Adam ist die Macht gegeben worden, dass er von den zweien sich *eins* auswähle, was er wollte, und er wählte das Licht und

streckte seine Hände aus [und nahm (es)][3] und liess die Finsternis und entfernte sich von ihr[4]. Ebenso ist jeder Mensch ermächtigt,

brachte seine Hand darauf, die Finsternis *aber* verliess er *und warf sie von sich. Also haben alle Menschen die Macht,*

zu glauben an das Licht,

dies ist das Leben des Vaters,

welches ist das ⟨*Leben*⟩ *und welches ist der Vater,*

der mich gesandt hat.

Und wer an mich geglaubt hat, wird leben, wenn er das Werk des Lichtes getan hat. Wenn er aber bekennt[5], dass das Licht ist, und tut das der Finsternis (Eigentümliche)[6], so hat er weder etwas, was er zur Verteidigung wird sagen können,

Ein jeder aber, der glaubt *(und) der tut die Werke* des Lichtes, wird leben *in ihnen. Wenn aber einer ist, der* bekennt, *dass er gezählt ist zum Lichte, indem er* tut *die Werke* der Finsternis, *ein solcher besitzt keine* Verteidigung *zu* sagen

noch wird er *sein* (das) Antlitz erheben können, und *um anzu*blicken den Sohn *Gottes,* der ich bin. Und ich werde *nämlich* zu ihm sagen:

---

[1] L „geschaffen". [2] S hat die Reihenfolge „Licht und Finsternis" wie Kopt. [3] Fehlt in A. [4] Würde man wa 'arḫaqa 'emnēhū (letzteres in S) = „entfernte er von sich" lesen, so wäre Übereinstimmung mit dem Kopt. da. [5] So nach AC. Die Lesart 'ī'amna = „nicht geglaubt hat" wäre in 'a'mana = „bekannt hat" zu korrigieren. [6] Nach ABC. S: weilt in Finsternis. Ähnlich L.

Du hast gesucht und gefunden, hast gebeten und empfangen. Was tadelst du uns? [1]	*Als du suchtest, hast du* gefunden, *und als du batest, hast du* empfangen. *Worin verurteilest du mich, Mensch?*

Weshalb hast du dich von mir [2] und meinem Reiche entfernt *und mich verleugnet? Weshalb* hast du mich bekannt und mich (dann doch) verleugnet?

Nun also sehet, dass jeder ermächtigt ist, sowohl zu leben als zu sterben! [3] Und wer mein Gebot tut und bewahrt,	*Hat denn nicht ein* jeder *die Macht,* zu leben *und* zu sterben? Wer *nun* bewahrt *hat meine Gebote,*

wird ein Sohn [4] des Lichtes d. i. meines *des* Vaters, *der sich in mir befindet,* sein.

Und denen, die bewahren und tun, und um des [5] willen bin ich herabgestiegen vom Himmel, ich [6], das Wort, bin Fleisch geworden und bin gestorben, indem ich lehrte und überführte, dass die einen [7] gerettet werden, die andern [7] aber ewiglich zugrunde gehen werden, indem [7] sie im Feuer gestraft werden am Fleisch und am Geist."	*Um derentwillen aber, die meine Worte verderben,* bin ich *herabgekommen* vom Himmel. Ich *bin* das Wort *(der Logos), ich bin* Fleisch geworden, indem ich *mich abmühte und* lehrte: *Die Geladenen werden* gerettet werden *und die Verlorenen werden verlorengehen* ewiglich. *Man wird sie ⟨peinigen⟩ lebendig und sie züchtigen an ⟨ihrem⟩* Fleisch und *ihrer Seele."*

40. (51.) Und wir *aber* sprachen zu ihm: „Herr, wir sind *tragen* wahrlich betrübt *Sorge* ihretwegen." Und er *aber* sprach zu uns: „Ihr tut *nämlich* gut (daran), denn also sind besorgt die Gerechten um die Sünder und beten *für sie* und flehen zu Gott [8] und *indem sie* bitten ihn *meinen Vater."* Und *wiederum* sprachen wir zu ihm: „Herr,

fleht dich niemand an?"	*Weshalb nun scheut sich niemand vor dir?"*

Und er *aber* sprach zu uns: „Ja, ich werde hören *auf* das Gebet der Gerechten, *das sie* für sie *tun."*

[1] Nach AC (S). L: Was (ist), das du nicht empfangen hast? [2] „Von mir" hat unter den äth. Hss nur A. [3] Nach AC. [4] welūda ist in walda zu verb. [5] Nach dem Kopt. sollte man ba 'entīahōmū = um ihrer willen statt ba 'entaze erwarten. [6] Nach S eingefügt. [7] Nach ABC ist 'ella vor jetḫagualū zu l. Dieses und das erste 'ella sind aber in 'ellū zu verbessern. Statt des 'ella vor westa ist 'enza zu schreiben. [8] AC+ dem Vater.

Und wir sprachen zu ihm: „Herr, dies alles hast du uns gelehrt und hast uns belebt [1] und dich uns gnädig erwiesen. Und wir werden es denen, denen es gebührt, predigen. Wird uns aber bei dir Lohn zuteil werden?“ [2]

Als er aber dies zu uns gesagt hatte, sprachen wir zu ihm: „Herr, *in allen Dingen nämlich hast du uns* gelehrt [3] *und dich unser erbarmt und uns gerettet, damit* wir predigen *denen, die würdig sind, gerettet zu werden, und wir uns einen* Lohn bei dir *erwerben.“*

41. (52.) Und er *antwortete aber und* sprach zu uns: „Gehet und prediget und

werdet gute Diener und Knechte!“ Und wir sprachen zu ihm: „Herr, du bist unser Vater.“ Und er sprach zu uns: „Sind alle Väter und alle Diener, alle Lehrer?“

ihr werdet werden Arbeiter .... und Diener.“ Wir *aber* sprachen zu ihm: *„Du bist es, der predigen wird durch uns.“ Da antwortete er uns, indem er sagte: „Nicht seid Väter alle noch seid Meister alle!“*

Und wir sprachen zu ihm: „Herr, hast du nicht gesagt: Nicht nennt (jemanden) auf Erden Vater und Meister, denn Einer ist euer Vater und Lehrer: der in den Himmeln? Jetzt sagst du uns, dass wir wie du vielen Kindern Väter werden sollen und auch Lehrer [4] und Diener.“

Wir sprachen zu ihm: „Herr, du *bist es, der zu uns* gesagt *hat:* Nicht nennet *euch* (jemanden) Vater auf Erden, einer *nämlich* ist euer Vater, der ⟨in⟩ den Himmeln, *und euer Meister — warum* sagst du uns jetzt: *Ihr werdet sein* Väter *vieler Kinder und* Diener *und Meister?“*

Uud er antwortete *aber* und sprach zu uns:

„Ihr habt recht gesagt. Wahrlich ich sage euch: alle, die auf euch gehört und an mich geglaubt haben, werden erhalten das Licht des Siegels, das in meiner Hand ist [5], und durch mich werdet ihr werden Väter und Lehrer.“

„Wie ihr es gesagt habt. Wahrlich *nämlich,* ich sage euch: *Wer* auf euch *hören wird und glauben wird* an mich, *der ⟨wird empfangen von⟩ euch* das Licht des Siegels *durch ⟨mich⟩ und die Taufe durch mich; ihr werdet ⟨werden⟩* Väter *und Diener und auch Meister.“*

[1] Oder: gerettet. [2] C ohne Frage: und es wird uns Lohn zuteil werden.
[3] Hier folgen zwei unverständliche Worte: „unsern Blick von weg“.
[4] Singular. Es ist wohl mit Wajnberg das ni in na zu verb. und der Plural zu l. [5] Kann auch einfach „durch mich“ übersetzt werden.

42. (53.) Und wir sprachen *aber* zu ihm: „Herr, wie *nun* ist es möglich, dass

dies drei an Einem werde?“

ein jeder von uns wird diese drei?“

Und er antwortete[1] und sprach *aber* zu uns: „Wahrlich, wahrlich ich sage euch: Väter werdet ihr *einmal* genannt werden, weil ihr

liebreich und barmherzig

mit nützlichem Herzen und in Liebe

ihnen offenbart habt,

was im[2] Himmelreich (ist),

die Dinge des Reiches der Himmel.

Und ihr werdet genannt werden Diener[3], weil sie durch meine Hand die Taufe des Lebens und die Vergebung der *ihrer* Sünden empfangen werden *durch euch.* Und *ihr werdet genannt werden* Lehrer *Meister,* weil ihr ihnen[4] mein *das* Wort überliefert[5] habt ohne Schmerz *Neid* und sie[4] zurechtgewiesen habt, und *als ihr sie zurechtgewiesen habt,* haben sie sich bekehrt in dem, worin ihr sie getadelt habt. Und vor ihrem Reichtum habt ihr euch nicht gescheut und habt nicht *ihrem* das Angesicht (= die Person) berücksichtigt, sondern ihr habt das Gebot[6] meines Vaters bewahrt und es *sie* getan. *Und* ⟨*ein*⟩ *grosser* Lohn *wird zuteil werden euch*[7] bei meinem himmlischen[8] Vater und jenen wird die Vergebung der Sünden zuteil werden und ewiges Leben und Anteil[9] am Reich *der Himmel.*“ Und wir sprachen *aber* zu ihm: „Herr, *selbst* wenn

sie einen zehntausendfachen Mund hätten, würden sie dir nicht danksagen können, wie es sich geziemt.“

ein jeder von uns hätte zehntausend *Zungen zum Reden, würden wir nicht* danksagen können dir, *der du uns derartiges verheissest.*“

Und *darauf* antwortete er und „Dies sage ich euch, damit ihr tut, wie ich euch getan habe;

indem er sagte zu uns: „*Nur das, was* ich euch sage, *tut,* wie ich *selbst auch es* getan habe;

---

[1] Und er antworette om. BC. [2] A einfacher: das des Himmelreiches. [3] Der Hauptsatz im Äth. ausgefallen. 'enta ist in 'esma = weil zu verb. [4] C. [5] Kopt. „gegeben“. [6] Der hier nicht erhaltene Kopt. hat den Plural gehabt. [7] C dafür: ist euer. [8] Kopt. wörtlich: „der in den Himmeln ist“. [9] Kopt. „und sie werden teilhaben“.

43. (54.) und seid **wie die klugen Jungfrauen,** die Licht angezündet und nicht geschlummert haben und die mit ihren Lampen **dem Herrn,** dem Bräutigam, entgegengegangen und mit ihm in das Bräutigamsgemach eingegangen sind.

und ihr werdet sein **wie die klugen Jungfrauen,** *die gewacht haben und nicht eingeschlafen sind, sondern heraus⟨gegangen⟩ sind zu (?)* **dem Herrn** *in das Brautgemach*[1].

Die Törichten aber, die mit ihnen redeten[2], **haben nicht zu wachen vermocht, sondern sind eingeschlafen.“** Und wir **sprachen** *aber* **zu ihm: „Herr, wer sind die Klugen und wer** *sind* **die Törichten?“** Und **er sprach zu uns:**

„Die Klugen sind diese fünf, die[3] der **Prophet** Töchter **Gottes** nennt, deren **Namen** man höre.“ **Wir aber** waren **traurig** und betrübt[4] und **weinten** über die, welche ausgeschlossen worden waren.

„Fünf Kluge und fünf Törichte, diese nämlich, hinsichtlich deren der **Prophet** *gesagt hat: Die Kinder* **Gottes** *sind sie. Höret nun ihre* **Namen!“ Wir aber weinten und waren traurig** *über die, welche eingeschlafen waren.*

Und **er sprach zu uns: „Die fünf Klugen** *nämlich,* **sind** diese[5]: **der Glaube** *und* **die Liebe** *und* **die** Freude[6] *Gnade,* **der Friede** *und* **die Hoffnung.**

Sobald die,

Die aber unter denen,

welche an mich **glauben,** *welche* **diese haben, werden** sie **denen, die an mich und an den, der mich gesandt hat,** glauben *geglaubt haben,* **Wegweiser**[7] **sein.** Ich *nämlich* **bin der Herr und ich bin der Bräutigam**[8],

mich **haben sie empfangen und** mit mir **sind sie hineingegangen in das Haus des Bräutigams** und **haben sich** mit dem Bräutigam (zu Tisch) **niedergelegt** und sich erfreut. **Die fünf Törichten aber**[9] haben

den **sie empfangen haben,** *und* **sie sind hineingegangen in das Haus des ⟨Bräutigams⟩** *und* **haben sich niedergelegt** *mit mir in meinem ⟨Braut⟩gemach ⟨und haben sich gefreut⟩.* **Die fünf Törichten aber** *als* ...

---

[1] Kopt hat hier einen verkürzten Text; es fehlt: eingegangen sind. [2] Um den entstellten äth. Worten einen Sinn abzugewinnen, sehe ich die Verbalform als im Geschlecht inkorrekt gebraucht an und zwar sowohl in der Person wie im Suffix (masc. pro fem.). [3] 'amēn mit ABC gestrichen. [4] „und betrübt“ fehlt in C. [5] 'ella in 'ellā verb. [6] Der Äth. setzt χαρά statt χάρις voraus (Schm.). [7] Äth. Führer. [8] So auch AC. [9] „aber“ mit S.

geschlafen und, als sie erwachten, kamen sie zum Haus des Bräutigams und klopften an die Türen[1], denn man hatte sie verschlossen; und sie weinten, weil sie geschlossen waren."

⟨*sie*⟩ ? *eingeschlafen waren ?*, erwachten *sie*, kamen *an die Tür des Brautgemaches und* klopften an, denn man hatte sie verschlossen[2]. *Da* weinten sie *und trauerten, dass ihnen nicht geöffnet wurde.*"

Und wir sprachen *aber* zu ihm: „Herr, *und* diese ihre klugen Schwestern nun, die *drinnen* im Hause *des Bräutigams*,

öffnen sie ihnen nicht und sind sie ihretwegen nicht traurig?" [Und er sprach zu uns: „Ja, sie sind traurig und bekümmert ihretwegen][3] und flehen den Bräutigam an und konnten noch[4] nicht (etwas) erlangen ihretwegen."

verharrten sie darin, ohne ihnen zu öffnen, *und trauerten sie* nicht ihretwegen *oder baten sie nicht den Bräutigam, dass er ihnen öffne?" Er antwortete, indem er zu uns sprach: „Sie konnten noch nicht Gnade finden ihretwegen.*"

Und wir sprachen zu ihm: „Herr, wann *an welchem Tage* werden sie um ihrer Schwestern willen[5] hineingehen?" Und ⟨*da*⟩ sprach er zu uns: „Wer ausgeschlossen ist[6], der ist ausgeschlossen." Und wir sprachen *aber* zu ihm: „Herr, ist diese Sache[7] entschieden? Wer sind nun diese Törichten?" Und er sprach zu uns: „Höret *ihre Namen:*

Einsicht, Erkenntnis, Gehorsam, Geduld, Barmherzigkeit.

die Erkenntnis *ist es und die* Einsicht, *der* Gehorsam, *die Langmut und die* Barmherzigkeit.

Diese[8] *nämlich* haben geschlafen unter denen[9], die geglaubt und mich bekannt haben[10].

44. (55.) Und indem die, welche geschlafen haben, mein Gebot nicht tun, werden sie ausserhalb des Reiches und der Hürde des Schafhirten

Es sind aber meine Gebote nicht getan worden durch die, welche geschlafen haben. *Sie werden daher bleiben* ausserhalb des Reiches *und der* Hürden

[1] Des folgenden Verbums wegen pluralisch übersetzt. hōhet ist manchmal Plural, vgl. z. B. 2 Esra 16_1. [2] Kopt. hat einiges Unklare. [3] Die eingeklammerten Worte fehlen in A, scheinbar in grösserer Übereinstimmung mit dem Kopt. Aber sie sind in der äth. Überlieferung ursprünglich. [4] „noch" in B. [5] Im Äth. ist ba'enta „um willen" mit A einzusetzen. [6] Mit S ist auch beim ersten Verb der Singular z. l. [7] Oder: Wort. Kopt. hat hier eine Lücke. [8] Lies 'ellā. [9] Lies badībĕhōmū. [10] Präterita nach AC.

stehen; und **wer ausserhalb** der Hürde stehen geblieben ist, den wird der Wolf **fressen.** Und obwohl er hört[1], wird er [gerichtet werden und][2] sterben, und **viel Leiden** und Mühsal und Geduld[3] wird auf ihm[4] sein, und obwohl er bös gequält wird und obwohl er zerstückelt und mit langer und schlimmer Strafe zerfleischt wird, wird er doch nicht schnell sterben können."

des ⟨**Hirten**⟩ *und seiner Schafe.* **Wer** *aber* **ausserhalb** *⟨der Hürde⟩ der Schafe bleiben wird, den werden die Wölfe* **fressen,** *und er wird ......, indem er stirbt in* **viel***em* **Leiden.** *Nicht wird ⟨Ruhe⟩ und Ausdauer ⟨in⟩ ihm sein, und man ⟨wird⟩ peinigen ihn böse (?), damit er ...., ⟨und man wird strafen⟩ ihn in grossen ⟨Strafen, und er wird⟩ sich befinden unter Martern."*

45. (56.) Und w i r sprachen zu ihm: „Herr, schön hast du uns alles offenbart."

Und er sprach zu uns: „Wisset und begreift diese Worte!"

Darauf antwortete er, indem er ⟨sagte⟩ zu uns: „Ihr begreifet nicht **diese Worte."**

Und **wir sprachen zu ihm:** *„Ja,* **Herr,**

diese fünf sind[5] es, durch die sie (fem.) die Aussicht haben, **in dein Reich** einzugehen, und fünf, die ausgeschlossen sind, wodurch sie[6] ausserhalb deines Reiches sein werden.

durch die fünf werden sie *hineinkommen* **in dein Reich.**

Jedoch die, welche gewacht haben und **mit** *dir,* **dem Herrn** *und* **dem Bräutigam,** hineingegangen sind *sind gewesen,* **werden sich** *dennoch* **nicht freuen wegen derer, die geschlafen haben."** Und **er sprach zu uns: „Sie** werden **sich freilich freuen, dass sie hineingegangen sind mit** *dem Bräutigam,* **dem Herrn, und werden betrübt sein**[7] **um derer willen, die geschlafen haben, weil sie ihre Schwestern sind.**

Und diese **Töchter Gottes, des Vaters**[8], sind **zehn."**

Die **zehn** *nämlich sind die* **Töchter Gottes, des Vaters."**

Und wir **sprachen zu ihm: „Herr,** deiner Grösse entspricht es, dass du ihren Schwestern

⟨*Da* **sprachen**⟩ **wir zu ihm:** „⟨**Herr**⟩, *das Deinige ist es, dass du ...." Er*

[1] „hört" fehlt in A. [2] Die eingeklammerten Worte nur in C. [3] A: Ungeduld. [4] „ihm" statt „ihnen" nach A B C. [5] Statt „sind es" hat A: Namen [6] L „diese" statt „wodurch sie". [7] Kopt. Präsens. [8] „des Vaters" nach A B C.

Gnade erzeigest." Und er sprach zu uns: „Diese Sache ist nicht die eurige, sondern dessen, der mich gesandt hat, indem auch[1] ich mit ihm übereinstimme.

sprach zu uns: „. . . . ., sondern *die* dessen, der mich gesandt hat, *und ich selbst* stimme mit ihm überein.

46. (**57.**) Ihr aber, indem ihr hingeht, prediget und lehret wahr und recht, indem ihr niemandes Person anseht und fürchtet, besonders aber (nicht) die Reichen, an denen (etwas) gefunden wird, die mein Gebot nicht[2] tun, die in ihrem Reichtum schwelgen." Und wir sprachen zu ihm: „Herr, sprichst du zu uns allein von den Reichen?" Und er sprach zu uns: „Auch von dem, der nicht reich ist[3]; sobald er gibt und nicht verweigert[4] dem, der nichts hat, von einem solchen[3]: er wird Täter bei den Menschen genannt werden.

Ihr aber prediget und lehret *in Geradheit (und) schön,* indem ihr *euch vor niemand scheuet* und *indem ihr euch vor niemand* fürchtet, *besonders aber (nicht vor) den Reichen, denn jene tun nicht meine Gebote, sondern schwelgen (?) in ihrem Reichtum."* Wir *aber* sprachen zu ihm: „Herr, *ob die* Reichen ⟨*es* allein⟩ *sind." Er antwortete, indem er* ⟨zu uns⟩ sprach: „⟨*Wenn*⟩ *jemand,* der nicht reich ist, *und der besitzt ⟨ein kleines⟩ Vermögen, den Bedürftigen* gibt ⟨*und den Armen*⟩, *so werden die* Menschen *ihn nennen Wohltäter.*

47. (**58.**) Wenn aber einer seine Last, d. i. die Sünde, die er [gegen die Person seines Nächsten][5] getan hat, tragend fällt, so soll sein Nächster (zur Vergeltung) für das, was er an seinem Nächsten (Gutes) getan hat, ihn zurechtweisen.

Wenn aber ⟨einer unter der⟩ Last fällt *wegen der Sünden,* die er ⟨getan⟩ hat, ⟨*so möge*⟩ sein Nächster ihn zurechtweisen für ⟨das *Gute, das*⟩ er getan hat seinem Nächsten.

Und wenn ihn ⟨*nun*⟩ sein Nächster zurechtgewiesen hat und er sich bekehrt hat, so wird er gerettet werden, und der, welcher ihn zurechtgewiesen hat, wird *Lohn empfangen und*

das ewige Leben erlangen. Wenn er aber sieht[6], wie

leben ⟨ewiglich⟩. Denn ein bedürftiger Mensch, wenn er sieht

[1] „auch" nach A B C (S). [2] A B C S. [3] scil. spreche ich. [4] Statt taʿabaja ist l'abaja zu konjizieren (Littmann). [5] Die eingeklammerten Worte fehlen in A C. [6] „sieht" nach S (A B C) eingesetzt.

dieser, der ihm (etwas) erweist, sündigt, und begünstigt ihn, ein solcher wird in einem grossen Gericht gerichtet werden [1].

den, der ihm Gutes erwiesen hat, sündigen, und er weist ihn nicht zurecht, so wird *er* gerichtet werden in einem *schlimmen* Gericht.

Denn ein Blinder *aber,* der einen Blinden führt, die beiden *pflegen* werden in eine Grube *zu* fallen.

Ebenso werden der Begünstiger, der die Person ansieht, und auch der, den [2] er begünstigt und dessen Person er ansieht, beide mit e i n e r Strafe gestraft werden, wie der Prophet gesagt hat: Wehe den Begünstigern, die den Sünder um eines Bestechungsgeschenkes willen gerecht sprechen, deren Gott ihr Bauch ist. Seht ihr, wie das Gericht ist? [3]

Und wer die Person ansieht, *um* ⟨*ihret*⟩*willen,* ⟨*der wird sein wie*⟩ *die beiden,*

wie der Prophet ⟨gesagt⟩ hat: *Wehe denen, die die Person ansehen und die* ⟨*rechtfertigen den Gottlosen*⟩ *um Geschenke willen, diese* deren ⟨Gott⟩ ist ihr Bauch. *Sehet nun, dass ein Gericht* [4] . . . . !

Denn wahrlich, ich sage euch: An jenem Tage werde ich mich *weder* vor dem *den* Reichen scheuen noch mit dem *den* Armen Erbarmen haben.

48. (59.) Wenn du mit deinen Augen gesehen hast, wie (einer) sündigt, so züchtige ihn unter vier Augen (wörtlich: allein). [Wenn er auf dich hört, so hast du ihn gewonnen.] [5] Wenn er aber nicht auf dich hört, so erscheine selbander oder höchstens selbdritt [6], züchtige deinen Bruder! Wenn er aber (auch dann) nicht auf dich hört, so soll er dir wie ein Heide und Zöllner sein.

Wenn du *siehst einen Sünder, so* ⟨*weise ihn zurecht*⟩ *zwischen dir und ihm.* Wenn er aber nicht auf ⟨dich⟩ hört, ⟨*so nimm*⟩ *einen andern mit dir* bis zu drei *und belehre* deinen Bruder! *Wiederum* wenn er nicht auf dich hört, *so stelle ihn hin vor dir wie*

---

[1] Kann auch heissen: mit großer Strafe gestraft werden. [2] „den" statt „die" (Plur.) mit ACS. [3] C add.: (mit dem) sie werden gerichtet werden. [4] uktisis muss mit Schmidt in ukrisis verbessert werden. [5] Das Eingeklammerte fehlt in AC. [6] Wörtlich: werde zu zweien und bis zu drei.

49. (60.) Wenn du eine Sache hörst, so schenke keinen Glauben gegen deinen Bruder und verleumde nicht und liebe es nicht, Verleumdung anzuhören. Denn so steht geschrieben: „Lass dein Ohr nichts gegen[1] deinen Bruder anhören, sondern (nur) wenn du gesehen hast, tadele, züchtige und bekehre ihn.“ Und wir sprachen zu ihm: „Herr, du hast uns in allem gelehrt und ermahnt. Aber, Herr, soll denn unter den Gläubigen, die unter ihnen an die Predigt deines Namens glauben werden, Entzweiung und Streit und Eifersucht und Verwirrung und Hass und Bedrängnis sein? Denn du hast doch gesagt: sie werden einander tadeln und haben die Person nicht angesehen[2]. Sündigen diese, die da den hassen, der sie gezüchtigt hat?“ Und er antwortete und sprach zu uns: „Warum nun[3] wird das Gericht stattfinden? Damit der Weizen in seine[4] Scheuern gelegt werde[5] und man seine Spreu ins Feuer werfe.

50. (61.) . . . die also hassen, und wer mich liebt und wer die tadelt, die mein Gebot nicht[6] tun, diese werden also gehaßt und verfolgt werden, und man wird (sie) verachten und verspotten[7]. Man wird auch absichtlich sagen, was nicht (wahr) ist, und es wird zu einer Zusammenrottung kommen gegen die, die mich lieben. Diese aber werden sie tadeln, damit sie gerettet werden. Und diejenigen, welche sie tadeln und züchtigen und ermahnen werden, wird man hassen und absondern und verachten, und die ihnen Gutes (zu tun) wünschen, werden (daran) gehindert werden. Aber diese, welche[8] so erduldet haben, werden beim Vater als Märtyrer gelten[9], weil sie um die Gerechtigkeit geeifert haben und nicht mit verderblichem Eifer geeifert haben.“[10] Und wir sprachen zu ihm: „Wird, Herr, solches auch in unserer Mitte eintreten?“ Und er sprach zu uns: „Fürchtet euch nicht vor dem, was nicht bei vielen, sondern (nur) bei wenigen eintreten wird!“ Und wir sprachen zu ihm: „Sage uns, auf welche Weise!“ Und er sprach zu uns: „Es wird eine andere Lehre und Streit kommen, und indem

---

[1] Statt ḫaba = bei l. mit S lā ʿela. [2] So nach ABC. Man erwartet den Subjunktiv = sollen. Die zweite Hälfte = ohne die Person angesehen zu haben. [3] „nun“ nach ABC. [4] Das Suffix kann aber auch einfach die Determination ausdrücken = in die Scheunen. [5] Das Passiv nach C. Sonst müßte man aus šernāj šernāja machen und übersetzen: dass man den Weizen lege. [6] „nicht“ nach S. [7] Aktiv nach S. [8] „welche“ nach ABC. [9] Wörtlich: sein. [10] Perf. nach ABC.

sie ihre eigene Ehre wollen und eine nichtsnutzige Lehre hervorbringen, wird dabei ein Ärgernis des Todes [1] eintreten, und sie werden lehren und selbst die, welche an mich glauben, von meinem Gebot abwenden und sie aus dem ewigen Leben herausbringen [2]. Wehe aber denen, die dies mein Wort und Gebot zum Vorwand benutzen, und auch denen, die auf sie hören, und denen, die sich entfernen vom Leben der Lehre, [denen, die sich entfernen vom Gebot des Lebens,] [3] sie werden mit ihnen ewiglich gestraft werden."

51. (62.) Und nachdem [4] er dies gesagt und das Gespräch mit uns beendigt hatte, sprach er wiederum zu uns: „Sieh, nach drei Tagen und drei Stunden wird der, der mich gesandt hat, kommen, damit ich mit ihm gehe." Und wie er sprach, trat Donner und Blitz und Erdbeben ein, und die Himmel spalteten sich, und eine lichte Wolke kam und nahm ihn fort. Und die Stimme vieler Engel [C: hörten wir], wie sie sich freuten und priesen und sprachen: „Versammle uns, Priester, im Licht der Herrlichkeit!" Und als sie sich dem Firmament des Himmels genähert hatten [5], hörten wir ihn sagen: „Gehet hin in Frieden!"

Das Explizit der Hs L lautet: Im Namen unseres Herrn Jesu Christi! Beendigt ist das Buch des Testamentes im Frieden Gottes. Er bewahre seinen Knecht Abēsēlōm in alle Ewigkeit! Amen.

---

[1] Statt „des Todes" hat C „der Hurerei". [2] Mit BC jāwaṣ" zu lesen. [3] Die Worte in Klammern fehlen mit Recht in AḂCS. [4] 'enza ist, wie oben geschehen, in 'emza zu verbessern oder sonst der Nachsatz mit wafaṣ̣ama zu beginnen. [5] BL haben den Singular.

Frommannsche Buchdruckerei (Hermann Pohle) in Jena. — 5305

33/34 SUPPLEMENTUM LYRICUM, neue Bruchstücke von Archilochus, Alcaeus, Sappho, Corinna, Pindar, Bacchylides ausgew. und erkl. von E. Diehl 3. Aufl. 83 S. 17. 2.60

35 LITURGISCHE TEXTE III: Die konstantinopolitanische messliturgie vor dem IX. jahrhundert, übersichtl. Zusammenstellung des wichtigsten Quellenmaterials v. A. Baumstark. 16 S. 09. 0.50

36 LITURGISCHE TEXTE IV: Martin Luthers Von ordnung gottesdiensts, Taufbüchlein, Formula missae et communionis 1523, hrsg. v. H. Lietzmann. 24 S. 09. 0.75

37 LITURGISCHE TEXTE V: Martin Luthers Deutsche Messe 1526, hrsg. v. H. Lietzmann. 16 S. 09. 0.50

38/40 ALTLATEIN. INSCHRIFTEN, hrsg. v. E. Diehl. 3. Aufl. 92 S. 21. 2.90

41/43 FASTI CONSULARES IMPERII ROMANI (30 v. Chr. – 565 n. Chr.) mit Kaiserliste u. Anhang bearb. v. W. Liebenam. 128 S. 10. 4.—, gbd. 6.90

44/46 MENANDRI reliquiae nuper repertae hrsg. v. S. Sudhaus. 2. Aufl. 103 S 14. 3.25, gbd. 5.—

47/49 LATEINISCHE ALTKIRCHLICHE POESIE ausgewählt v. H. Lietzmann. 64 S. 10. 2.—

50/51 URKUNDEN ZUR GESCHICHTE DES BAUERNKRIEGES UND DER WIEDERTÄUFER hrsg. v. H. Böhmer. 2. Aufl. 36 S. 21. 1.10

52/53 FRÜHBYZANTINISCHE KIRCHENPOESIE I: Anonyme hymnen des V.—VI. jahrhunderts ed. Dr. Paul Maas. 32 S. 10. 1.—

54 KLEINERE GEISTLICHE GEDICHTE DES XII. JAHRHUNDERTS hrsg. v. A. Leitzmann. 30 S. 10. 1.—

55 MEISTER ECKHARTS BUCH D. GÖTTLICHEN TRÖSTUNG U. VON DEM EDLEN MENSCHEN (liber benedictus), hrsg. v. Ph. Strauch. 2. Aufl. 51 S. 22 1.70

56 POMPEIANISCHE WANDINSCHRIFTEN UND VERWANDTES, ausgew. v. E. Diehl. 60 S. 10. 1.90

57 ALTITALISCHE INSCHRIFTEN, ausgew. v. H. Jacobsohn. 32 S. 10. vergr.

58 ALTJÜDISCHE LITURG. GEBETE, ausgew. und mit Einleitung hrsg. v. W. Staerk. 32 S. 10. 1.—

59 DER MIŠNATRAKTAT BERAKHOTH IN VOKALISIERTEM TEXT, m. sprachl. u. sachl. Bemerkungen v. W. Staerk. 18 S. 10 0.60

60 EDWARD YOUNGS GEDANKEN ÜBER DIE ORIGINALWERKE in einem Schreiben an Samuel Richardson, übersetzt von H. E. v. Teubern, hrsg. v. K. Jahn. 46 S. 10. 1.50

61 LITURGISCHE TEXTE VI: Die Klement. liturgie a. d. Const. apost. VIII nebst anhängen hrsg. v. H. Lietzmann. 32 S. 10. 1.—

62 VULGÄRLATEIN. INSCHRIFTEN hrsg. v. E. Diehl. 180 S. 10. 5.60

63 GOETHES ERSTE WEIMARER GEDICHTSAMMLUNG mit varianten hrsg. v. A. Leitzmann. 35 S. 10 1.10, gbd. 3.—

64 DIE ODEN SALOMOS aus dem syrischen übersetzt mit Anmerkungen von A. Ungnad und W. Staerk. 40 S. 10. vergr.

65 AUS DER ANTIKEN SCHULE. Samml. griech. texte auf papyrus holztafeln, ostraka ausgew. u. erklärt v. E. Ziebarth. 2. Aufl. 33 S. 13. 1.—

66 DIE FRÖSCHE DES ARISTOPHANES, mit ausgewählten antiken scholien, hrsg v. W. Süss. 90 S. 11. 2.90, geb. 4.50

67 DIETRICH SCHERNBERGS Spiel von Frau Jutten (1480). Nach der einzigen Ueberlieferung im Druck des Hieronymus Tilesius (Eisleben 1565), hrsg. v. E. Schröder 56 S. 11. 1.—

68 LATEIN. SACRALINSCHRIFTEN ausg. v. F. Richter. 45 S. 11. 1.40

69 POETARVM ROMANORVM VETERVM reliquiae selegit E. Diehl. 165 S. 11. 2.50, geb. 4.50

70 LITURGISCHE TEXTE VII: Die Preussische Agende im Auszug hrsg v. H. Lietzmann. 42 S. 11. 0.50, geb. 2.40

71 CICEROS REDE FÜR T. ANNIUS MILO mit dem commentar des Asconius und den Bobienser Scholien, hrsg. v. P. Wessner. 74 S. 11. 1.20, geb. 3.—

72 DIE VITAE VERGILIANAE u. ihre antiken Quellen, hrsg. v. E. Diehl. 60 S. 11. 0.90

73 DIE QUELLEN VON SCHILLERS UND GOETHES BALLADEN zusammengest. v. A. Leitzmann. 51 S. 3 Abb. 2. Aufl. 23 1 70

74 ANDREAS KARLSTADT VON ABTUHUNG DER BILDER und das keyn bedtler vnther den christen seyn sollen 1522 und die Wittenberger beutelordnung hrsg. v. H. Lietzmann. 32 S. 11. 1.—

75 LITURGISCHE TEXTE VIII: Die Sächsische Agende im Auszug hrsg. v. H. Lietzmann. 36 S. 11. 1.10, geb. 3.—

76 AUSWAHL AUS ABRAHAM A S. CLARA hrsg. v. K. Bertsche. 47 S 11. 1 50

77 HIPPOCRATIS de aere aquis locis mit der alten lateinischen Übersetzung hrsg. v. G. Gundermann. 50 S. 11. 1.70

78 RABBINISCHE WUNDERGESCHICHTEN des neutestamentlichen zeitalters in vokal. text mit sprachl. und sachl. bemerkungen v. P. Fiebig. 28 S. 11. 1.—

79 ANTIKE WUNDERGESCHICHTEN zum studium der wunder des Neuen Testaments zusammengest. v. P Fiebig. 27 S. 2. Aufl. 21. 1.—

80 VERGIL AENEIS II mit dem commentar des Servius herausgeg. von E. Diehl. 131 S. 11. 2.—, geb. 4.—

81 ANTI-XENIEN in auswahl hrsg. v. W. Stammler. 68 S. 11. 1.—, geb 2.80

82 APOLLONIUS DYSCOLUS: De pronominibus pars generalis edidit Dr. Paulus Maas. 44 S. 11. 1.25

83 ORIGENES, EUSTATHIUS V. ANTIOCHIEN, GREGOR V. NYSSA über die Hexe von Endor hrsg. v. Erich Klostermann. 70 S. 12. 1.20

84 AUS EINEM GRIECHISCHEN ZAUBERPAPYRUS v. Richard Wünsch. 31 S. 11. 1.—

85 DIE GELTENDEN PAPSTWAHLGESETZE, Pii X constitutio vacante sede apostolica — Pii X constitutio commissum nobis — Leonis XIII constitutio praedecessores nostri — Leonis XIII. instructio (regolamento) hrsg. v. F. Giese. 56 S. 12. 1.—

86 ALTE EINBLATTDRUCKE hrsg. v. Otto Clemen. 77 S. 11. 2 50

87 UNTERRICHT DER VISITATOREN 1528 an die pfarrherrn im kurfürstentum zu Sachsen hrsg. v. H. Lietzmann. 48 S. 12. 1 50

88 JOHANNES BUGENHAGENS BRAUNSCHWEIGER KIRCHENORDNUNG 1528, hrsg. von H. Lietzmann. 152 S. 12. 2.25

89 EURIPIDES MEDEA mit scholien herausg. von Ernst Diehl. 116 S. 11. 1.75, geb. 3.60

90 DIE QUELLEN VON SCHILLERS WILHELM TELL zusammengestellt v. Albert Leitzmann. 47 S. 12. 1.50, geb. 3.20

91 SCHOLASTISCHE TEXTE I: Thomas v. Aquin, Texte zum Gottesbeweis, ausgew. u. chron. geord. v. E. Krebs. 2. Aufl. 64 S. 12. 2.—

92 MITTELHOCHDEUTSCHE NOVELLEN I: Die heidin (IV. Redaktion), hrsg. v. L. Pfannmüller. 51 S. 12. 1.50

93 SCHILLERS ANTHOLOGIE-GEDICHTE kritisch hrsg. v. W. Stammler. 71 S. 12. 1.20, geb. 3.—

94 ALTE UND NEUE ARAMÄISCHE PAPYRI übersetzt und erklärt von W. Staerk. 73 S. 12. 2.25

95 MITTELHOCHDEUTSCHE NOVELLEN II: Rittertreue. Schlegel hrsg. v. L. Pfannmüller. 63 S. 12. 2.—

96 Der Franckforter („eyn deutsch theologia") hrsg. v. W. Uhl. 64 S. 12. 2.—
97 Diodors römische Annalen bis 302 a. Chr. samt dem Ineditum Vaticanum hrsg. v. A. B. Drachmann. 72 S. 12. 2.25
98 Musaios, Hero u. Leandros m. ausgew. Varianten u. Scholien hrsg. v. A. Ludwich. 54 S. 12. 1.70
99 Authentische Berichte über Luthers letzte Lebensstunden herausgegeben von Dr. J. Strieder. 42 S. 12. 1.25
100 Goethes römische Elegien nach der ältesten Reinschrift hrsg v. A. Leitzmann. 56 S. 12. 1.70, geb. 3.50
101 Frühneuhochdeutsches Glossar von Alfred Götze. 2. Aufl. 136 S. 20. 4.25
102 Die Generalsynodal-Ordnung hrsg. v. A. Uckeley. 20 S. 12. 0.30
103 Die Kirchengemeinde- und Synodalordnung f. d. Provinzen Preußen, Brandenburg, Pommern, Posen, Schlesien u. Sachsen hrsg. v. A. Uckeley. 36 S. 12. 0.50
104 Die rheinisch-westfäl. Kirchenordnung hrsg. v. A. Uckeley. 64 S. 12. 2.—
105 Mystische Texte aus dem Islam. Drei Gedichte des Arabi 1240. Aus d. Arab. übers. u. erläutert v. M. Horten. 18 S. 12. 0.60
106 Das niederdeutsche Neue Testament nach Emsers übersetzung. Rostock 1530. Eine Auswahl aus den Lemgoer Bruchstücken mit einer Einleitung, hrsg. v. E. Weissbrodt. 32 S. 12. 1.—
107 Herders Shakespeare-Aufsatz in dreifacher gestalt mit Anm. hrsg. v. F. Zinkernagel. 41 S. 12. 1.20
108 Konstantins Kreuzesvision in ausgew. texten vorgelegt v. J. B. Aufhauser. 26 S. 12. 0.90
109 Luthers kleiner Katechismus, der deutsche text in seiner geschichtlichen entwicklung v. J. Meyer. 32 S. 12. 1.—
110 Historische attische Inschriften ausgewählt u. erklärt von E. Nachmanson. 82 S. 13. 2.60
111 Auswahl aus den Iliasscholien zur Einführung in die antike Homerphilologie, ausgew. u. geordn. v. W. Deecke. 92 S. 12. 2.90
112 Supplementum Euripideum, bearb. v. H. v. Arnim. 80 S. 13. 2.50
113 Svpplementvm Sophoclevm hrsg. v E. Diehl. 33 S. 13. 1.—
114 Die Verfassung des deutschen Reiches vom Jahre 1849 mit Vorentwürfen, Gegenvorschlägen u. Modifikationen bis zum Erfurter Parlament, hrsg v. L. Bergsträsser. 104 S. 13. 3.25
115 Griechische Inschriften zur griechischen Staatenkunde ausgewählt v. F. Bleckmann. 79 S. 13. 2.50
116 Die Quellen zu Heinrich von Kleists Michael Kohlhaas. Hrsg. von Rudolf Schlösser. 14 S. 13. 0.50
117 Meister Eckhart: Reden der Unterscheidung, hrsg. v. E. Diederichs. 45 S. 13. 1.40
118 Oratorum et rhetorum Graecorum nova fragmenta ed. K. Jander 42 S. 13. 1.40
119 Texte zu dem Streite zwischen Glauben und Wissen im Islam. Die Lehre vom Propheten und der Offenbarung bei dem islamischen Philosophen Farabi, Avicenna u. Averroes, dargest. v. M. Horten. 43 S. 13. 1.40
120 Hippokrates über Aufgaben und Pflichten des Arztes in einer Anzahl auserlesener Stellen aus dem Corpus Hippocraticum hrsg. v. Th. Meyer-Steineg u. W. Schonack. 27 S. 13. 0.50
121 Historische griechische Inschriften b. auf Alexander d. Grossen. ausgewählt u. erklärt v. E. Nachmanson. 60 S. 13. 1.90

122 URKUNDEN ZUR ENTSTEHUNGSGESCHICHTE DES DONATISMUS hrsg. v. Hans von Soden. 56 S. 13. 1.75
123 HUGO VON ST. VICTOR SOLILOQUIUM DE ARRHA ANIMAE UND DE VANITATE MUNDI hrsg. v. K. Müller. 51 S. 13. 1.65
124 DEUTSCHE LYRIK DES SIEBZEHNTEN JAHRHUNDERTS in Auswahl hrsg. v. P. Merker. 53 S. 13. 1.65
125 LITURGISCHE TEXTE IX: Die Hannoversche Agende im Auszug hrsg. v. J. Meyer. 30 S. 13. 1.—
126 ANTIKE JESUS-ZEUGNISSE vorgel. v. J. B. Aufhauser. 51 S. 13. vergr.
127 AUSGEWÄHLTE PREDIGTEN JOHANN TAULERS hrsg. v. L. Naumann. 62 S. 14. 2.—
128 BÜRGERS GEDICHT DIE NACHTFEIER DER VENUS hrsgeg. v. W. Stammler. 56 S. 14. 1.75
129 TEXTE ZUR GESCHICHTE DES MONTANISMUS hrsg. v. D. N. Bonwetsch. 32 S. 14. 1.—
130 DER TOSEPHTATRAKTAT ROŠ HAŠŠANA in vokal. Text m. sprachl. Textkrit. u. sachl. Bem. v. Lic. Paul Fiebig. 16 S. 14. 0.50
131 DIE LINDISCHE TEMPELCHRONIK neu bearb. v. Chr. Blinkenberg. 59 S. 15. 1.90
132 DIE RÖMISCHEN KRÖNUNGSEIDE DER DEUTSCHEN KAISER von H. Günter. 51 S. 15. 1.65
133 LAMELLAE AVREAE ORPHICAE edidit commentario instruxit Alex. Olivieri. 28 S. 15. 0.90
134 P. VERGILI MARONIS BVCOLICA cum auctoribus et imitatoribus in usum scholarum edidit C. Hosius. 64 S. 15. 2.—, geb. 3.60
135 NOVAE COMOEDIAE FRAGMENTA in papyris reperta exceptis Menandreis edidit O. Schroeder. 77 S. 15. 2.25
136 TÜRKISCHE NACHRICHTEN für Uebungen im Türkischen in der Originalschrift von Arthur Ungnad. 39 S. 16. 1.25
137 VITAE HOMERI ET HESIODI in usum scholarum edidit Vdalricvs de Wilamowitz-Moellendorff. 58 S. 16. 1.80
138 CRATIPPI Hellenicorum fragmenta Oxyrhynchia scholarum in usum ed. J. H. Lipsius. 35 S. 16. 1.10
139 MUHAMMEDANISCHE GLAUBENSLEHRE. Die Katechismen des Fudāli u. d. Sanussi übers. u. erl. v. Max Horten. 57 S. 16. 1.75
140 TÜRKISCHE VOLKSTÜMLICHE UND VOLKSPOESIE. Für Uebungen in türkischer Originalschrift zusammengestellt u. m. Anmerkungen vers. von Dr. Karl Lokotsch. 23 S. 17. 0.75
141 LITURGISCHE TEXTE X: Einführung in das römische Brevier von H. Lietzmann. 48 S. 17. 1.50
142 MARTIN LUTHERS 95 Thesen nebst dem Sermon von Ablaß und Gnade 1517. Hrsg. v. O. Clemen. 34 S. 17. 0.50
143 VULGÄRLATEIN. UEBUNGSBUCH hrsg. von F. Slotty. 64 S. 18. 2.—
144 LITURGISCHE TEXTE XI: Taufe und Firmung. Nach dem röm. Missale, Rituale und Pontificale, hrsg. v. I. Herwegen. 46 S. 20. 1.50
145 CALLIMACHI FRAGMENTA NUPER REPERTA, ed. R. Pfeiffer. 94 S. 21. 3.—
146 PROBEN HOCH- UND NIEDERDEUTSCHER MUNDARTEN, hrsg. v. Alfred Goetze. 110 S. 22. 3.50
147 OCTAVIA PRAETEXTA cvm elementis commentarii, ed. Carolus Hosius. 72 S. 22. 2.25
148 DIE DEUTSCHEN LIEDER DER CARMINA BURANA. Herausgegeben von Dr Friedrich Lüers 34 S. 22. 1.—
149 DIE KLEINEREN HISTORIKERFRAGMENTE AUF PAPYRUS. Bearbeitet von Fr. Bilabel. 63 S. 23. 2.—

Frommannsche Buchdruckerei (Hermann Pohle) in Jena. 5305.

Kolibri 360 GmbH
Ferdinand-Jühlke-Straße 7
99095 Erfurt, Deutschland
produktsicherheit@kolibri360.de